The Income Approach tc Valuation

A classic textbook that has guided generations of students through the intricacies of property valuation receives a twist and a makeover in the hands of two new authors and under continuity editorship of one of the previous authors.

The Income Approach to Property Valuation has always been a firm favourite with undergraduate and postgraduate students and teachers. This new edition has a refreshingly contemporary focus and a more international reach, thoroughly updated and with new directions to meet the increasingly international perspectives of current real estate students. The links between theory and practice are further reinforced with updated and new case studies and exhaustive examples and problems with worked examples.

The Income Approach to Property Valuation facilitates understanding in:

- How to fully analyse market rents, sales prices, and trends to derive market evidence to support an opinion of market value.
- The investment method(s) of valuation and how to apply them to a wide variety of situations.
- How specific internal and external factors impact on risk, value, and security of income.
- A range of valuation methods and how and when to apply them.
- The use of Excel formulae and how to apply them to *the income approach*.
- The expectations of valuers and the professional frameworks they operate within.
- And contemporary issues and their impact on valuations.

This extensively revised and extended new edition is perfect for both real estate students worldwide and professional candidates working towards career enhancement, including the assessment of professional competence (APC) for the Royal Institution of Chartered Surveyors.

Nick Nunnington is a visiting professor of corporate real estate at Nottingham Trent University and an assistant professor at Neapolis University Paphos. He worked for many years at Sheffield Hallam University, before moving to Abu Dhabi to become an associate vice provost for the 17 colleges of the Higher Colleges of Technology across the United Arab Emirates. He also developed and taught real estate courses at the University of Adelaide. Nick is now a founder of the PropTech start-up Workplaced (www.workplaced.com), which delivers solutions to maximise productive hybrid workplace strategies.

Thomas Dimopoulos holds a PhD from Cyprus University of Technology, a MEng in rural and surveying engineering from Aristotle University of Thessalonica, and an MSc in real estate from Oxford Brookes University. He is an assistant professor and the head of the Real Estate Department at Neapolis University Paphos. He is also the director and founder of AXIA Chartered Surveyors, a firm that is practicing property valuations in Cyprus, and the director and founder of Real Geosolutions, a firm that is offering GIS, AVM, and PropTech solutions. He was appointed to the European Board of IVSC (International Valuation Standards Council) in January 2020 and remains an active participant. He was the chairman of RICS Cyprus from 2017 until 2022 and is a member of the board of the Association of Property Valuers in Cyprus. Before AXIA, he worked in several companies in Cyprus and abroad. He has also published several articles in accredited international journals. His research and publications focus on property valuation methodologies, property taxation, geospatial data analysis, and mass appraisals with the use of artificial intelligence and machine learning techniques.

Fernanda Antunes Batista da Silva holds a PhD in real estate from the University of Aberdeen and an MSc in real estate from the University of Reading. She is an associate lecturer at the University College of Estate Management and a visiting lecturer at Neapolis University Paphos. Her research focuses on flexible workspaces, commercial real estate investment, property valuation, and corporate real estate. Previously, she was the course leader for the MSc in property development and planning at Nottingham Trent University, where she led valuation modules and contributed to investment-related modules at both undergraduate and postgraduate levels. Earlier in her career, she worked as a valuer for financial institutions in Brazil.

The Income Approach to Property Valuation

8th Edition

Nick Nunnington, Thomas Dimopoulos and
Fernanda Antunes Batista da Silva

Routledge
Taylor & Francis Group
LONDON AND NEW YORK

Designed cover image: Nick Nunnington

Eighth edition published 2026
by Routledge
4 Park Square, Milton Park, Abingdon, Oxon, OX14 4RN

and by Routledge
605 Third Avenue, New York, NY 10158

Routledge is an imprint of the Taylor & Francis Group, an informa business

First edition published by Routledge & Kegan Paul 1979
Seventh edition published by Routledge 2018

British Library Cataloguing-in-Publication Data
A catalogue record for this book is available from the British Library

ISBN: 978-1-032-79070-1 (hbk)
ISBN: 978-1-032-77868-6 (pbk)
ISBN: 978-1-003-49035-7 (ebk)

DOI: 10.1201/9781003490357

Typeset in Times New Roman
by Apex CoVantage, LLC

Contents

Acronyms

Acronyms are used throughout the text to simplify the presentation of valuation terms in regular use and to avoid the unnecessary repetition of the names of organisations, regulations, and standards. Whilst this is mainly provided for the valuation examples, we have included all acronyms for completeness.

It will help readers who like to chapter-hop rather than reading from beginning to end, as acronyms are used consistently after the first introduction of the relevant expression. We start this process at the beginning of Chapter 1.

ACP	aluminium composite panels
ACQSC	Aged Care Quality and Safety Commission (Australia)
ACV	asset of community value
ADGM	Abu Dhabi Global Market
ADR	average daily rate, or average room rate
AED	United Arab Emirates dirham
AER	annual equivalent rate
AI	Appraisal Institute
AI	artificial intelligence
ALMR	Association of Licensed Multiple Retailers
AML	anti–money laundering
ANN	artificial neural networks
APAC	Asia Pacific
APC	Assessment of Professional Competence
API	Australian Property Institute
APR	annual percentage rate
AR	augmented reality
ARY	all risks yield
ASA	American Standards Association
ASF	annual sinking fund
ASIC	Australian Securities and Investments Commission
ASIC	application-specific integrated circuit
AST	assured shorthold tenancy
AT	assured tenancy
AVM	automated valuation models
AWP	annuity £1 will purchase
BBPA	British Beer and Pub Association

BCIS	Building Cost Information Service
BIM	building information modelling
BIS	Bank for International Settlements
BO	beneficial ownership
BOMA	Building Owners and Managers Association
BREEAM	Building Research Establishment Environmental Assessment Method
BVI	Bundesverband Investment und Asset Management
CAMA	computer-assisted mass appraisals
CAPM	capital asset pricing model
CBD	central business district
CBV	Chartered Business Valuator Institute (Canada)
CE	certainty equivalent
CGT	capital gains tax
CLRA	Commonhold and Leasehold Reform Act
CMBS	commercial mortgage-backed securities
CMS	Centres for Medicare and Medicaid Services (US)
CNAREA	Canadian National Association of Real Estate Appraisers
CNN	convolutional neural network
CoMP	Code of Measuring Practice
COL	cost of living
COV	coefficient of variation
CPD	continuing professional development
CPI	customer price index
CQC	Care Quality Commission
CRL	Contribution sur les Revenus Locatifs
CRREM	Carbon Risk Real Estate Monitor
CS	cognitive systems
CSR	corporate social responsibility
CUB	Custo Unitário Básico (Brazil)
CV	capital value
CV	computer vision
CVA	company voluntary arrangements
CVT	computer vision technology
CZ	commercial zone
DCF	discounted cash flow
DGNB	Deutsche Gesellschaft für Nachhaltiges Bauen/German Sustainable Building Council
DIFC	Dubai International Finance Centre
DLNN	deep learning neural network
DRC	depreciated replacement cost
EBITDA	earnings before interest, taxation, depreciation, and amortisation
EBITDAR	earnings before interest, taxation, depreciation, amortisation, and rent
EC	European community
ECV	effective capital value
ENFIA	unified property ownership tax (Cyprus)
EPBD	Energy Performance of Buildings Directive
EPC	energy performance certificate

EPOS	electronic point of sale system
ERT	emotion recognition technology
ERV	estimated rental value
ESG	environmental, social, and governance
ET	eye tracking
ETEK	Scientific and Technical Chamber (Cyprus)
EU	European Union
EVS	European Valuation Standards
F&B	food and beverage
FAR	floor area ratio
FASB	Financial Accounting Standards Board
FDI	foreign direct investment
FF&E	furniture, fixtures, and equipment
FIRE	finance, insurance, real estate
FL	fuzzy logic
FMOP	fair maintainable operating profit
FMT	fair maintainable turnover
FOMO	fear of missing out
FRI	full repairing and insuring (terms) (of a lease)
FV	future value
FVIP	financial viability in planning
FW	flexible workspace
GCC	Gulf Cooperation Council
GDP	gross domestic product
GDV	gross development value
GEA	gross external area
GFA	gross floor area
GFC	global financial crisis
GIA	gross internal area
GIM	gross income multiplier
GIS	geographic information system
GNN	graph neural network
GOEF	German open-ended funds
GOP	gross operating profit
GPA	Government Property Agency
GPU	graphics processing unit
GRETI	Global Real Estate Transparency Index
GRM	gross rental multipliers
GRNN	general regression neural networks
GST	goods and services tax
HA	Housing Act
HBU	highest and best use
HMO	house in multiple occupation
HQE	Haute Qualité Environmentale
HVAC	heating, ventilation, and air conditioning
IAAO	International Association of Assessing Officers
IAS	International Accounting Standard

IASB	International Accounting Standards Board
IBFC	income before fixed charges
ICMS	International Construction Measurement Standards
IES	International Ethics Standards
IFRS	International Financial Reporting Standards
IMS	information management systems
IoT	Internet of Things
IPMS	International Property Measurement Standards
IPMSC	International Property Measurement Standards Coalition
IPT	immovable property tax
IPTI	International Property Tax Institute
IR	internal repairing lease
IRFY	inflation risk-free yield
IRI	internal repairing and insuring lease
IRR	internal rate of return
IRRV	Institute of Revenues Rating and Valuation
ITZA	in terms of Zone A
IVS	International Valuation Standards
IVSC	International Valuation Standards Council
IWG	International Workplace Group
LED	light-emitting diode
LEED	leadership in energy and environmental design
LLC	limited liability company
Lmi	Leesman Index
LRA	Leasehold Reform Act
LRHUDA	Leasehold Reform Housing and Urban Development Act
LTA	Landlord and Tenant Act
MAE	mean absolute error
MDR	multiple dwelling relief
MEES	minimum energy efficiency standards
MENA	Middle East and North Africa
MICE	meetings, incentives, conferences, and exhibitions
ML	machine learning
MLT	machine learning technologies
MLV	mortgage lending value
MOU	memorandum of understanding
MR	market rent (IVS)
MRA	multivariate regression analysis
MRICS	member of the Royal Institution of Chartered Surveyors
MV	market value (IVS)
NABBERS	National Australian Built Environment Rating
NDV	net development value
NER	net effective rent
NIA	net internal area
NIY	net initial yield
NN	neural network
NOI	net operating income

NPL	non-performing loan
NPV	net present value
OAR	overall rate models
OECD	Organisation for Economic Co-operation and Development
OLS	ordinary least squares
OMV	open market value
pa	per annum (per year)
PCA	principal component analysis
PE	price earnings
PESTEL	political, economic, social, technological, environmental, and legal
PII	professional indemnity insurance
PLC	public limited company
PMT	regular payment
POM	property operations and maintenance
PPM	planned preventative maintenance
PRS	private rented sector
PS	professional standards
PV	present value
Pv£1pa	present value £1 per annum
QS	quantity surveyor
RA	Rent Act
RADR	risk-adjusted discount rate
RCA	reinstatement cost assessment
Red Book	*RICS Valuation – Global Standards*
REITs	real estate investment trusts
REO	reasonably efficient operator
RERA	Real Estate Regulatory Authority (Dubai)
REV	Recognised European Valuer
RevPAR	revenue per available room
RF	fandom forest
RFR	risk-free rate
RG	robust geo-estimation
RICS	Royal Institution of Chartered Surveyors
RMSE	root mean square error
RP	risk premium
RPI	retail price index
RST	rough set technology or theory
RV	rateable value
SCORFA	special commercial or financial advantage
SDLT	Stamp Duty Land Tax (replaced by a transaction tax in parts of the UK)
SEE	standard error of the estimate
SF	sinking fund
SFR	sinking fund return
SIA	Security Industry Authority
SNN	spatial neural network
SOE	state-owned enterprises
SOM	self-organising maps

SPV	special-purpose vehicle
SREA	standards for real estate appraisals
SSSI	site of special scientific interest
SVM	support vector machines
TEGoVA	The European Group of Valuers' Associations
TEY	true equivalent yield
TPU	tensor processing unit
TR	total return
TUPE	transfer of undertakings (protection of employment)
TV	terminal value
TVM	time value of money
UAE	United Arab Emirates
UBR	Uniform Business Rate
UK	United Kingdom
UKVS	UK Valuation Standards
ULT	unexpired lease term
UN	United Nations
US	United States
USPAP	Uniform Standards of Professional Appraisal Practice
VAT	value-added tax
VIC	value influence centres
VP	vacant possession
VPGA	*RICS Global Valuation Practice Guidance – Applications* (RB)
VPO	valuation professional organisations
VPS	valuation technical and performance standard (RB)
VR	virtual reality
VRS	valuers registration scheme
WAAC	weighted average cost of capital
WALE	weighted average lease expiry
WAULT	weighted average unexpired lease term
WAULTC	weighted average unexpired lease term certain
WELL	WELL Building Standard
WGBC	World Green Building Council
XAI	explainable AI
XGBoost	Extreme gradient boosting
YP	Years' purchase (UK valuers' term for PV £ pa)
YPp	Years' purchase in perpetuity

Preface

For the eighth edition, with the retirement of David Mackmin and the withdrawal of Andrew Baum, we feel obliged to recognise the enormous contribution that these originators have made over the years since the first edition in 1979. Andrew and David were very conscious of the fact that much of the methodology of the *income approach* was, when they set out to write the early editions, rote taught and applied in a very simplistic way.

Over the years, so much has changed, and Andrew and David, and later Nick, evolved the book to keep ahead of the shifts in practice, scale, regulation, internationalisation, and market sophistication. Over the period of the later editions, the property market has grown into a truly international market, with ever-sophisticated products that now focus on branding, amenities, and services, not simply a rental income. At the same time, investors must deal with national and international shocks, serious geopolitical tension, and risks requiring the services of qualified professional valuers with strong ethical beliefs and a global commercial awareness. Clients demand greater global consistency in valuation standards, value definitions, and basis of property measurement and application of the *income approach*.

With two new authors, we have a remit to change and move things in a slightly different direction than all the previous editions, but with the weight of maintaining the legacy of *the income approach*. With this edition, we see greater significance on the themes of environmental, social, and governance (ESG); the application of technology and artificial intelligence (AI); and a significant need to qualify the validity and reliability of big data.

We also recognise that university education continues to evolve dramatically, with a much greater emphasis on application and employability. Students need to be prepared for placements/internships or graduate jobs that throw them in at the deep end. Those placements are increasingly international, and expectations of what an intern can do are much higher than before. They need to be able to manage the application of valuation theory, not just regurgitate it, and be able to balance consistency of purpose and professional standards with local, national, and international contexts and specific client requirements.

Investment advisors now usually analyse property purchases using the internal rate of return (IRR) and look for valuations and appraisals which are discounted cash flow (DCF) based, with the potential net operating income clearly identified over, say, a ten-year period and then discounted using present value factors. Income flows can be with or without assessment of income (rental) growth over the holding period. As we explore throughout the book, the market is evolving increasingly complex, occupier-focused property products with greater volatility and multiple income streams. These reinforce DCF as the only valuation tool capable of delivering appropriate results and analysis.

Thomas's relationship with this text began long before he ever imagined contributing to it. The first edition was published in 1979, the year he was born. Throughout his MSc studies, he encountered subsequent iterations, each unveiling new insights and reinforcing a commitment to property valuation. However, it was the fourth edition, published in 1997, that first captivated his attention through its rigorous blending of theory and practice. For him, each new edition represented a milestone in his intellectual journey, reflecting the evolution of an industry that never ceases to fascinate. Now, he feels privileged to participate in shaping this eighth edition. He hopes that readers, whether practicing professionals or students new to the subject, will find the eighth edition causing the same spark of curiosity and comprehensive guidance that inspired him so many editions ago. In revising and refining this work, we all continue a legacy more than four decades in the making, striving to preserve the precision and passion that define it, whilst always looking forward to the changing demands of our field.

The most significant change, given we now have Greek and Brazilian-born authors and one who travels constantly to teach and work, was to make the book much more international in its orientation and not so UK-centric. We have tried to find international case studies, and examples, which will help readers from any country feel more affinity with the text. You will find examples in dollars, pounds, and euros to reflect this approach. We also took the decision when using US-based examples to use square feet, which is their adopted unit of measurement, and metres for the UK and Europe, to give the examples authenticity.

We have also decided to strip out many of the references to UK law, or where we do use it, to try to contextualise it in terms of its impact, rather than the specifics of case law, and to provide contrasting examples for other countries. Our opinion is that law is frequently country-specific and is constantly changing; in the life of this book, many laws, especially case laws, may change or become obsolete. We feel that you should rely on your law courses for the latest and local legal advice which you need to integrate with your valuation studies.

We decided to take a greater focus on the *income approach* for commercial property and removed many of the UK residential elements that are too narrow a focus for our global perspective. Residential valuation was a bit of an obsession with David, and he has a separate text on it, and so we hope he will forgive us.

We have divided the new book into four parts. In **Part I**, we start with the core frameworks, concepts, mathematics, and techniques. We start in a similar logical way to the previous edition, with an examination of what we do mean by *property* and *property markets*, their characteristics and operations. However, in this edition, we have taken a more philosophical approach. We move on to an initial examination of property as an investment asset class.

We then set out expanded perspectives of the frameworks, both global and local, within which the *income approach* operates. These are not only the regulatory and professional standards but also the economic and legal contexts within which valuation takes place.

The remainder of Part I of the book is an expanded consideration of professional and ethical issues. We introduce a greater global reach and integrate perspectives not only from the latest edition of the *Royal Institution of Chartered Surveyors (RICS) Red Book* but also from the European Group of Valuation Associations (TEGoVA) European Valuation Standards, the *Blue Book*, and the International Valuation Standards (IVS). We also integrate the updated International Property Measurement Standards (IPMS).

This is followed by familiar material refreshed, reset, and extended to cover the basic financial concepts and techniques that underpin the *income approach*.

For **Part II** of the book, we return to a valuation methods approach and focus on applications of the income approach within the comparison, investment, profits, development (residual), and cost methods. We start with the most fundamental of all, the comparable method. We have spent a lot of time expanding this section to recognise that data, specifically big data, is playing an increasing role in valuations and how we can use techniques, including regression analyses, to take a more advanced approach to this method.

Throughout Part II, our objective is to set valuations in context, using some existing, but with many new, case studies and examples, to help those new to the subject see how a professional valuer applies the concepts and methods and works within international and national standards. We have introduced a new step-by-step approach to explain the application of the method(s), either manually or using Excel, to arrive at an assessment of market value.

Our new Chapter 8 is a departure from past editions, incorporating both freehold and leasehold investment valuations in the same place. This is followed by the profits method, the residual method (development appraisals), and the cost method. In all these chapters, we have new examples and case studies designed to extend the application of the methods, and examine some of the nuances, of how the *income approach* works in practice.

In **Part III**, rather than looking back as in previous editions, we look forward and examine many contemporary issues, including automated valuation methods and artificial intelligence, ESG and sustainability and its impacts on value, and how to value more volatile incomes, such as co-working.

We conclude with an evaluation of valuers, examining uncertainty and how practice is evolving in managing disruptive changes in the industry and markets.

Part IV contains a number of elements to demonstrate the application of the *income approach* in practice with two extended case studies, example valuation reports, and solutions to the large numbers of problems set at the end of many chapters. We hope this allows readers to explore *the income approach* at their own pace.

Fernanda Antunes Batista da Silva
Thomas Dimopoulos
Nick Nunnington
Brazil, Cyprus, and Portugal

Note: This eighth edition is based on:

International Valuation Standards Council (IVSC) (2024). *International Valuation Standards* (IVS). Available at: www.ivsc.org.

Royal Institution of Chartered Surveyors (RICS) (2021). *RICS Valuation – Global Standards* (*Red Book*). London: RICS.

International Financial Reporting Standards (IFRS) (2023). International Accounting Standards Board (IASB). Available at: www.ifrs.org.

Basel Committee on Banking Supervision (BCBS) (2022). *Basel III: Finalising post-crisis reforms*. Bank for International Settlements (BIS)

The European Group of Valuers' Associations (TEGoVA) (2020). European Valuation Standards (EVS).

The Appraisal Foundation (2021). Uniform Standards of Professional Appraisal Practice (USPAP). Washington, DC: Appraisal Standards Board.

Financial Accounting Standards Board (FASB) (2019). Conceptual Framework for Financial Reporting (2021).

Acknowledgements

As we have considerably edited and removed significant amounts of the last edition, we do not feel it is appropriate to list all the contributors from previous editions. They are acknowledged in the last edition and know who they are. We are grateful for the contributions of all previous contributors who have shaped the evolution of this important text. Again, we are obviously mindful of the continuing legacy of David and Andrew and their influence as main writers of the past seven editions.

We do, however, want to acknowledge the contributions to the previous edition of Howard Day, BSc, FRICS, MAE, MCIArb, FAVLP, in respect of the profits method, where we have retained some examples and narrative. For the same reason, we recognise the contribution of Adam Baldwin, BA (Hons), MSc, MRICS, who contributed and edited parts of the seventh edition and whose hand can be seen in some elements of this edition.

We have retained the use of a number of software packages and are very grateful to both EstateMaster and Argus for allowing an illustration of how their products support valuers.

This year we are also indebted to Soraya Hemming, James Gibson-Flowith, and Gareth Thomas from Harworth for providing a detailed strategic development acquisition and decision-making scenario, and to Philip Bacon, Dorus Helsen, and Joris Blommaert from Horwath HTL for working with us to provide a detailed boutique and restaurant appraisal using their ForValue system.

Last but not the least, Nick's wife, Mandy, for the many hours spent proofreading, on acronyms, and consistency checking.

Part I

Core frameworks, concepts, mathematics, and techniques

Chapter 1

The historical foundations of property as a concept, and how to value it

What is property, and what is real estate?

Property is a broad legal and economic concept, whilst *real estate* specifically refers to land and immovable structures. Both concepts are integral to individual livelihoods, business operations, and the functioning of economies and societies. *Property* refers to the legal rights or interests that persons or entities have in assets, which can be tangible (e.g. land, buildings, vehicles) or intangible (e.g. intellectual property, stock). It contains both real property, which comprises land and anything permanently tied to it, and personal property, which refers to movable assets.

Real estate, a specific type of property, refers to things that can't be moved, like land, buildings, and structures, as well as the rights that come with owning, using, and selling them. *Real estate* is focused on land and the permanent changes that are made to it, whereas *property*, on the other hand, refers to all types of ownership.

The concept of property has evolved significantly over time, shaped by the diverse philosophical perspectives of influential thinkers. Each of these historical figures – whether advocating for private ownership, communal systems, or revolutionary approaches – has contributed to our modern understanding of property and its role in society. Their opinions explore the ethical, social, and financial aspects of property, providing perspectives that still have an impact on discussions about justice, wealth distribution, and economic systems today. We will examine how these fundamental concepts have shaped how property is viewed and managed in the modern era in the paragraphs that follow.

Aristotle: property as natural and ethical

Aristotle (384–322 BC) was an ancient Greek philosopher who regarded property as a natural and necessary institution that plays a central role in individual well-being and societal harmony.

He emphasised the importance of private ownership, arguing that it incentivises productivity and fosters responsibility. However, he also maintained that property should serve the common good, not just the interests of its owner. For Aristotle, property had both practical and moral dimensions. It was a means to achieve self-sufficiency (*autarkeia*) and a tool for virtuous living. He cautioned against greed and advocated for moderation, viewing property as essential for a balanced and just society. His views laid the foundation for later discussions on the ethical responsibilities of property owners and the importance of equity.

DOI: 10.1201/9781003490357-2

Plato: communal ownership for harmony

Plato (427–347 BC) was a Greek philosopher who advocated for communal property to ensure unity and prevent corruption among the ruling class and critiqued private property for its potential to create division and selfishness. In his work *The Republic* (*Πολιτεία*), he proposed communal ownership for the guardian class, ensuring that those responsible for governance and protection were not distracted by personal wealth. Plato believed that property ownership could corrupt individuals and undermine societal unity. In another work, *The Laws* (*Νόμοι*), Plato adopted a more pragmatic stance, allowing for limited private property but with strict regulations to prevent excessive wealth accumulation. His vision prioritised equality and social cohesion, emphasising the moral and ethical use of property over its accumulation.

Adam Smith: property and economic prosperity

Adam Smith (1723–1790) was a Scottish economist who championed private property as essential for economic growth and individual liberty. He is often referred to as the Father of Economics and viewed property as a cornerstone of economic development and individual liberty. He argued that secure private property rights were essential for fostering innovation, investment, and the division of labour. Property for Smith was a natural extension of labour, and its protection was a primary role of government. Smith was also aware of the disparities that come with property ownership. Despite appreciating the financial advantages of property, he cautioned against monopolies and the concentration of wealth, supporting policies like taxes to lessen inequality. For Smith, owning property was both a moral obligation and an economic necessity.

Karl Marx: property as exploitation

Karl Marx (1818–1883) was a German philosopher who advocated for collective ownership and took a radical stance on property, defining it as a tool of exploitation under capitalism. In his works, such as *The Communist Manifesto*, Marx critiqued private property in the means of production, arguing that it enabled the bourgeoisie to profit from the labour of the proletariat. He famously described this relationship as the root of class struggle. According to Marx, private property alienates workers by separating them from their labour, its results, and other people. In his vision of a classless society where property served the general good rather than the increase of personal wealth, he argued for its eradication in favour of collective ownership.

Pierre-Joseph Proudhon: property is theft

Pierre-Joseph Proudhon (1809–1865) was a French theorist known for his provocative declaration 'Property is theft!', which encapsulates his critique of private property as a source of inequality and exploitation. He distinguished between 'property' (exclusive ownership) and 'possession' (use-based ownership). He opposed private property that allowed individuals to profit from resources they did not create, but supported possession as a just form of ownership. Proudhon supported mutualism, which is a system of equal resource allocation and cooperative ownership. In his ideal society, property relations would be based on usage and group benefit rather than exclusion and profit, and it would be decentralised and self-managed.

Other influential historical figures with notable views on property

John Locke (1632–1704). An English philosopher who argued that property is a natural right derived from labour and essential for individual liberty, advocating for its protection under social contracts.

Jean-Jacques Rousseau (1712–1778). A French philosopher who critiqued private property as the source of inequality and social corruption, famously stating, 'The first man who enclosed a plot of land and declared it his own was the true founder of civil society'.

Thomas Hobbes (1588–1679). An English political philosopher who viewed property as a creation of the state necessary to maintain order and prevent the chaos of a 'state of nature'.

Jeremy Bentham (1748–1832). An English philosopher and jurist who saw property as a legal construct designed to maximise utility and happiness within society.

David Ricardo (1772–1823). A British economist who explored the relationship between property, rent, and economic inequality, emphasising the role of land ownership in wealth distribution.

*Henry Ge*orge (1839–1897). An American economist and social reformer who criticised land monopolisation, advocating for a 'single tax' on land to address inequality and promote fairness.

John Stuart Mill (1806–1873). An English philosopher and economist who supported private property but argued for limitations to ensure social justice, including redistribution and taxation.

Georg Wilhelm Friedrich Hegel (1770–1831). A German philosopher who linked property to personal freedom and self-realisation, viewing it as an extension of the individual.

Thomas Paine (1737–1809). An American revolutionary and writer who proposed a system of universal income funded by taxes on property to counteract inequality.

Friedrich Engels (1820–1895). A German philosopher and collaborator with Marx who extended Marxist critiques of property, emphasising its role in class oppression and the exploitation of labour.

Private ownership vs collective systems

The philosophical perspectives on property reveal a spectrum of views, ranging from the defence of private ownership to its outright critique. Thinkers like Aristotle and Adam Smith champion private property as a natural extension of human labour and a cornerstone of societal and economic stability. For Aristotle, property supports personal virtue and serves the common good, while Smith emphasises its role in fostering innovation, investment, and market efficiency. Similarly, John Locke argued that property is a natural right derived from labour, essential for liberty and protected by social contracts. In contrast, David Ricardo and Jeremy Bentham focus on the economic and utilitarian dimensions of property, linking it to wealth distribution and societal happiness. Together, these thinkers recognise private property as an essential institution for progress but acknowledge its potential for inequality and societal imbalance.

On the other hand, philosophers like Karl Marx and Pierre-Joseph Proudhon critique private property for perpetuating inequality and exploitation. Marx viewed property, particularly in the means of production, as the root of class struggle, advocating for its abolition in favour of collective ownership. Proudhon famously declared, 'Property is theft', proposing mutualism as a system of equitable resource distribution based on use rather than ownership. Similarly, Jean-Jacques Rousseau identified private property as the source of inequality and social corruption, advocating for a return to more natural, egalitarian systems. Henry George and Thomas Paine also critiqued land monopolisation and proposed solutions like land taxation or universal income to address property-related disparities. These perspectives underscore that property has a dual role as a source of opportunity and of conflict, inspiring calls for more equitable systems.

Plato offers a unique middle ground, envisioning communal ownership for the ruling class to prevent corruption and ensure unity while allowing limited private property under strict regulations for others. This echoes the ideas of Thomas Hobbes, who saw property as a creation of the state necessary for societal order, and John Stuart Mill, who advocated for private ownership tempered by redistributive measures to promote social justice. Hegel extended these ideas by linking property to personal freedom and self-realisation, viewing it as an extension of the individual. Together, these views highlight property as not merely an economic construct but a deeply social and moral institution, shaping modern debates on resource distribution, social justice, and economic systems.

Following this important review of the philosophical origins of property as a concept, we can now focus on the real-world impacts of property ownership in contemporary society. Property valuation, a field that connects theory and practice by establishing the monetary value of property, is one important component.

To ensure fair transactions and well-informed policymaking, property valuation is essential in legal, investment, and economic decision-making situations. The concept of property valuation, its importance, and the methods used to determine the value of real estate are covered in detail in the following section.

What is property valuation?

The applied field of valuation works to support that decision by examining the market worth of real estate and the potential returns on real estate investments. *Valuers* are the people who give advice on the market value of real estate as well as on its purchase and sale. *Value* in this context typically refers to *market value* (value in exchange), but it can also indicate *investment value* or *worth*. A property needs to be useful to be valuable. It might be valuable to the current owner but can have no market worth if that usefulness is limited to that person.

Who is a valuer/chartered surveyor?

In the context of this book, a *valuer* would normally be a member of the Royal Institution of Chartered Surveyors (RICS), or one of the other valuation professional organisations (VPOs) approved with the agreement of the RICS, who meet the qualification requirements set out in the RICS Valuation Global Standards (*Red Book*) Professional Standard (PS) 2 2.1–2.8 and who acts with independence, objectivity, and confidentiality as specified in PS 2 3. But in other countries, similar professional bodies, such as the European Group of Valuers' Associations (TEGoVA), and/or licensing requirements (as in Singapore) exist to control or monitor the work of valuers. An overarching and global definition of valuation standards and expectations is set out in the International Valuation Standards (IVS).

The following definition of 'valuer' is given in the IVS (International Valuation Standards Council [IVSC] 2024, p. 11):

> An individual, group of individuals or individual within an entity . . . possessing the necessary qualifications, ability and experience to execute a valuation in an objective, unbiased, ethical and competent manner.

Why do we need property valuations?

Secured lending

Property valuation for secured lending ensures that the value of a property used as collateral aligns with the loan amount provided by a financial institution. Lenders use this valuation to mitigate risk, ensuring they can recover their funds by selling the property in case of borrower default. Accurate valuations are critical for establishing trust between lenders and borrowers and maintaining financial stability in the lending market.

Financial statements

Property valuations play a vital role in preparing accurate financial statements by reflecting the fair value of real estate assets owned by an organisation. This ensures transparency and compliance with accounting standards, such as the International Financial Reporting Standards (IFRS). By incorporating reliable valuations, companies provide stakeholders with a true picture of their asset base, which is essential for investor confidence and regulatory compliance.

Inform buy/sell decisions

Property valuations are essential for individuals and businesses making buy-or-sell decisions, offering an informed perspective on a property's current market value. This helps buyers avoid overpaying and sellers set realistic prices. With a reliable valuation, stakeholders can negotiate confidently and make strategic financial decisions, ensuring fair transactions for all parties involved.

Development appraisal

Valuations for development appraisal assess the feasibility and profitability of proposed real estate projects. By estimating the cost of development, potential sales, or rental income, developers can gauge whether a project will yield the desired returns. This analysis is crucial for obtaining financing, securing investor confidence, and minimising the financial risks associated with large-scale developments.

Statutory valuations

Statutory valuations are conducted to comply with legal requirements, such as for probate, divorce settlements, or compulsory purchase orders. These valuations ensure equitable distribution or compensation by providing an objective assessment of a property's value. They play a key role in legal proceedings, offering clarity and fairness in resolving disputes or fulfilling statutory obligations.

Assessment for taxation purposes

Accurate property valuations are essential for calculating tax liabilities, including property tax, inheritance tax, or capital gains tax. These valuations ensure that individuals and businesses pay the correct amount of tax, avoiding disputes with tax authorities. By providing a clear basis for taxation, they help governments maintain revenue streams while ensuring taxpayers meet their obligations fairly.

Compensation claims

Valuations for compensation claims are critical in determining the fair value of properties affected by external factors, such as compulsory acquisition, damage, or loss. These valuations ensure that property owners are adequately reimbursed for their losses. By providing objective assessments, they uphold the principles of fairness and equity, particularly in cases involving government acquisitions or disputes with third parties.

What are the methods *of*, and the approaches *to*, valuation?

Various valuation methods and approaches have been developed to cater to the diverse characteristics of properties and the purposes of valuation, such as buying, selling, financing, taxation, and development. Each method offers a unique perspective, ranging from market-based comparisons to income potential and construction costs, ensuring a comprehensive assessment tailored to the property type and market conditions. Selecting the appropriate method is essential to provide a fair, reliable, and objective valuation that aligns with industry standards and stakeholder needs. While bibliography suggests three approaches – comparison, income, and cost – the widely used methods are the following five.

Sales comparison method (market approach)

The sales comparison method determines the value of a property by comparing it to similar properties that have recently sold in the same area. Adjustments are made for differences, such as location, size, condition, and features. This method reflects real market conditions, making it ideal for residential properties and other assets with sufficient market activity. However, its accuracy depends on the availability of reliable and up-to-date sales data.

Income capitalisation method (the income approach)

The income capitalisation method values a property based on its ability to generate income. It calculates the value by dividing the annual net operating income (NOI) by a capitalisation rate that reflects market expectations for return on investment. This method is widely used for income-producing properties, such as rental apartments and commercial assets. Its strength lies in evaluating the property's investment potential, though it requires precise income and expense data.

Profits method

The profits method evaluates properties whose value is closely tied to the income they generate through business operations, such as hotels, theatres, or gas stations. The valuation is based on the net profit attributable to the property after deducting operating expenses and applying industry norms. This method is highly specialised and is valuable for unique properties with no comparable sales, though it relies on accurate financial data and market assumptions.

Residual method

The residual method is used for properties with development potential. It calculates the value by estimating the potential gross development value (GDV) after redevelopment, minus development costs and the developer's profit. This approach highlights the potential of undeveloped or underutilised land, particularly in urban areas. It can also be used to identify the highest and best use of the land through the examination of different potential developments. Its effectiveness depends on detailed knowledge of construction costs, market conditions, and planning regulations.

Cost method

The cost approach (or depreciated replacement cost method, DRC) estimates a property's value by calculating the cost of rebuilding it, minus depreciation, plus the value of the land. This method is most effective for new or unique properties where comparable sales are scarce, such as schools or hospitals. It provides a practical estimate of replacement cost but may not always align with market value if replacement costs differ significantly from what the market is willing to pay.

While other methods, such as the cost approach and the sales comparison approach, are explicitly analysed in this book, the primary focus remains on the *income approach* due to its technical depth and direct applicability to income-generating properties. Unlike simpler valuation techniques that rely on comparable sales data or replacement costs, the *income approach* involves complex financial modelling, requiring a deep understanding of investment principles, risk assessment, and market dynamics.

The *income approach* is inherently more analytical because it integrates key investment concepts, such as discounted cash flow (DCF) analysis, yield capitalisation, and sensitivity analysis. It allows valuers to quantify income potential, account for risk variations, and adjust for factors like inflation, rental growth, and occupancy trends. These elements are particularly crucial for institutional investors, lenders, and asset managers, who require precise financial evaluations to inform their investment decisions. Furthermore, the *income approach* is essential in the valuation of leasehold and freehold interests, where rental income projections and investment returns dictate market value, ensuring that valuers can assess complex investment scenarios with precision.

Through a focus on the *income approach*, this book equips valuers with the necessary skills to navigate modern property markets, where investment-grade real estate assets require sophisticated valuation techniques. Whether applied to commercial office buildings, retail centres, or long-leasehold properties, this approach remains the most relevant for professionals engaged in real estate investment, financing, and asset management.

What are the bases of value?

Bases of value describe the fundamental premise on which reported values are determined. It is critical that the basis of value aligns with the terms and the purpose of the valuation, as it directly influences the valuer's choice of methods, inputs, assumptions, and ultimately, the reported value (IVSC 2024).

Whilst many bases of value exist, the IVS define six primary bases of value, as outlined in **Table 1.1**.

Most valuations are conducted on the basis of establishing the market value or market rent, reflecting the property's position within the market. For example, when valuing a residential property for secured lending, the market value estimation is used, as the financial institution providing the loan needs to assess the potential sale price of the property in the event of borrower

Table 1.1 IVS-defined bases of value

Market value (MV)	'[T]he estimated amount for which an asset and/or liability should exchange on the valuation date between a willing buyer and a willing seller in an arm's length transaction, after proper marketing and where the parties had each acted knowledgeably, prudently and without compulsion' (IVSC 2024, p. 24).
Market rent (MR)	'[T]he estimated amount for which an interest in real property should be leased on the valuation date between a willing lessor and a willing lessee on appropriate lease terms in an arm's length transaction, after proper marketing and where the parties had each acted knowledgeably, prudently and without compulsion' (IVSC 2024, p. 26).
Equitable value	'[T]he estimated price for the transfer of an asset or liability between identified knowledgeable and willing parties that reflects the respective interests of those parties' (IVSC 2024, p. 27).
Investment value (worth)	'[T]he value of an asset to a particular owner or prospective owner for individual investment or operational objectives' (IVSC 2024, p. 27).
Synergistic value	'[T]he result of a combination of two or more assets or interests where the combined value is more than the sum of the separate values. If the synergies are only available to one specific buyer, then synergistic value will differ from market value, as the synergistic value will reflect particular attributes of an asset that are only of value to a specific purchaser' (IVSC 2024, pp. 27–28).
Liquidation value	'[T]he amount that would be realised when an asset or group of assets are sold from a liquidation sale, with the seller being compelled to sell as of a specific date' (IVSC 2024, p. 28).

default. Similarly, when a landlord seeks to relet a vacant commercial warehouse, they would rely on an estimate of market rent, as the asking rent must be competitive with other similar properties in the market to attract potential tenants.

On the other hand, when assessing a distressed asset requiring a forced sale outside the normal parameters of the open market, a liquidation value would be determined.

The significance of the eighth edition

In the quickly changing real estate investment landscape of today, which is influenced by shifting market conditions, economic uncertainty, and technological breakthroughs, the eighth edition of *Income Approach to Property Valuation* is extremely important. The difficulties presented by shifting interest rates, inflation, and international uncertainties – all of which have increased the demand for accurate and trustworthy valuation methodologies – make this edition especially pertinent. The book gives practitioners the tools they need to accurately evaluate income-generating assets during times of economic turbulence, such as post-COVID-19 recovery or geopolitical upheavals, by providing updated guidelines.

The eighth edition offers frameworks for evaluating non-traditional revenue streams as the real estate market expands to encompass new asset classes, like data centres, logistical hubs, and flexible workspaces. It also considers developments like digitalisation; environmental, social, and governance (ESG) compliance; and hybrid working, bringing valuation procedures in line with shifts in society and contemporary investment goals. Its usefulness is increased by adding sophisticated methods for income capitalisation and DCF analysis, particularly when assessing properties with erratic income trends or long-term growth potential. To account for uncertainties in income projections, a critical component in today's unpredictable markets, this edition highlights the significance of risk adjustments and sensitivity analysis.

The eighth edition's technological integration is yet another highlight. It investigates how to improve valuation inputs and spot market trends using automated valuation models (AVMs), PropTech solutions, and big data analytics. These tools increase property appraisal accuracy in addition to efficiency. Additionally, as cross-border investments grow more common, the book's relevance in international real estate transactions is ensured by its alignment with IVS.

In recognition of the rising demand for energy-efficient and green-certified properties that fetch higher prices, the eighth edition also covers sustainability and ESG issues. It guarantees that valuers are prepared to manage the intricacies of contemporary valuation scenarios by integrating frameworks to take these elements into consideration. Updated case studies and real-world examples further support the practical applicability, connecting abstract ideas with useful information for investors, property valuers, and other stakeholders.

In conclusion, the eighth edition of *Income Approach to Property Valuation* is a vital tool for understanding the intricacies of today's real estate markets. It is a fundamental resource for obtaining accurate and trustworthy appraisals in a constantly shifting environment because of its all-encompassing approach, which considers new trends, cutting-edge techniques, and international standards. It continues to be a crucial resource for both academics and practitioners to comprehend and successfully use the revenue method to property assessment.

Questions

1. What is the difference between *property* and *real estate*?
2. How did Aristotle and Plato differ in their views on property?
3. Why did Karl Marx oppose private property ownership?
4. What role does property valuation play in economic and financial decision-making?
5. What are the main methods used in property valuation?
6. How do different bases of value impact property valuations?
7. What are the ethical considerations in property valuation?
8. How do property valuation standards contribute to market stability?

Reference

International Valuation Standards Council (IVSC) (2024). *International valuation standards*. London: IVSC.

Chapter 2

Legal and taxation frameworks

The legal framework

Introduction

Land and *buildings*, *property*, and *real property* are all terms used loosely to describe what it is that valuers and appraisers value. However, in law, it is the legal interest in land and/or buildings that is the subject matter of the valuation; this may also be referred to as a 'bundle of legal rights'.

Legal interests in land vary across jurisdictions, but broadly, ownership can be classified into freehold (or fee simple) and leasehold interests. *Freehold* represents absolute ownership, often considered perpetual, subject to legal and regulatory constraints. A freeholder can transfer, subdivide, or lease their land, provided it complies with relevant laws and any necessary permissions. In some jurisdictions, if a freeholder dies without a will or heirs, ownership may revert to the state.

Leasehold, in contrast, is limited in time to the term specified in the lease. It is created out of a freehold, and the rights granted within the lease may be limited – that is, the freeholder may limit what the leaseholder can or cannot do with the land. The length of leases varies globally, with some countries permitting long lease terms, such as 99 or even 999 years, which is the maximum allowed in the United Kingdom (UK) and Ireland. Some countries, like Australia, do not specify any minimum or maximum lease term. Others, such as Argentina, have laws that restrict the maximum term to 20 years for residential leases and 50 years for any other kind of lease. In addition to any legal permissions or licences, the leaseholder will generally have to obtain the freeholder's consent or licence to do anything other than that permitted within the lease.

In some countries, national legislation may limit the length of leases, for specific policy reasons, such as transparency and money laundering. The Economic Crime (Transparency and Enforcement) Act 2022 applies to all UK properties. In simple terms, it prevents overseas entities from granting leases longer than seven years.

Freehold ownership

A freehold title may still be subject to a wide range of limitations, and so in that sense, the rights are not completely absolute. In many countries, land ownership generally includes everything below the surface and everything above the surface into infinity. Imagine a 1 square hectare or

DOI: 10.1201/9781003490357-3

square acre of land with the surface boundaries extended below to the centre of the earth and above to the far reaches of space.

However, there are many restrictions and limitations which may need to be considered by valuers, including thus:

- **Adjoining property rights.** Owners must not interfere with neighbouring properties by restricting access to water, causing pollution, or undermining structural support. Some jurisdictions recognise rights of light, air, or access.
- **Easements and servitudes.** Certain legal rights may be granted to third parties, such as rights of way, or to extract resources like timber, fishing, or hunting.
- **Mineral and resource rights.** In many countries, valuable underground resources, such as oil, gas, and precious metals, are owned by the state rather than the landowner.
- **Airspace regulations.** While a property owner has rights over the airspace above their land, these are subject to aviation laws, particularly near airports or military zones.
- **Restrictive covenants.** In some countries, a restrictive covenant clause may be inserted as part of a land transfer, whereby the seller restricts activity on the land sold generally for the benefit of land retained. For instance, in the UK in the 19th century, land put up for sale for housing was often sold with a restriction to a single-family home. Land sold today may be subject to future payments should more valuable development be permitted at a future date, such as grazing land being granted permission for housing development.
- **Land use planning and development control.** Most jurisdictions require landowners to obtain planning or zoning approvals before undertaking significant development. Building regulations govern construction standards, including fire safety, energy-efficiency, and sustainability.

Leasehold occupation

The rights of a leaseholder are similarly restricted. A lease will generally include a number of specific conditions, plus other restrictions or covenants. A *covenant* can be defined as a formal agreement or promise, usually included in a contract or deed. They may set out what *can be done*, such as peaceful enjoyment of the property, or *not be done*, such as change a shopfront. The most significant lease covenants relate to the payment of rent, the use of any land and buildings, the repair of any buildings, the freeholder's consent for any development or alterations and improvements, and the yielding up at the end of the lease.

In many countries, tenants have substantial rights to occupy the building leased to them without interference from the landlord. In the UK, this is known as *quiet enjoyment*, which means the leaseholder has a right not just to occupy the property, but to use it in a lawful way, without interference.

However, in certain circumstances, this may be overridden. For example, in the UK, the police have powers of access exercisable if they believe there is a criminal activity taking place or if they have reason to believe someone they are seeking is inside a property. As in many countries, the state has certain powers to use property during times of conflict, and they and other authorities may have or may obtain powers of compulsory acquisition when land (property) is required for specific uses. Examples might be social housing, education, new highways, or highway improvements. This may be referred to as *eminent domain*.

Types of leases

The nature of a lease is often qualified with a prefix. The most common lease types are *occupation* leases, *periodic* tenancies, *ground* leases, leases *for life* (which are now by law leases for 90 years, terminable on death), *reversionary* leases, and *sub*leases.

Occupation lease

An *occupation lease* is the typical lease that can be found in relation to all types of property. It will be for a fixed term of years. Lease durations vary significantly across markets. We provide a summary table of lease lengths and other lease covenants in different countries later in this chapter.

Historically, institutional leases in the UK and some other countries were for 25 years, with rent reviews to market rent (upward only) every five years. The term developed in the 1960s, when institutions such as insurance/assurance companies, pension funds, and property developers sought tenants willing to take a long lease of 25 years with regular reviews of rent. This provided the newly emerging institutional investors with a secure, but not fixed, income and some certainty of continued occupation. Tenants, especially retailers, banks, and public sector tenants, also saw these leases as opportunities to secure space for a favourable length of time without any immediate need to find alternative premises. Retailers in particular sought (and still do) a long enough lease to spread the substantial costs of fit out over.

Lease lengths in the UK have fallen, in line with many other countries, and currently, occupation leases are often for three to five years; if over five years, there will generally be a tenant's break clause[1] and a rent review.

In France, the standard commercial lease is known as a bail commercial, lasting 9 years, with a tenant break option every 3 years. In the United States, lease terms vary by sector, with office leases often ranging from 3 to 10 years and retail leases sometimes exceeding 20 years in prime locations. Some markets, such as Hong Kong, favour shorter leases, often around 3 years. Similarly, in Brazil, commercial leases typically range from 2 to 5 years, with 5-year terms ensuring automatic renewal rights for tenants, while shorter durations give landlords flexibility to avoid renewal obligations (DLA Piper 2024).

Lease structures also differ widely: Whilst some jurisdictions prefer full repairing and insuring (FRI) leases, where tenants assume insurance and maintenance responsibilities (common in the UK and Australia), others favour gross leases, where landlords cover property costs and include them in the rent (common in parts of Europe and Asia).

The use of formal rent reviews at specified periods, for example, at years 3 and 6 in a 9-year lease, is common in the UK, Commonwealth countries, and some others. However, in many European countries, rent is increased using some form of index, and in some cases using an index plus an uplift to create real returns above inflation. In retail leases, rents may also use an element of turnover, at an agreed percentage, in the rental agreement.

Periodic tenancy

A *periodic tenancy* is usually a right to occupy on a recurring basis, such as weekly, monthly, or quarterly, until either the landlord or the tenant gives notice to terminate. Unlike a fixed-term lease, which has a predetermined end date, periodic tenancies automatically renew at the end of each rental period, unless proper notice is given. There is an expectation that provided the tenant occupies the property in a tenant-like manner, then the right to occupy will be renewed week to week or month to month. The rent is the market rent for that property let on those terms. These periodic tenancies do not have a value per se. A tenant moving out may, however, be able to charge an incoming tenant for any tenant's fixtures and fittings.

Long leaseholds

Long leaseholds, sometimes referred to as *ground* leases or *building* leases, often granted for terms of 99, 125, or even 999 years, are a type of lease used in several countries. Originating in the Roman Empire, long leaseholds evolved differently across markets; while Napoleonic reforms largely replaced them with private land ownership in continental Europe, they remained a significant form of tenure in other regions (JLL 2019).

The use of long leaseholds in the UK dates back to the Industrial Revolution and to a time when much of the land in the UK was owned by large estates, the owners of which held entailed title – that is, the title had to pass down the male line and could not be sold free of the entail. This restriction on inheritance and rights to sell meant that, as more land was needed to provide for the new factories and housing, landowners were unwilling or unable to sell the freehold interest. So instead, they sold very long leases or combined a sale, which generated almost the same sale price as a freehold disposal, with a ground rent. These ground rents were sometimes only peppercorn, that is, at a nominal sum, but if they were let at a rent reflecting the value of the land for housing or industrial use, then that rent would be many times the level of rent previously paid for the land for farming. In a period of non-inflation, no one thought to include provision for rent reviews. The custom of granting leases rather than selling freeholds for development continues today; owners in this case may be public sector or private, but with a wish to retain ultimate ownership. In some cases, a more complex form of lease will be encountered, where the freeholder wishes to participate in any growth in occupation rent that may occur, be it real or inflationary. These are geared rents, so that the relationship between the initial ground rent and the occupation rent secured on completion of the development becomes a fixed ratio, and as the occupation rent increases, so too does the ground rent.

In Germany, leaseholds were reintroduced after the First World War through the 1919 Leaseholds Ordinance, aimed at facilitating homeownership for lower-income households (JLL 2019). In Hong Kong, nearly all land is leasehold, typically for 75 or 99 years, with ownership retained by the government. Introduced after the British occupation in 1841, the Hong Kong model has influenced ground lease systems in mainland China and Singapore, where most land is also leasehold, and the sale of long-term ground leaseholds remains a key source of government funding (Pope and Young 2024).

Lease for life

A *lease for life* is a lifetime tenancy agreement between the new or current owner of a property and a tenant who wants to lease the property until they pass away. They are often used in equity release situations and by investors who are looking for a long-term capital growth investment, as the property itself may not be income-producing. The valuation of life interests is outside the scope of this book.

Reversionary lease

A *reversionary lease* is, as its name implies, a lease of property which begins after a current lease terminates through the effluxion of time. These could have an impact on the value of a freehold depending on the lease terms but are too specialised to be considered here.

Sublease

Where permitted by a lease, a leaseholder may create *subleases*. Clearly, a developer creating a new distribution site on a 125-year lease will need to sublet the new units on occupation leases so will expect to have that as a right written into the building lease (sometimes referred to as a head lease). A tenant on an occupation lease may find that they have surplus accommodation or may wish to vacate and sublet the property.

Most UK leases will have a covenant (clause) in the lease which prohibits the subletting in whole or in parts and a prohibition on the assignment of the lease (selling on the lease to another occupant). In current leases, this will typically be 'without the landlord's consent, whose consent should not be unreasonably withheld'. But if a tenant assigns or sublets, they may still be liable under the lease, as the freeholder can require the leaseholder to enter into a guarantee under the Landlord and Tenants Covenant Act 1995.

Ownership rights worldwide may go by different names but fundamentally will be recognised as equivalent to freehold and leasehold ownership. In some countries, sovereignty is with the state, and in some, native or communal rights will be found. For the valuer, an initial stage of due diligence is the clarification of the ownership rights and the extent to which those rights are encumbered. The process for verifying ownership differs by country. In many markets, this involves reviewing land registries or equivalent official records. Where registration is incomplete, valuers may rely on legal documentation provided by property owners or their representatives, including title deeds and lease agreements.

Statutes affecting rights to let property

Although freeholders have the right to let their property, in whole or in part, on whatever terms they want or are able to negotiate, government can interfere and has interfered with these rights, particularly in relation to residential lettings. This interference has been both in terms of the right to recover possession at the end of a lease and in respect of the rent that might be legally recoverable, that is, the amount that may be recoverable at law may be more or less than the market rent, or the rent negotiated, and entered into the lease. Legislation has generally been seen by government as necessary at certain times to protect tenants who are seen as being at risk or vulnerable. We provide a summary of such statutory influences in our comparative table at the end of this section.

Residential property

The residential sector is the most likely sector of the property market to come under scrutiny by governments around the world. This is because, in periods of excess demand, market rents can be perceived to be rising too high and too fast. In other circumstances, governments may see residential landlords as unscrupulous. Landlord-tenant regulations vary widely across jurisdictions, but certain themes are common. The following examples illustrate different approaches.

Rent control and security of tenure

Rent control refers to government-imposed regulations that limit the amount landlords can charge for rent, often to protect tenants from excessive increases. Security of tenure, which ensures tenants have the right to remain in a property under certain conditions, is often linked to such regulations. Some jurisdictions retain legacy rent-controlled tenancies, such as '*regulated tenancies*' in the UK where the lease started before 15 January 1989. In these cases, maximum rents, referred to as 'fair rents', are defined very precisely in the respective legislation, including the need to disregard market scarcity. Regulated tenants have security of tenure, that is, when the protected tenancy comes to an end, it continues indefinitely as a statutory tenancy. In Germany, federal states can impose rent control in municipalities with tight rental markets, limiting rent increases in new contracts to 10% above the local comparative rent index (Breidenbach et al. 2022).

Leasehold enfranchisement

In some countries where long leaseholds exist, leaseholders may have a legal right to purchase the freehold of their property or extend their lease. For instance, in the UK, tenants of long leases of residential property (21 years or longer) will typically have the right to enfranchise, that is, to require the owner to sell the freehold to the tenant. This is a complex area of law and outside the scope of this text. Provisions will vary by country, but in England and Wales, it was introduced by the Leasehold Reform Housing and Urban Development Act 1993. Recently, the Leasehold and Freehold Reform Act 2024, approved but not yet enacted, will extend the opportunities for enfranchisement, making it potentially easier and cheaper to undertake. In the UK, the Leasehold and Commonhold Reform Bill is set to bring the feudal leasehold system to an end and prevent long leaseholds and improve the commonhold system of land tenure.

Energy-efficiency requirements

Many governments impose minimum energy-efficiency standards for rental properties. In the UK and the European Union (EU), properties with low energy-performance ratings may not be legally lettable. Similar provisions apply in Australia, where certain states enforce energy-efficiency disclosure for rental properties.

Commercial tenancies

In many countries, laws establish the rights and responsibilities of landlords and tenants, influencing lease negotiations and market dynamics. Key themes related to commercial (business) tenancies include what follows.

Security of tenure and lease renewal

In some jurisdictions, business tenants benefit from statutory protection, ensuring that leases continue beyond their expiry date unless terminated through a legally defined process. For instance, in the UK, the Landlord and Tenant Act 1954 grants security of tenure, requiring landlords to serve formal notice and demonstrate specific grounds for possession, such as tenant breaches, redevelopment plans, or owner occupation. In Brazil, the *Lei do Inquilinato* (Tenancy Law) allows commercial tenants to request judicial lease renewal if they have occupied the premises for at least five years under a fixed-term written contract and have conducted the same business for at least three years. If the landlord lacks valid grounds for refusal – such as personal use or redevelopment – the lease is generally extended under similar terms.

Compensation for tenant improvements

In certain jurisdictions, tenants may claim compensation for authorised improvements upon lease termination, particularly if these improvements were made under a formal agreement with the landlord. The criteria for compensation often depend on whether the improvements were required under the lease terms, their impact on the property's value, and the existence of reinstatement clauses. In the UK, the Landlord and Tenant Act 1927 contains provisions for such compensation, allowing tenants to recover costs for improvements that enhance rental value.

Energy-efficiency requirements

As with residential property, environmental regulations influence commercial leasing. In the UK, the Minimum Energy Efficiency Standard (MEES) prohibits landlords from letting commercial properties with an energy performance certificate (EPC) rating below 'E' (UK Government 2015). In the EU, the Energy Performance of Buildings Directive (EPBD) sets efficiency targets, requiring member states to enforce stricter energy performance standards and encourage sustainable retrofits (EU 2024).

A global comparison of lease lengths and significant terms

It is not possible to provide a comprehensive overview of all lease terms across all countries, and this is outside the scope of this text. However, we present a summary of lease terms for some countries to illustrate the complexities and contrasts between them in the following table. We do this to illustrate the dangers of assuming the world is the same and how essential it is for valuers operating internationally to be fully aware of the legal and other contexts within which they are operating. Readers should consider the wide variety of lease types, lease lengths, and rights to renew as set out in Table 2.1.

Comparison of lease arrangements across four continents

In addition to the many variations in lease length, renewal options, and the balance of the use of licences versus leases, a number of other differences can be observed globally in terms of principal lease covenants.

These are set out in Table 2.2.

Planning control and/or zoning

Throughout the world, the use and development of land is controlled to an extent by central and local government. Areas may be zoned for specific uses, or permission may have to be obtained for the actual development of a site for a specific use or change of use.

In the UK, local authorities are responsible for planning policy; planning legislation requires owners and occupiers seeking to develop or alter property or to change the use of property to apply for, and secure, the necessary planning permission.

Table 2.1 A comparison of lease types and lengths across a number of countries

Europe	

France	• France has a minimum obligatory term for a commercial lease of nine years. Tenants have the right to terminate the lease at the end of each third year. The parties are free to negotiate a notice period longer than six months.
	• Short-term leases can be granted for a maximum term of three years (several short-term leases can be granted consecutively, provided that, in aggregate, their duration does not exceed three years), but in this case, legislation relating to commercial leases does not apply.
	• In specific cases, the tenant may waive its right to terminate the lease at the end of the relevant three-year period. The parties can agree whatever termination options/mechanisms they wish.
	• For example, they may agree that there will be a financial penalty for exercising a termination.
	• The cases in which the tenant may waive its legal right at the end of each three-year period are:
	  • Leases relating to premises for office use
	  • Leases relating to premises only suitable for a specific business activity due to layout or design
	  • Leases relating to storage premises
	  • Leases with an initial duration of more than nine years
	• Whilst there is no maximum term for a commercial lease, any lease of more than 12 years must be notarised, and additional fees are payable.

Germany	• No minimum or maximum period for general leases of commercial and/or residential premises, and they may be entered into for periods of more than 30 years. However, long-term arrangements like this may be terminated by either party after 30 years. • Hereditary building rights (*Erbbaurechten*) that entitle the holder to use the leased property for up to 99 years and to erect buildings on the land can be created for longer periods of time, typically 99 years.
Poland	• The term of a **tenancy** known as a *dzierżawa* can be definite or indefinite. A tenancy is regarded as having an indefinite term, where the parties agree on a period of over 30 years and the initial term of the lease has come to an end. The tenancy continues indefinitely, with either party having the right to terminate within a relatively short statutory time of up to 1 year. • The term of a **lease** known as a *najem* can also be indefinite or definite, with a maximum term of 10 years, or in the case of a lease concluded between 'entrepreneurs' (which, under Polish law, are defined as natural or legal persons, or non-corporate organisational units, conducting economic activity on their own behalf), 30 years. A lease is regarded as having an indefinite term where the parties agree on a period of over 10 years (or 30 years for a lease between entrepreneurs), and the initial term of the lease has come to an end. The lease continues indefinitely, with either party having the right to terminate within a relatively short statutory time of up to 3 months.

(*Continued*)

Table 2.1 (Continued)

Asia Pacific (APAC)	
Australia	• No minimum or maximum term for a commercial lease. The duration of the lease term can be for any number of years or up to a fixed date. • Generally, a commercial office lease for a larger tenant will be for 8 to 10 years, with one or two option terms (exercisable by the tenant) of 3 to 5 years each. In most Australian jurisdictions, leases for a period which exceeds 3 years must be registered on a public register with the relevant government authorities for the lease to be legally effective. The main benefit of registering the lease on a public register is to provide third parties with notice of the tenant's leasehold interest in the property. • Retail leases, usually in shopping centres, are governed by specific legislation, and in most jurisdictions, retail leases have a minimum term (inclusive of options for renewal included in the lease) of 5 years, unless the tenant agrees to waive that right.
Hong Kong	• Whilst having many origins in UK commercial law and practice, there are increasing differences. There is no minimum or maximum term for a commercial lease. Leases tend to be shorter than 5 years; 10 years or more, whilst possible, are uncommon. • In the office, market terms are typically 2 to 3 years with or without an option to renew for another 2 to 3 years. In the retail market, leases are comparatively longer, with three to five years being usual, with or without an option to renew for another three to five years. The current market trend is toward ever-shorter leases.
New Zealand	• No minimum or maximum term for a commercial lease. The duration of the lease term can be for any number of years or up to a fixed date and can include rights for the tenant to renew the lease (often subject to certain preconditions being met). These are commercial matters for the parties to agree.
North and South America	
Canada	• No minimum or maximum term for a commercial lease. The duration of the lease term can be for any number of years or up to a fixed date. • Most office or retail leases for a larger tenant will be for 5 to 10 years, with one or more option terms (exercisable by the tenant), usually of 1 to 5 years each. Ground leases tend to be for longer terms, such as 20 to 99 years, and in many cases, the rent is prepaid for the entire term. • In most Canadian jurisdictions, leases for a period greater than 3 years may be registered on a public register maintained by the relevant government authorities subject to certain statutory requirements. The main benefit of registering the lease on a public register is to provide third parties with notice of the tenant's leasehold interest in the property. In some Canadian jurisdictions, registration of long-term leases (for example, for a term greater than 30 years in British Columbia) will attract land transfer tax in the same manner as a transfer of freehold title.

(*Continued*)

Table 2.1 (Continued)

North and South America	
United States (US)	• The duration of commercial leases is case-determined and can be of any length. Typically, they range from 3 to 10 years. Longer terms may apply to ground leases and 'net leases'. • Net leases are where the tenant is responsible for all the costs of operating, insuring, and paying taxes on the real property.
Brazil	• The usual term of commercial lease agreements depends on the parties' intention to secure the right of automatic renewal or not. If the tenant wants to secure its right of renewal, they should negotiate the term of the contract for 5 years. On the other hand, if it's not in the landlord's interest to automatically renew the contract, they should accept a term shorter than 5 years. Accordingly, it is common that commercial leases have initial terms ranging from 2 to 5 years.
Middle East	
United Arab Emirates (UAE)	• It is important to recognise that in some countries, whilst some federal laws may apply across the country, there may be specific property legislation and regulations applying to a particular state, or in this case, an emirate. • In ***Abu Dhabi***, the duration of commercial leases is case-determined and can be any length of time but typically ranges from 3 to 10 years. Longer-term leases tend only to be encountered in the context of 'ground leases' and 'net leases'. In 'net leases', the tenant is responsible for the cost of operating, insuring, and paying taxes on the real property. A 'ground lease' is a subspecies of net lease where the tenant is given rights to develop a parcel of land leased to it. • ***Dubai*** has a more developed real estate market, and it is common to see a commercial lease for a term of between 1 and 5 years.

Source: This table has been produced using information from a number of sources, but most specifically the DLA Piper website.

A principle often referred to in valuation books is 'highest and best use'. This requires or expects a valuer undertaking a valuation to have regard to both the current use to which a property is put and to its potential use, which may give rise to a higher market value. The point is that if buyers see a half-empty office property that has the potential to be converted into luxury apartments and used in this way, it will have a higher market value, taking into account all costs of securing vacant possession and of conversion. This enhanced value, due to the development potential, then may be the market value. So as part of due diligence, a valuer must explore the current permitted use of a property and its potential use, for which permission can be assumed or obtained. In this respect, in England, a wide range of permitted use changes, including essential alterations, are allowed (unless specifically excluded by a planning authority in, say, a conservation area) under The Town and Country Planning (General Permitted Development etc) (England) (Amendment) Order 2024.

In many countries, the planning and zoning documentation provides specific and detailed information, not only on the permitted uses for a site, but also on the maximum gross floor area

Table 2.2 A comparison of commercial property lease types and lengths across a number of countries (as of 2024)

Country	*Lease length*	*Legal right to renew*	*Do rent reviews apply?* Period	*Basis of rent review*	*Index basis*	*Repair*	*Value-added tax (VAT)* Chargeable?
APAC							
Australia	3–6 years	No	2 years	Negotiable; both open market value (OMV) and indexation in use	3–4% or another national index	Tenant for internal repair only.	**10%** applied as goods and services tax (GST) and applicable to commercial and retail leases only.
China	2–3 years	Yes	On expiry	OMV	N/A	Unless otherwise agreed, the lessor is responsible for the maintenance and repair of the premises.	After 2016: **9%**.
Hong Kong	2–3 years	No	1–3 years	OMV	N/A	Tenant for internal repair only.	NO VAT but stamp duty payable on rent and term of the lease.
Japan	2–5 years	In some cases	6 months, prior to lease expiry		N/A	Landlord responsible for the maintenance and repair of the building structure. In many commercial leases, the lessees are responsible for internal repairs, maintenance, and repair of wear and tear.	NO VAT but consumption tax at **10%**.
Singapore	1–5 years	Yes	Not used	Not used	N/A	Keep the premises in a 'tenantable condition'. Required to inform the landlord if repairs are needed and give access to view the state of repairs and carry out the repairs.	**7%** applied as GST and applicable to commercial and retail leases only.

(Continued)

Table 2.2 (Continued)

Country	*Lease length*	*Legal right to renew*	*Do rent reviews apply?* Period	*Basis of rent review*	*Index basis*	*Repair*	*Value-added tax (VAT)* Chargeable?
AMERICAS							
Argentina	3–5 years	No guarantee	2–3 years	Consumer Price Index (CPI)	CPI	If repairs are needed for the habitability of the premises, the landlord pays.	Yes, but some exceptions for residential.
Canada	5–10 years	Yes	Negotiable	Negotiable; both OMV and indexation in use	If index: CPI or specific %	Generally, the tenant for internal repair only. The landlord is typically responsible for maintaining the exterior, structure, and common areas of the property, subject to reimbursement by tenants through a shared operating expense mechanism.	Applied as goods and services tax (GST) and applicable to commercial and retail leases at **5%** (additional state taxes may apply).
United States	3–10 years	No	Both periodic reviews and indexation used			Tenants responsible for maintaining and repairing non-structural, non-building system portions of the leased premises, landlords for repairing and maintaining the property exterior, structure, systems, and common areas, subject to reimbursement, through a shared operating expense mechanism. Some leases (known as 'full-service gross' leases) require tenants only to pay increases above those operating expenses that the landlord incurs in a pre-determined *base year.*	There is no value added tax on base rents in the United States. However, some city and other local governmental authorities do impose sales or occupancy taxes on base rents.

(*Continued*)

Table 2.2 (Continued)

Country	*Lease length*	*Legal right to renew*	*Do rent reviews apply?* Period	*Basis of rent review*	*Index basis*	*Repair*	*Value-added tax (VAT)* Chargeable?
EUROPE							
France	3, 6, 9 years	Yes	On expiry	OMV	Construction index	Under the French Civil Code, expenses relating to the maintenance of the premises and major repairs are payable by the landlord. It is market practice for the landlord to recover all, or a major part, of these expenses from the tenant in the form of service charges.	**20%** or 2.5%; landlord can opt for one of two taxes to apply: VAT (at 20%) or Contribution sur les Revenus Locatifs (CRL) amounting to 2.5%.
Germany	5–10 years	No	Not used	Not used	Cost of living (COL)	The tenant is responsible for minor repairs. Under statutory law, the landlord is responsible for the maintenance and repair of the leased property. Most leases contain an apportionment of responsibility, by which the landlord will be responsible for the structure, external appearance, and common areas (Dach und Fach).	Generally, rental payments exempt from VAT; **19%** if the landlord can waive the exemption and the lease is to an entrepreneur who uses the land for business purposes.
Nether lands	5–10 years	Market-driven	5–10 years	OMV	COL	The tenant is responsible for minor repairs. Landlord obliged to carry out major maintenance, enhancement, and repairs. Parties may agree a different arrangement.	**21%**, but only payable if parties agree to charge it and the premises are used for taxable purposes.

(*Continued*)

Table 2.2 (Continued)

Country	*Lease length*	*Legal right to renew*	*Do rent reviews apply?* Period	*Basis of rent review*	*Index basis*	*Repair*	*Value-added tax (VAT)* Chargeable?
Poland	3–7 years	No	Not used	Not used	COL	The tenant is responsible for minor repairs. The landlord must maintain the leased property so that it is fit for its purpose throughout the period of the lease. This obligation cannot be contracted out of in the agreement. The landlord is liable for defects in the leased property which, wholly or partially, prevent it from being used for its intended purpose according to the lease. In such cases, the tenant may request a rent reduction or, if the defects are irreparable or the landlord refuses to repair the property, may even withdraw from the agreement.	**23%** on commercial property (including agricultural real estate); **8%** on short-term leases.
Spain	3–10 years	No	3–5 years	OMV	COL	Under the Urban Leases Act, the landlord is responsible for the maintenance and repair of leased property; however, the parties to a commercial lease are free to allocate this responsibility to the tenant. Small repairs due to ordinary wear and tear will be borne by the tenant.	**21%** on commercial premises.

(*Continued*)

Table 2.2 (Continued)

Country	*Lease length*	*Legal right to renew*	*Do rent reviews apply?* Period	*Basis of rent review*	*Index basis*	*Repair*	*Value-added tax (VAT)* Chargeable?
UK	Up to 25 years, but typically 3–7 years	Yes	3–5 years	OMV (some on an upward-only basis)	N/A	Tenant for internal repair only. In multi-let developments, the landlord repairs and maintains common areas and the structure of the building and recovers the cost from the tenants via a service charge.	**20%**, but VAT not payable unless landlord opts to do so.

Source: This table has been produced using information from a number of sources and databases, including the DLA Piper website.

(GFA), height of buildings, density of construction, and other parameters which inform what could be constructed. In some countries, this detailed information is highly transparent and available online, and so consideration of the *highest and best use* is straightforward. In other countries, it may be much more opaque and subject to complex negotiations, which make it more challenging to assess the development potential of a site.

In some cases, planning-based restrictions may have a direct impact upon the investment potential of a development. For example, it is becoming common in many countries for restrictions to be placed upon the proportion of residential apartments in a large block that can be sold to investors (in some cases, specifically overseas investors). This is because large volumes of empty apartments impact negatively upon the vitality of a development scheme and the ability to energise supporting services and create a vibrant community.

A detailed understanding of country-specific planning and zoning practices has to be part of a valuer's knowledge base and kept current through continuing professional development (CPD).

Taxation of property

Property taxes are one of the traditional revenue sources for most governments of the world and have existed for a very long time. The tax is collected at different administration levels, ranging from the national to the federal to the state/county/region to the municipality. According to a United Nations Human Settlements Programme report, property tax is widely implied and generally accepted to be a combined assessed value of land, the buildings that it contains, and any improvements made thereto (UN-HABITAT 2011). Excluding the application of taxation to land values, property tax attempts to generate revenues from structures and land improvements, although the latter part has given rise to many debates on its fairness. In many mature property markets, taxation is usually related to some form of valuation, since this is considered to be fairer than other methods. Often, taxation on capital is in price bands with a minimum price, below which there is no tax. This allows for taxation to be progressive if that is the policy of a country.

The Organisation for Economic Co-operation and Development (OECD) classifies property taxes into six basic types: recurrent taxes on immovable property; recurrent taxes on net wealth; estate, inheritance, and gift taxes; taxes on financial and capital transactions; other non-recurrent taxes on property; and other recurrent taxes on property (OECD 2014). Some of the aforementioned six types are subdivided into further categories – for example, recurrent net wealth taxes are divided into individual and corporate taxes.

Property tax in Europe is mainly concerned with the collection of annual taxes on immovable property, the latter comprising both real property and real estate. Approximately 66% of property taxes in the EU are of a recurring nature, contributing to 1.5% of the gross domestic product (GDP) and collected annually on an ad valorem basis. There are significant variations in these recurrent property tax rates. Rates are highest in the UK (3.4% of GDP), Denmark (2.4% of GDP), and France (2.1% of GDP), whilst they are not applied at all in Croatia, Luxembourg, and Malta. For the group of 28 EU countries (EU28), property taxes accounted for 2.3% of GDP in 2012, with recurrent taxes on immovable property accounting for 1.5% and other property taxes accounting for the balance 0.8%. These taxes together generated €293.5 billion in 2012.

As with the legal framework, it is not possible to provide a comprehensive overview of property taxation across all sectors and countries, and it is outside the scope of this text. However, we present a summary of some countries' approach to the tax on residential properties to illustrate the complexities and contrasts between them. Again, we remind readers of the dangers of assuming everything is the same the world over, how essential it is for valuers operating internationally to be fully aware of the taxation and other contexts within which they are operating, and to take specialist advice. Where appropriate, we make reference to taxation issues and their direct impact on the *income approach* within the relevant chapters.

Readers should consider the wide variety of taxation approaches and rates set out in Table 2.3.

Table 2.3 A comparison of residential property taxation rates (as of 2024)

Country	*Capital gains tax (CGT)* (on residential other than primary residence)	*Stamp duty on sale of residential property*	*Income tax on residential rental income*
Australia	25–45%	Varies by State but around 3–4%	32.5–45%
Canada	50%	No stamp duty, but additional taxes on occupation	25%
Dubai	No CGT	2% (registration fee)	0%
France	2–5 years	2–12% (transfer tax)	20–30%
Hong Kong	No CGT	Up to 4.25% for permanent residents, 15% for non-permanent residents	0–15%
Italy	26%	2–9%, minimum of €1,000	21%
Japan	40%	3% plus 60,000 yen plus a consumption tax	20.42%
Portugal	14.5–48% of 50% of the gain	0.8% of fiscal value	28%
Singapore	No CGT	Citizens, 17%; foreigners, 30%	24%
Spain	26% for residents, 15% for non-residents	1.5% of the purchase price	19%
Thailand	15%	Yes, 0.1%	0–35%
UK	10–20%	From 5% (on homes £40,000–£250,000) up to 15% (on homes £925,001–£1.5 million)	0–45%
United States	0%, 15%, or 20%	No stamp duty equivalent	30%

Source: This table has been produced using information from a number of online resources.

It can be seen that there are wide variations, and this may influence property markets indirectly in different ways. For example, stamp duty or transaction-based taxes set in value bands will influence both market activity and pricing.

Market activity. When an increase in tax percentages is announced, especially if it is a significant increase, there will be a flurry of transactions to beat the deadline, often temporarily pushing prices upwards.

Market pricing. Prices will be influenced by the value at which bands are set. If a band is up to, say, £250,000, a lot of properties around this price point will be set at £249,999. To price a property at, say, £255,000 makes little sense if there are several thousand pounds more tax to be paid.

Other taxes, such as inheritance tax, are part of an estate and often subject to specific taxation planning.

The application of VAT, especially where it is optional, may influence rental values and create a kind of two-tier pricing situation in commercial markets.

What is probably most significant are the wide variations in taxes, for example, between 0 and 45% income tax on residential investment income. Global investors will consider both taxation policies and the rate of tax applied when evaluating which countries to invest in, as will those seeking a second home or investment with rental potential.

Note

1 A *break clause* is a lease agreement provision that allows the tenant, the landlord, or both to terminate the lease before its expiry date.

References

Breidenbach, P., Eilers, L., Fries, J. (2022). Temporal dynamics of rent regulations – The case of the German rent control. *Regional Science and Urban Economics*, 92. Available at: https://doi.org/10.1016/j.regsciurbeco.2021.103737

DLA Piper (2024). *Realworld Law: Commercial Leases*. Available at: www.dlapiperrealworld.com/law/

EU (2024). *Energy Performance of Buildings Directive (EU/2024/1275)*. Available at: https://eur-lex.europa.eu/legal-content/EN/TXT/?uri=OJ:L_202401275&pk_keyword=Energy&pk_content=Directive

JLL (2019). *Leaseholds in Major German Cities: Analysis of their Value Drivers and Significance on the Investment Market*. Available at: www.jll.de/content/dam/jll-com/documents/pdf/research/emea/germany/en/de-leaseholds-in-major-german-cities-march-2019-jll.pdf

OECD (2014). *Revenue Statistics 2014*. Paris: OECD Publishing. Available at: https://doi.org/10.1787/rev_stats-2014-en-fr

Pope, A.R., Young, M.R. (2024). Urban ground leases: a cross-country comparison. *Journal of Property Research*, 41(2). Available at: www.tandfonline.com/doi/full/10.1080/09599916.2024.2308908#d1e1106

UK Government (2015). *The Energy Efficiency (Private Rented Property) (England and Wales) Regulations 2015*. Available at: www.legislation.gov.uk/uksi/2015/962/made

UN-HABITAT (2011). *Global Report on Human Settlements 2011: Cities and Climate Change*. Available at: https://unhabitat.org/global-report-on-human-settlements-2011-cities-and-climate-change

Chapter 3

Investment property

Investment property

Moles and Terry (1997, p. 20) define an *asset* as 'any item owned or right possessed by a firm or individual which has an economic, commercial, or exchangeable value'. In this sense, *property* qualifies as an asset, given its significant economic and commercial value. However, property differs from other major asset classes, such as bonds and stocks.

Investment property is defined by RICS (2021, p. 10) as

> property that is land or a building, or part of a building, or both, held by the owner to earn rentals or for capital appreciation, or both, rather than for:
>
> - use in the production or supply of goods or services, or for administrative purposes, or
> - sale in the ordinary course of business.

Examples of common types of investment property include thus:

Residential property	Houses, apartments, or multi-family units rented out to tenants
Commercial property *(property used solely for business purposes)*	Office buildings, shopping centres, retail warehouses, co-working spaces, data centres, warehouses, etc. rented to businesses
Land	Purchased for future development or resale

Investment property is considered a tangible asset, meaning it has intrinsic value in its physical form, unlike financial assets. The most obvious difference is its ***fixed geographical location***, hence the importance of the quality of that location for the land's current or alternative uses as determined by its general and special accessibility and its interrelationship with other competing and complementary buildings, locations, and land uses.

In recent years, the importance of physical location has been somewhat eroded by the growing significance of connectivity. The quality of a property's internet infrastructure, particularly the capacity and speed of its internet connectivity, has increasingly become a crucial factor in assessing real estate value. In some cases, robust connectivity has reduced the reliance on traditional locational advantages, particularly in sectors where remote work and digital operations have gained traction. Post–COVID-19 pandemic, the idea of vitality has emerged, with hybrid workers seeking areas with many vibrant amenities rather than the sometimes-sterile traditional office locations.

DOI: 10.1201/9781003490357-4

Another significant distinction between property and stocks or bonds as investments lies in the inherent ***heterogeneity*** of property. Beyond location, properties differ in various aspects, including the quality of permitted planning use, the quality and state of repair of the building, the nature of the legal title, and for tenanted properties, the covenant strength of the tenant. These factors will be explored in the following paragraphs.

Once a property is developed, its investment quality is largely shaped by two key factors: the quality of the permitted planning use and the quality of the physical improvements (such as buildings) on the site. The permitted planning use refers to zoning and planning permissions that define how a property can be utilised. This affects its flexibility for future changes in use, such as converting from commercial to residential. Properties with adaptable uses may offer higher long-term value, and compatibility of the planning use with the surrounding area is crucial. The quality of physical improvements includes the design, construction, and condition of buildings, which directly affect the property's value. High-quality materials, energy-efficient systems, and modern amenities reduce maintenance costs and improve tenant appeal. Flexible layouts that can accommodate changing needs increase demand.

Equally important factors in assessing a property investment are the quality and the duration of the legal title. The nature of the legal title can significantly impact the rights and obligations associated with the property, influencing its marketability and value. The owner of a UK-style freehold (or equivalent) title effectively owns all the land described in the title deeds in perpetuity, including everything below it to the centre of the earth and everything above. Freehold rights may be restricted by covenants in the title and/or by the rights of others, such as rights of way, or by law, whereby certain mineral and other rights may be vested in the state. On the other hand, a leasehold title limits ownership by time. The leaseholder holds rights to the property for a specified period, and these rights are governed by the terms of the lease agreement, which could include restrictions (covenants) negotiated between the landlord and tenant or imposed by law. The quality of an investment property will also be affected by the standing of the tenant in terms of what is sometimes referred to as their covenant strength. This largely relates to the risk of default. More established companies, with positive trading accounts, will be regarded as a stronger covenant. The quality of the lease, in terms of its length and the appropriateness of the lease conditions relevant to the use, location, sector, and construction, will also be a consideration.

Real estate investments are also influenced by government intervention, with the legal framework varying significantly between countries. In many countries, the legal framework evolves slowly and has a long history and tradition. In other countries, for example, the United Arab Emirates (UAE), laws can and do change overnight, and this introduces the concept of the risk framework within which an asset sits. In the UK, for example, the risk of fundamental land tenure rights being changed is low, but in other countries, these can change almost overnight, making the risks associated with the legal framework a very different proposition.

In addition to the core principles of the investment method, real estate as an asset class possesses several distinctive characteristics that differentiate it from other investment types. These include the following.

Lumpiness

Real estate investments are characterised by substantial capital requirements and lack the capacity for division into smaller parts for incremental investment. In contrast to stocks or bonds, which can be acquired in minimal quantities, real estate transactions generally encompass substantial, indivisible assets. This attribute affects liquidity, as an investor is unable to divest a fraction of a building to obtain capital. The complete asset must be sold, refinanced, or fractionalised via structured vehicles, like real estate investment trusts (REITs).

Depreciation and obsolescence

In contrast to financial assets, real estate is susceptible to physical deterioration and functional obsolescence. Structures deteriorate over time due to attrition, necessitating continuous upkeep and financial investment. Furthermore, fluctuations in market demand, alterations in regulations, or technical innovations might make properties obsolete, requiring expensive renovations or adaptive reuse approaches. Investors must incorporate these risks into their cash flow estimates and value models.

Active management requirement

Real estate investment often demands active management, distinguishing it from passive financial assets, like stocks or bonds. Property owners must oversee leasing, maintenance, tenant relationships, and regulatory compliance. The degree of management intensity varies by property type, with hotels and retail spaces requiring more hands-on involvement compared to stabilised office buildings or long-term leased industrial properties. The need for active management introduces operational risks and costs that must be factored into investment decisions.

Illiquidity

In contrast to other investment vehicles, real estate is by nature less liquid. Long-term due diligence, legal processes, and substantial transaction costs – such as taxes, agency fees, and legal bills – are all part of transactions. During downturns, it can be challenging to sell properties rapidly at fair market value due to market conditions that further worsen illiquidity. Publicly traded securities, on the other hand, can be sold at market values practically instantly.

Leverage and financing risks

Leverage, or the use of borrowed funds to finance acquisitions, is a common feature of real estate transactions. Leverage raises financial risk even if it can boost returns. Changes in lending terms, interest rate swings, and refinancing risk can all have an impact on an investor's capacity to repay debt and asset values. When assessing the viability of an investment, the cost of borrowing and the availability of funding are important considerations.

Inflation hedge

A common belief is that real estate is a good inflation hedge. Property values and rentals typically rise with rising prices, maintaining purchasing power, in contrast to fixed-income instruments, which may depreciate in inflationary circumstances. This advantage is dependent on supply-demand relationships and lease arrangements that permit rent modifications over time.

Tax considerations

Tax laws have an impact on real estate investments, which may have consequences for total returns. Investors may benefit from depreciation allowances, capital gains taxes, and deductions for mortgage interest payments. Taxation is another instrument that governments employ to control real estate markets, which affects incentives for investment. When assessing real estate prospects, investors need to take jurisdiction-specific tax effects into account.

These characteristics play a crucial role in shaping the investment strategy, risk assessment, and valuation methodology within the income approach to property valuation. Properly accounting for these factors ensures a more comprehensive analysis of real estate assets as investment vehicles.

Comparing property to other asset classes

An *investment* can be described as an exchange of capital today for future benefits, generally in the form of income (dividends, rent, etc.), and sometimes in the form of capital. The investment income from property is the net rent paid by tenants. The market price of an investment is determined in the market by the competitive bids of buyers for the available supply at a given point in time under market conditions prevailing at that time. Short supply and high demand – that is, scarcity – will lead to price (value) increases, whilst low demand and high supply will lead to falls in prices and values.

The definition of *market value* requires the valuer to express an opinion of the market price that the valuer believes would have been achieved if that property had been sold at that time under the market conditions at that time, hence the importance of market knowledge and market comparables.

The unique characteristics of property make property investment valuation a more complex combination of *art and science* than that exercised by brokers and market makers who are in the market for stocks and shares. In the stock market, sales volume generally allows for price (value) comparison to be made minute by minute. As stocks, shares, and property are the main investments available, there is bound to be some similarity between the pricing (valuation) methods used in the various markets and some relationship between the investment opportunities offered by each. A basic market measure is the yield, or capitalisation rate, calculated as the ratio of net operating income (or net rent) to property value or purchase price:

$$Capitalisation\ rate\ (yield) = \frac{Net\ operating\ income}{Property\ value\ or\ purchase\ price}$$

The assessment of the capitalisation rate allows comparison to be made between investments in each market and between different investments in different markets. There is a complex interrelationship of yields and patterns of yields within the whole investment market. In turn, these yields reflect market perceptions of risk and become key to pricing and valuation methods.

The capitalisation rate is often explained using the Gordon growth model (Gordon 1959). This model values an asset as a perpetuity that pays a stream of income (dividends) growing at a constant rate:

$$P = \frac{X}{r - g}$$

where P is asset value and X is net income, g is the growth rate of income, and r is the required rate of return (opportunity cost of capital).

Solving the equation for the capitalisation rate (y) yields:

$$y = \frac{X}{P} = r - g$$

Following Fisher's classic work on interest (Fisher 1930), and as Baum and Crosby (2008) note, the required rate of return r can be broken down into the risk-free rate r_f plus the risk premium r_p. Additionally, Ball et al. (1998) also include an annual rate of property depreciation d to account for the diminishing rent-earning capacity of a building over time due to aging and obsolescence. Therefore, the capitalisation rate formula can be expressed as:

$$y = r_f + r_p - g + d$$

where r_f and r_p represent risk factors that are offset by the prospects of rental growth g adjusted by depreciation d.

The ***risk-free rate*** is usually assessed through a proxy, often using government bond yields. This compensates investors for the time value of money and theoretically represents a guaranteed return for investors without any risk of loss of liquidity and anticipated inflation (Geltner et al. 2010). On the other hand, the ***risk premium*** compensates the investor for the risk associated with the cash flows and is harder to quantify, as it includes both exogenous and endogenous factors (Geltner et al. 2010; Hutchison et al. 2011). Exogenous factors include broader economic conditions, capital market trends, and long-term urban market factors, such as market size, employment levels, and economic volatility. Endogenous factors relate to the individual property and include factors such as location, building condition, tenant quality, and lease terms.

Baum and Crosby (2008) explore risk in relation to property in detail, but as a framework for the *income approach*, the following are possibly the risks most relevant to later discussions:

- **Property risk.** How important are the issues of illiquidity in relation to resales and the fixed location of each parcel of land compared to owning, say, stocks and shares?
- **Sector risk.** Are there other risks that relate specifically to the property under consideration and its use type, such as the ever-increasing growth of online shopping and the impact on traditional retail centres or possible location issues, leading to potential falls in demand?
- **Building risk.** Is the building ageing faster due to poor sustainability, which might have been noted? But the overall risk is of the unknown defects that might appear and render the building of no use, such as increased flooding, affecting foundations.
- **Tenant risk.** Will the tenant pay the rent? Will voids occur? Will the tenant abide by the terms of the lease?
- **Legal and legislative risk.** Could there be undiscovered issues, such as rights of way, or changes in property laws which have a disproportionate effect on property values in general or to specific sectors compared to other investments?
- **Planning risk.** Are there emerging new planning policies which impact on the previously assumed location advantages and permissible uses, such as the construction of a new bypass?
- **Taxation risk.** The possibility of future taxes on property, or changes in taxation of property.

Whilst the preceding framework helps illustrate the factors influencing capitalisation rates, in practice, valuers rely on the comparison method, which considers market transactions to determine appropriate capitalisation rates for specific properties. This approach will be discussed in detail in Chapter 7.

The four quadrants of real estate investment

The real estate investment market can be analysed through four quadrants which illustrate the interplay between capital markets, the space market, and the development industry. These quadrants provide a structured approach to understanding how supply, demand, pricing, and returns are interconnected. The quadrants include:

1. **Public equity market.** REITs and real estate stocks.
    - Represents real estate investments that trade in public exchanges, such as REITs and publicly listed property companies.
    - Investors gain exposure to real estate without direct ownership responsibilities.
    - Prices are influenced by investor sentiment, broader stock market trends, and capital market conditions.
2. **Private equity market.** Direct property ownership.
    - Includes direct ownership of income-producing properties, such as office buildings, retail centres, industrial warehouses, and residential complexes.
    - Investors derive returns from rental income and capital appreciation.
    - Valuation is influenced by property-specific factors, market conditions, and long-term economic trends.
3. **Public debt market.** Commercial mortgage-backed securities (CMBS).
    - Represents securitised mortgage products that trade in public bond markets.
    - CMBS investors receive fixed income payments from pools of real estate–backed loans.
    - Valuation and yields are influenced by interest rates, credit risk, and macroeconomic conditions.
4. **Private debt market.** Whole loans, direct mortgage lending.
    - Consists of direct lending to real estate owners through mortgages and loans.
    - Typically involves banks, insurance companies, and private lenders.
    - Returns depend on loan terms, interest rates, and the borrower's ability to service debt.

Each quadrant serves a distinct role in real estate investment, with varying risk-return profiles and market dynamics. Investors may allocate capital across these quadrants to diversify their exposure to real estate assets.

Property returns

Real estate returns are generated through two primary components:

- **Income return (yield).** The rental income received from a property, expressed as a percentage of its **value**.

- **Capital return (appreciation).** The change in a property's value over time due to market conditions, improvements, or economic factors.

Total property return is typically measured using the ***total return formula***, which combines income and capital appreciation:

Total return *= (net operating income (NOI) + capital gain)/initial property value*

Alternatively, in percentage terms:

Total return *= income return + capital return*

Where:

Income return (%) = net rental income/property value
Capital return (%) = change in property value/initial property value

Example 3.1 Property return calculation

An investor purchases a commercial office building for €2,000,000. The building generates an annual NOI of €150,000. After five years, the investor sells the building for €2,500,000.

Step 1: Calculate the income return

Income return (%) is calculated as:

$$\textit{Income return} = (\textit{NOI/initial property value}) \times 100\%$$
$$= (150{,}000/2{,}000{,}000) \times 100 = 7.5\%$$

Step 2: Calculate capital return

Capital return (%) is calculated as:

$$\textit{Capital return} = (\textit{sale price} - \textit{initial purchase price})/\textit{initial purchase price}) \times 100\%$$
$$= ((2{,}500{,}000 - 2{,}000{,}000)/2{,}000{,}000) \times 100 = 25\%$$

Since the property was held for five years, we calculate the annualised capital return:

$$\textit{Annual capital return} = 25/5 = 5\% \textit{ per year}$$

Step 3: Calculate total return

Total return (%) is the sum of income return and capital return:

$$\textit{Total return} = \textit{income return} + \textit{annual capital return}$$
$$= 7.5\% + 5\% = 12.5\% \textit{ per year}$$

This investment provided the investor with a **total return of 12.5% per year**, consisting of a **7.5% annual yield from rental income** and a **5% annual appreciation in property value** over five years. This example demonstrates how property returns are derived from both income (rents) and capital appreciation (property value increase).

The structural nature of property investment

Property investment is inherently structured around several key economic, financial, and physical components. Following are some critical perspectives on the structural nature of property investment.

Capital intensity and long-term investment

Property investment requires substantial initial capital and is typically a long-term asset due to its slow appreciation and extended holding periods. The capital structure often includes equity, mortgages, and institutional funding sources. Pivo and McNamara (2005) highlight that property investment is experiencing structural changes, with increased interest in sustainable and responsible investment. Capital flows into energy-efficient real estate mitigate investment risks.

Market liquidity and investment cycles

Real estate investments are inherently illiquid, meaning that they cannot be quickly converted to cash without potential value loss. Property markets follow cycles of boom and bust, influenced by macroeconomic factors. Amidu et al. (2021) argue that market disruptions significantly impact investment decision-making due to liquidity constraints and regulatory structures.

Leverage and debt financing

Real estate investments often use leverage (debt financing) to increase returns, though high leverage increases risks in downturns. Legarza (2025) explores how investment structures shape risk perception and decision-making in property development. Asiedu and Esfahani (2001) examine the ownership structure in foreign direct investment projects, noting that property investments require optimised capital structures for profitability.

Regulatory and legal structures

Regulatory frameworks heavily impact property investments through zoning laws, tax policies, and compliance requirements. Haila (1991) identifies four types of investment structures in land and property that are shaped by regulatory frameworks. Ball et al. (1996) examine the cyclical nature of property investment, showing how legal and economic conditions affect long-term investment stability.

Risk management and diversification strategies

Property investors diversify across sectors (residential, commercial, industrial) and locations to mitigate risks. Hartzell (2012) describes global property investment structures, emphasising how investors adjust strategies based on financial innovations. Chan et al. (2003) analyse REITs as a structured investment vehicle that provides liquidity and risk diversification.

Sustainable and technological innovations

Modern property investment is being reshaped by PropTech (property technology), smart buildings, and blockchain applications. McNamara (2017) discusses how sustainability and digital innovations are altering the structural composition of property investment. Baum et al. (2021) argue that property investments are evolving towards a more data-driven structure, integrating financial modelling and AI-based risk assessment.

The structural nature of property investment involves capital planning, debt structuring, regulatory frameworks, risk management, and technological advancements. Investors need strategic financial planning and market adaptation to navigate economic cycles, legal changes, and emerging technologies.

The cyclical nature of property investment and property cycles

Real estate investment is inherently cyclical, meaning that property values, rents, and investment returns fluctuate over time due to changing economic conditions, market sentiment, and supply-demand dynamics (Quigley 2002). Unlike stocks or bonds, real estate operates on long development timelines, leading to lag effects in supply adjustments, which can prolong market imbalances and exacerbate volatility (Roulac 1996). These cycles influence investment strategies, financing conditions, and asset pricing.

The cyclical nature of property investment is influenced by several key factors, including what follow.

Economic growth and contraction

Real estate performance correlates with GDP growth, employment rates, and consumer spending. Economic expansion leads to increased demand for space, higher occupancy rates, and rising rents, whereas downturns reduce occupancy and rental incomes, causing asset values to decline (Marcato & Nanda 2022). Historical data has shown that real estate cycles are closely tied to broader economic fluctuations, making them susceptible to recessionary pressures (Nguyen et al. 2020).

Interest rate fluctuations

Changes in interest rates play a significant role in shaping real estate cycles. Higher interest rates increase borrowing costs, reducing investment activity and limiting access to capital, whereas lower rates encourage property purchases, refinancing, and development activity (Kurucz 2010). The inverse relationship between interest rates and property valuations highlights the impact of monetary policy on real estate market dynamics (Quigley 2002).

Capital market conditions

Investor appetite for real estate fluctuates based on alternative investment opportunities, risk perception, and liquidity in financial markets. During periods of financial stability and economic optimism, capital inflows into real estate increase, driving property prices up. Conversely, economic uncertainty and liquidity constraints reduce real estate investment activity (Özbaş et al. 2014). Moreover, global financial markets have shown increased synchronisation, meaning that shocks in one region can have significant spillover effects on real estate cycles worldwide (Huangfu et al. 2024).

Supply and demand imbalances

The long development cycle in real estate often leads to supply-demand mismatches. Overbuilding during economic booms results in excessive inventory, which depresses prices during downturns, whereas construction slowdowns during recessions can create housing shortages when demand rebounds (Kader et al. 2022). This lag effect is a critical determinant of real estate price fluctuations, as the market takes time to correct itself after periods of over- or undersupply (Torstensen & Roszbach 2019).

Government policies and regulations

Tax incentives, zoning laws, monetary policies, and land-use regulations significantly impact real estate cycles by shaping investment behaviour and development feasibility. Policy interventions, such as interest rate adjustments, tax relief for real estate investors, and rent control measures, can either stabilise or amplify market fluctuations (Lee et al. 2013). Additionally, regulatory uncertainty can deter investment, as developers and investors require stable policy environments to make long-term decisions (Quigley 2002).

Phases of the property cycle

The real estate market cycle follows a repeating pattern driven by economic conditions, supply-demand dynamics, and investor behaviour. Investors' emotional responses often mirror these fluctuations, influencing decision-making and market trends. Many researchers argue that the period of such cycle can be between 10 and 15 years. Set out in Figure 3.1 is an expanded breakdown of each phase of the market quadrant cycle, incorporating both market fundamentals and the psychological emotions experienced by the investor.

Recovery phase: *emotional response of hope and optimism*

Market characteristics

- The economy begins to recover from a downturn, but demand for real estate is still weak.
- Vacancy rates remain high, but they stop increasing.
- Rents stabilise, and property prices remain low.
- New construction is minimal as developers remain cautious.
- Early investors begin acquiring undervalued properties, anticipating future growth.

Investor emotions

- *Cautious optimism*. Investors see early signs of recovery but remain sceptical.
- *Hope*. Opportunistic investors recognise potential deals and start investing.

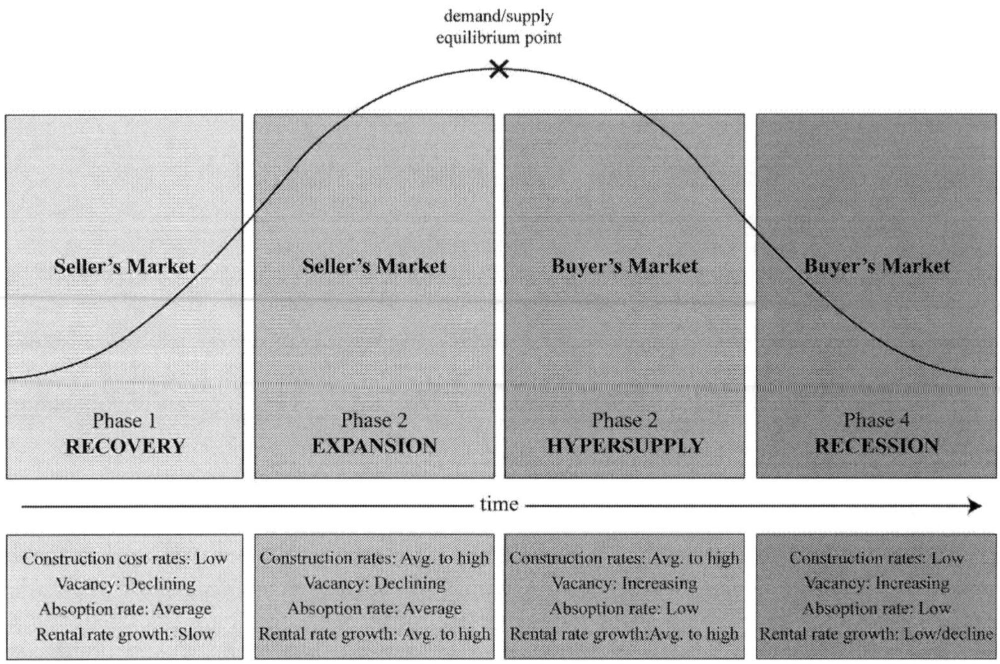

Figure 3.1 The real estate market cycle.

Best investment strategies

- Acquire undervalued properties before prices rise.
- Focus on assets in recovering locations with growth potential.
- Secure long-term leases to ensure cash flow stability.

Expansion phase: *emotional response of excitement and euphoria*

Market characteristics

- Economic growth accelerates, leading to rising employment and business expansion.
- Demand for real estate increases, lowering vacancy rates.
- Rents rise, improving property cash flows.
- New construction begins as developers respond to growing demand.
- Investor confidence surges, pushing property values higher.

Investor emotions

- *Excitement.* Rising property values and rental income make real estate highly attractive.
- *Euphoria*. Investors rush into the market, believing prices will continue to rise.

Best investment strategies

- Buy properties in prime locations before prices peak.
- Secure long-term tenants at favourable lease rates.
- Leverage financing carefully to maximise returns while minimising risk.

Risk warning ⚠

- Overconfidence can lead to speculative buying at inflated prices.
- Developers may overbuild, creating a future supply surplus.

Hypersupply phase: *emotional response of anxiety and denial*

Market characteristics

- New developments flood the market, leading to increased competition.
- Demand slows as affordability declines.
- Vacancy rates begin rising, putting downward pressure on rents.
- Property values stop growing, and some investors start exiting the market.
- Lending conditions tighten as banks become more cautious.

Investor emotions

- *Anxiety*. Investors worry about declining demand and falling rental income.
- *Denial*. Some refuse to believe that the market is slowing, holding on to properties despite warning signs.

Best investment strategies

- Avoid overpaying for properties, as prices may decline.
- Monitor market fundamentals closely, and be prepared to exit risky investments.
- Focus on stable, income-generating assets with strong tenants.

Risk warning ⚠

- Many investors remain overleveraged, leading to financial distress.
- Those who bought at peak prices may face negative equity.

Recession phase: *emotional response of panic and despair*

Market characteristics

- Economic conditions weaken, and businesses struggle, reducing demand for space.
- Vacancy rates peak, and landlords offer concessions to attract tenants.
- Rents decline, affecting property cash flows.
- Property values drop, forcing some investors to sell at losses.
- Foreclosures and distressed sales become common.

Investor emotions

- *Panic*. Fear spreads as investors see asset values plummet.
- *Despair*. Many investors regret overleveraging and selling at a loss.

Best investment strategies

- Look for distressed opportunities, as motivated sellers offer discounts.
- Secure financing cautiously, as banks may have stricter lending criteria.
- Hold cash reserves to take advantage of buying opportunities at the market bottom.

Risk warning ⚠

- Selling during panic phases locks in losses – patience is crucial.
- The real estate market takes time to recover; avoid rushed decisions.

Connecting the cycle with market psychology

Investor behaviour, or sentiment, significantly amplifies market fluctuations due to the following:

- *Fear of missing out (FOMO).* Drives investors to overpay during booms.
- *Panic selling.* Causes unnecessary losses during downturns.
- *Contrarian investors.* Those who buy in downturns and sell near peaks often achieve the best returns.

Through an understanding of the real estate cycle and its psychological effects, investors can make more informed, rational decisions and avoid common pitfalls.

The current state of global real estate investment

The worldwide real estate market has experienced significant oscillations in recent years, influenced by economic changes, variations in interest rates, and evolving investment techniques. This section analyses the valuation of the worldwide market, investment patterns, regional insights, sector-specific developments, and investor sentiment at the time of writing. Markets are obviously dynamic, so these are a snapshot, and readers are encouraged to explore the resources provided at the end of this chapter to establish the current market position and forecasts.

In the year 2024, the global real estate industry was estimated to be worth around US$634.90 trillion. According to forecasts, the market is expected to increase at a compound annual growth rate (CAGR) of 2.77%, potentially reaching approximately US$727.80 trillion in the year 2029. It is anticipated that the residential sector will be the most dominant, accounting for around US$516.80 trillion in the United States in 2024 (Statista 2024).

Research by Savills (2023) indicates that global real estate investment fell to US$699 billion, representing the lowest level since 2012. The current downturn is primarily due to cyclical factors, particularly a notable increase in interest rates in major markets. Early indicators of recovery are apparent, with projections indicating a 7% increase in investment by the end of 2024, totalling US$747 billion, followed by an additional 27% increase to US$952 billion in 2025.

The United States is anticipated to be a key contributor, with an estimated real estate value of around US$132.0 trillion in 2024. The market is experiencing increased demand for suburban properties, attributed to the growth of remote work (Statista 2024).

Conversely, investment activity in Europe has been limited, as evidenced by a 64% decline in transaction volumes in the UK during Q2 2023 relative to the same period in the prior year. The region is anticipated to experience a recovery as economic conditions stabilise (Abrdn 2023).

Markets in Singapore and Tokyo (Asia-Pacific) continue to attract significant investment, supported by strong economic fundamentals and urbanisation trends, as indicated by PwC research in 2023.

Comparative analysis of commercial and residential property investments

Commercial and residential property investments each offer unique advantages and challenges, making them essential components of a diversified real estate portfolio. Residential properties,

Table 3.1 Comparison of key aspects between residential and commercial property investments

Aspect	*Residential property*	*Commercial property*
Rental yields	2–6%	5–12%
Capital requirement	Moderate	High
Lease duration	1–2 years	5–10 years
Market sensitivity	Less sensitive	More sensitive
Vacancy risk	Moderate	High

including single-family homes and apartments, are favoured for their accessibility, lower capital requirements, and steady demand, as they fulfil a basic human need. Global residential property is valued at $516.80 trillion, accounting for 81.5% of total real estate value in 2024 (Statista 2024). Rental yields range between 2 and 6%, with urban areas like New York and London offering stable, though lower, returns compared to emerging markets, such as Dubai, where yields exceed 5% (National Association of REALTORS 2024). In contrast, commercial properties, including office buildings, retail spaces, and industrial facilities, deliver higher income potential, with yields typically ranging from 5 to 12%, and benefit from long-term leases averaging five to ten years (Savills 2024). However, they require substantial capital investment and are more sensitive to economic fluctuations, as seen during the COVID-19 pandemic, when global retail property values fell by 20% (PwC 2023).

Factors affecting global property investment

Global property investment is influenced by a wide range of macroeconomic, political, and social factors. Events such as Brexit and the COVID-19 pandemic have significantly shaped investment trends and market dynamics.

Economic uncertainty and geopolitical events

Brexit created significant uncertainty in the property market, particularly in London. The UK's exit from the EU raised concerns about trade, financial regulations, and London's role as a global financial hub. In 2019, foreign investment in UK commercial property fell by 30% compared to pre-referendum levels (CBRE 2019). Global trade tensions, especially between the United States and China, also affected property markets by disrupting investment flows into regions dependent on global supply chains. Brexit, and the resulting restrictions on time that British ex-pats could remain in the EU, had a serious impact on the holiday and second home residential market in many Mediterranean countries.

Public health crises

The COVID-19 pandemic caused a significant shift in the demand for real estate across all businesses. Remote work reduced demand for office space in 2024, when global vacancy rates reached 18% (Cushman & Wakefield 2024). The retail sector suffered tremendously because of lockdowns and the growth of e-commerce; in 2020, retail property prices fell by 20% globally (PwC 2023). On the other side, the epidemic raised demand for logistical and industrial facilities as e-commerce expanded. Investors turned their attention from traditional assets, like retail stores and office buildings, to thriving sectors, like biological sciences, logistics, and data centres.

Inflation and interest rate fluctuations

Interest rate fluctuations and inflation have a significant impact on global real estate investment, affecting borrowing costs, yields, and asset values. Rising inflation usually leads to increased interest rates, which make it more expensive to finance building projects and real estate acquisitions, because central banks aim to preserve price stability. For example, global real estate investment volumes dropped to $699 billion in 2023–2024, the lowest level since 2012, because of rising borrowing rates (Savills 2023). These increasing rates have the greatest effect on commercial properties, where leveraged financing is common, impeding new developments and decreasing profitability. However, inflation can raise property values and rental revenue since real estate is sometimes seen as a hedge against price increases. Usually, when inflation is strong, investors focus on sectors that have consistent cash flows, such as logistics and real estate. Fixed-rate financing becomes more attractive and shields borrowers from rising rates, whereas variable-rate loans are riskier. The correlation between inflation and interest rates highlights the need for careful planning when making real estate investments, balancing risks and rewards in a shifting economic landscape.

Technological advancements

Global real estate investing has undergone a transformation due to technological breakthroughs that have brought efficiency, transparency, and new opportunities. Blockchain, artificial intelligence (AI)–powered analytics, and smart contracts are examples of PropTech advancements that improve risk assessment, expedite transactions, and lessen the need for intermediaries. International investors can make well-informed decisions by using virtual and augmented reality tools to remotely tour and visualise properties. Cloud computing and 5G have fuelled the growth of digital infrastructure assets like data centres and telecom towers, opening up profitable investment opportunities. IoT and energy management systems are two examples of smart building technologies that improve operational efficiency and match characteristics with ESG objectives. Crowdfunding platforms democratise access to real estate markets, while AVMs offer quick, precise valuations. Whilst cybersecurity measures are becoming more and more important to secure sensitive data, digital twin technology enhances asset management by virtually mimicking real assets. These developments give investors a competitive edge in the digital age by making international real estate markets more accessible, data-driven, and varied.

A changing relationship between the end user and the suppliers of property products

The dynamics between the end user (or tenant) and the supplier (the landlord) have, in the commercial market, shifted dramatically over the past two decades. Shorter lease periods, greater flexibility, and agile business practices have largely signalled the end of 20-year leases, with strict upward-only rent review clauses even in the UK. Companies demand flexibility not only of space but also of lease terms, and the exponential growth of flexible space operators and co-working creates a different focus on tenants as customers. This has serious implications for the security and volatility of income flows and creates a need to retain tenants (increasingly seen as customers) by a better understanding of their requirements and the provision of a wide range of amenities which extend beyond the simple provision of, say, an office space. When an asset manager of a prestigious office tower in Dubai says 'We are actually curators of customer experience', this signals an entirely different relationship in terms of management, communication, and cash flows from a traditional landlord and tenant. All of this needs to be considered in the new world of the income approach.

Climate change and ESG factors

Due to their impact on asset prices, market demand, and operational strategies, climate change and ESG considerations are becoming more and more important in international real estate investing. Often known as the 'green premium', environmentally friendly properties with energy-efficient features, green certifications, and climate-resilient designs are drawing higher rentals, valuations, and investor interest. Growing tenant preferences for environmentally and socially conscious environments are in line with societal factors, including affordable housing and community-focused developments. Attracting ethical investors requires governance features, such as openness in ESG reporting and anti-corruption initiatives. The move toward net-zero carbon standards is driven by financial concerns for non-compliant properties, such as becoming stranded assets, while regulatory pressures, such as tougher building codes and carbon levies, encourage sustainable development and retrofitting. The need to match investment strategies with ESG principles to increase asset value, lower risks, and satisfy changing market expectations is highlighted by the convergence of environmental, social, and governance issues.

Urbanisation and demographic shifts

The need for residential, commercial, and industrial spaces is being reshaped by urbanisation and demographic changes, which are significant forces behind global real estate investment. Rapid urbanisation has raised demand for housing, infrastructure, and mixed-use urban developments, especially in developing nations like Asia and Africa. With 68% of the world's population expected to reside in cities by 2050, there will be plenty of opportunity for real estate investors to profit from growing metropolitan regions (United Nations (UN) 2022). As the need for these specialist property types increases, aging populations in industrialised countries are turning their attention to retirement communities, health-care facilities, and senior housing. The desire for contemporary, amenity-rich urban living spaces and adaptable co-working spaces is being driven by younger, tech-savvy generations joining the workforce. Diversified investment options have also been brought about by the growth of megacities and secondary cities with reasonably priced real estate markets. Long-term investment strategies are being shaped by these demographic changes and urbanisation, which highlight the necessity of sustainable, flexible, and community-focused developments to satisfy the changing demands of urban populations.

Government policies and regulations

Government regulations have a big impact on how people invest in real estate. Whilst higher taxes on luxury real estate restrict investment in markets like Vancouver and Singapore, tax incentives for infrastructure and affordable housing promote investment in other parts of the world. Residential investment profits are also impacted by rent limits and more stringent tenant protection legislation in places like New York and Berlin.

Government policy may dictate that public sector occupiers have to occupy office buildings with a minimum environmental certification. This can overnight remove a huge section of the market from even being able to consider buildings below that threshold.

In a country such as the UAE, these changes can happen very quickly, as policy becomes law in a very short space of time. Policies include restricting the ability of someone to live in one emirate and work in another, having obvious impacts on the residential market. Countries can literally create new markets through, for example, the creation of new visas, again in the UAE, that allow digital nomads to enter and work with much reduced restrictions, which suddenly can open up new markets such as co-living.

Changes in the economy, society, and geopolitics have a significant impact on international real estate investment. The COVID-19 epidemic and Brexit are two examples of how external shocks can change market priorities and compel investors to modify their approach. Navigating the constantly changing global real estate market requires an understanding of these dynamics.

Questions

1. What are the primary characteristics that make real estate a distinctive investment choice?
2. How does the definition of *investment property* by RICS (2021) distinguish it from other property types?
3. Why is location crucial in property investment, and how has its importance evolved in recent years?
4. What are the four main phases of a property cycle?
5. How do property cycles impact investment decisions?
6. What is the capitalisation rate, and how is it used in property valuation?
7. What factors contribute to property market cycles?
8. Explain the concept of 'lumpiness' in real estate investment.
9. How does leverage affect real estate investment, and what are the risks?
10. Why is liquidity a challenge in real estate investment?

References

Abrdn (2023) *Global real estate market outlook: Q3 2023*. Available at: www.abrdn.com/en-au/institutional/insights-and-research/global-real-estate-market-outlook-q3-2023 (Accessed: 27 February 2025).

Amidu, A.R., Filippova, O. and Levy, D. (2021) Property investment decision-making behaviour amidst market disruptions: An institutional perspective. *Property Management* 39, 1–21.

Asiedu, E. and Esfahani, H.S. (2001) Ownership structure in foreign direct investment projects. *Review of Economics and Statistics* 83, 647–662.

Ball, M., Colin, L. and MacGregor, B.D. (1998) *The economics of commercial property markets*. London: Routledge.

Ball, M., Morrison, T. and Wood, A. (1996) Structures investment and economic growth: A long-term international comparison. *Urban Studies* 33, pp. 1687–1706.

Baum, A. and Crosby, N. (2008) *Property investment appraisal*. Oxford: Blackwell.

Baum, A.E., Crosby, N. and Devaney, S. (2021) *Property investment appraisal*. 4th ed. Chichester: Wiley-Blackwell.

CBRE (2019) *UK commercial property market report*. Available at: www.cbre.com

Chan, S.H., Erickson, J. and Wang, K. (2003) *Real estate investment trusts: Structure, performance, and investment opportunities*. Oxford: Oxford University Press.

Cushman & Wakefield (2024) *Global office space trends*. Available at: www.cushmanwakefield.com

Fisher, I. (1930) *The theory of interest*. New York: Macmillan. Available at: http://oll.libertyfund.org/Home3/EBook.php?recordID=0219http://oll.libertyfund.org/Home3/index.php (Accessed: 27 December 2022).

Geltner, D.M., Miller, N.G., Clayton, J. and Eichholtz, P. (2010) *Commercial real estate: Analysis & investment*. 2nd ed. Mason, OH: Thompson South-Western.
Gordon, M.J. (1959) Dividends, earnings, and stock prices. *The Review of Economics and Statistics* 41(2), 99–105. Available at: https://about.jstor.org/terms (Accessed: 27 December 2022).
Haila, A. (1991) Four types of investment in land and property. *International Journal of Urban & Regional Research* 15.
Hartzell, D. (2012) *Global property investment: Strategies, structures, decisions*. Chichester: Wiley-Blackwell.
Huangfu, Y., Yu, H., Dong, Z. and Wang, Y. (2024) Research on the risk spillover among the real economy, real estate market, and financial system: Evidence from China. *Land* 13, 890.
Hutchison, N.E., Adair, A.S. and Findlay, N. (2011) The impact of covenant strength on property pricing. *Journal of Property Research* 28(2), 167–188. Available at: https://doi.org/10.1080/09599916.2010.538479
Kader, S.A., Zayed, N.M., Salah Uddin, M. and Nitsenko, V. (2022) Factors affecting demand and supply in the housing market: A study on three major cities in Turkey. *Computation* 10, 196.
Kurucz, A. (2010) The relationship between general economic cycle and the real estate cycle. *Regional and Business Studies* 2, 41–56.
Lee, S.H., Kim, J.H. and Kim, J.J. (2013) Dynamic cycles of unsold new housing stocks, investment in housing, and housing supply-demand. *Mathematical and Computer Modelling* 57, 2094–2105.
Legarza, A. (2025) Revisiting structure and agency: The perception and materialisation of risk in property development and investment. *Planning Theory & Practice* 1–9.
Marcato, G. and Nanda, A. (2022) Asymmetric patterns of demand-supply mismatch in real estate. *The Journal of Real Estate Finance and Economics* 64, 440–472.
McNamara, P. (2017) The changing nature of property investment: Implications for urban planning. In Hillier, J. and Metzger, J. (eds.) *Connections: Exploring contemporary planning theory and practice with patsy healey*. Abingdon: Routledge.
Moles, P. and Terry, N. (1997) *The handbook of international financial terms*. Oxford: Oxford University Press.
National Association of REALTORS®. (2024) *Quick real estate statistics*. National Association of REALTORS®. https://www.nar.realtor/research-and-statistics/quick-real-estate-statistics (Accessed: 11 June 2025).
Nguyen, M., Xuan, P. and Bui, T. (2020) Causal relationship between banking system development and real estate market. *Management Science Letters* 10(1), 41–52.
Özbaş, B., Özgün, O. and Barlas, Y. (2014) Modeling and simulation of the endogenous dynamics of housing market cycles. *Journal of Artificial Societies and Social Simulation* 17, 19.
Pivo, G. and McNamara, P. (2005) Responsible property investing. *International Real Estate Review* 8(1), 128–143.
PwC (2023) *Impact of COVID-19 on retail property values*. Available at: www.pwc.com
Quigley, J.M. (2002) *Real estate prices and economic cycles*. eScholarship, University of California.
RICS (2021) *RICS valuation – Global standards*. London: Royal Institution of Chartered Surveyors (RICS).
Roulac, S. (1996) Real estate market cycles, transformation forces, and structural change. *Journal of Real Estate Portfolio Management* 2, 1–17.
Savills (2023) *Global real estate investment outlook*. Available at: www.savills.com
Savills (2024) *Global real estate investment: 2025 and beyond*. Available at: www.savills.com/impacts/market-trends/global-real-estate-investment-2025-and-beyond.html (Accessed: 27 February 2025).
Statista (2024) *Real estate worldwide*. Available at: www.statista.com/outlook/fmo/real-estate/worldwide (Accessed: 27 February 2025).
Torstensen, K.N. and Roszbach, K. (2019) Housing markets in Scandinavia: supply, demand, and regulation. In Nijskens, R., Lohuis, M., Hilbers, P. and Heeringa, W. (eds.) *Hot Property*. Cham: Springer. https://doi.org/10.1007/978-3-030-11674-3_11
United Nations (UN) (2022) *World urbanization prospects: The 2022 revision*. Department of Economic and Social Affairs, Population Division. Available at: https://population.un.org/wup

Chapter 4

Market analysis

Markets within a real estate system

The real estate system consists of three major components, namely, the space market, the asset market, and the development industry (Geltner et al. 2010). Understanding these components is essential for accurate property valuation because each component plays a unique and interconnected role in influencing property values.

The space market, also known as the user or occupier market, is where properties are leased or rented by individuals or businesses seeking space (the *demand side*) from property owners looking to rent out their premises (the *supply side*). This market directly impacts cash flow, which is a key determinant of property value. Rental rates, occupancy levels, and lease terms collectively shape the revenue potential of a property and provide insight into its current income-generating ability. For instance, high rental rates and low vacancy rates indicate strong demand and can positively influence property values. Conversely, a decline in occupancy levels or rental rates can signal reduced demand, potentially diminishing property values. Analysing the space market helps valuers understand tenant preferences, rental trends, and the broader economic factors affecting both demand and supply. This market is particularly relevant for estimating income and forecasting future property performance, as it reflects how the property will be utilised, and the profile and scale of its tenants, who ultimately generate the revenue(s) flowing to the landlord.

In contrast, the asset market is where properties are bought by investors wanting to buy real estate (the *demand side*) from investors wanting to sell real estate (the *supply side*). This market captures investor sentiment, which is shaped by expectations of future income, risks, and returns. In the asset market, real estate competes with other assets, such as bonds and stocks; therefore, valuers also need to consider broader global and local economic indicators, such as interest rates, inflation, and capital flows, which impact investor decisions and consequently affect yields and cap rates. The asset market plays a crucial role in translating the property's income-generating potential into a market value, which is influenced by investment trends and the availability of capital. For example, low interest rates typically make real estate investments more attractive, leading to increased demand and potentially higher property values. Understanding this market allows valuers to gauge how anticipated income aligns with investor expectations and market conditions, providing a foundation for determining the property's value.

The development industry translates the financial capital from the asset market into physical capital in the space market, thereby governing the supply side of real estate (Geltner et al. 2010). The industry is driven by factors such as land availability, construction costs, regulatory frameworks, and development timelines, all of which shape the potential for new developments. The development industry influences the supply of properties in the market, which, in turn,

DOI: 10.1201/9781003490357-5

affects competition and demand. For instance, if new developments saturate the market with similar properties, this could lead to increased vacancy rates and potentially reduced rental rates for existing properties. Conversely, limited development activity can constrain supply, potentially driving up demand and values for existing properties. Additionally, new developments can set benchmarks for modern amenities and building standards, which can impact the attractiveness and value of older properties. By considering the development industry, valuers can anticipate shifts in supply dynamics and assess how new projects might alter the competitive landscape, affecting both the subject property's value and its long-term marketability.

Table 4.1 summarises the three components of the real estate system. Together, they create a comprehensive framework for analysing the real estate market. The interplay between the space market, asset market, and development industry helps valuers anticipate changes in demand, supply, and investment sentiment, all of which are crucial for accurate property valuation. Each market provides unique insights that, when combined, offer a holistic view of the factors driving property values. Understanding these dynamics allows valuers to make informed assessments that consider both current market conditions and future trends, ensuring a well-rounded and accurate property valuation.

Table 4.1 The real estate system

Real estate market/industry	*Definition*	*Key indicators*
Space market	Market where properties are leased or rented by individuals or businesses seeking space from property owners looking to rent out their premises	Rental levels, vacancy rate, occupancy rate, take-up levels
Asset market	Market where properties are bought by investors wanting to buy real estate from investors wanting to sell real estate	Price, capitalisation rate, investment volumes
Development industry	Industry that converts cash from the asset market into buildings to the space market	Construction activity, development pipeline

Market analysis in property valuation

Market analysis in property valuation focuses on identifying the factors influencing property value (Fanning et al. 1994). At a macro level, it examines the subject property within the broader context of market dynamics, analysing how it competes with other properties based on supply and demand conditions. At a micro level, it evaluates the subject property itself, focusing on its location, physical characteristics, and legal attributes.

Macro-level analysis – analysing the market where the subject property exists

The subject property's market

The definition of the relevant real estate market constitutes a foundational step in any real estate market analysis (Geltner et al. 2010). Such analyses are typically undertaken at both the metropolitan level and the submarket level in which the subject property is situated. For instance, when valuing an office building in Marylebone, London, the primary geographical scope of the analysis would be the Marylebone submarket. However, the dynamics of demand and supply in other submarkets, as well as within the broader metropolitan area, may exert an influence on the Marylebone submarket. Consequently, a comprehensive market analysis necessitates consideration of both the specific submarket and the wider metropolitan context, in this example, the London office market.

As seen in the aforementioned example, the market must also be segmented in terms of property sector, for example, retail, office, industrial, hotel, residential, and others. Further segmentations are also important to consider, such as location, scale, quality, and sustainability. Table 4.2 provides an overview of common segmentations made in different real estate sectors, and while the list is not exhaustive, it illustrates the diversity of criteria that investors, valuers, and market analysts use to classify properties.

Beyond geographical and sectoral considerations, the temporal scope of market analysis is also important. Most valuation reports emphasise recent market conditions, examining data from the past quarters to the previous year. However, structural shifts in the market may necessitate a broader temporal analysis. For example, the Global Financial Crisis (GFC) and, more

Table 4.2 Property sectors, property types, and further segmentations

Sector	*Examples of property types*	*Further segmentations*
Retail	Shopping centres, retail warehouse, standard retail	**Location:** regional, neighbourhood, strip/convenience centre **Quality:** class A, B, or C **Function:** store, supermarket, restaurant
Office	High-rise, mid-rise, low-rise; business park; medical office	**Tenancy:** multi- or single-tenant **Location:** central business district (CBD), urban or suburban location **Quality:** class A, B, or C **Sustainability:** Building Research Establishment Environmental Assessment Method (BREEAM) (UK), Leadership in Energy and Environment Design (LEED) (US), Haute Qualité Environmentale (HQE) (France), Deutsche Gesellschaft für Nachhaltiges Bauen/German Sustainable Building Council (DGNB) (Germany) certification
Industrial	Warehouse, manufacturing, data centre	**Specialisation:** storage, distribution, logistic, manufacturing **Location:** proximity to transportation hubs, urban or rural settings
Hotel	Hotel, motel, bed and breakfast	**Building structure:** resort, boutique, conference centre **Brand position:** luxury, upper upscale, upscale, upper midscale, midscale, economy, and independent **Service:** full service, limited service
Residential	Apartments/multi-family, houses	**Building structure:** detached house, semi-detached house, terraced house, apartment **Demographic**: senior housing, student housing

Source: Adapted from MSCI (2021), Moody's (2023), STR (2024).

recently, the COVID-19 pandemic significantly altered office space usage patterns, underscoring the need to assess both short-term fluctuations and long-term trends.

By incorporating these dimensions – geographical scope, property type, and temporal range – valuers and analysts can produce comprehensive market analyses that accurately reflect the forces shaping real estate values.

Real estate demand

Real estate demand is 'the quantity of a particular type of real estate product or service that will be purchased or leased' (Fanning et al. 1994, p. 137). Figure 4.1 illustrates the fundamental law of demand, which asserts that as prices rise, the quantity demanded decreases. In the context of real estate, this means that at higher prices or at higher rents, less space is demanded.

When examining the impact of price or rent fluctuations, it is crucial to differentiate between *actual* and *expected* changes (Mourouzi-Sivitanidou & Sivitanides 2020). An *actual* increase in price or rent reduces the quantity of space demanded, as previously noted. In the retail market, for instance, higher rents may deter retailers from occupying properties, leading to a decline in demand for retail space. This reflects a movement *along* the demand curve. Actual prices or rents are endogenous determinants of demand – meaning, they arise within the system and are influenced by the interaction of supply and demand in the market. These factors play a direct role in shaping market behaviour.

In contrast, the *expectation* of future price or rent increases can lead to a different outcome. Anticipation of rising rents may prompt retailers to accelerate their decision to enter the market, aiming to secure space before rents escalate further. As a result, *expected* price or rent increases can stimulate demand, producing an effect opposite to that of *actual* increases. This behaviour illustrates the phenomenon of rising demand during periods of anticipated price or rent growth. Unlike *actual* changes, *expectations* about future prices or rents cause a *shift* in the demand

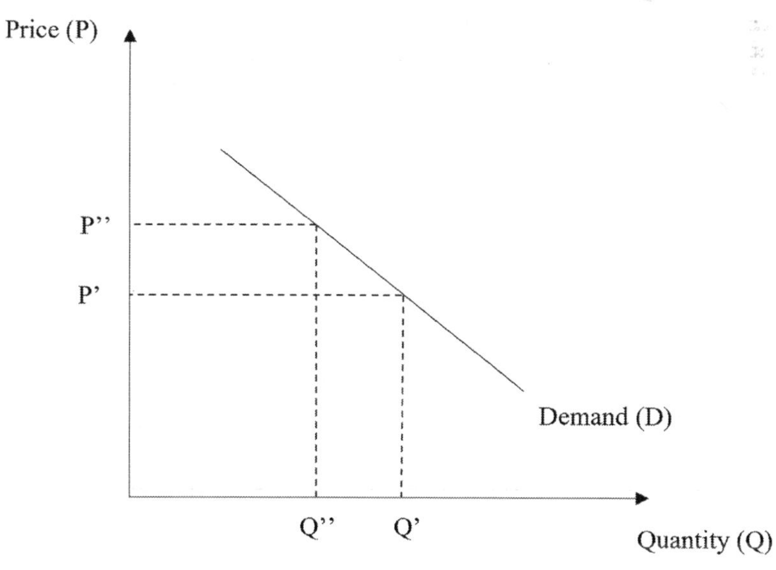

Figure 4.1 The fundamental law of demand.

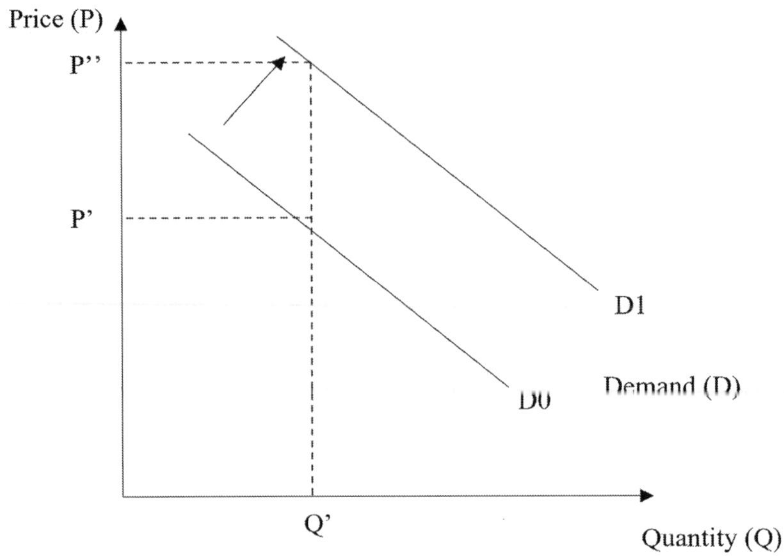

Figure 4.2 The shift in the demand curve.

curve and serve as exogenous determinants of demand. Figure 4.2 illustrates this shift. In the real estate market, where the quantity of space is inelastic in the short term (represented by the vertical Q line), such shifts in demand result in price or rent increases.

While *actual* prices and rents are endogenous determinants of real estate demand, the quantity demanded is also influenced by non-price or exogenous factors, which cause shifts in the demand curve. These exogenous determinants play a crucial role in market analysis, as they provide insights into the future prospects of the real estate market. According to Mourouzi-Sivitanidou and Sivitanides (2020), these exogenous drivers of demand in the user market can be categorised into five areas: market size, income, price of substitutes, consumer or firm expectations, and interest rates and availability of credit. Table 4.3 provides an explanation of each area and includes important factors to consider in different real estate sectors.

When conducting a market analysis, it is essential to account for both endogenous and exogenous factors influencing real estate demand. In the space market, key indicators include thus:

- *Average rents.* The average rent tenants pay for leasing space. Average rents might be broken down further into more specific categories, such as Average Grade A rent or Average Rent by Building Research Establishment Environmental Assessment Method (BREEAM).
- *Take-up levels.* Sometimes known as absorption rates (square metres or square feet). The amount of space being leased by tenants in a specific time frame, such as a quarter or a year. Take-up levels may also be categorised by the occupier's sector. For example, in the office sector, financial services were responsible for the largest share of take-up at 30% in 2023.
- *Vacancy rates (%).* The amount of vacant space divided by the existing rentable area in a submarket or in a building.
- *Occupancy rates (%).* The amount of space occupied by a tenant divided by the existing rentable area in a submarket or in a building.

Table 4.3 Exogenous Demand Factors

Key area	*Explanation*	*Specific factors/considerations for real estate sectors*
Market size	The market size, including variables such as population and employment, has a positive effect on demand; that is, as the population grows and employment levels rise, the demand for real estate increases.	**Retail:** number of households, foot traffic **Office:** office employment (e.g. employment in finance, insurance, real estate (FIRE)) **Industrial:** distribution employment, manufacturing employment **Hotel:** air passenger volume, number of visitors **Residential:** total employment, number of households, and demographic structure
Income and wealth	Income and wealth directly impact purchasing power; as income rises, so does the demand for real estate, while lower income can reduce demand.	**Retail:** disposable income **Office:** average wages, corporate profitability **Industrial:** industrial output, infrastructure investment **Hotel:** tourism expenditure, discretionary spending **Residential:** real income, median household income
Price or rent of substitutes	The price or rent of substitutes affects demand, as a decrease in the price of alternative properties can lead to reduced demand for the original property.	**Retail:** high retail rents possible to make the retailer sell online, shifting the demand of real estate downward **Office:** high rents in standard offices possible to encourage businesses to adopt remote work policies or move to flexible workspaces, reducing demand for traditional office space **Industrial:** high industrial rents possible to push companies to relocate to areas with lower costs, decreasing demand for industrial properties in expensive locations **Hotel:** high hotel room rates potentially leading travellers to consider alternatives like vacation rentals (e.g. Airbnb), lowering demand for hotel accommodations **Residential:** high house prices potentially making prospective buyers opt for renting an apartment instead, shifting demand to the rental market

(Continued)

Table 4.3 (Continued)

Key area	*Explanation*	*Specific factors/considerations for real estate sectors*
Consumer or firm expectations	Anticipated changes in prices, rental levels, and corporate performance can influence current demand, with positive expectations driving up demand and negative expectations dampening it.	**Retail:** retailer's expectation on consumer's disposable income **Office:** corporation's expectations on office employment growth and rental rates, square-feet-per-employee requirements **Industrial:** expectations about e-commerce growth **Hotel:** hotel's expectations on tourism levels **Residential:** prospective buyer's expectation on house price appreciation
Interest rates and the availability of credit	When interest rates are low, borrowing becomes more affordable, encouraging property investment, while higher interest rates make financing more expensive and can dampen demand. Similarly, the accessibility of credit enables more buyers to enter the market, but tighter credit conditions can limit purchasing power and reduce demand.	**Retail:** availability of business loans for retailers **Office:** corporate access to credit for lease commitments or workspace expansions **Industrial:** financing options for supply chain infrastructure investments **Hotel:** availability of financing for developers and operators **Residential:** mortgage interest rates, lending criteria, and credit availability for homebuyers

- *Obsolescence and fit for purpose.* Increasingly with changing technology, social change, and evolving demographics, the gap between the characteristics, amenities, and quality of supply and the specific demands of occupiers is becoming more critical.

In contrast, the asset market – where properties are bought and sold – relies on indicators such as the following:

- *Long-term trends.* As assets may need to be held for some years to cover acquisition costs, an investor must look at longer-term trends affecting the sector(s).
- *Sustainable demand.* Again, as assets may be of help for a significant period, an estimate of how stable demand will be for that investment.
- *Average prices.* The prices at which properties are bought and sold.
- *Capitalisation rates or net initial yield (%).* Initial expected return of a real estate asset, calculated by dividing the expected NOI of the first year by the property sales price.
- *Investment volumes.* The total value of property transactions within a specific time frame, such as a quarter or a year.
- *Notable transactions.* Key transactions that highlight significant market movements, such as record-breaking sales or acquisitions by major players.

While numerous data sources are typically available in major markets to facilitate the analysis of demand for a specific property type, the valuer must critically evaluate the data to ensure it is free from bias or manipulation intended to serve particular interests. These sources are explored further in Chapter 7.

Example 4.1 Demand analysis: Maple City office market

Disclaimer: The examples provided in this chapter are based on a fictitious office building in Maple City, referred to as the subject property. The analyses presented are simplified snippets, intended to offer a brief illustration of the type of information typically included in a full valuation report for an office property.

Space market

Leasing activity in Maple City saw a notable increase in Q4 2024, with take-up reaching 1 million square feet by the end of the quarter, reflecting a 5% rise compared to the previous quarter. For the entire year (Q1–Q4), total take-up stood at 5 million square feet across 400 transactions, a 10% decline from the long-term average.

This decrease is likely a result of the ongoing shift towards hybrid working models in the post-pandemic environment. In the central business district (CBD), the average rent for Grade A office space is currently $35/ft² per annum, marking a 3% increase from the previous year, a sign of sustained demand for modern, high-quality office environments.

Rents for BREEAM Outstanding/Excellent–certified office buildings in the CBD are even higher, averaging $55/ft² per annum, highlighting the premium for energy-efficient and sustainable office spaces. On the other hand, rental growth for secondary office space in the CBD has remained modest, with rents stabilising at around $20/ft² per annum. The overall vacancy rate for office space in Maple City remains relatively low at 5.2%, indicating strong demand. However, there is a clear distinction between prime and secondary office spaces, with older, less energy-efficient buildings experiencing higher vacancy rates.

Asset market

In the investment market, the outlook for office assets in Maple City remains positive. Capitalisation rates for prime office assets in the CBD have compressed slightly, with rates averaging 4.5%, down from 5.0% last year. This reflects strong investor interest in high-quality office properties, particularly those with long-term leases in place with multinational tenants.

Secondary office assets, however, are seeing less investor interest, with cap rates for these properties averaging 6.5%, indicating a risk premium for less-modern buildings. Investment volumes in the office market have remained strong, with over $2.5 billion in transactions recorded in 2024, primarily driven by institutional investors and REITs.

Notable transactions include the sale of the Skyline Tower, a 95,000 ft² fully leased Grade A office building in the CBD with a weighted average unexpired lease term (WAULT) of 10.5 years, which traded for $50 million, reflecting a cap rate of 4.3%. Another significant deal was the acquisition of Maple City Central, a mixed-use development with 120,000 ft² of office space, which was sold for $70 million, at a 5% cap rate.

Real estate supply

The real estate supply to be investigated in an appraisal context includes the competitive properties within the subject property's market area (Fanning et al. 1994). These properties may either belong to the existing stock/inventory or represent new supply currently under construction and expected to enter the market in the near future. Therefore, the quantity of this competitive supply depends on the prices being paid in the asset market and the cost of building new units in the development industry (Fanning et al. 1994).

Traditionally, the supply curve is portrayed as an upward-sloping curve that demonstrates that greater quantity is supplied at higher prices in real estate. However, as highlighted by Mourouzi-Sivitanidou and Sivitanides (2020), in real estate, it is important to consider the behaviour of the curve in the short term (Figure 4.3) and in the long term (Figure 4.4). In the short term, construction lags and other constraints render the supply curve vertical, reflecting the inelasticity of supply. Over the long term, these constraints are alleviated, allowing the supply curve to assume its conventional upward-sloping form.

A variety of secondary data is available to analyse the supply of a given property type. Basic information to be gathered include thus:

- *Inventory*. The total stock of existing space, typically measured in square metres or square feet, categorised by property type.
- *Development pipeline*. The volume of space under development in new construction projects, including details on their current status (e.g. planned, under construction, or nearing completion), expected delivery timelines, and classification as either speculative – developments undertaken without a secured tenant or buyer – or pre-let, where the property is leased to a tenant before construction or completion.
- *Competitive projects*. Information on similar developments currently under construction within the market area.

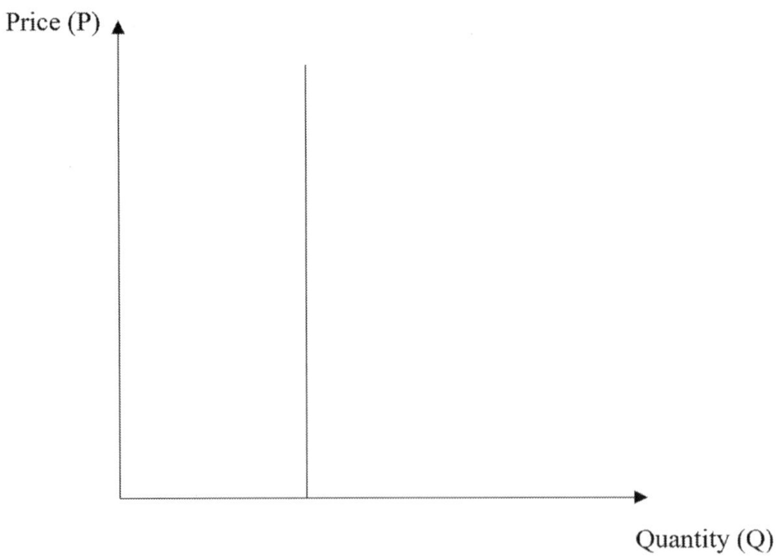

Figure 4.3 Short-term real estate supply.

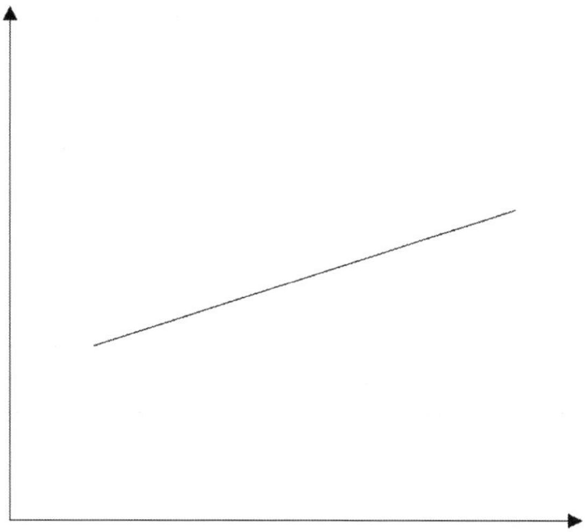

Figure 4.4 Long-term real estate supply.

Example 4.2 Supply analysis

The office market in Maple City currently holds a total inventory of approximately **6 million m²**, with a significant concentration in the CBD.

As of Q4 2024, **200,000 m²** of office space are under construction, primarily within the CBD. The focus of these developments is on high-quality (Grade A), sustainable office buildings (LEED Gold and above), catering to the increasing demand for modern, energy-efficient workspaces. Around 70% of the new office space under construction has already been pre-let to large multinational firms and professional services companies.

The most significant developments currently under construction are two large-scale projects:

- **Maple Tower**, a 35-storey office building offering **8,500 m²** of space
- **CityPoint Plaza**, a 20-storey mixed-use development circa **5,000 m²** of office space

Both projects are targeting large multinational firms, with over 80% of their space pre-let to corporate tenants.

Micro-level analysis – the subject property

The micro-level analysis focuses on the subject property and its immediate environment. This calls for careful evaluation of the location amenities, user preferences and their willingness to pay for specific location, and subject property attributes. With an understanding of these elements, the analysis can determine the property's competitive position in the market and its ability to attract tenants or buyers willing to pay premium prices or rents.

Location

The location of a property is often one of the most critical factors influencing its value. Extensive research underscores the significant impact of location attributes on both property prices and rental levels, making location analysis a cornerstone of any property market analysis. For instance, in the residential market, Benson et al. (1998) observed that ocean views can increase the market price of houses in Bellingham, Washington, by nearly 60%. Similarly, Morancho (2003) found that proximity to green spaces affects housing prices, with every 100 m increase in distance resulting in a decrease of €1,800. Nguyen-Hoang and Yinger's review (2011) provides evidence that school quality translates into price premiums in the residential market across several geographical regions. In the office market, Kopczewska and Lewandowska (2018) reported that in London in 2015, every additional 100 m from the nearest metro station reduced office rents by £0.7/ft² per year, while a central business district (CBD) location increased office rents by approximately £20/ft² per year.

Wyatt (2013) highlights the importance of agglomeration economies – benefits arising from the proximity of firms and individuals to one another – when evaluating the location of a subject property. For instance, offices often cluster in CBD locations, close to shopping and hotel facilities (Wyatt 2013; Liusman et al. 2017), to benefit from enhanced accessibility, knowledge sharing, access to a diverse talent pool, and proximity to both clients and complementary businesses. Similarly, retail units tend to concentrate in shopping centres or high-street retail districts, forming compact clusters that broaden their catchment areas and attract a higher volume of customers (Zhang et al. 2020). Table 4.4 provides an overview of location factors relevant to different property sectors.

Table 4.4 Key location factors across property sectors

Sector	*Location factors to consider*
Retail	Footfall, proximity to complementary businesses, visibility, accessibility for customers (e.g. pedestrian flow, car access), competition within the area, demographic characteristics of the surrounding population, proximity to transport hubs
Office	CBD/urban/suburban location, accessibility by private transport (e.g. car parking, traffic congestion), accessibility by public transport (e.g. train, bus), proximity to retail (e.g. supermarket, restaurant), leisure amenities (e.g. parks, public gardens), local amenities (e.g. gyms, cafes), incidence of crime, quality of nearby infrastructure, the concentration of other office users (e.g. finance, legal services)
Industrial	Accessibility to major roads, highways, and transportation networks; proximity to suppliers and markets; proximity to port or rail facilities; proximity to waste disposal or recycling facilities; labour availability
Hotel	Proximity to retail centres, popular tourist destinations, and local attractions; proximity to transportation hubs (e.g. airports, train stations); visibility and accessibility
Residential	Proximity to beaches or other desirable natural features, distance to schools, quality of schools within the district, proximity to public transport and local amenities (e.g. shopping centres, parks), crime rates, access to healthcare and emergency services, proximity to employment hubs, overall safety and liveability of the area

- *Deferred annuity.* In this case, a lump sum is exchanged today for an annuity that starts after a specified deferral period, typically longer than a year. For example, an individual aged 50 might purchase a life annuity that begins at age 65, making it a deferred annuity.

Ordinary annuity (also called annuity in arrears)

An *ordinary annuity* is an annuity in which the cash flows occur at the end of the period. This means that the first annual sum will be paid (received) 12 months after the purchase or taking out of the policy. The payments could be weekly, monthly, quarterly, or for any period, provided they are in arrears. For example, an ordinary annuity that pays $1,000 per year for five years can be visually represented like this:

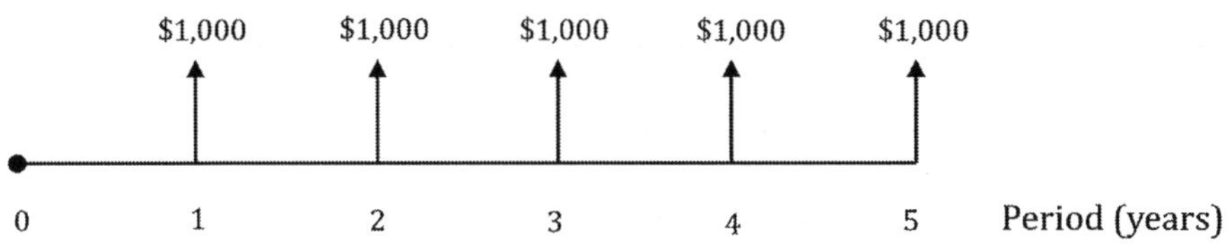

Figure 6.6 Example of an ordinary annuity.

Present value of an ordinary annuity

To determine the present value of an ordinary annuity, we sum the present values of each payment. Considering the preceding annuity, how much money should we allocate into our fund today to receive this payment stream if our remaining balance continually earns 4% annually?

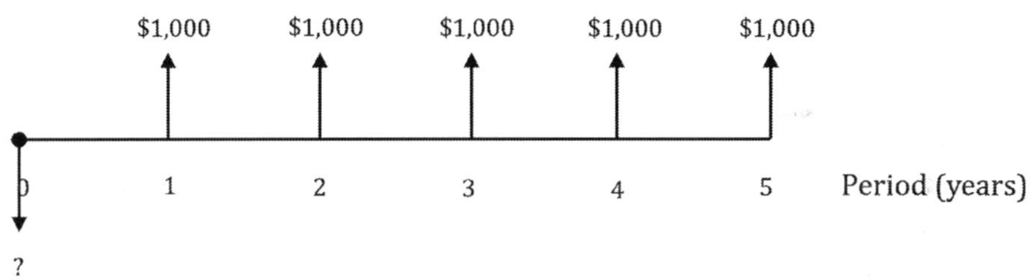

Figure 6.7 Finding the PV of an ordinary annuity.

Given that all our cash flows are in the future, we can use Equation 6.2 to bring these future values to present values:

Table 6.4 PV of an ordinary annuity

Year	*Cash flow ($)*	*Calculation*	*Present value*
1	1,000	$PV = \frac{1,000}{(1+0.04)^1}$	961.54
2	1,000	$PV = \frac{1,000}{(1+0.04)^2}$	924.56

(*Continued*)

Example 4.3 Location analysis

The subject property is located within the CBD of Maple City, an area with a population of approximately 1.2 million people that hosts numerous corporate headquarters, financial institutions, and law firms, making it a significant business hub in Maple City.

Public transport connectivity is excellent, with the Maple City metro station located **250 m** to the east, offering access to multiple metro lines that connect the CBD to other key areas of the city.

The nearest bus stop, situated just 100 m from the property, further enhances public transport options for tenants and visitors.

The subject property is well-served by retail and leisure amenities. A range of restaurants, cafés, and retail stores are located within walking distance, with a supermarket just **200 m** away, providing convenient access to daily essentials.

For leisure, nearby parks, including Maple Park (**500 m** away), and public gardens offer outdoor spaces for relaxation. Additionally, several gyms and fitness centres are located within the vicinity, providing convenient options for health and wellness. Local amenities also include medical clinics, banks, and convenience stores, ensuring that tenants have easy access to essential services. In terms of safety, the CBD of Maple City has a relatively low incidence of crime, supported by strong police presence and surveillance systems in place throughout the district. The quality of nearby infrastructure is high, with modern utilities, reliable internet connectivity, and well-maintained roads and public spaces. However, traffic congestion is a significant issue in the CBD, particularly during peak hours.

The area experiences heavy vehicular flow, which can lead to delays and reduced accessibility by private transport.

Despite the availability of underground parking at the property, the limited parking space in the surrounding area exacerbates the issue, making commuting by car less convenient for tenants and visitors during busy periods.

Site

The site characteristics of the subject property include its size, shape, topography, geology, drainage, orientation, and any other feature that might influence the property's value. Whilst in the past, access to parking and cars was essential, increasingly, environmental factors will come into play, such as access to sustainable public transport. Access for safety, fire, and waste disposal is another important issue. Aside from the physical features, it is also important to consider legal restrictions, such as zoning, and planning constraints, which may restrict the site's potential uses. Other risks to be considered include flooding and changes of use around the site, which may have a negative impact.

Example 4.4 Site analysis

The subject property is situated on a 2,215.87 m^2 urban plot along Maple Road, in a well-connected commercial district. The plot is bordered to the north by 65 m of land owned by Greenfield Enterprises, to the south by 70.50 m of land owned by Oakwood

Developments, to the east by 36 m of land owned by Redwood Corporation, and to the west by 63 m of land owned by Mr Birchwood. The site features level topography with good drainage and faces north, offering optimal natural light. Access to the site is provided by Maple Road, a well-maintained concrete-paved road that connects the area to key parts of the city. The site's location ensures good connectivity with public transport options, including nearby bus and metro stations, enhancing accessibility for future office tenants.

The subject property is situated within Commercial Zone 1 (CZ1), which supports a variety of commercial uses typically found in central areas. This includes corporate headquarters, large hotels, department stores, and entertainment facilities, particularly in high-rise mixed-use buildings.

The commercial floor area ratio (FAR) in CZ1 is set at 6.0, with the potential for increases through incentives, such as the inclusion of public plazas or pedestrian-friendly amenities. Residential and community facilities are also permitted in CZ1.

The area benefits from strong public transportation links, though due to the central location's congestion, off-street parking is required at a rate of one space per 40 m^2.

Physical attributes

The physical attributes of a property include its size, age, condition, design, internal layout, amenities, and sustainability features. Modern, well-maintained buildings with high-specification finishes, energy-efficient systems, and flexible layouts are often preferred by users. In residential properties, attributes such as the number of bedrooms, natural lighting, adequate storage solutions, and outdoor amenities, such as balconies, gardens, and pool areas, can significantly enhance a property's appeal. Additionally, energy-efficient features, such as high-performance insulation, smart home technology, and renewable energy systems, increasingly influence buyer and tenant preferences.

In commercial properties, considerations such as material handling, product flow, and maintenance arrangements are critical. Retail properties are often evaluated based on their internal configuration, including frontage length, depth, ground floor area, and facilities for display, sales, and storage. For office properties, tenants might prioritise open or flexible layouts. In industrial properties, uninterrupted ground-floor areas, high load-bearing capacities, generous eaves heights, and well-designed loading and access points are typically favoured, along with additional amenities, like car parking, ventilation, and on-site canteen facilities.

Across all property types, sustainability features, such as energy certifications and green credentials, can command price and rental premiums, reflecting their growing importance in modern property valuation. For instance, Porumb et al. (2020) found that office buildings with green certifications (BREEAM, LEED, and DGNB) in Finland, France, and Germany commanded a price premium of approximately 20% between 2010 and 2015. In the United States, Fuerst and McAllister (2011) identified rental premiums of around 5% for LEED-certified properties and 4% for Energy Star–certified buildings. The impact on sales prices was even more pronounced, with LEED-certified buildings achieving an average premium of 25% and Energy Star–certified buildings reaching 26%. This topic will be explored in greater detail in Chapter 13, 'Contemporary impacts on markets and values', where the implications of sustainability on property markets will be discussed further.

Example 4.5 Physical attribute analysis

The subject property is a ten-storey office building with a total internal area of 65,900 ft^2. Completed in April 2024, the building meets modern construction standards and has achieved a BREEAM Excellent certification. The exterior incorporates contemporary cladding materials, including aluminium composite panels (ACP) and high-performance glass, both of which offer superior thermal insulation and soundproofing. Double-glazed windows provide abundant natural light while contributing to overall energy-efficiency. Inside, the building features two high-speed lifts serving all ten floors, ensuring accessibility and convenience. Thirty-six underground parking spaces offer an added advantage in a central urban location where parking is often constrained. Additional specifications include a centralised heating, ventilation, and air conditioning (HVAC) system; light-emitting diode (LED) lighting throughout; and a building management system designed for efficient operations and ease of maintenance.

Tenancy profile

The tenancy profile of the property is another critical factor in micro-level analysis. Properties with strong covenants – tenants with reliable financial standing and long lease terms – tend to attract investors due to their income stability. A diverse tenant mix, with limited exposure to any single sector or tenant, reduces risk and enhances the property's appeal. On the other hand, high vacancy rates or short unexpired lease terms may signal operational challenges, requiring further investigation into market dynamics and property management practices. Key tenancy factors, including lease length, rent reviews, repairing and insurance obligations, user restrictions, and alienation rights, all play a significant role in shaping the overall risk and return profile of a property, ultimately influencing its value. These aspects will be examined in detail in the following sections.

Lease term/length

As discussed in Chapter 2, a *lease term* refers to the length of time a lease remains valid. The duration of the lease directly impacts the value of the subject property, as it ensures income stability. In other words, the rent agreed upon for the term of the lease is considered secure because it is contractually guaranteed.

Short leases increase the potential for voids, which can diminish cash flow and reduce market value, due to the associated income risk. In contrast, longer leases may provide greater security of income for the owner but could reduce ability to respond to changing market requirements, leading to an ageing building which may not meet current user specifications. There may also be a reduced opportunity to secure increases of rent due to inflation or market movement. While there is no optimum lease length, market observations will indicate what tenants are requesting for a particular use and type of building. For instance, properties such as retail units, restaurants, and leisure facilities which require significant and expensive fit-outs may require longer leases to justify the spreading of the cost of the fit-out over enough years of operation to be viable commercial propositions.

The unexpired lease term is a critical consideration for investors, as clearly the longer the term – provided it is to a tenant of good covenant (that is, one deemed to be financially secure and able to pay the rent for that term) – the more secure the investment will be.

Rent indexation and reviews

When analysing the tenant(s) in the subject property and the leases that they occupy the building under, it is important to assess the mechanisms for increasing the rent. This varies globally and can take the form of indexation to align the rent to a relevant price index, or to add an additional uplift to that index. A lease may provide for a specified rent review, or a number of reviews at specified dates. These may be to open market value, and historically in the UK, these have been upward only, ensuring that rental decreases cannot occur. Whatever the provision in the lease or license for adjustment of rents and when and to what basis these occur, it will have potential, sometimes significant, implications for the income stream. Will the rent increase, maintaining or enhancing income stability, or will it decrease, potentially leading to reduced revenue? The timing and structure of rent reviews are essential to understanding the future trajectory of rental income and its impact on the property's overall financial performance.

In markets such as the UK, where five-yearly rent reviews are still prevalent, rents fixed for a set period can expose the property to inflation risk. As rents may not adjust until the review date, they can lag behind inflation, reducing the rent's real value over time. In contrast, many European markets commonly link rents to inflation indices, such as the retail price index (RPI) or CPI. This practice helps safeguard rental income against inflationary pressures, offering a more stable and predictable income stream. Such arrangements ensure that the rent increases in line with economic conditions, reducing the risk of the property's income falling behind the market.

Rent review, that is, the process that adjusts the rent being paid by a tenant to reflect the current market conditions, also has a significant impact on the property's value because it directly influences future income stability and the potential for revenue growth.

In markets like the UK, where five-yearly rent reviews remain common, rents that are fixed for a set period may be exposed to inflation risk. In contrast, rents linked to inflation, a practice more common in many European countries, ensure that the rent increases in line with inflation or other economic indicators.

The actual terms for payment may also be important. Typically, rent is payable in advance, but should this be monthly or quarterly? Monthly payments will ensure earlier awareness of any payment problems but will be more costly administratively. Quarterly means that three or even four months could pass before inability to pay becomes clear, and action must be taken to recover rent and to repossess the space.

Repairing and insurance obligations

Analysing repairing and insurance obligations is essential for understanding the financial implications and risk profile of a property, as these obligations directly affect the NOI and, consequently, the property's value. If the landlord is responsible for repairs or insurance, these costs must be deducted from the gross rental income to determine the property's net income, which affects its valuation. In contrast, when tenants are responsible for repairs and insurance, the income stream becomes more predictable, and the financial and managerial risk to the landlord is reduced.

From a valuation perspective, it is generally more favourable for tenants to bear the costs of repairs and insurance obligations, as this reduces the landlord's ongoing financial liabilities and enhances the property's net income. However, there is an argument that, as lease terms get shorter, landlords might prefer to take on this responsibility to ensure the property is kept in the best condition to command the best market rent and is always fully insured. Where a landlord is responsible, then there is an annual average cost to be covered. This might be possible through a service charge, but in addition, there will be a greater amount of hands-on management, which may have a cost. Where the tenant is responsible, work can be done to suit the tenant, but if it is the landlord's responsibility, arrangements must be made to undertake work around the tenant.

User restrictions and alienation rights

Another important aspect of tenancy profile to analyse is the presence of user restrictions and alienation rights, as these directly impact tenant flexibility, financial viability, and the property's long-term value. *User restrictions* refer to limitations imposed by the landlord on how the tenant can use the leased space. Alienation rights, on the other hand, concern the tenant's ability to assign (sell) or sublet the lease to another party.

Strict user restrictions, such as limitations on business activities (e.g. 'to use only for the sale of cheese') or operating hours (e.g. 'to use by the tenant only between the hours of 08:30 and 18:00 hours'), can reduce a tenant's ability to generate revenue, making the lease less attractive and potentially leading to lower rents. These factors influence not only the tenant's willingness to pay rent but also the property's capital value, as they limit the number of potential tenants who may find the space suitable.

Tenants taking on a lease cannot be certain that they will need all or part of the space for the duration of the lease. Much will depend upon the length of the lease and the nature of the tenants' use of the property. If business is better than expected, then the tenant may need more space; if business is not as good as initially forecast, the tenant may need less space or may need to extricate themselves from all the lease obligations. For this reason, a tenant will prefer a lease which gives them complete freedom to assign (sell), or sublet, the whole or part of the demised premises.

The landlord will prefer a total restriction on assignment and subletting, so that they can maximise control over any future change in the tenant. The former will enhance tenants' rental bids, and the latter will restrict tenants' bids. For this reason, covenants or clauses covering assignment and subletting are usually permissive, but subject to first offering to surrender the lease to the landlord and/or only with the landlord's written consent.

The covenant strength of the tenant(s)

The *covenant strength* of a tenant refers to the financial stability, creditworthiness, and reliability of a tenant in fulfilling their lease obligations. The covenant strength of a tenant might be assessed through credit ratings in markets where this is available. Some of the major rating agencies are Standard & Poor, Dun & Bradstreet, Fame, Experian, and Moody's Investor Service. Other methods for assessing covenant strength include examining financial statements to assess the tenant's liquidity, profitability, and leverage ratios and checking historical payment records. For tenants that are subsidiaries, the financial health and creditworthiness of the parent company play a crucial role, with a parent guarantee strengthening the tenant's covenant. Publicly available data, such as press releases, market performance reports, and regulatory filings, can also enhance understanding of the tenant's financial position.

Other considerations

It is important to recognise that a wide and diverse range of other regulatory issues can impact upon the property market, and therefore the value of buildings. Also, valuers should be aware that these are constantly changing, and in some cases, their introduction will have wide-ranging and rapid impacts upon property markets. For instance:

- It is common in the UK and many countries to require public tenants to only occupy buildings of a specified sustainability ranking. For the owner of a building outside this requirement, it could find 30–40% of its potential occupiers removed from the market overnight, having a significant impact upon its value.
- In the UAE, the sudden introduction of a law requiring those who worked in Abu Dhabi emirate to reside in that emirate and not commute from adjoining Dubai or other emirates had a profound impact upon the value of residential property in each emirate.

Contemporary approaches to market analysis

Increasingly, property agents, developers, and other professionals are applying strategic marketing tools to real estate markets as property products become more sophisticated and end user–orientated and have a complex arrangement of amenities and brand positions.

As an example of this approach, we refer to a JLL study based in Portugal which focused on the rapidly expanding co-living market. *Co-living* is a solution to enable affordable yet high-quality accommodation through an element of sharing. It is not only for digital nomads and the millennial generation but has also been applied to more mature communities, impacted by loneliness during the pandemic. In many countries, demand is high, supply is low, and growth has been in double digits.

Co-living in the asset market would be allocated to what is known as the *alternatives sector* investment, associated with broader, longer-term trends with much less volatility and uncertainty, such as age demographics, urbanisation, technology, or macro growth in consumer markets and household wealth. Investors are always looking for opportunities such as co-living, where demand is much bigger than supply and where it is a new concept that enables them to obtain asset diversification and enhanced risk-adjusted returns.

Defining micro segments

But even the co-living market is not one-size-fits-all, and as the JLL (2019) study demonstrates, property segments now have identifiable 'micro segments' which may have to be considered in an appraisal context. The JLL study identified a number of micro segments of co-living across Europe, of which five of the most robust were classified by location, building characteristics, and end user needs. These are set out in Table 4.5.

Defining target segment profiles

In addition to micro segments, increasingly developers are examining the characteristics and behaviours of target segments to ensure they provide the right kind of product, its shape, design, finishes, and increasingly, the many amenities that are expected by lifestyle-orientated customers. In the same report, JLL identified the demographic, behavioural, and lifestyle characteristics of the different customers associated with the micro segments identified earlier. In the report, they were given names and photographs to make it more personal.

Whilst a little on the margins of our valuation focus in this book, the earlier example illustrates the sophistication and detail of market analysis that property professionals are increasingly engaging in.

Table 4.5 Micro segments of the co-living market

Co-living market analysis		
Micro segment	*Characteristics*	*Active brands (as at 2020)*
Community-orientated	Value- and/or purpose-driven communities with a strong emphasis on shared values and contributing to a positive social impact on the local community	**Treehouse** **Techfarm** **Embassy Network**
Urban lifestyle–orientated	Integrate multiple-community residential solutions and shared services into a single ecosystem; focused on differentiation through attention to detail and unique designs with enhanced personality; provide the comfort of a hotel with the feeling of being at home	**Quarters** **We** **Circles** **Habyt**
Travel-orientated	Promote balance between work and leisure through the creation of a new 'workstations' concept; integrate a global co-living and co-working space network in attractive locations, where the same kinds of experiences and standards are guaranteed	**Selina** **Outsite**
Work-orientated	Hybrid hotel, residence, and co-working models; aimed at ensuring productivity and a service offering that meets the target's core needs, access to entrepreneurs	**Urban Campus** **Zoku** **Aviato Club**
Cost-orientated	Affordability – hacker-house evolution for all-inclusive experiences, with basic benefits and services for its residents	**Hacker House** **StartUp House**

It demonstrates above all the need for those involved in property markets (and therefore those who value properties in these markets) to fully understand their end users more than ever before. This is not an example from a marketing textbook but from a real estate consultant (JLL) report, illustrating the characteristics of co-living for European investors.

We encourage readers to explore the websites of leading property companies, such as JLL, CBRE, Cushman & Wakefield, and Colliers, wherever you live, to see for yourselves this kind of approach, especially in residential markets.

Summary

- The real estate system comprises the space market, the asset market, and the development industry.
- The space market, also known as the user or occupier market, is where properties are leased or rented by individuals or businesses seeking space (demand side) from property owners looking to rent out their premises (supply side).
- The asset market is where properties are bought by investors wanting to buy real estate (demand side) from investors wanting to sell real estate (supply side).
- The development industry translates the financial capital from the asset market into physical capital in the space market, thereby governing the supply side of real estate.

Table 4.6 Customer profiles associated with the micro segments of the co-living market

Lifestyle	*Profile*	*Statement*	*Motivations*
Digital nomad	33 years old; single, with no kids; from Spain; web developer (remote worker)	'The things you own end up owning you – that's why I prefer flexibility to enjoy life'.	• Freedom; a non-9-to-5 routine • Healthy lifestyle • Minimalist, 'non-materialistic' • Professional growth and personal development • 'Live intensely'
(No) mad	30 years old; single, with no children; from the Netherlands; interior designer (remote worker)	'Now I want to slow down and stay longer to work, learn, and feel the local vibe'.	• Work–life balance • Stability, routine, and focus • Meaningful relationships • Income management/savings • Local immersive experience • 'Slow down'
Expat	35 years old; married, with no children; from the UK; creative business consultant	'I want to feel welcome abroad and find a home away from home'.	• Looking for authentic daily life • Aim for better quality (cost of living, safety, climate) • Soft landing with comfort of 'home, away from home' • Integration in local culture
Business traveller	42 years old; single, with no children; from Germany; hedge fund manager	'Sayonara, boring hotel room and lonely drink at the bar. I'm looking for a homely atmosphere and new inspiring people'.	• Maximum convenience • *Bleisure* (business + pleasure) • Humanised experience (home vs hotel) • Get out of routine and open new horizons • Learn, be inspired, and network
Local	29 years old; single, with no children; from Aveiro, Portugal; photographer and content creator	'I want to come home and have someone to have dinner [with] and talk to, but with all the privacy and comfort I always had'.	• Buy house at 31–35 years old • Personal and professional balance • Maintain same conditions as parents' home • Stable employment and income situation

Source: JLL Portugal (2019), 'Portugal Co-living: Follow the Trend'.

- Market analysis in property valuation focuses on identifying the factors influencing property value.
- At a macro level, it examines the subject property within the broader context of market dynamics, analysing how it competes with other properties based on supply and demand conditions.
- At a micro level, it evaluates the subject property itself, focusing on its location, physical characteristics, and legal attributes.

Questions

1. What are the components of the real estate system, and how do they interact with one another?
2. What indicators should valuers focus on when analysing the space market versus the asset market?

3. How do endogenous and exogenous factors affect real estate demand?
4. How might economic crises or pandemics reshape market analysis approaches?
5. What are the implications of short-term supply inelasticity for real estate valuation?

References

Benson, E.D.; Hansen, J.L.; Schwartz, Jr., A.L.; Smersh, G.T. (1998). Pricing Residential Amenities: The Value of a View. *Journal of Real Estate Finance and Economics*, 16: 1, 55–73.

Fanning, S.; Grissom T.; Pearson T. (1994). *Market Analysis for Valuation Appraisals*. Chicago: Appraisal Institute.

Fuerst, F.; McAllister, P. (2011). Green Noise or Green Value? Measuring the Effects of Environmental Certification on Office Values. *Real Estate Economics*, 39, 45–69.

Geltner, D.M.; Miller, N.G.; Clayton, J.; Eichholtz, P. (2010). *Commercial Real Estate: Analysis & Investment*. 2nd ed. Mason, OH: Thompson South-Western.

JLL. (2019). Portugal Co-living: Follow the Trend. Available at: https://www.jll.pt/pt/estudos-e-tendencias/estudos-de-mercado/portugal-coliving-follow-the-trend-2019 [Accessed on 10/03/2020].

Kopczewska, K.; Lewandowska, A. (2018). The Price for Subway Access: Spatial Econometric Modelling of Office Rental Rates in London. *Urban Geography*, 39: 10, 1528–1554.

Liusman, E.; Ho, D.C.W.; Lo, H.C.; Lo, D.Y.F. (2017). Office Rents, Mixed-Use Developments, and Agglomeration Economies: A Panel Data Analysis. *Journal of Property Investment & Finance*, 35: 5, 455–471.

Moody's. (2023). *Types of Retail*. Available at: https://www.moodyscre.com/insights/cre-news/types-of-retail/ [Accessed on 27/05/2022].

Morancho, A.B. (2003). A Hedonic Valuation of Urban Green Areas. *Landscape and Urban Planning*, 66, 35–41.

Mourouzi-Sivitanidou, R.; Sivitanides, P. (Eds.). (2020). *Market Analysis for Real Estate*. London: Routledge.

MSCI (2021). *MSCI Property Indexes Methodology*. [Online]. Available at: www.msci.com/documents/1296102/1311232/MSCI+Property+Indexes+Methodology.pdf/a33f5a6c-a5e4-4d18-f460-d8c537ded968 [Accessed on 27/05/2022].

Nguyen-Hoang, P.; Yinger, J. (2011). The Capitalization of School Quality into House Values: A Review. *Journal of Housing Economics*, 20, 30–48.

Porumb, V.; Maier, G.; Anghel, I. (2020). The Impact of Building Location on Green Certification Price Premiums: Evidence from Three European Countries. *Journal of Cleaner Production*, 272, 1–11.

STR (2024). *Glossary*. [Online]. Available at https://str.com/data-insights/resources/glossary [Accessed on 11/12/2024].

Wyatt, P. (2013). *Property Valuation*. 2nd ed. Chichester: Wiley-Blackwell.

Zhang, S.; van Dujin, M.; van der Vlist, A.J. (2020). The External Effects of Inner-city Shopping Centers: Evidence from the Netherlands. *Journal of Regional Science*, 60: 4, 583–611.

Chapter 5

Expectations of valuers

Introduction

The public frequently becomes perplexed when they obtain two valuations with disparate valuation outcomes. This misunderstanding frequently results from a lack of knowledge about the factors and procedures involved in the appraisal process. Significant disparities can raise questions regarding accuracy and dependability, even while valuation variances are normal and expected. Valuers must provide a clear explanation of the foundation and goal of their assessments, together with the data sources, underlying assumptions, and valuation techniques used, in order to prevent such misunderstandings.

Extreme differences should be closely explored, even though varying appraisals are a normal feature of the profession. Significant differences between two values could be a symptom of inconsistent data sources, approaches, or underlying presumptions. If the appraisals are being used for important financial choices, like lending, acquisitions, or legal issues, this can be reason for concern. To resolve the discrepancies in these situations, a review by an impartial expert or regulatory agency might be required. In the end, differences in valuation should not be concerning but rather viewed as a reflection of various viewpoints and analytical techniques, provided that valuers conduct themselves openly and in accordance with professional norms.

Stakeholders can better understand value discrepancies if a clear explanation of the reasoning behind the conclusions is given. Furthermore, professional associations and regulatory agencies are essential in establishing standards to improve uniformity and comparability in real estate appraisals. The industry can endeavour to decrease disparities and boost public trust in valuation results by promoting improved communication, standardising valuation procedures, and making sure valuers follow professional guidelines.

The main reasons for different valuation results

There are several key reasons that valuation results may differ, even when conducted by qualified professionals:

- *Purpose of the valuation.* The intended use of the valuation can lead to different assumptions and methodologies. For example, a valuation for investment purposes may use a different approach than one for mortgage lending or financial reporting.
- *Choice of methodology.* Valuations can be conducted using different approaches, such as the market approach, income approach, or cost approach. The choice of methodology depends on factors such as property type, availability of data, and the valuer's expertise.

DOI: 10.1201/9781003490357-6

- *Market conditions and timing.* Valuations conducted at different times may yield different results due to fluctuations in market conditions, interest rates, economic trends, and investor sentiment.
- *Data availability and quality.* The sources of data used in a valuation, including comparable sales, rental income, and expenses, can vary; some valuers may have access to more comprehensive or recent data, leading to discrepancies in results.
- *Professional judgement and assumptions.* Valuers rely on their professional experience and judgement when making assumptions about key factors, such as depreciation, risk, and future market trends; these subjective elements can contribute to differences in valuation outcomes.
- *Regulatory and legal factors.* Different jurisdictions and regulations may influence valuation practices, leading to variations in how properties are assessed.

By understanding these factors, stakeholders can better interpret valuation differences and ensure that the most appropriate valuation approach is used for their specific needs.

The role of professional bodies and valuation standards

The property valuation profession is governed by a range of standards designed to ensure accuracy, consistency, and transparency. These standards help in maintaining the credibility of valuation reports and ensuring that they are legally defensible. Key valuation standards include thus:

- *Royal Institution of Chartered Surveyors (RICS Red Book).* The RICS (2021) Valuation – Global Standards provides a comprehensive framework for ethical and professional valuations, ensuring credibility and consistency across international markets. RICS establishes best practices and guidelines for valuers to follow.
- *International Valuation Standards (IVS).* Published by the IVSC, IVS establishes globally recognised valuation principles applicable to various asset types, ensuring comparability and reliability. These standards provide guidance on methodologies, reporting, and ethical considerations.
- *The European Group of Valuers' Associations (TEGoVA) European Valuation Standards (EVS).* TEGoVA sets the EVS, which are widely used for regulatory and commercial valuation purposes across Europe. These standards focus on compliance with European Union directives and ensure alignment with regional regulations.
- *International Financial Reporting Standards (IFRS).* IFRS (2023) requires valuation methods that align with financial reporting regulations, ensuring that property valuations accurately reflect an entity's financial position. These standards are crucial for businesses and financial institutions that rely on accurate property valuations for balance sheets and investment decisions.
- *Appraisal Institute (AI).* A leading global professional association of real estate appraisers, the AI establishes ethical standards, education programmes, and certification requirements for valuers.
- *Financial Accounting Standards Board (FASB).* FASB (2019) sets financial reporting standards in the United States, affecting how property values are reported in financial statements.
- *Uniform Standards of Professional Appraisal Practice (USPAP).* Managed by the Appraisal Foundation, USPAP provides widely accepted standards for ethical conduct and professional appraisal practice in the United States.

- *International Accounting Standards Board (IASB).* The IASB sets the International Financial Reporting Standards (IFRS), which influence valuation practices for financial reporting globally.
- *Bank for International Settlements (BIS).* Through Basel Accords, BIS affects risk assessment and property valuations used in the financial and banking sectors.
- *Other local regulatory authorities.* Many countries have their own valuation regulatory bodies, such as the Australian Property Institute (API) and the Canadian National Association of Real Estate Appraisers (CNAREA), which set valuation practices tailored to their respective markets.

These professional bodies and standards collectively shape the valuation profession, ensuring that valuations are conducted fairly, consistently, and transparently. Adhering to these standards is critical in fostering public trust, supporting financial transparency, and facilitating investment confidence in the real estate sector. Compliance with international and regional standards also ensures that valuations are accepted across different jurisdictions.

Additionally, professional bodies play a crucial role in maintaining and advancing standards in the valuation profession, as they provide guidance, education, and ethical frameworks to ensure that valuations are conducted accurately and consistently. Some of the key roles of professional bodies include:

- *Setting standards.* Organisations such as RICS, IVSC, and TEGoVA establish globally recognised valuation standards that professionals must adhere to, ensuring uniformity and reliability in property assessments.
- *Regulatory oversight.* Professional bodies monitor the compliance of valuers with ethical and professional guidelines, ensuring accountability and integrity in valuation practices.
- *Education and certification.* Some organisations offer training, accreditation, and continuous professional development programmes, keeping valuers updated with the latest methodologies, regulatory changes, and market trends.
- *Dispute resolution.* In cases where valuation discrepancies lead to disputes, professional bodies provide mechanisms for resolving conflicts through mediation or disciplinary actions against non-compliant valuers.
- *Promoting public trust.* By enforcing strict ethical codes and professional conduct guidelines, these institutions help enhance public confidence in valuation outcomes and the integrity of the profession.

By fostering high standards and ensuring compliance with established guidelines, professional bodies play an essential role in improving the reliability, credibility, and transparency of property valuations.

Inconsistencies in Measurement

Another major issue in property valuation is measurement inconsistencies that lead to discrepancies in reported property sizes and, consequently, valuation results. These inconsistencies arise due to variations in measurement methods, standards, and interpretation of guidelines. Some common causes include:

- *Different measurement standards.* Various countries and organisations use different property measurement standards, such as the International Property Measurement Standards (IPMS),

RICS Code of Measuring Practice, and the Building Owners and Managers Association (BOMA) Standards, leading to variations in reported property areas and the treatment of comparables.
- *Methodology variations.* Some valuers include external areas, internal partitions, or common areas differently, depending on whether they follow gross internal area (GIA), net internal area (NIA), or gross external area (GEA) methodologies.
- *Lack of clarity in definitions.* The interpretation of terms like 'usable space', 'rentable area', and 'common area' can differ across markets, leading to inconsistent property sizes in valuation reports.
- *Human error and subjectivity.* Measurement techniques depend on the valuer's expertise, the tools used, and the precision applied during calculations, all of which introduce the potential for variation.
- *Technology and equipment differences.* The use of laser measurement tools, digital mapping, and traditional tape measures can yield different results, affecting the accuracy and consistency of measurements.

To mitigate these inconsistencies, valuers should adhere to recognised international measurement standards, like IPMS, ensuring transparency and uniformity in property assessments. Proper training, regulatory oversight, and clear guidelines can also help minimise variations in reported measurements and improve valuation reliability.

IPMS and their importance

The IPMS is another set of standards, this time focused on measurement, that plays a crucial role in promoting consistency and transparency in property measurement worldwide. These standards, developed by the International Property Measurement Standards Coalition (IPMSC; www.ipmsc.org), aim to standardise the way property areas are measured across different markets, ensuring that valuations and appraisals are based on consistent and comparable data.

Consistency in property measurement is fundamental to ensuring fair and accurate valuations, especially in a global environment. Standardised measurement practices help reduce discrepancies, enhance transparency, and provide confidence in valuation results. When measurement practices are inconsistent, it becomes challenging to compare properties accurately, leading to confusion among investors, developers, and regulatory bodies. Ensuring that all valuers adhere to the same measurement guidelines allows stakeholders to make informed decisions based on reliable data. Furthermore, maintaining consistency reduces the risk of overvaluation or undervaluation, which can impact investment decisions, lending approvals, and legal agreements. International measurement standards such as the IPMS promote uniformity across different markets, ensuring smoother cross-border transactions and aligning valuation practices worldwide. Adhering to these recognised standards also supports regulatory compliance, minimising the likelihood of disputes and legal complications. By implementing and enforcing standardised measurement guidelines, valuers can enhance credibility, contribute to market stability, and ensure a more transparent and fair property assessment process.

Because it removes disparities brought up by different measurement techniques employed in different locations and for different kinds of properties, IPMS is especially crucial in valuation. IPMS promotes cross-border real estate transactions, improves market transparency, and guarantees that developers, investors, and other stakeholders may make well-informed decisions based on standardised and trustworthy property data by offering a uniform approach to property measurement.

Valuation profession and transparency

Transparency plays a critical role in the valuation profession, as outlined by the IVS. According to IVS, transparency ensures that valuation practices are conducted consistently and fairly, fostering trust among stakeholders, such as investors, lenders, and regulatory authorities. Key aspects of transparency in valuation under IVS include:

1. *Standardised methodologies*. IVS emphasises the use of globally recognised valuation approaches – market approach, *income approach*, and cost approach – to ensure uniformity and comparability across different valuations.
2. *Disclosure of assumptions and inputs*. Valuers must disclose all significant assumptions, methodologies, and limitations in their valuation reports. This ensures that users of valuation reports can understand and assess the credibility of the conclusions.
3. *Compliance with ethical standards*. Transparency is closely linked to professional ethics. IVS mandates that valuers act independently, objectively, and free from conflicts of interest, thereby enhancing the credibility of valuation conclusions.
4. *Market evidence and data transparency*. Valuers are encouraged to base their assessments on verifiable market data and comparable transactions. Where market data is limited, IVS recommends clear disclosure of how estimates were derived.
5. *Regulatory compliance*. IVS-aligned valuation reports support regulatory compliance, particularly in financial reporting, investment analysis, and lending. This contributes to market stability and investor confidence.
6. *Consistency with accounting and financial standards*. IVS works alongside global financial reporting standards, such as IFRS, to ensure valuations are relevant for financial statements and investment decision-making.

Overall, IVS establishes transparency as a cornerstone of professional valuation, ensuring valuations are reliable, are comparable, and contribute to informed decision-making in the global financial and real estate markets.

Ethics and professional standards

Valuers must adhere to the highest standards of professional behaviour since they play a critical role in real estate markets. To ensure impartial and equitable property evaluations, which in turn promote confidence in the valuation process, ethical issues are crucial. Valuers must practice with independence, impartiality, and honesty, according to the RICS (RICS 2021). According to the IVSC (IVSC 2024), valuers should also follow guidelines that guarantee openness, dependability, and professionalism.

Integrity and objectivity are central to the profession, as valuers must provide unbiased assessments based on factual data and sound professional judgement. They should avoid any conflicts of interest that might compromise their impartiality (The European Group of Valuers' Associations [TEGoVA] 2020). Confidentiality is another crucial aspect, as valuers often handle sensitive client information that must be protected and disclosed only when legally required. Ensuring competency and due diligence is fundamental, as professionals need to possess the necessary expertise and stay updated with evolving industry practices (FASB 2019).

Valuers are also accountable to a variety of stakeholders, such as the public, financial institutions, regulatory agencies, and clients. Their valuation studies emphasise the significance of moral behaviour and regulatory compliance and have an impact on credit approvals, investment decisions, and court cases. To ensure uniformity and accountability in their work, valuation practitioners must adhere to stringent rules set forth by, for example, in the United States, the USPAP (The Appraisal Foundation 2021).

By adhering to these ethical and professional standards, valuers contribute to the credibility of the valuation industry, promote transparency in real estate transactions, and support informed decision-making across various sectors. Ethical lapses can lead to significant financial and reputational damage, underscoring the need for strict adherence to professional codes of conduct.

Valuers play a crucial role in property markets, requiring them to uphold the highest standards of professional conduct. Overall, and across different bodies and standards, ethical considerations from professional bodies (the general principles are similar) could be summarised as:

- *Integrity and objectivity.* Valuers must remain impartial and provide unbiased assessments based on factual data and professional judgement. Their analyses should be free from external influences that could impact the reliability of their reports.
- *Confidentiality.* Sensitive client information should be protected and disclosed only when legally required. Valuers must ensure that their work does not compromise client confidentiality or proprietary information.
- *Competency and due diligence.* Professionals must ensure they possess the requisite knowledge, skills, and experience to conduct accurate valuations. Ongoing professional development and training are necessary to stay updated on evolving industry practices.
- *Avoiding conflicts of interest.* Valuers should disclose any potential conflicts and refuse assignments where impartiality cannot be guaranteed. Ethical guidelines require them to act transparently and in the best interest of all stakeholders.
- *Professional scepticism.* A questioning mindset should be applied throughout the valuation process to verify sources and ensure robust conclusions. Valuers must critically assess market data, assumptions, and projections to produce well-supported valuation reports.
- *Responsibility to stakeholders.* Valuers have a duty not only to their clients but also to regulatory authorities, financial institutions, and the general public. Their reports influence lending decisions, investment strategies, and financial disclosures, making their ethical responsibilities paramount.
- *Legal and regulatory compliance.* Valuers must adhere to national and international legal requirements, ensuring that their reports meet the necessary legal and professional criteria. Compliance with regulations helps mitigate risks and enhances credibility in the industry.

Maintaining high ethical and professional standards ensures that valuations remain credible and contribute to the integrity of the property market. Ethical lapses can lead to significant financial and reputational damage, underscoring the need for strict adherence to professional codes of conduct.

By following these principles, valuers can enhance public trust, promote transparency in real estate transactions, and support informed decision-making across various sectors.

The role of a competent valuer in financial stability

A competent valuer plays a crucial role in financial stability by ensuring accurate, reliable, and transparent property and asset valuations. Their work supports sound investment decisions, risk management, and regulatory compliance, thereby reducing financial system vulnerabilities. Here are key ways a competent valuer contributes to financial stability.

Providing accurate and reliable valuations

Competent valuers apply standardised methodologies (market, income, and cost approaches) to ensure that asset values reflect real market conditions.

Reliable valuations prevent asset overvaluation or undervaluation, which could lead to financial instability when assets are used as collateral for loans.

Enhancing market transparency and investor confidence

By following IVS and relevant ethical guidelines, valuers ensure that investors, lenders, and stakeholders receive clear, unbiased, and well-documented valuations. Transparent valuations reduce speculation and prevent asset bubbles, which can lead to financial crises.

Supporting risk management for lenders and financial institutions

Financial institutions rely on valuations for mortgage lending, credit risk assessments, and investment appraisals.

Competent valuers help prevent excessive lending against inflated property values, reducing the risk of NPLs.

Ensuring regulatory compliance and financial reporting accuracy

Valuers contribute to financial stability by ensuring compliance with financial reporting standards, such as IFRS and Basel regulations (BCBS 2022) for capital adequacy in banks. Accurate valuation reports ensure fair representation of assets in company balance sheets, helping investors make informed decisions.

Mitigating systemic risks in real estate and capital markets

A competent valuer identifies market trends, risks, and vulnerabilities, which policymakers and financial institutions can use to implement corrective measures. Mispriced assets due to poor valuation practices can trigger economic downturns, as seen in the 2008 financial crisis, where property overvaluation contributed to a banking collapse.

Advising on sustainable and long-term asset value

Valuers assess long-term asset sustainability, including factors like climate risks, market volatility, and obsolescence. This helps investors, banks, and governments adopt more resilient financial policies that support economic stability.

A competent valuer is more than just a technical expert; they act as gatekeepers of financial stability, ensuring that valuations support prudent lending, responsible investment, and regulatory oversight. Their work helps prevent asset bubbles, reduces credit risks, and ensures a stable financial system that can withstand economic shocks.

Questions

1. Why is the role of valuers crucial in financial stability?
2. What factors contribute to variations in valuation results?
3. What are the three primary approaches to property valuation?
4. How do valuation standards contribute to consistency and transparency?
5. What role do professional bodies play in the valuation industry?
6. Why is transparency important in property valuation?
7. How do inconsistencies in property measurement affect valuations?
8. What ethical responsibilities do valuers have in their profession?

References

The Appraisal Foundation (2021). *Uniform Standards of Professional Appraisal Practice (USPAP)*. Washington, D.C.: Appraisal Standards Board.

Basel Committee on Banking Supervision (BCBS) (2022). *Basel III: Finalising Post-Crisis Reforms*. Bank for International Settlements (BIS).

The European Group of Valuers' Associations (TEGoVA) (2020). *European Valuation Standards (EVS)*.

Financial Accounting Standards Board (FASB) (2019). *Conceptual Framework for Financial Reporting*.

International Financial Reporting Standards (IFRS) (2023). *International Accounting Standards Board (IASB)*. Available at: www.ifrs.org

International Valuation Standards Council (IVSC) (2024). *International Valuation Standards (IVS)*. Available at: www.ivsc.org

Royal Institution of Chartered Surveyors (RICS) (2021). *RICS Valuation – Global Standards (Red Book)*. London: RICS.

Chapter 6

Valuation mathematics

This book explores the process of property valuation, with particular focus on investment properties, that is, those acquired to generate returns through rental income, future resale, or both. Evaluating an investment property requires a valuer to grasp how the future financial gains derived from owning a freehold or leasehold interest in a property can be quantified in terms of a present value (PV). To do this, a valuer must have knowledge of the mathematics of finance and the theory of compounding and discounting, as it relates to savings and investments. This chapter sets out the financial or investment-based arithmetic tools utilised in the 'income approach'. They are based on the concept of compound interest and are illustrated with both property- and non-property-based examples.

The time value of money (TVM)

TVM is a fundamental concept in financial mathematics; the basic principle is that a sum of money is worth more now than in the future. This principle stands on several pillars. Firstly, the opportunity cost of money in hand today, rather than being simply held in a safe. It can be invested, yielding interest and appreciating in value over time. Secondly, there is a constant threat of inflation; money might lose purchasing power, making it worth less in the future compared to its current value. Lastly, the element of uncertainty underscores this principle; until money is actually obtained, there is no certainty it will remain available in the future (Hussain 2008).

Quite simply, we would all prefer money today in preference to the same amount of money at a future date. But for investors, the question to be considered is what sum of money today and what sum of money in the future might be considered by them to be of equal value? Is there a relationship between money today and money in the future, and if so, what is that relationship?

Individuals and organisations with surplus money, that is, money not required for the satisfaction of immediate purchases of goods and services, can save or invest that surplus, or even leave it under a mattress! If saved, the savings bank will normally pay interest on the amount saved. The amount of interest will vary from bank to bank, from state to state, and may be very small. But the larger the sum, the more likely it is that some interest will accrue. This is because the bank uses that saved money; it lends it to those who have a borrowing need. The amount of saved money that can be lent as a percentage of money deposited in a bank is generally limited by central banks or by government.

An investor looking at an *investment*, loosely defined as 'an exchange of money today for future benefits which can be conveniently limited to income or capital growth', must weigh the potential returns from an investment against the opportunity cost of not earning interest

DOI: 10.1201/9781003490357-7

elsewhere. To justify the risks involved, investors seek returns that are equivalent to or exceed the interest they could have earned through safer alternatives, such as savings.

There is a pay-off between money today and money in the future.

If £/$/€100 can be saved to earn interest at 5% over a year (5% per annum), then it will grow to £/$/€105. So no investor would pay £/$/€100 today for £/$/€100 in a year's time; the future amount will have to be discounted.

Taking a minimum requirement to be a 5% return on investment, then that future amount can be found by dividing (in this case using £'s) £100 by 1.05 (100 ÷ 1.05) = **£95.238**.

Why?

Because £95.238 + 5% of £95.238 = £95.238 + £4.762 = **£100**.

TVM plays a pivotal role in financial decisions. For example, it enables individuals to determine the regular savings required in an interest-bearing account to achieve a financial goal. TVM also underpins key financial analyses, such as capital budgeting, investment appraisal, loan structuring, retirement planning, and business and property valuation.

Cash flow diagram and table

A very important tool to use whenever we are dealing with TVM is a cash flow diagram, as it helps in visualising the timing of the cash flows. Cash flow diagrams exhibit two key characteristics: the amount and the timing. The *amount* denotes the transaction's magnitude and its direction. Financial *gains* (cash inflows from receipts, or income) are positive and represented by upward arrows. Financial *losses* (cash outflows from expenses or disbursements) are denoted with a negative sign and a downward arrow. *Timing* indicates when the amount is either received or spent.

For example, imagine you are a property investor. Your property generates $20,000 in rent one year from now, followed by $30,000 in rent two years from now. By year 3, the property becomes vacant, incurring $10,000 in business rates. Subsequently, in years 4 and 5, the property is fully occupied again, yielding $40,000 annually. This scenario can be represented visually as:

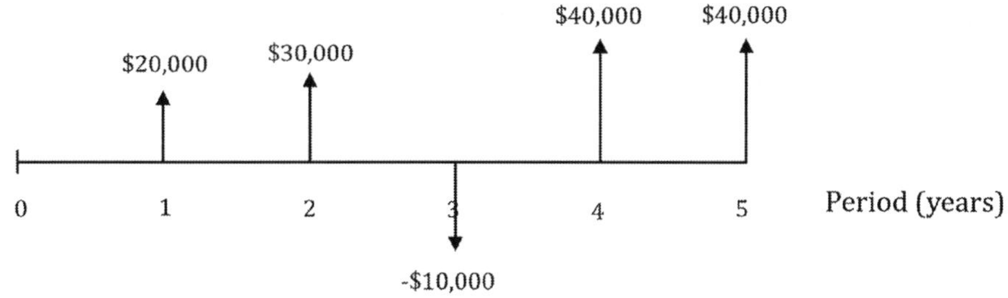

Figure 6.1 Example of a cash flow diagram.

Another valuable tool, as highlighted by Dahlquist and Knight (2022), is a timeline or a cash flow table, which provides a visual and structured breakdown of cash flows over time.

Table 6.1 Example of a cash flow table

Year	*0 (today)*	*1*	*2*	*3*	*4*	*5*
Principal added/withdrawn ($)	0	20,000	30,000	−10,000	40,000	40,000

Single-payment problems

A *single* or *lump sum payment* is a monetary sum paid in one single point in time, present or future, rather than being spread across instalments. If the lump sum is in the present, it is called present value (PV). For instance, consider you have $1,000 and you deposit that today in a bank with an annual interest rate of 3%. Imagine that you wish to know how much you will accumulate, the future value (FV) after five years, without making any additional deposits or withdrawals during this time frame. Visually, this can be represented as follows:

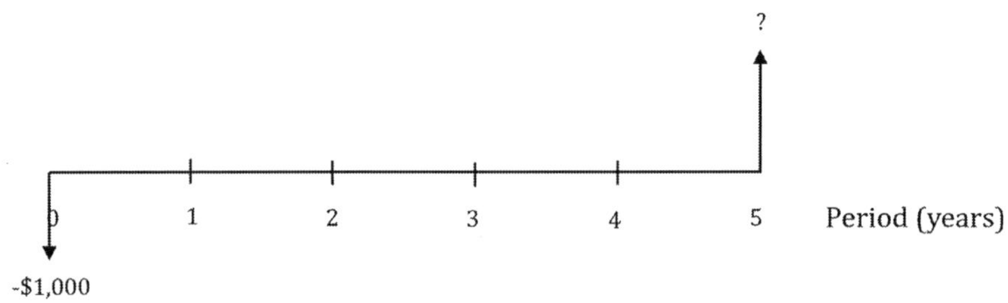

Figure 6.2 Example of a single-payment problem – PV known, FV unknown.

After one year, you will have $1,000 plus the interest earned over that time.

That is, $\$1,000 + \$1,000 \times 0.03 = \mathbf{\$1,030}$.

Then, one year later, you will have $\$1,030 + \$1,030 \times 0.03 = \mathbf{\$1,060.90}$.

If you repeat this action over five years, you will have:

Year 1: $\$1,000 + \$1,000 \times 0.03 = \mathbf{\$1,030}$
Year 2: $\$1,030 + \$1,030 \times 0.03 = \mathbf{\$1,060.90}$
Year 3: $\$1,060.90 + \$1,060.90 \times 0.03 = \mathbf{\$1,092.73}$
Year 4: $\$1,092.73 + \$1,092.73 \times 0.03 = \mathbf{\$1,125.51}$
Year 5: $\$1,125.51 + \$1,125.51 \times 0.03 = \mathbf{\$1,159.27}$

These dynamics can be seen in Table 6.2. In year 0 (today), an initial deposit of $1,000 is made, with no interest accrued, since the payment occurs on the last day of the year. Year 1 begins with the same balance as year 0, but by the year's end, interest is earned on the previous balance $(\$1,000 \times 0.03 = \$30)$, resulting in a new balance comprising the sum of the previous balance and the interest earned $(\$1,000 + \$30 = \$1,030)$. This pattern continues for years 2, 3, 4, and 5, with interest being earned on the previous balance each year and no additional principal being added or withdrawn. Consequently, the final balance at the end of each year showcases the cumulative effect of interest accumulating over time on the principal sum.

Table 6.2 Cash flow table, single-payment problem – PV known, FV unknown

Year	*0 (today)*	*1*	*2*	*3*	*4*	*5*
Previous balance ($)		1,000	1,030	1,060.90	1,092.73	1,125.51
Interest earned ($)		30	30.90	31.83	32.78	33.77
Principal added/withdrawn ($)	1,000					
New balance ($)	1,000	1,030	1,060.90	1,092.73	1,125.51	1,159.27

Note that the following formulas were used, and these can be written as:

Year 1: Present Value + Present Value × Interest rate, or Present Value × (1 + Interest Rate)
Year 2: Present Value × (1 + Interest Rate) + Present Value × (1 + Interest Rate) × Interest Rate, or Present Value × (1 + Interest Rate)^2
Year 3: Present Value × (1 + Interest Rate)^2 + Present Value × (1 + Interest Rate)^2 × Interest Rate, or Present Value × (1 + Interest Rate)^3
Year 4: Present Value × (1 + Interest Rate)^3 + Present Value × (1 + Interest Rate)^3 × Interest Rate, or Present Value × (1 + Interest Rate)^4
Year 5: Present Value × (1 + Interest Rate)^4 + Present Value × (1 + Interest Rate)^4 × Interest Rate, or: Present Value × (1 + Interest Rate)^5

The preceding formulas can be expressed more succinctly using time value of money notation as follows:

$$FV = PV \times (1+r)^n$$

Where:
$\boldsymbol{FV}$ = future value
$\boldsymbol{PV}$ = present value
$\boldsymbol{r}$ = interest rate or discount rate
$\boldsymbol{n}$ = number of periods

Equation 6.1 Future value formula

Using Equation 6.1 in our example, we have:

$$FV = 1{,}000 \times (1+0.03)^5$$
$$FV = \mathbf{\$1{,}559.27}$$

Interest in the preceding equation was expressed as a decimal, for example, 3% = 3 ÷ 100 = 0.03. Also, it is important to emphasise that the interest rate $\boldsymbol{r}$ should correspond to the rate for a single period of time, aligned with the interval, measured by $\boldsymbol{n}$. For example, if the interest rate is stated annually, then the number of periods should be in years. Similarly, if the interest rate is provided quarterly, the number of periods should be in quarters, and so forth.

The component of the formula denoted as $(1+r)^n$, which multiplies the PV, is commonly called the **amount of £1** in the UK. The term is another name for compound interest, that is, the future worth of £1 if invested today, allowing for compound interest at a given rate.

The transition from present value to future value involves a process known as **compounding**, which is set out in Figure 6.3. *Compounding* refers to the growth of money over time as interest accumulates on both the initial principal (present value) and any previously earned interest. This process relies on an interest rate, expressed as a percentage, which represents either the cost of borrowing money or the return earned on invested funds.

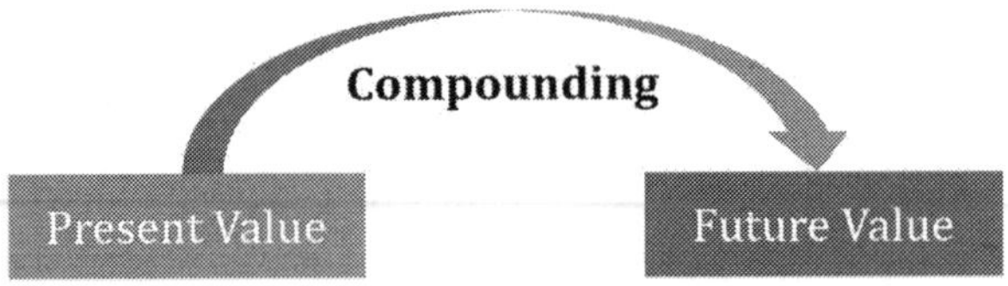

Figure 6.3 Compounding.

Spreadsheets, such as Excel,[1] have built-in functions and formulas that can be used to solve TVM problems. The syntax for the FV formula function is:

= FV(rate,nper,pmt,[pv],[type])

Where:

rate = (required) the interest rate per period.
nper = (required) the total number of periods interest is earned.
pmt = (required) the periodic annuity payment (we will use this later).
Note: If PMT is omitted, you must include the PV argument.
pv = (optional) the present value. If PV is omitted, the PMT argument must be included.
type = (optional) the type of annuity (we will use this later).

Excel function 6.1 FV function

Hence, with the aid of Excel, one can arrive at the solution either by manually implementing the formula or by utilising a built-in function. To set up our example in Excel, the following steps can be undertaken:

Excel formula	*Excel* function
=1000*(1 + 0.03)^5	=FV(rate,nper,pmt,[pv],[type]) =FV(0.03,5,,-1000) *Please note:* 1. The presence of a double comma within the function was used to indicate the absence of a payment (PMT). 2. The argument [type] was not used.

Note that in the function, payment details and the annuity type are excluded, as they are unnecessary for lump sum scenarios. Furthermore, the present value was inputted with a negative sign, as **−$1,000**. This convention aligns with the cash flow sign convention commonly used in spreadsheets, enabling them to interpret cash flow direction. While omitting the negative sign is not critical, the future value result would have been returned as negative (though still accurate in absolute terms).

Example 6.1

You are contemplating leasing a property that demands €50,000 for renovations, which is beyond your current financial means, especially with the added expense of rent. However, the landlord offers to lend you the €50,000 today, under the condition that you return the loan in two years, with a 5% annual interest rate applied. How much will you be required to repay the landlord at the conclusion of the second year?

Solution

The task is to calculate the FV of the loan after two years.

From the enunciate we have:

PV = €50,000 (the loan amount)
n = 2 years
r = 5% (0.05 in decimal form)
FV =?

$$FV = PV \times (1 + r)^n$$
$$FV = 50{,}000 \times (1 + 0.05)^2$$
$$FV = €55{,}125.000$$

In Excel:

Excel formula	*Excel* function
=50000*(1 + 0.05)^2	=FV(0.05,2,,-50000)

Thus, you will need to repay €55,125 to the landlord at the end of the second year.

Example 6.2 Compound interest in the form of the amount of £1

NOTE: For completeness, and in case you read other UK valuation textbooks, we include an example of compounding using the amount of £1 approach.

Calculate the amount of £1 after four years at 10%.

Solution

$$\text{Amount of £1} = (1 + r)^n$$
$$\text{Amount of £1} = (1 + 0.1)^4 = 1.4641$$

The calculation shows that £1 will accumulate to **£1.4641** (£1.47 to the nearest pound and pence) after four years at 10% pa if left to collect compound interest annually.

At times, it is necessary to determine the present value of a future lump sum of money or a series of cash flows expected to be received in the future. This process, known as **discounting**, allows us to assess the value today of amounts that will be received or paid at a later date. Discounting involves applying a discount rate, which is a percentage that reflects the cost of capital, or the rate of return required by an investor.

Figure 6.4 Discounting.

Calculating the PV of a future lump sum stands as the inverse operation of determining an FV:

$$PV = \frac{FV}{(1+r)^n}$$

Where:

FV = future value
PV = present value
r = interest rate or discount rate
n = number of periods

Equation 6.2 Present value formula

The component of the formula denoted as

$$\frac{1}{(1+r)^n},$$

which multiplies the FV, is commonly called the '**PV of £1**' in the UK. The term is the PV of £1 to be received in the future discounted over a given period at a given rate of interest.

The Excel built-in function for the PV formula is:

=PV(rate,nper,pmt,[fv],[type])

Where:

rate = (required) the interest rate per period.
nper = (required) the total number of periods interest is earned.
pmt = (required) the periodic annuity payment (we will use this later).
Note: If PMT is omitted, you must include the PV argument.
fv = (optional) the future value. If FV is omitted, the PMT argument must be included.
type = (optional) the type of annuity (we will use this later).

Excel function 6.2 PV function

Now, suppose we aim to ascertain the required deposit amount today into a bank account with a 4% annual interest rate to accumulate \$50,000 in five years from now. The cash flow diagram would look like this:

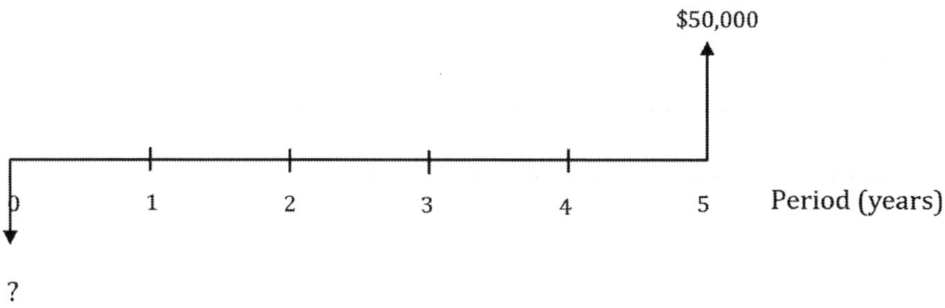

Figure 6.5 Example of a single-payment problem – PV unknown, FV known.

Applying the formula, we have:

$$PV = \frac{FV}{(1+r)^n} = \frac{50,000}{(1+0.04)^5} = \$41,096.36$$

Using Excel, the following steps can be undertaken:

Excel formula	*Excel* function
=50000/((1 + 0.04)^5)	=PV(rate,nper,pmt,[fv],[type]) =PV(0.04,5,,-50000) *Please note:* 1. The presence of a double comma within the function was used to indicate the absence of a payment (PMT). 2. The argument [type] was not used.

To understand what is happening underneath the formula, let us explore the behaviour of the present value of \$41,096.36 and how it evolved to the future value of \$50,000 in year 5.

In year 0 (present), an initial deposit of \$41,096.36 is made, with no interest accrued, since the payment occurs on the last day of the year.

Year 1 begins with the same balance as year 0, but by the year's end, interest is earned on the previous balance $(0.04 \times 41,096.36 = 1,643.85)$, resulting in a new balance comprising the sum of the previous balance and the interest earned $(41,096.36 + 1,643.85 = 42,740.21)$.

This pattern continues for years 2, 3, 4, and 5, with interest being earned on the previous balance each year and no additional principal being added or withdrawn. Consequently, the final balance at the end of each year showcases the cumulative effect of interest accumulating over time on the principal sum.

Table 6.3 Cash flow table, single-payment problem – PV unknown, FV known

Year	*0 (today)*	*1*	*2*	*3*	*4*	*5*
Previous balance ($)	0	41,096.36	42,740.21	44,449.82	46,227.81	48,076.92
Interest earned ($)	0	1,643.85	1,709.61	1,777.99	1,849.11	1,923.08
Principal added/ withdrawn ($)	41,096.36	0	0	0	0	0
New balance ($)	41,096.36	42,740.21	44,449.82	46,227.81	48,076.92	50,000

Example 6.3

If Mr X requires a return of 10% pa, how much would you advise Mr X to pay for the right to receive £200 in five years' time?

Solution

We know that:

$\boldsymbol{FV} = £200$

$\boldsymbol{r} = 10\%$

$\boldsymbol{n} = 5$ years

PV = ?

$$PV = \frac{FV}{(1+r)^n}$$

$$PV = \frac{200}{(1+0.1)^5}$$

$$PV = \mathbf{124.18}$$

In Excel:

Excel formula	*Excel* function
=200/((1 + 0.1)^5)	=PV(0.1,5,,-200)

What this means is that if **£124.18** is invested now and earns interest at **10% per annum (pa)**, then it will accumulate to **£200** in five years' time.

Or **£124.18** is the value today of **£200** due in five years' time if discounted at **10% pa**.

Example 6.4

You are a real estate investor intending to buy a property three years from now. You estimate that the price of the property will be €1,000,000. However, you plan to secure a mortgage, which requires a down payment of 25%. How much money are you required to deposit today into a bank account with a 5.5% annual interest rate to be able to make the down payment three years from now?

Solution

In this case, *FV* is the down payment to be paid in three years (***n***), and **r** is the interest rate of 5.5%.

The down payment is $0.25 \times 1{,}000{,}000 = 250{,}000$

To find the amount required to put into a bank account today so we're able to pay the down payment in three years, we apply the PV formula:

$$PV = \frac{FV}{(1+r)^n}$$

$$PV = \frac{250{,}000}{(1+0.055)^3}$$

PV = **€ 212,903.42**

In Excel:

Excel formula	*Excel* function
=250000/((1 + 0.055)^3)	=PV(0.055,3,,-250000)

In this case, you will need to put €212,903.42 into the bank account to make the down payment three years from now.

Example 6.5

An investor is offered five separate investment opportunities on five separate occasions; each will produce a certain cash benefit of £10,000, the first in exactly one year's time and the other four at subsequent yearly intervals. The investor is seeking a 10% return from his money. What price should be paid for each investment?

Solution

$$PV = \frac{FV}{(1+r)^n}$$

Option 1: $PV = \dfrac{10{,}000}{(1+0.1)^1} = £9{,}090.91$

$$\textbf{Option 2: } PV = \frac{10{,}000}{(1+0.1)^2} = £8{,}264.46$$

$$\textbf{Option 3: } PV = \frac{10{,}000}{(1+0.1)^3} = £7{,}513.15$$

$$\textbf{Option 4: } PV = \frac{10{,}000}{(1+0.1)^4} = £6{,}830.13$$

$$\textbf{Option 5: } PV = \frac{10{,}000}{(1+0.1)^5} = £6{,}209.21$$

The figures of PV show the individual prices to be paid today for each investment in order that the investor can achieve a 10% return on capital invested and obtain the return of the capital invested. For example, £7,513.15 is paid for a future benefit of £10,000 in three years' time; in three years' time, the receipt of £10,000 repays the investment of £7,513.15; the difference represents the equivalent compound interest at 10% on the £7,513.15.

This is what is meant by a 10% return.

In other words, the investor is exchanging a sum of money today for a known future sum which will be equal to the sum of money today plus the interest forgone if the capital had been invested elsewhere at the same rate of interest. In each case, the receipt of £10,000 at the due date returns the respective capital sum or price paid, and the difference between the price today and the £10,000 is equivalent to the 10% annual compound interest that would have been earned on the purchase sums if they had been saved at the investor's 10% required rate of return.

Example 6.6

You want to be able to contribute $50,000 for the down payment on a residential property. You currently have $25,000 in a savings account that is earning 4% interest every year. How long will it take for this account to grow into the targeted $50,000, assuming no additional deposits or withdrawals will be made?

Solution

In this case we have the following:

$$\begin{aligned} PV &= 25{,}000 \\ FV &= 50{,}000 \\ r &= 4\% \\ n &= ? \end{aligned}$$

Since the $\boldsymbol{n}$ is unknown, we need to rearrange Equation 6.1:

$$FV = PV \times (1+r)^n$$

$$\frac{FV}{PV} = (1+r)^n$$

$$ln\left(\frac{FV}{PV}\right) = ln(1+r)^n$$

$$ln\left(\frac{FV}{PV}\right) = n \times \ln(1+r)$$

$$n = \frac{ln\left(\frac{FV}{PV}\right)}{\ln(1+r)}$$

$$n = \frac{ln\left(\frac{50,000}{25,000}\right)}{\ln(1+0.04)}$$

$$n \approx 17.67\ years$$

In Excel:

Excel formula	*Excel* function
=LN(50000/25000)/LN(1 + 0.04)	=NPER(0.04,,25000,-50000)

If you deposit \$25,000 today, it will take approximately 17 years and 8 months to accumulate the \$50,000 required for the down payment.

Example 6.7

You want to be able to contribute £50,000 for the down payment of a property 10 years from now. You currently have £25,000 in a savings account that is earning 4% interest every year.

What rate of interest will you need to grow your saved money from £25,000 to £50,000 in this time period, again with no other deposits or withdrawals?

Solution

This question is very similar to Example 3; however, now the $\boldsymbol{r}$ is unknown, and the $\boldsymbol{n}$ is 10 years. Therefore, we need to rearrange Equation 6.1 again, this time:

$$FV = PV \times (1+r)^n$$

$$\frac{FV}{PV} = (1+r)^n$$

$$\left(\frac{FV}{PV}\right)^{\frac{1}{n}} = 1+r$$

$$r = \left(\frac{FV}{PV}\right)^{\frac{1}{n}} - 1$$

$$r = \left(\frac{50,000}{25,000}\right)^{\frac{1}{10}} - 1$$

$$r \approx 0.0718 \text{ or } \mathbf{7.18\%}$$

In Excel:

Excel formula	*Excel* function
=((50000/25000)^(1/10))-1	=RATE(nper,pmt,pv,[fv]) =RATE(10,,-25000,50000) *Please note:* 1. The presence of a double comma within the function was used to indicate the absence of a payment (PMT).

To achieve a down payment of £50,000 within a ten-year time frame, you would need to locate a savings account offering an annual interest rate of 7.18%, given that you deposit £25,000 today.

Multiple-payment problems

At times, we must deal with a stream of payments, or cash flows, made or received periodically. In real estate, typical examples include rent and mortgage payments. Other practical examples include a car loan being repaid with equal monthly instalments and a life insurance policy being purchased with monthly payments (Chan and Tse 2017). These examples configure annuities. An ***annuity*** is a stream of ***fixed periodic payments***, either to be made or received in the future. By applying time value of money formulae to each payment, we can calculate the present or future values of these payment streams.

Before proceeding with further consideration and analysis of annuities, there are a number of common terms which need to be explained. They are:

- *Ordinary annuity or annuity in arrears*. This means that the first annual sum will be paid (received) 12 months after the purchase or taking out of the policy. The payments could be weekly, monthly, quarterly, or for any period, provided they are in arrears.
- *Annuity due or annuity in advance.* Payments in this type of annuity are made at the start of the chosen period, whether it be weekly, monthly, annually, or otherwise.
- *Immediate annuity.* The word 'immediate' is used to distinguish a normal annuity from a deferred annuity. An immediate annuity is one where the income commences immediately either 'in advance' or 'in arrears'.

Table 6.4 (Continued)

Year	*Cash flow ($)*	*Calculation*	*Present value*
3	1,000	$PV = \frac{1,000}{(1+0.04)^3}$	889.00
4	1,000	$PV = \frac{1,000}{(1+0.04)^4}$	854.80
5	1,000	$PV = \frac{1,000}{(1+0.04)^5}$	821.93
Present value of the ordinary annuity			**$4,451.82**

Since annuities are fixed periodic payments, we can calculate the present value of an annuity in one step using the following equation:

$$PV = PMT \times \frac{\left[1 - \frac{1}{(1+r)^n}\right]}{r}$$

Where:

PMT = periodic payment, that is, inflow or outflow amount that occurs at each period of a financial stream

PV = present value

r = interest rate

n = number of periods

Equation 6.3 Present value of an ordinary annuity

Applying the formula for the preceding example, we have:

$$PV = 1,000 \times \frac{1 - \frac{1}{(1+0.04)^5}}{0.04} = 4,451.82$$

Using Excel, the following steps can be undertaken:

Excel formula	*Excel* function
=1000*(1–1/((1 + 0.04)^5))/0.04	=PV(rate,nper,pmt,[fv],[type]) =PV(0.04,5,-1000)

To understand what is happening underneath the formula, let us explore the behaviour of this stream of payments over time in Table 6.5.

Table 6.5 Cash flow table – PV of ordinary annuity problem

Year	*0 (today)*	*1*	*2*	*3*	*4*	*5*
Previous balance ($)	0	4,451.82	3,629.89	2,775.09	1,886.09	961.54
Interest earned ($)		178.07	−45.20	111.00	75.44	38.46
Principal added/ withdrawn ($)	4,451.82	−1,000	−1,000	−1,000	−1,000	−1,000
New balance ($)	4,451.82	3,629.89	2,775.09	1,886.09	961.54	0

Starting from year 0 (today), we have determined that we need to deposit $4,451.82 into our fund. In year 1, before any withdrawals are made, a year's worth of interest at a rate of 4% is compounded and added to our balance. Then, the first $1,000 is withdrawn, leaving us with $3,629.89. Similarly, in year 2, another year's worth of interest is added to our balance before the second $1,000 is withdrawn, leaving us with $2,775.09. This pattern continues until year 5, when the fund has served its purpose and is depleted, reaching a balance of zero.

As the previous example highlighted, annuities differ from savings accounts in their structure and the composition of returns. When investing in a savings account, the principal remains intact, and interest accumulates on this untouched capital. In contrast, purchasing an annuity involves converting the initial outlay (purchase price) into periodic payments. The return from an annuity consists of two parts:

- *Return* **on** *capital.* The interest earned on the remaining capital. It is calculated by multiplying the interest by the capital outstanding.
- *Return* **of** *capital.* The gradual repayment of the original purchase price over the annuity's term. It is calculated by subtracting the interest from the annuity payment.

The distinction between 'return *on*' and 'return *of*' capital is important in the case of life annuities, because the capital element may be held in some countries for taxation purposes to be the return of the annuitant's capital and therefore is exempt from income tax. In practice, the tax authority will have to indicate how the distinction is to be made. Clearly a precise distinction is not otherwise possible because of the uncertain nature of the annuitant's life.

The distinction in the case of certain property valuations undertaken on an annuity basis is important because normally there is no distinction between 'return *on*' and 'return *of*' capital other than for life annuities. Thus, tax may be payable on the whole income from the property, and the desired return may not be achieved unless this factor is accounted for in the valuation.

The return ***on*** and a return ***of*** capital can be demonstrated by constructing a table:

Table 6.6 Return *on* and return *of* capital

Year	*Capital outstanding*	*Interest (return on capital)*	*Income*	*Return of capital*
1	$4,451.82	$4,451.82 × 0.04 = $178.07	$1,000	$1,000 – $178.07 = $821.93
2	$4,451.82 – $821.93 = $3,629.90	$3,629.89 × 0.04 = $145.20	$1,000	$1,000 – $145.20 = $854.80

(Continued)

Table 6.6 (Continued)

Year	*Capital outstanding*	*Interest (return on capital)*	*Income*	*Return of capital*
3	\$3,629.90 − \$854.80 = \$2,775.09	\$2,775.09 × 0.04 = \$111	\$1,000	\$1,000 − \$111 = \$889
4	\$2,775.09 − \$889 = \$1,886.09	\$1,886.09 × 0.04 = \$75.44	\$1,000	\$1,000 − \$75.44 = \$924.56
5	\$1,886.09 − \$924.56 = \$961.54	\$961.54 × 0.04 = \$38.46	\$1,000	\$1,000 − \$38.46 = \$961.54

In the context of property valuation, the *PV* of an ordinary annuity formula is used to assess the current value of finite and regular cash flows, including rental payments. Therefore, the valuation of finite and regular rental income streams involves capitalising – that is, converting to present value – these income streams using a suitable discount rate ***r*** for the duration ***n*** that the rental income *PMT* is received.

The component of the formula denoted as $\frac{\left[1-\frac{1}{(1+r)^{n}}\right]}{r}$, which multiplies the PMT, is commonly called the **'PV of £1 per annum' (PV£1 pa)** by valuers in the UK. The Latin 'per annum', or as shortened, pa, is the expression normally used for annual receipts or payments. The PV of £1 per annum is the PV of a series of payments of £1 due annually for a given period of time, discounted at a given rate. PV£1 pa is also called **years' purchase (YP)** in the UK and some Commonwealth countries and in some software packages.

The *YP* multiplier represents the number of years over which the periodic payment must be received to recoup the present value, akin to the payback period but taking the time value of money into consideration. In the preceding example, it will take approximately $\frac{\left[1-\frac{1}{(1+0.04)^{5}}\right]}{0.04} \approx 4.45$ years of receiving \$1,000 to recoup the original present value of \$4,451.82.

Example 6.8

If you're considering investing in an annuity that offers you **£10,000** annually for the next 20 years and the annuity earns an annual interest rate of 4.5%, how much should you invest today to secure this stream of payments?

Solution

In this case, the periodic payment (PMT) is **£10,000**; the number of years (n) is 20 years, and r is the interest rate of 4.5%.

To find the amount required to invest today to secure this annuity, we apply the following formula:

$$PV = PMT \times \frac{\left[1 - \frac{1}{(1+r)^n}\right]}{r}$$

$$PV = 10000 \times \frac{\left[1 - \frac{1}{(1+0.045)^{20}}\right]}{0.045}$$

PV = **£ 130,079.36**

In Excel:

Excel formula	*Excel* function
=10000*(1–1/((1 + 0.045)^20))/0.045	=PV(rate,nper,pmt,[fv],[type]) =PV(0.045,20,-10000)

Example 6.9

How much should an investor pay for the right to receive an income of £675 for 64 years if the investor requires a 12% return on the investment?

Solution

For this question, we know that:

$$\boldsymbol{r} = 0.12$$
$$\boldsymbol{n} = 64$$
$$\boldsymbol{PMT} = £675$$
$$\boldsymbol{PV} = \textbf{?}$$

$$PV = PMT \times \frac{\left[1 - \frac{1}{(1+r)^n}\right]}{r}$$

$$PV = 675 \times \frac{\left[1 - \frac{1}{(1+0.12)^{64}}\right]}{0.12}$$

$$PV = 5{,}621.02$$

Excel formula	*Excel* function
=675*(1–1/((1 + 0.12)^64))/0.12	=PV(rate,nper,pmt,[fv],[type]) =PV(0.12,64,-675)

Example 6.10

If an investor is offered a single investment opportunity which generates £10,000 at the end of each year for the next five years and wishes to achieve a 10% return, what price should they pay today?

Solution

This is similar to Example 6.5. The answer can be found by adding the five previously calculated PVs, for example, £9,090.91 + £8,264.46 + £7,513.15 + £6,830.13 + £6,209.21, which equals a sum of £37,907.87.

However, it is quicker to use Equation 6.3:

$$PV = PMT \times \frac{\left[1 - \frac{1}{(1+r)^n}\right]}{r}$$

$$PV = 10{,}000 \times \frac{\left[1 - \frac{1}{(1+0.1)^5}\right]}{0.1}$$

$$PV = \mathbf{£37{,}907.87}$$

Excel formula	*Excel* function
=10000*(1–1/(1 + 0.1)^5)/0.1	=PV(rate,nper,pmt,[fv],[type]) =PV(0.1,5,-10000)

The investor must achieve a return of capital (the initial £37,908) and a return of 10% on the capital (that is, the interest forgone). The proof is shown in the following table.

Year	*0 (today)*	*1*	*2*	*3*	*4*	*5*
Previous balance (£)	0	37,907.87	31,698.65	24,868.52	17,355.37	9,090.91
Interest earned (£)	0	3,790.79	3,169.87	2,486.85	1,735.54	909.09
Principal added/ withdrawn (£)	37,907.87	−10,000	−10,000	−10,000	−10,000	−10,000
New balance (£)	37,907.87	31,698.65	24,868.52	17,355.37	9,090.91	0

Example 6.11

Calculate the PV of £1 pa at 5% for 20 years, given that the PV of £1 in 20 years at 5% is 0.3769.

Solution

For the question, we know that:

$$r = 0.05$$
$$n = 20$$
$$PV\ of\ £1 = 0.3769$$

Since PV of £1 $= \dfrac{1}{(1+r)^n}$, we can rewrite the formula of the PV£1 pa ***from***

$$PV\ of\ £1\ pa = \frac{\left[1 - \dfrac{1}{(1+r)^n}\right]}{r}$$

to

$$PV\ of\ £1\ pa = \frac{[1 - \text{PV of £1}]}{r}$$

Now we can apply the figures provided in the question:

$$PV\ of\ £1\ pa = \frac{[1 - 0.3769]}{0.05}$$

$$PV\ of\ £1\ pa = 12.462$$

A common application of Equation 6.3 in property relates to mortgages. When a property purchaser borrows money by way of a legal mortgage, it is usually agreed between the parties that the loan will be repaid in full by a given date in the future. Like anyone else lending money, the lender or bank, known as the mortgagee, will require the capital sum to be repaid and will require interest on any outstanding amounts of the loan until such time as the loan and all interest is recovered.

This is comparable to the purchase of an annuity certain, that is, an investment that pays a set amount of money for a specified period of time. Indeed, conceptually, from the point of view of the mortgagee, it is the purchase of an annuity. It follows that the mortgagee will require a return on capital and the return of capital.

The repayment mortgage allows the mortgagor (borrower) to pay back a regular sum each year, often by equal monthly instalments. This sum is made up of interest on the capital owed and the partial return of capital. In the early years of the mortgage, most of the payment represents interest on the loan outstanding, and only a small amount of capital is repaid.

However, over the period of the loan, as more and more capital is repaid, the interest element becomes smaller, and the capital repaid larger.

For the purpose of explanation, interest is assumed fixed throughout the term of the loan. In practice, lenders offer a wide range of mortgage deals, but the underlying concept is the same – a requirement that capital borrowed be repaid and that interest be paid on money owed. The rates of interest and pattern of payment can be infinitely variable and often involve lower interest payments during the early years.

To determine the regular payment (*PMT*), Equation 6.3 can be rearranged as follows:

$$PMT = PV \times \frac{r}{\left[1 - \frac{1}{(1+r)^n}\right]}$$

Equation 6.4 Ordinary annuity payment formula when PV is known

In the UK, the multiplier $\frac{r}{\left[1 - \frac{1}{(1+r)^n}\right]}$ is called **the annuity that £1 will purchase**.

This function shows the amount that will be paid back at the end of each year for n years at r% in return for £1 invested. It calculates what is known as the annuity that £1 will purchase. It can be used to calculate the annual equivalent of a capital sum, although valuers tend to prefer to divide a capital sum by the PV of £1 pa to calculate an annual equivalent.

The Excel built-in function for the PMT is as follows:

=PMT(rate,nper,pv,[fv],[type])

The PMT function formula has the following arguments:

rate = (required) the interest rate per period.
nper = (required) the total number of periods interest is earned.
PV = (required) if PV is omitted, you must include the FV argument.
FV = (optional) if FV is omitted, you must include the PV argument.
type = (optional) the type of annuity – it can be 0 or omitted for payments due at the *end* of the period (ordinary annuities) or 1 for payments due at the *beginning* of the period (annuities due).

Excel function 6.3 PMT function

The valuer who can understand the concept of the normal repayment mortgage and can solve standard problems that face mortgagors and mortgagees should readily understand most investment valuation problems. The following examples are indicative of such mortgage problems.

Example 6.12

What annuity will a capital sum of £4,918 purchase over six years at 6%?

Solution

PV = £4,918
n = 6 years
r = 0.06
PMT = ?

$$PMT = PV \times \frac{r}{\left[1 - \frac{1}{(1+r)^n}\right]}$$

$$PMT = 4{,}918 \times \frac{0.06}{\left[1 - \frac{1}{(1+0.06)^6}\right]}$$

$$PMT = \mathbf{£1{,}000.14}\ \boldsymbol{pa}$$

In Excel:

Excel formula	*Excel* function
=4918*0.06/(1-(1/((1 + 0.06)^6)))	=PMT(rate,nper,pv,[fv],[type]) =PMT(0.06,6,-4918)

Example 6.13

A mortgage of €250,000 is obtained over a 25-year term from a bank at an interest rate of 4.98%. Considering the loan is to be repaid in equal repayments at the end of each month, find the repayment amount.

Solution

From the question, we have:

PV = €250,000

$$\boldsymbol{n} = 25 \text{ years} = 25\ years \times 12\ months = 300\ months$$

$$\boldsymbol{r} = 0.0498 \text{ per annum} = \frac{0.0498}{12} = 0.00415\ per\ month$$

PMT =?

$$PMT = PV \times \frac{r}{\left[1 - \frac{1}{(1+r)^n}\right]}$$

$$PMT = 250{,}000 \times \frac{0.00415}{\left[1 - \frac{1}{(1+0.00415)^{300}}\right]}$$

$$\boldsymbol{PMT = €1{,}458.56\ per\ month}$$

In Excel:

Excel formula	*Excel* function
=250000*0.00415/(1-(1/((1 + 0.00415)^300)))	=PMT(rate,nper,pv,[fv],[type]) =PMT(0.00415,300,-250000)

Example 6.14

What annuity will a capital sum of £5,000 purchase for five years if a 10% yield is required by the purchaser?

Solution

PV = £5,000
n = 5 years
r = 10%
PMT =?

$$PMT = PV \times \frac{r}{\left[1 - \frac{1}{(1+r)^n}\right]}$$

$$PMT = 5{,}000 \times \frac{0.10}{\left[1 - \frac{1}{(1+0.10)^5}\right]}$$

$$PMT = \boldsymbol{£1{,}319\ pa}$$

In Excel:

Excel formula	*Excel* function
=5000*0.1/(1-(1/((1 + 0.1)^5)))	=PMT(rate,nper,pv,[fv],[type]) =PMT(0.1,5,-5000)

The £1,319 has two constituent parts.
The return ***on*** capital is 10% of the outlay, that is, £500 pa, which leaves £819 extra. This is the return ***of*** capital.

A *perpetuity*, defined as an annuity with no end or a stream of cash flows continuing indefinitely, allows for the simplification of Equation 6.3 into:

$$PV = PMT \times \frac{1}{r}$$

Equation 6.5 Present value of a zero-growth perpetuity

The component of the formula denoted as $\frac{1}{r}$, which multiplies the PMT in the perpetuity formula, is commonly called the **years purchase in perpetuity (YPp)** by valuers in the UK.

Now, let us consider that we are thinking about buying a perpetuity that pays $1,000. Visually, it can be represented as follows:

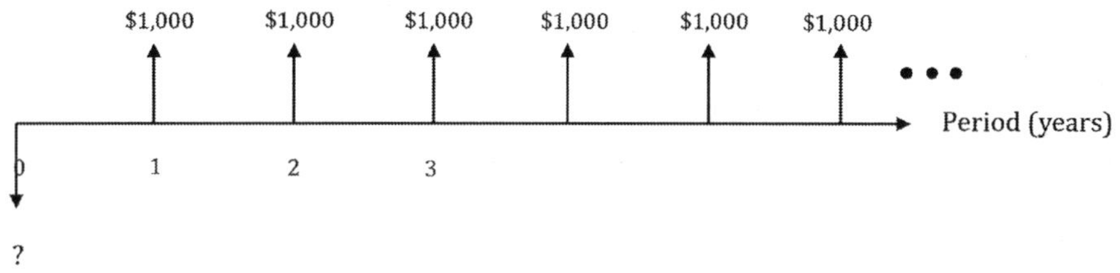

Figure 6.8 Example of a perpetuity.

Considering that the interest rate is 4%, how much money should we allocate into our fund today?

$$PV = PMT \times \frac{1}{r}$$

$$PV = 1{,}000 \times \frac{1}{0.04}$$

$$\boldsymbol{PV = \$\ 25{,}000}$$

Thus, we should allocate $25,000 into our fund today for the right of receiving $1,000 forever, at 4% interest rate.

Using Excel, the following steps can be undertaken:

Excel formula	*Excel* function
=1000*1/0.04	=PV(rate,nper,pmt,[fv],[type]) =PV(0.04,1000,-1000)

Note: When using Excel for perpetuities, it is necessary to set a high number of periods, such as a thousand years, because Excel defaults to Equation 6.3 instead of the simplified Equation 6.4. This is illustrated in the following with Equation 6.3 for a thousand-year period, yielding the same result as Equation 6.4.

$$PV = 1{,}000 \times \frac{\left[1 - \dfrac{1}{(1+0.04)^{1000}}\right]}{0.04} = \$25{,}000$$

Example 6.15

Considering the following equations:

$$PV\ of\ £1 = \frac{1}{(1+r)^n}$$

$$PV\ of\ £1\ \text{pa} = \frac{\left[1 - \dfrac{1}{(1+r)^n}\right]}{r}$$

What happens to the PV of £1 as the time period increases?

What effect does this have on the value of the multiplier YP or PV £1 pa?

To answer these questions, assume that the interest rate is 10%.

Solution

Years	*PV of £1*	*YP or PV of £1 pa*
1	0.909091	0.9090909
2	0.826446	1.7355372
3	0.751315	2.486852

(Continued)

(Continued)

Years	*PV of £1*	*YP or PV of £1 pa*
4	0.683013	3.1698654
5	0.620921	3.7907868
6	0.564474	4.3552607
7	0.513158	4.8684188
8	0.466507	5.3349262
9	0.424098	5.7590238
10	0.385543	6.1445671
90	0.000188	9.9981178
91	0.000171	9.9982889
92	0.000156	9.9984445
93	0.000141	9.9985859
94	0.000129	9.9987145
95	0.000117	9.9988313
96	0.000106	9.9989376
97	0.0000966	9.9990342
98	0.0000878	9.999122
99	0.0000798	9.9992018
100	0.0000726	9.9992743
Perpetuity	0	10

The PV of £1 decreases over time, and the PV £1 pa increases over time, reaching a maximum value at infinity.

In addition, it can be seen that the PV £1 pa is the accumulation of the PV of £1. For example, the PV £1 pa for two years at 10% equals the PV of £1 for one year (0.90909) plus the PV of £1 for two years (0.82645), namely, 1.736.

As ***n*** approaches perpetuity, the PV of £1 tends towards 0; the PV of £1 to be received such a long time in the future is reduced to virtually nothing.

As shown in Example 6.10, the formula of the *PV of* £1 pa can be written as:

$$PV\ of\ £1\ pa = \frac{[1 - \text{PV of £1}]}{r}$$

Therefore, if the PV of £1 tends to 0 at perpetuity, and if n is infinite, then given that:

$$PV\ £1\ pa = \frac{[1-0]}{r}$$

$$PV£1\ pa = \frac{[1-0]}{r}$$

Example 6.16

A freehold property produces a net income (annual net rent) of €120,000 a year. If an investor requires a return of 8%, what price should be paid? The income is perpetual.

Solution

In this case, the periodic payment (PMT) is €120,000, and ***r*** is the target rate of return of the investor, 8%.

To find the value of this property, we need to capitalise this income that is in perpetuity. Therefore, we can use Equation 6.4.

$$PV = PMT \times \frac{1}{r}$$

$$PV = 120,000 \times \frac{1}{0.08}$$

$$PV = \textbf{€1,500,000}$$

In Excel:

Excel formula	*Excel* function
=10000*1/0.08	=PV(rate,nper,pmt,[fv],[type]) =PV(0.08,1000,-120000)

The simplest way to assess the present value of an income in perpetuity is to divide the sum by the decimal equivalent of the required rate of return, or yield, for example, €120,000 ÷ 0.08 = **€1,500,000**.

A common property valuation application of the present value of an ordinary annuity formula (Equation 6.5, Equation 6.6, and Equation 6.7) involves calculating the PV of a perpetual income stream starting at a future date. This is referred to as a ***deferred perpetuity***. The calculation involves (1) calculating the PV of a perpetuity in the future and (2) discounting this FV to a PV in year 0 (today).

This is particularly relevant in property valuation, as properties are often expected to revert to a higher perpetual income after an initial period, such as following a rent review or the renewal of a lease after an under-renting period.

Example 6.17

Calculate the PV of a perpetual income of £60,000 pa beginning in five years' time using a discount rate of 12%.

Solution

This can be visually represented in Figure 6.9:

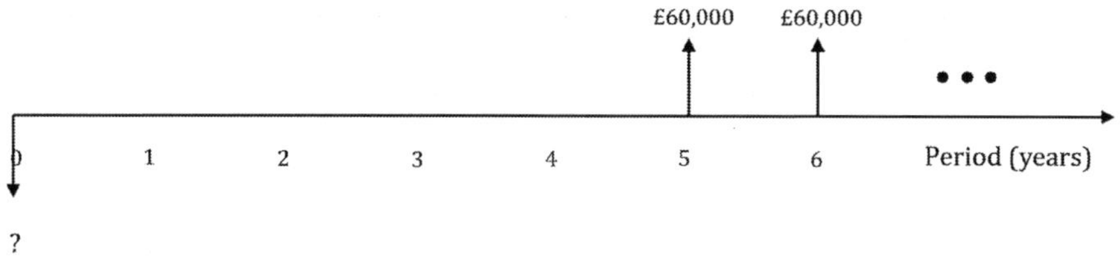

Figure 6.9 Example 6.14: cash flow (deferred perpetuity).

The first step is to calculate the PV of the perpetuity in year 5.

$$PV = PMT \times \frac{1}{r}$$

$$PV = 60,000 \times \frac{1}{0.12}$$

$$PV = £500,000$$

Note that this will result in a PV in year 5.

Now, we have a situation like Figure 6.9, where the PV in year 5 of £500,000 is the FV in year 0 (today). So the second step to solve this exercise is to apply Equation 6.2.

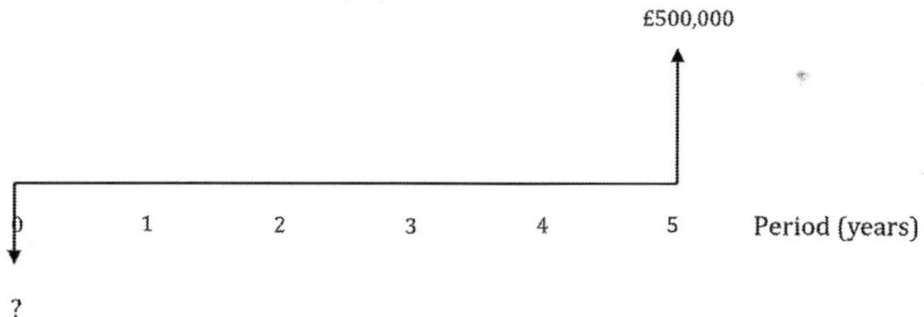

Figure 6.10 Example 6.17: cash flow (FV known).

$$PV = \frac{FV}{(1+r)^n}$$

$$PV = \frac{500,000}{(1+0.12)^5}$$

$$PV = £283,713.43$$

Thus, the PV of a perpetual income of £60,000 pa beginning in five years' time using a discount rate of 12% is £283,713.43.

In Excel:

Excel formula	*Excel* function
Step 1: =60000/0.12 **Step 2:** =500000/((1 + 0.12)^5)	**Step 1:** =PV(rate,nper,pmt,[fv],[type]) =PV(0.12,1000,-60000) **Step 2:** =PV(rate,nper,pmt,[fv],[type]) =PV(0.12,5,,-500000)

The use of this PV, or discounting technique, to assess the price to be paid for an investment or the value of an income-producing property ensures the correct relationship between future benefits and present value, namely, that the investor will obtain both a return on capital and a return of capital at the market-derived rate of interest (yield or discount rate) used in the calculation.

In property valuation in the UK, the preceding calculations would be demonstrated using a different structure:

Income	£60,000
YP perp @12%	$\frac{1}{r} = \frac{1}{0.12} = 8.333$
Deferred 5 years (PV£1 5 years @ 12%)	$\frac{1}{(1+r)^n} = \frac{1}{1.12^5} = 0.567$
PV	£283,713.43

Example 6.18

How much would it cost today to purchase an annuity of €1,000 for three years followed by an annuity of €1,200 for three years on a 10% basis?

Solution

This can be visually represented by the following figure:

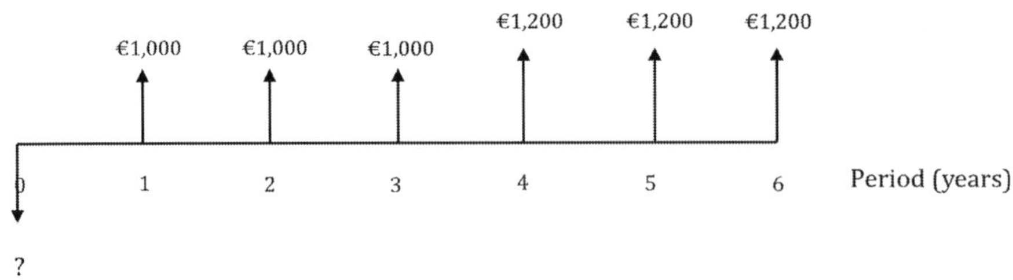

Figure 6.11 Example 6.18: cash flow.

Note that we have two groups of annuities:

- *Immediate annuity.* Paying €1,000 for three years, in years 1, 2, and 3.
- *Deferred annuity.* Paying €1,200 for three years, in years 4, 5, and 6.

Step 1: Calculate the PV of the immediate annuity (year 0).
The PV of an immediate annuity is calculated as:

$$PV = PMT \times \frac{\left[1 - \frac{1}{(1+r)^n}\right]}{r}$$

$$PV = 1{,}000 \times \frac{\left[1 - \frac{1}{(1+0.1)^3}\right]}{0.1}$$

$PV =$ **€2,486.85**

Step 2: Calculate the PV of the deferred annuity (year 3).

$$PV = PMT \times \frac{\left[1 - \frac{1}{(1+r)^n}\right]}{r}$$

$$PV = 1{,}200 \times \frac{\left[1 - \frac{1}{(1+0.1)^3}\right]}{0.1}$$

$PV =$ **€2,984.22**

The two present values cannot be added directly because they are calculated for different points in time. The PV from step 1 represents year 0 (today), while the PV from step 2 corresponds to year 3, effectively making it a future value relative to today. Therefore, we need to do step 3.

Step 3: Discount FV of the deferred annuity.

$$PV = \frac{FV}{(1+r)^n}$$

$$PV = \frac{2{,}984.22}{(1+0.1)^3}$$

$PV =$ **€2,242,09**

Step 4: Add the two PVs.

Now we can add the PV from step 1 with the PV from step 3 to find the cost to purchase today an annuity of €1,000 for three years followed by an annuity of €1,200 for three years on a 10% basis:

$PV = €2,486.85 + €2,242.09 = \mathbf{€4,728.94}$

In Excel:

Excel formula	*Excel* function
Step 1: =1000*(1-(1/(1+0.1)^3))/0.1 **Step 2:** =1200*(1-(1/(1+0.1)^3))/0.1 **Step 3:** =2,984.22/(1+0.1)^3	**Step 1:** =PV(rate,nper,pmt,[fv],[type]) =PV(0.1,3,-1000) **Step 2:** =PV(rate,nper,pmt,[fv],[type]) =PV(0.1,3,-1200) **Step 3:** =PV(rate,nper,pmt,[fv],[type]) =PV(0.1,3,,-2,984.22)

In property valuation in the UK, the preceding calculations would be demonstrated using a different structure:

Immediate annuity	€1,000	
×YP 3 years @10%	2.48685	€2,486.85
Deferred annuity	€1,200	
×YP 3 years @10%	2.48685	
×PV £1 3 years @10%	0.75131	€2,242.09
PV		**€4,728.94**

Example 6.19

Calculate the PV of a property that generates $100,000 in annual net rent for the first two years, followed by $150,000 annually in perpetuity, assuming a 6.5% discount rate.

Solution

This can be visually represented by the following figure:

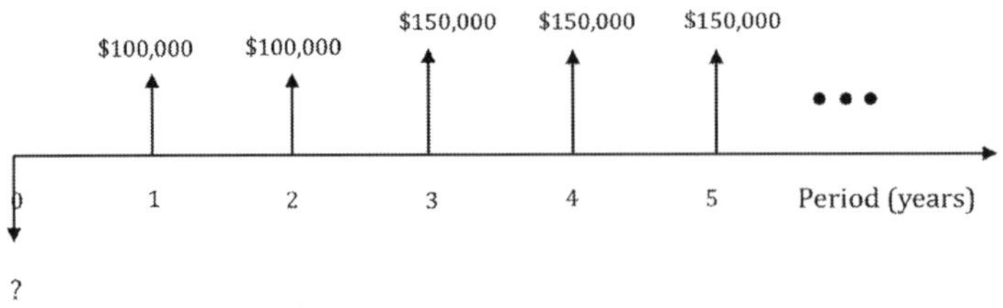

Figure 6.12 Example 6.19: cash flow.

In this question, we have:

- *Immediate annuity.* Rent of $100,000 pa for two years – years 1 and 2.
- *Deferred perpetuity.* Rent of $150,000 from year 3 to infinity.

Step 1: Calculate the PV of the immediate annuity (year 0).

$$PV = PMT \times \frac{\left[1 - \frac{1}{(1+r)^n}\right]}{r}$$

$$PV = 100{,}000 \times \frac{\left[1 - \frac{1}{(1+0.065)^2}\right]}{0{,}065}$$

$$PV = \mathbf{\$182{,}062.54}$$

Step 2: Calculate the PV of the deferred perpetuity (year 2).

$$PV = PMT \times \frac{1}{r}$$

$$PV = 150{,}000 \times \frac{1}{0.065}$$

$$PV = \mathbf{\$2{,}307{,}692.31}$$

Step 3: Discount FV of the deferred perpetuity.

$$PV = \frac{FV}{(1+r)^n}$$

$$PV = \frac{2{,}307{,}692.31}{(1+0.065)^2}$$

$$PV = \mathbf{\$2{,}034{,}598.35}$$

Step 4: Add the PVs.

$$PV = \$182{,}062.54 + \$2{,}034{,}598.35 = \mathbf{\$2{,}216{,}660.99}$$

In Excel:

Excel formula	*Excel* function
Step 1: =100000*(1-(1/(1 + 0.065)^2))/0.065 **Step 2:** =150000/0.065 **Step 3:** =2307692.31/((1 + 0.065)^2)	**Step 1:** =PV(rate,nper,pmt,[fv],[type]) =PV(0.065,2,-100000) **Step 2:** =PV(rate,nper,pmt,[fv],[type]) =PV(0.065,1000,-150000) **Step 3:** =PV(rate,nper,pmt,[fv],[type]) =PV(0.065,2,,-2307692.31)

It will be shown later that this approach is the same as that used by valuers when valuing a variable income flow from a property investment.

Future value of an ordinary annuity

There will be times when we will be interested to know how to project the future value of a series of payments. For instance, let us consider that we plan to invest \$1,000 at the end of each year into a real estate investment trust (REIT) with an annual rate of return of 4% and want to know how much money we will have at the end of five years. Visually, this can be represented as:

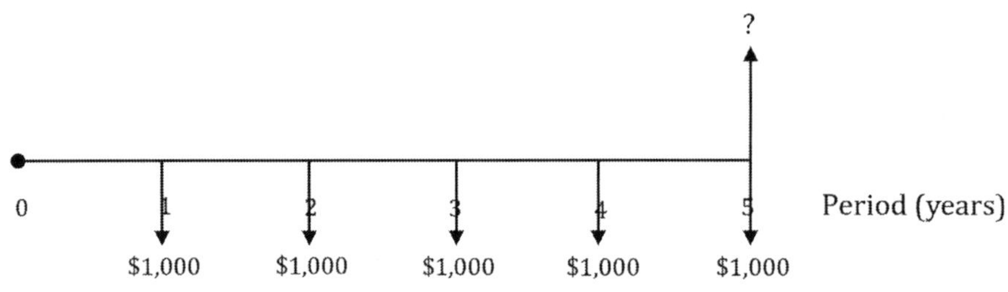

Figure 6.13 Finding the FV of an ordinary annuity.

Before exploring the future value of an ordinary annuity formula, let us understand how the cash flows and compounding occur over time. Table 6.7 illustrates the progression of the investments, the interest earned, and the growing balance.

Note that the first \$1,000 is not invested today (year 0) but on the last day of year 1, so it does not earn interest initially. In year 2, the \$1,000 invested in year 1 accrues 4% interest, generating \$40, and another \$1,000 is added, resulting in a balance of \$2,040. This process repeats annually, with each balance earning interest and a new \$1,000 payment added at the end of the year. By year 5, the final \$1,000 payment does not accrue interest, culminating in a total balance of \$5,416.32.

Table 6.8 illustrates how each annual cash flow contributes to the future value of the annuity by compounding at a rate of 4%. Each \$1,000 payment grows based on the number of years it

Table 6.7 Cash flow table – FV of ordinary annuity problem

Year	*0 (today)*	*1*	*2*	*3*	*4*	*5*
Previous balance ($)	0	0	1,000	2,040	3,121.60	4,246.46
Interest earned ($)		0	0.04 × 1,000 = 40	0.04 × 2,040 = 81.60	0.04 × 3,121.60 = 124.86	0.04 × 4,246.46 = 169.86
Principal added ($)		1,000	1,000	1,000	1,000	1,000
New balance ($)		1,000	2,040	3,121.60	4,246.46	5,416.32

Table 6.8 FV of an ordinary annuity

Year	*Cash flow ($)*	*Calculation*	*Future value*
1	1,000	1,000 × $(1+0.04)^4$	1,169.86
2	1,000	1,000 × $(1+0.04)^3$	1,124.86
3	1,000	1,000 × $(1+0.04)^2$	1,081.60
4	1,000	1,000 × $(1+0.04)^1$	1,040.00
5	1,000	1,000 × $(1+0.04)^0$	1,000.00
Future value of the annuity			**5,416.32**

remains invested: The first payment compounds for four years, the second for three years, and so on. The future value of each payment is calculated using Equation 6.1.

Since annuities are equal periodic payments, we can calculate the future value of an annuity in one step using the following equation:

$$FV = PMT \times \frac{(1+r)^n - 1}{r}$$

Where:

PMT = periodic payment, that is, inflow or outflow amount that occurs at each period of a financial stream.
FV = future value.
r = interest rate.
n = number of periods.

Equation 6.6 Future value of an ordinary annuity

The component of the formula denoted as $\frac{(1+r)^n - 1}{r}$, which multiplies the PMT, is commonly called the **amount of £1 pa** in the UK and represents the future worth of £1 invested at the end of each year accruing compound interest at a given rate.

Applying the formula for the preceding example, we have:

$$FV = 1{,}000 \times \frac{(1+0.04)^5 - 1}{0.04} = 5{,}416.32$$

Using Excel, the following steps can be undertaken:

Excel formula	*Excel* function
=1000*((1 + 0.04)^5–1)/0.04	=FV(rate,nper,pmt,[pv],[type]) =FV(0.04,5,-1000)

Example 6.20

You decide to invest €200 per month for ten years in a fund that earns an annual interest rate of 6%. What will be the value of the investment at the end of ten years?

Solution

$PMT = €200$ per month

$r = 6\%$ pa $= 6\%/12 = 0.5\%$ or 0.005 per month

$n = 10$ years $= 10 \times 12 = 120$ *months*

$$FV = PMT \times \frac{(1+r)^n - 1}{r}$$

$$FV = 200 \times \frac{(1+0.005)^{120} - 1}{0.005}$$

$FV =$ **€32,775,87**

In Excel:

Excel formula	*Excel* function
=200*((1 + 0.005)^120–1)/0.005	=FV(rate,nper,-pmt,[pv],[type]) =FV(0.5%,120,-200)

Example 6.21

Calculate the amount of £1 pa for five years at 10%.

Solution

$$\begin{aligned} PMT &= £1 \text{ pa} \\ r &= 10\% \text{ pa or } 0.10 \\ n &= 5 \text{ years} \\ FV &= ? \end{aligned}$$

$$FV = PMT \times \frac{(1+r)^n - 1}{r}$$

$$FV = 1 \times \frac{(1+0.1)^5 - 1}{0.1}$$

$$FV = £6.1051$$

Multiplying this figure by 100 to calculate the sum that £100 invested at the end of each year will accumulate to after five years produces the same figure as in the savings passbook.

In Excel:

Excel formula	*Excel* function
=1*((1 + 0.1)^5–1)/0.1	=FV(rate,nper,-pmt,[pv],[type]) =FV(0.1,5,-1)

Example 6.22

If a student invests £600 in a savings account at the end of each year and at the end of 20 years has £43,230, at what rate of interest has this sum accumulated?

Solution

$$\begin{aligned} \textbf{PMT} &= £600 \text{ pa} \\ n &= 20 \text{ years} \\ \textbf{FV} &= £43{,}230 \\ r &= ? \end{aligned}$$

$$FV = PMT \times \frac{(1+r)^n - 1}{r}$$

$$\frac{43{,}230}{600} = \frac{(1+r)^{20} - 1}{r}$$

$$72.05 \times r = (1+r)^{20} - 1$$

Note that this equation involves ***r*** on both sides, meaning that solving it directly would typically require numerical methods or iterative solutions. Another way is to use the rate function in Excel, as previously demonstrated in Example 6.6.

=RATE(nper,pmt,pv,[fv])
=RATE(20,600,,-43,230)

The Excel function returns a rate of **12%.**

Proof:

Substitute **r** = 0.12 into the equation:

$$72.05 \times 0.12 = (1 + 0.12)^{20} - 1$$

$$\mathbf{8.65 = 8.65}$$

Property investors may need to make provision from today for some future expenditure, and this may be for some form of maintenance – for example, a building may require a new roof, or service roads may require resurfacing.

Such obligations may be passed on to the tenant in the form of service charges payable in addition to rent, which may include the provision of a fund to be used for major works in the future. However, in some cases, the landlord may be responsible for major repairs, which cannot be recovered from the tenant. Such obligations should be reflected in the purchase price of the investment. Such expenditure can be budgeted for by investing a lump sum today, or by investing a regular sum from today, both of which will grow due to compound interest. A property owner may need to know how much should be invested now or on an annual basis.

The first step is to estimate the probable amount needed at an expected future date. The investor could then either invest a lump sum immediately which, with the accumulation of interest, will meet the estimated outlay when it arises. This can be calculated using Equation 6.2 or the PV of £1 multiplier. For instance, if we anticipate needing $50,000 for renovations in five years and the interest rate of the savings account is 4%, then we will need to save $41,096.36[2] today.

However, it may be more effective to set aside part of the income received from the investment (i.e. rent) regularly in a sinking fund account which is planned to accumulate to the required sum by the date in the future when the expenditure is required. To calculate this amount that should be regularly set aside to achieve a future expenditure, we must rearrange Equation 6.6 as follows:

$$PMT = FV \times \frac{r}{(1+r)^n - 1}$$

Equation 6.7 Ordinary annuity payment formula when FV is known

Note that in the UK, the multiplier $\frac{r}{(1+r)^n - 1}$ is called **the annual sinking fund (ASF)** and is the fraction of £1 which must be invested at regular intervals to produce £1 at a given point in the future with compound interest accruing at a given rate.

Referring to our previous example, to accumulate $50,000 in five years with a 4% interest rate, we would need to set aside $9,231.36[3] per year in the savings account.

Example 6.23

An investor intends to buy a shop in which the window frames are deteriorating. It is estimated that they will require complete replacement in four years at a cost of £18,500. The shop produces a net income of £17,500 pa.

How much of this income should be set aside each year to meet the expense, assuming the money is invested with a guaranteed fixed return of 3% per year?

Solution

FV = 18,500
r = 3% or 0.03
n = 4 years
PMT = ?

$$PMT = FV \times \frac{r}{(1+r)^n - 1}$$

$$PMT = 18,500 \times \frac{0.03}{(1+0.03)^4 - 1}$$

$$\boldsymbol{PMT = £4,422}$$

In Excel:

Excel formula	Excel function
=18500*0.03/((1 + 0.03)^4–1)	=PMT(0.03,4,,-18500)

Example 6.24

The sum of £10,000 has been borrowed on a repayment mortgage at 6% for 25 years.

1. Calculate the annual repayment of interest and capital.
2. Calculate the amount of interest due and the amount of capital repaid in the first year.
3. Calculate the amount of interest due and the amount of capital repaid in the tenth year.

Solution

1.

$$PMT = PV \times \frac{r}{\left[1 - \frac{1}{(1+r)^n}\right]}$$

$$PMT = 10,000 \times \frac{0.06}{\left[1 - \frac{1}{(1+0.06)^{25}}\right]}$$

$$\boldsymbol{PMT = £782.27\ pa}$$

In Excel:

Excel formula	*Excel* function
=10000*0.06/(1-(1/((1 + 0.06)^25)))	=PMT(rate,nper,pv,[fv],[type]) =PMT(0.06,25,-10000)

2. As interest in the first year is added immediately to capital borrowed, then:

   £10,000 × 0.06 = £600 interest due in first year

   As the total to be paid is £782.27, the amount of capital repaid will be:

   £782.27 − £600 = £182.27 in the first year.

3. The amount of interest in the tenth year will depend upon the amount of capital outstanding at the beginning of the tenth year, that is, after the ninth annual payment. Assuming that £182.27 a year is a notional sinking fund, after nine years, the capital repaid will be:

$$FV = PMT \times \frac{(1+r)^n - 1}{r}$$

$$FV = 182.27 \times \frac{(1+0.06)^9 - 1}{0.06}$$

$$FV = £2,094.52$$

In Excel:

Excel formula	*Excel* function
=182.27*((1 + 0.06)^9–1)/0.06	=FV(rate,nper,-pmt,[pv],[type]) =FV(0.06,9,-182.27)

Capital outstanding is £10,000 less £2,094.52, which is £7,905.48, or calculate the value at 6% of the right to receive 16 more payments of £782.27.

$$PV = PMT \times \frac{\left[1 - \frac{1}{(1+r)^n}\right]}{r}$$

$$PV = 782.27 \times \frac{\left[1 - \frac{1}{(1+0.06)^{16}}\right]}{0.06}$$

$$PV = \mathbf{£7,905.54}$$

In Excel:

Excel formula	*Excel* function
=782.27*(1–1/((1 + 0.06)^16))/0.06	=PV(rate,nper,-pmt,[fv],[type]) =PV(0.06,16,-782.27)

Capital outstanding is **£7,905.54.**

Therefore, interest will be £7,905.54 × 0.06 = £474.33, and capital repaid in year 10 is £782.27 – £474.33 which is **£307.94.**

In passing, it may be observed that the capital repaid in year 1 plus compound interest at 6% will amount to £307.94 after nine years:

$$FV = PV \times (1+r)^n$$

$$FV = 182.27 \times (1+0.06)^9$$

$$FV = £307.94$$

In Excel:

Excel formula	*Excel* function
=182.27*(1 + 0.06)^9	=FV(rate,nper,-pmt,[pv],[type]) =FV(0.06,9,,-182.27)

As interest on a mortgage is based on capital outstanding from year to year, the change in capital repaid from year to year must be at 6%.

$$FV = PV \times (1+r)^n$$
$$FV = 182.27 \times (1+0.06)^{24}$$
$$\boldsymbol{FV = £738}$$

In Excel:

Excel formula	*Excel* function
=182.27*(1 + 0.06)^24	=FV(rate,nper,-pmt,[pv],[type]) =FV(0.06,24,,-182.27)

As the annual payment is £782.27, if we deduct the capital repaid in the last year, that will represent the interest due in the last year of the mortgage: £782.27 − £738 = £44.27. In many cases, the rate of interest will change during the mortgage term. If it does, the mortgagor will usually have the choice of varying the term or repaying at an adjusted rate.

Example 6.25

Given the figures in the previous example, advise the borrower on the alternatives available if the interest rate goes down to 5% at the beginning of year 10.

Solution

The borrower can either (1) continue to pay £782.30 a year and repay the mortgage earlier or (2) reduce the annual payment.

1. **No change in annual payment.** In this alternative, the amount of capital outstanding at beginning of year 10 (i.e. the end of year 9) is (as before) £7,905.54. So PV = **£7,905.54**.
   The repayment is also, as before, £782.27, so PMT = **£782.27**.
   The rate now is 0.05, so r = **0.05**.
   $\boldsymbol{n = ?}$

$$PV = PMT \times \frac{\left[1 - \frac{1}{(1+r)^n}\right]}{r}$$

$$\frac{PV}{PMT} \times r = 1 - \frac{1}{(1+r)^n}$$

$$\frac{1}{(1+r)^n} = 1 - \frac{PV \times r}{PMT}$$

$$(1+r)^n = \frac{1}{1 - \frac{PV \times r}{PMT}}$$

$$\ln((1+r)^n) = ln\left(\frac{1}{1 - \frac{PV \times r}{PMT}}\right)$$

$$n \times ln(1+r) = ln\left(\frac{1}{1 - \frac{PV \times r}{PMT}}\right)$$

$$n = \frac{\ln\left(\frac{1}{1 - \frac{PV \times r}{PMT}}\right)}{\ln(1+r)}$$

$$n = \frac{\ln\left(\frac{1}{1 - \frac{7{,}905 \times 0.05}{782.27}}\right)}{\ln(1+0.05)}$$

$$n = \mathbf{14.42}\ \boldsymbol{years}$$

In Excel:

Excel formula	*Excel* function
=LN(1/(1-(7905.54*0.05)/782.27))/ LN(1 + 0.05)	=NPER(rate,pmt,[-pv],[type]) =NPER(0.05,782.27,-7905.54)

The loan in this alternative is repaid before the due date.

2. **Change in annual payment.** In this alternative, the amount of capital outstanding at beginning of year 10 (i.e. the end of year 9) is (as before) £7,905.54, and the number

of years remaining is 16 years. So PV = £7,905.54, and n = 16 years, as a change in interest at the beginning of year 10 means that, including the payment in year 10, there are 16 more payments due. Since the rate now is 0.05, we need to find the new PMT.

$$PMT = PV \times \frac{r}{\left[1 - \frac{1}{(1+r)^n}\right]}$$

$$PMT = 7{,}905.54 \times \frac{0.05}{\left[1 - \frac{1}{(1+0.05)^{16}}\right]}$$

$$\boldsymbol{PMT = £729.44}$$

In Excel:

Excel formula	*Excel* function
=7905.54*0.05/(1-(1/(1 + 0.05)^16))	=PV(rate,nper,-pmt,[fv],[type]) =PMT(0.05,16,-7905.54)

Summary of formulae

Table 6.9 summarises the formulae discussed in this chapter and is designed to assist in identifying and solving financial calculations effectively. The first column, **Find**, identifies the variable to be determined, representing the unknown in the calculation. The second column, **Given**, outlines the known variable required to perform the calculation.

The **UK multiplier** column clarifies the terminology used in UK-based calculations, while the **Formula** column presents the various formulas for present value, future value, and ordinary annuities. The final **Excel** column provides the corresponding Excel function for practical application.

Table 6.9 TVM formulae

Find	*Given*	*UK multiplier*	*Formula*	*Excel*
Future value	Present value	Amount of £1 $(1+r)^n$	$FV = PV \times (1+r)^n$	=fv(r,n,,-pv)
Future value	Periodic payment	Amount of £1 pa $\frac{(1+r)^n - 1}{r}$	$FV = PMT \times \frac{(1+r)^n - 1}{r}$	=fv(r,n,-pmt)
Present value	Future value	The PV of £1 $\frac{1}{(1+r)^n}$	$PV = \frac{FV}{(1+r)^n}$	=pv(r,n,,-fv)

(Continued)

Table 6.9 (Continued)

Find	*Given*	*UK multiplier*	*Formula*	*Excel*
Present value	Periodic payment	PV of £1 pa or year's purchase (YP) $\frac{\left[1-\frac{1}{(1+r)^n}\right]}{r}$	$PV = PMT \times \frac{\left[1-\frac{1}{(1+r)^n}\right]}{r}$	=pv(r,n,-pmt)
Present value	Periodic payment in perpetuity	Year's purchase in perpetuity (YPp) $\frac{1}{r}$	$PV = \frac{PMT}{r}$	=pv(r,1000,-pmt)
Periodic payment	Present value	Annuity that £1 will purchase	$PMT = PV \times \frac{r}{\left[1-\frac{1}{(1+r)^n}\right]}$	=pmt(r,n,-pv)
Periodic payment	Future value	Annual sinking fund $\frac{r}{(1+r)^n-1}$	$PMT = FV \times \frac{r}{(1+r)^n-1}$	=pmt(r,n,,-fv)

Annuity due

Until now, all the formulae and explanations have focused on ordinary annuities – annuities where cash flows occur at the end of each period. A variation of this is the annuity due, in which cash flows occur at the beginning of each period. For example, an annuity due with cash inflows of $1,000 per year over five years can be represented as follows:

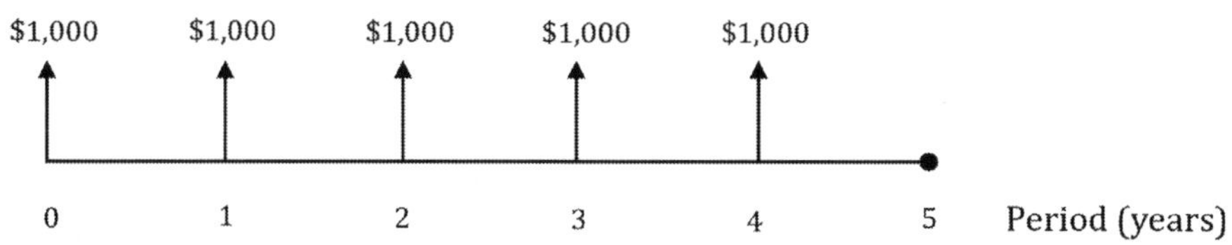

Figure 6.14 Example of an annuity due.

It is common practice in commercial properties to require the tenant to pay rent in quarterly instalments at the beginning of each quarter. In lower-value areas and for residential properties, the expectation is that rent is paid monthly in advance. As highlighted by Wyatt (2013), since the income is received earlier than it would be with annual payments in arrears, these arrangements slightly enhance the value of the investment.

Present value of an annuity due

To determine the present value of an annuity due, we calculate the sum of the present values of each payment, considering that payments are made at the beginning of each period. Considering the example of an annuity due with \$1,000 payments for five years (Figure 6.14) and a 4% annual interest rate, how much money should we allocate today to receive this payment stream? Visually this question can be represented in the cash flow diagram as:

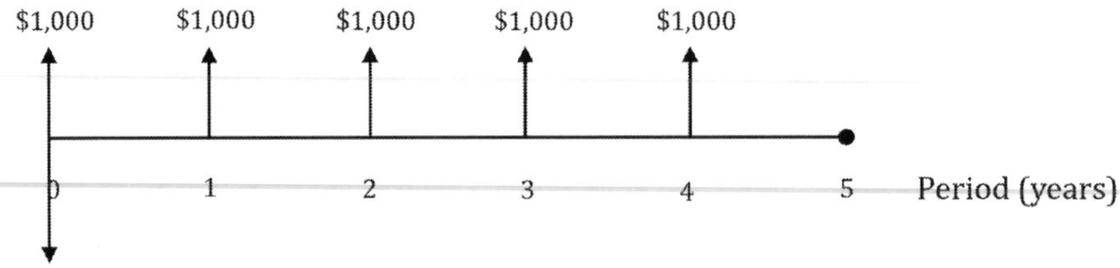

Figure 6.15 Finding the PV of an annuity due.

Since the first cash flow is received immediately, there is no need to discount it. For the remaining cash flows, which occur in future periods, we use Equation 6.2 to calculate their present values.

Table 6.10 Present value of an annuity due

Year	*Cash flow (\$)*	*Calculation*	*Present value*
0	1,000	$PV = \frac{1,000}{(1+0.04)^0}$	1,000
1	1,000	$PV = \frac{1,000}{(1+0.04)^1}$	961.54
2	1,000	$PV = \frac{1,000}{(1+0.04)^2}$	924.56
3	1,000	$PV = \frac{1,000}{(1+0.04)^3}$	889.00
4	1,000	$PV = \frac{1,000}{(1+0.04)^4}$	854.80
Present value of the annuity due			**\$4,629.90**

Alternatively, the present value of the annuity due can be calculated directly using this formula:

$$PV = PMT_{Year\ 0} + PMT \times \frac{\left[1 - \frac{1}{(1+r)^{n-1}}\right]}{r}$$

Equation 6.8 PV of an annuity due

$$PV = 1{,}000 + 1{,}000 \times \frac{\left[1 - \frac{1}{(1+0.04)^{5-1}}\right]}{0.04}$$

$$PV = \$4{,}629.90$$

Using Excel, the following steps can be undertaken:

Excel formula	*Excel* function
=1000 +1000*(1–1/(1+0.04)^4)/0.04	=PV(rate,nper,pmt,[fv],[type]) =PV(0.04,5,1000,,1) **Note:** Since this represents an annuity due, it is essential to specify 1 for the 'type' argument to reflect that payments occur at the beginning of each period.

Note that the figures used in this example are identical to those in the present value calculation for an ordinary annuity, except for the timing of the cash flows. The annuity due assumes payments are made at the beginning of each period, resulting in a higher present value ($4,629.90) compared to the ordinary annuity ($4,451.82) due to the reduced effect of discounting. This difference arises because payments are received sooner in an annuity due – meaning they are subject to less discounting, thus increasing their present value.

Starting from year 0 (today), we have determined that we need to deposit $4,629.90 to fund the annuity due. Unlike an ordinary annuity, the first $1,000 are immediately withdrawn after the deposit. By the start of year 1, the remaining balance of $3,629.90 accrues interest of $145.20 (at a rate of 4%), resulting in a total of $3,775.10 before the second $1,000 withdrawal is made. After this withdrawal, the balance decreases to $2,775.09. This pattern continues for each subsequent year: The remaining balance earns interest at the start of the year, and then $1,000 is withdrawn. By year 4, the final interest amount of $38.46 is added to the balance of $961.54, bringing the total to exactly $1,000. This amount is then withdrawn, leaving the account balance at zero.

The key distinction between an annuity due and an ordinary annuity lies in the timing of withdrawals or payments. In an annuity due, they occur at the start of each period, making the payments more valuable because they are received sooner and subject to less discounting.

Table 6.11 Cash flow table – PV of annuity due problem

Year	*0 (today)*	*1*	*2*	*3*	*4*
Previous balance ($)	0	3,629.90	2,775.09	1,886.09	961.54
Interest earned ($)	0	145.20	111.00	75.44	38.46
Principal added/ withdrawn ($)	4,629.90–1,000	−1,000	−1,000	−1,000	−1,000
New balance ($)	3,629.90	2,775.09	1,886.09	961.54	0

This requires a larger initial deposit to fund the payments. Conversely, in an ordinary annuity, withdrawals are made at the end of each period – meaning that the funds remain in the account longer, earning additional interest before each payment, but the delayed timing reduces the present value of each payment.

Example 6.26

Revisit Example 6.9 and determine the amount the investor should pay for the right to receive the income of £675 yearly in advance over 64 years. The investor requires a 12% return on the investment. Compare this figure with the result of Example 6.9 – which PV is higher, and why?

Solution

$$r = 0.12$$
$$n = 64$$
$$\mathbf{PMT} = £675 \text{ in advance}$$
$$\mathbf{PV} = ?$$

$$PV = PMT_{Year\ 0} + PMT \times \frac{\left[1 - \frac{1}{(1+r)^{n-1}}\right]}{r}$$

$$PV = 675 + 675 \times \frac{\left[1 - \frac{1}{(1+0.12)^{64-1}}\right]}{0.12}$$

$$PV = \$6,295.54$$

In Excel:

Excel formula	*Excel* function
=675 + 675*(1–1/(1 + 0.12)^63)/0.12	=PV(rate,nper,pmt,[fv],[type]) =PV(0.12,64,-675,,1)

In Example 6.9, the present value of £5,621.02 was calculated for an ordinary annuity, where payments are made at the end of each period. In contrast, in Example 6.26, the present value of £6,295.54 was calculated for an annuity due, where payments are made at the beginning of each period. As a result, the present value in Example 6.26 is higher. This is because each payment in an annuity due is invested for an additional period, allowing it to accumulate more interest over time, which increases its present value.

Future value of an annuity due

Let us examine how much money will accumulate over five years if we invest $1,000 annually in a REIT starting today, assuming an annual rate of return of 4%. In visual terms we have:

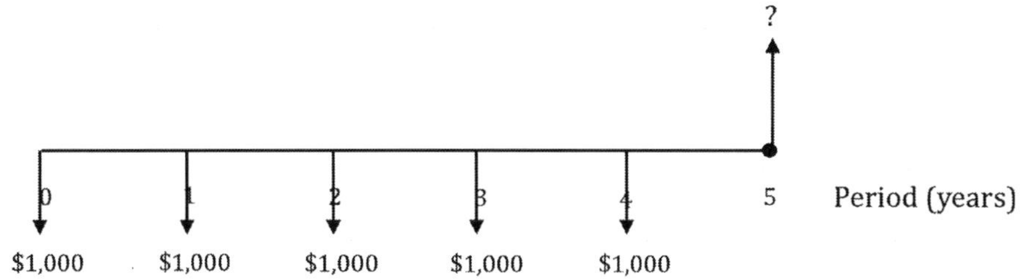

Figure 6.16 Finding the FV of an annuity due.

Table 6.11 illustrates how each annual cash flow contributes to the future value of the annuity by compounding at a rate of 4%. Each $1,000 payment grows based on the number of years it remains invested: The first payment compounds for five years, the second for four years, and so on. The future value of each payment is calculated using Equation 6.1.

Table 6.11 FV of an ordinary annuity

Year	*Cash flow ($)*	*Calculation*	*Future value*
0	1,000	$1,000 \times (1 + 0.04)^5$	1,216.65
1	1,000	$1,000 \times (1 + 0.04)^4$	1,169.86
2	1,000	$1,000 \times (1 + 0.04)^3$	1,124.86
3	1,000	$1,000 \times (1 + 0.04)^2$	1,081.60
4	1,000	$1,000 \times (1 + 0.04)^1$	1,040.00
Future value of the annuity due			**5,632.98**

We can calculate the future value of an annuity due in one step using the following equation:

$$FV = (1+r) \times PMT \times \frac{(1+r)^n - 1}{r}$$

Equation 6.9 Future value of an annuity due

Applying the formula to the previous example, we have:

$$FV = (1+0.04) \times 1{,}000 \times \frac{(1+0.04)^5 - 1}{0.04} = 5{,}632.98$$

Using Excel, the following steps can be undertaken:

Excel formula	*Excel* function
=(1 + 0.04)*1000*((1 + 0.04)^5–1)/0.04	=FV(rate,nper,pmt,[pv],[type]) =FV(0.04,5,-1000,,1) **Note:** Since this represents an annuity due, it is essential to specify 1 for the 'type' argument to reflect that payments occur at the beginning of each period.

Note that the figures used in this example are identical to those used in the future value calculation for an ordinary annuity (pages 108–110), except for the timing of the cash flows. The annuity due assumes payments are made at the beginning of each period, resulting in a higher future value ($5,632.98) compared to the ordinary annuity ($5,416.32) due to the earlier compounding effect of each payment.

Example 6.27

You decide to invest €200 at the beginning of each month for ten years in a fund that earns an annual interest rate of 6%. What will be the value of the investment in year 10? Compare your result with Example 6.20.

Solution

PMT = €200 per month
r = 6% pa = 6%/12 = 0.5% or 0.005 per month
n = 10 years = 10×12 = 120 months

$$FV = (1+r)\times PMT \times \frac{(1+r)^n - 1}{r}$$

$$FV = (1+0.005)\times 200 \times \frac{(1+0.005)^{120} - 1}{0.005}$$

$$FV = €32{,}939.75$$

In Excel:

Excel formula	*Excel* function
=(1 + 0.005)*200*((1 + 0.005)^120–1)/0.005	=FV(rate,nper,-pmt,[pv],[type]) =FV(0.5%,120,-200,,1)

In Example 6.20, the €200 was invested at the end of each period – meaning it did not earn interest during the first month (month 0). In the current example, the €200 is invested at the beginning of each month, allowing each payment to start earning interest immediately, including the first payment in month 0 (today). This timing difference results in the current example accumulating more interest, leading to a higher final value of €32,939.75 compared to €32,775.87 when invested at the end of each period.

Incomes receivable quarterly or monthly in advance

It has been noted that almost all property is let quarterly or monthly in advance terms. Although the market maintains the tradition of valuing on an in arrears assumption (and of comparing market capitalisation rates on the same basis), it is a simple process, if mathematically more complex, to find the PV on a quarterly or monthly basis. Valuers using standard investment valuation software can readily provide investors with the effective or net effective rate of return, as well as the nominal rate of return – that is, whilst not valuing on a quarterly or monthly income basis using appropriate rates, they will provide the investor with details of their return, which reflect the actual frequency of income payment.

Most valuation software includes options to switch assumptions, that is, from annual in arrears to, say, quarterly in advance. However, to do so can be dangerous if the in arrears rates are simply substituted in the formula to assess in advance PVs.

The present value of an annuity received in a basis other than yearly can be expressed as:

$$PV = PMT \times \left[1 - \left(1 + \frac{r}{f}\right)^{-n \times f}\right] \times \left[\frac{\left(1 + \frac{r}{f}\right)}{\frac{r}{f}}\right]$$

Where:

- ***PMT*** = periodic payment
- ***r*** = interest rate
- ***f*** = frequency of occurrence in a year
- ***n*** = number of years

Equation 6.10 Present value of an annuity due with different frequency of occurrence

For instance, let us say that a landlord expects to receive \$5,000 at the beginning of each month as rent for the next ten years. To calculate the present value of this income stream, considering a discount rate based on the ongoing market rate of 5% pa, we have:

$$PV = 5{,}000 \times \left[1 - \left(1 + \frac{0.05}{12}\right)^{-10 \times 12}\right] \times \left[\frac{\left(1 + \frac{0.05}{12}\right)}{\frac{0.05}{12}}\right]$$

Excel formula	*Excel* function
=5000*(1-(1+(0.05/12))^(-10*12)) *((1+(0.05/12))/(0.05/12))	=PV(rate,nper,pmt,[fv],[type]) =PV(0.05/12,120,-5000,,1) **Note:** Since the rents are paid in months, we need to transform the rate and the period to months. Since this represents an annuity due, it is essential to specify 1 for the 'type' argument to reflect that payments occur at the beginning of each period.

Switching to quarterly or monthly without amending *r*% to reflect the change will produce different opinions of PV. This would be strictly true if the comparison was genuinely with an income in arrears. However, the rates used by valuers, whilst derived from sales on the assumption of annual in arrears and applied in the same way, are known to be the market yields on investments where the income is actually paid quarterly or monthly in advance. To find the PV of a sum or PV of a quarterly or monthly sum payable in advance, one needs the discount rate to be derived from comparable quarterly or monthly sales analysis. It is dangerous to simply switch as it can imply a higher value where no higher value would be achieved in the market.

Nominal and effective rates of interest

The rates of return provided by different forms of saving, and from different forms of investment, are usually compared by means of the annual rate of interest being paid or earned. For example, banks and saving organisations usually quote the annual rate of interest that will be earned in savings accounts, such as 10% pa, representing the amount of interest that will be added to each £100 of savings left in the account for a full 12 months. Sometimes, however, it is necessary to check whether the rate of interest quoted is nominal or effective. The quoted or stated interest rate is also known as the nominal interest rate, and it does not incorporate the compounding periods into effect. In contrast, the effective interest rate considers the compounding periods and reflects the actual interest rate (Mazur and Moshenko 2021). The difference in annual interest can be significant, depending on the frequency with which interest is added and the rate.

For example, if a savings bank pays interest at 10% pa and it is added at the end of each complete year, with the effect that £10 is added to the account for every £100 held for a year in that account, then the rate of 10% is both the nominal rate of interest and the effective rate of interest.

On the other hand, if savings bank B pays 10% pa with 5% interest paid half-yearly, then that 10% will produce a higher effective rate than the nominal 10% because the interest paid after six months will itself earn interest over the second half of the year. The total accumulation of £1 invested in an account for one year will therefore be $£(1.05)^2 = £1.1025$. Interest is 10.25 pence accumulated on £1 invested. The effective annual rate (EAR) of interest is therefore 10.25%. The effective annual rate formula can be written as:

$$EAR = \left(1 + \frac{r}{f}\right)^f - 1$$

Where:

EAR = effective annual interest rate

r = nominal interest rate per year

f – = frequency of occurrence in a year (e.g. 12 for monthly, 4 for quarterly)

Equation 6.11 Effective annual rate (EAR) formula

Applying the formula to the earlier example, we have:

$$EAR = \left(1 + \frac{0.1}{2}\right)^2 - 1 = 10.25\%$$

The built-in Excel function for the EAR is as follows:

=EFFECT(nominal_rate,npery)

The EFFECT functions formula has the following arguments:

- **Nominal rate** = (required) the nominal interest rate.
- ***Npery*** = (required) the number of compounding periods per year.

Excel function 6.4 Effect function

However, if the 10% is quoted as the annual percentage rate (APR), or as an effective annual rate (EAR) of interest, but is actually credited every six months, then it is important to recognise that 5% or £5 for every £100 will not be added after the first six months. Instead, the savings provider will calculate the effective rate for every six months.

By rearranging Equation 6.8, we interpret $\frac{r}{n}$ as the nominal rate per compounding period (periodic rate), in the case of this example, the six-monthly rate of interest. To find this value, we use the known EAR of 10% and $n = 2$ compounding periods per year:

$$EAR = \left(1 + \frac{r}{f}\right)^f - 1$$

$$\frac{r}{f} = \sqrt[f]{EAR + 1} - 1$$

$$\frac{r}{f} = \sqrt[2]{1.1} - 1$$

$$\frac{r}{f} = 0.0488 \; or \; 4.88\% \text{ per semestre}$$

Proof:

$$EAR = \left(1 + \frac{r}{f}\right)^f - 1$$

$$EAR = (1 + 0.0488)^2 - 1 = 0.1 \; or \; 10\%$$

So 4.88% is the six-monthly rate of interest. If interest is to be paid quarterly, the rate of interest per quarter becomes 2.411% – that is, in the preceding calculation, n is replaced by 4:

$$EAR = \left(1 + \frac{r}{f}\right)^f - 1$$

$$\frac{r}{f} = \sqrt[4]{1.1} - 1$$

$$\frac{r}{f} = 0.02411 \; or \; 2.411\% \text{ per quarter}$$

Investments are therefore best compared by means of the annual effective rate of interest. Savings accounts and banks and all other finance companies are required in the UK to disclose this annual equivalent rate (AER) on savings and the annual percentage rate (APR) on loans and mortgages. The APR may be based on sums which include arrangement costs.

This principle leads to the conclusion that given a specific effective rate of interest per year, the effective rate per period will get ever smaller as the interest-earning period gets smaller. For many purposes, a daily rate is used, and interest is earned and credited daily. Similarly, certain loans and mortgages will have interest charged on the amount of the loan or debt outstanding from day to day or from month to month.

Example 6.26

Find the effective annual rate if the nominal rate of interest is 12% per year and interest is added monthly.

Solution

$r = 12\%$ per year
$f = 12$ months in a year
EAR = ?

$$EAR = \left(1+\frac{r}{f}\right)^{f} - 1$$

$$EAR = \left(1+\frac{12\%}{12}\right)^{12} - 1$$

$$EAR = 0.1268 \text{ or } 12.68\%$$

Excel formula	*Excel* function
=(1 + 0.12/12)^12–1	=EFFECT(nominal_rate,npery) =EFFECT(0.12,12)

Example 6.27

The APR on a credit card is 36% per year. The bank does not charge interest if the balance is paid in full by the due date each month. However, if the balance is not paid in full, interest is charged daily. What is the EAR in this case?

Solution

$r = 36\%$ per year
$f = 365$ days in a year
EAR = ?

$$EAR = \left(1 + \frac{0.36}{365}\right)^{365} - 1$$

$$EAR = 0.4331 \text{ or } 43.31\%$$

Excel formula	*Excel* function
=(1 + 0.36/365)^365–1	=EFFECT(nominal_rate,npery) =EFFECT(0.36,365)

Summary

- TVM asserts that a sum of money today is more valuable than the same amount in the future due to opportunity cost, inflation risks, and uncertainty.
- *Compounding* is the process of converting PV to FV.
- *Discounting* is the process of converting FV to PV.
- In *ordinary annuity* (annuity in arrears), payments are made at the end of the period.
- In *annuity due*, payments are made at the beginning of the period.
- The application of TVM concepts and formulae is crucial for the income approach in property valuation.
- In property valuation, we are often concerned with determining the present value of a series of rents, which, although typically paid in advance, are assumed to follow the structure of ordinary annuities.
- *Nominal interest rate* is the stated rate – it does not take the compounding periods into effect.
- *Effective interest rate* is the interest rate that considers the compounding periods.

Questions

1. How much would you have to deposit now to have \$100,000 in ten years if interest is 5% per annum?
2. You have won the lottery, and you were given two choices: to receive \$1,000,000 today or to receive \$1,200,000 two years from now. Which one should you take? Your bank pays a 5% rate of interest.
3. A UK government gilt in the amount of £10,000 will mature in five years, has no coupon payments, and carries an interest rate of 5.75%. What is the value of this gilt today?
4. If a real estate investor requires a return of 12% pa, how much would you advise the investor to pay for the right to receive \$1,500,000 in five years' time?
5. If you invest \$10,000 today, \$12,000 one year from now, and \$15,000 two years from now, how much money will you have at the end of the third year if you invest in a fund paying 5%?
6. You bought a piece of land in 2014 for \$200,000 which appreciated in value at the rate of 5% per year for the first five years and then at the rate of 6% for the next five years. Find its value after 10 years.

7. Alice is looking to set up a fund today, offering a guaranteed compounded annual rate of 6.78%. Her withdrawal plan involves taking out €10,000 in one year and €30,000 in two years before depleting the fund. How much should she invest today to meet her goal?
8. If an investor is offered a single investment opportunity which generates €150,000 at the end of each year for the next ten years and wishes to achieve a 10% return, what price should they pay today?
9. Calculate the PV of a property that generates R$200,000 in annual net rent for the first five years, followed by R$250,000 annually in perpetuity, assuming a 5.5% discount rate.
10. You estimate that £70,000 is required five years from now to refurbish your property. An annual sinking fund is to be set up that will return an interest rate of 4.32% per annum. What should be the annual sinking fund instalments?
11. You decide to invest €500 per month for ten years in a fund that earns an annual interest rate of 6%. What will be the value of the investment at the end of ten years?
12. How long will it take to accumulate $1,000,000 if you decide to invest $10,000 per year into a fund paying 9.85% per year?
13. If an investor is offered a single investment opportunity which generates R$150,000 at the beginning of each year for the next 10 years and wishes to achieve a 10% return, what price should they pay today?
14. If you invest £600 in a savings account paying 3% pa at the beginning of each year, how much will you accumulate over 20 years?
15. You are comparing two bank offers to determine which one would give you the highest return on your investment. Bank A offers an interest rate of 7% compounded monthly, while Bank B offers 7.2% compounded semi-annually. Which bank offers the higher EAR?

Notes

1 A list of Excel functions is available from Microsoft (2024).

2 $PV = \frac{FV}{(1+r)^n} = \frac{50,000}{(1+0.04)^5} = 41,096.36$

3 $PMT = FV \times \frac{r}{(1+r)^n - 1} = 50,000 \times \frac{0.04}{(1+0.04)^5 - 1} = 9,231.36$

References

Chan, W.-S. and Tse, Y.-K. (2017). *Financial Mathematics for Actuaries*. 2nd ed. World Scientific Publishing Company.
Dahlquist, J. and Knight, R. (2022). *Principles of Finance*. Houston: OpenStax.
Hussain, R. (2008). *Introduction to Finance: A Workbook*. Scranton, PA: Kania School of Management, University of Scranton.
Mazur, I. and Moshenko, K. (2021). *Business/Technical Mathematics*. Victoria: BCcampus.
Microsoft (2024). *Excel Functions*. Available at: https://support.microsoft.com/en-us/office/excel-functions-alphabetical-b3944572-255d-4efb-bb96-c6d90033e188#bm16
Wyatt, P. (2013). *Property Valuation*. 2nd ed. Chichester: Wiley-Blackwell.

Part II

Valuation methods

Chapter 7

The comparison method

Introduction

The comparison method involves evaluating the subject property by comparing it with similar, recently sold properties and adjusting for any differences. This process of adjusting based on market evidence is either explicitly or implicitly integrated into many valuation approaches, making it a cornerstone of property valuation.

Imagine you are considering buying a house. Your first step might be browsing online for properties that meet your needs. But how can you accurately determine a property's value? While online listings provide a helpful starting point, they are not always reliable; some may be priced above or, perhaps less frequently, below the market rate. So what would be an indicator of market value?

Let us say that you know similar homes on a particular street recently sold between €300,000 and €350,000. These prices provide strong evidence of the market value. However, real estate is inherently heterogeneous; no two properties are identical. Even similar houses on the same street can differ in many ways, including size, age, number of bedrooms, amenities such as gardens or swimming pools, architectural style, orientation (the direction the front of the house faces), and views. In properties occupied by tenants, such as commercial real estate, differences extend beyond physical characteristics to include the lease term (or the lease length), the tenant covenant strength, the occupancy rate in multi-let properties, the amenities and services provided for tenants, and more.

Other than heterogeneity, another challenge in real estate is the fragmented nature of information. While platforms provide details on recent transactions and market trends (Table 7.1), the data available is often limited and varies across geographies around the world. For instance, according to JLL Global Transparency Index 2024, the most transparent markets are the United Kingdom, France, and the United States. On the other hand, the least transparent countries are Iraq, Ethiopia, and Guatemala. This disparity in transparency makes it difficult for investors and stakeholders to access reliable data in certain regions, leading to potential inefficiencies in decision-making and higher risks in less transparent markets. As a result, accurate property valuation and market assessments may be skewed, particularly in regions with limited or unreliable information.

Market rental evidence is obtained from various sources, as seen in Table 7.1. The importance of each source has led the RICS to derive a hierarchy of evidence, detailed in Table 7.2. This hierarchy is divided into three categories: (a) direct comparables, (b) general market data, and (c) other sources. Category A includes recent transactions of near-identical or similar properties with information available, prioritising this as the strongest evidence possible. Asking prices and properties with active offers are also included in category A, but they only should

DOI: 10.1201/9781003490357-9

Table 7.1 Examples of online platforms for real estate information

Continent	*Country*	*Platforms*
North America	USA	Redfin, Realtor, Zillow, CoStar, Compstak
	Canada	RealMaster, Redfin, Realtor, Zillow, CoStar, Honest Door
	Mexico	Inmuebles24, MercadoLibre Inmuebles, Lamudi
South America	Brazil	Viva Real, ZAP Imoveis, OLX Imoveis, Secovi, FipeZap Index
	Argentina	Zonaprop, Properati, MercadoLibre Inmuebles, Reporte Inmobiliario
	Peru	Urbania, Adondevivir, Properati, Cámara Peruana de la Construcción
Europe	UK	HM Land Registry, Zoopla, Right Move, Zillow, CoStar, EG Radius
	Germany	ImmoScout24, Immowelt, Engel & Völkers
	France	MeilleursAgents, SeLoger, Bien'ici, Green acres
Africa	Nigeria	PropertyPro, Private Property Nigeria, Nigeria Property Centre
	South Africa	Property24, Private Property, Remax, Zoopla
	Egypt	Aqarmap, OLX Egypt, Property Finder Egypt, Coldwell Banker Egypt,
Asia	China	Anjuke, Fang, Beike, Lianjia,
	India	MagicBricks, 99acres, PropTiger
	Japan	Suumo, Homes.co.jp, AtHome, Real Estate Japan, Tokyo Kantei
Australia/ Oceania	Australia	CoreLogic, Domain, Realestate.com.au, PriceFinder
	New Zealand	Realestate.co.nz, Trade Me Property, OneRoof, CoreLogic NZ,
	Papua New Guinea	Hausples, Marketmeri

Table 7.2 RICS hierarchy of evidence

Category	*Description*
Category A: direct comparables	Direct transactional evidence, including: • Contemporary, completed transactions of near-identical properties for which full and accurate information is available; this may include data from the subject property itself. • Contemporary, completed transactions of other, similar real estate assets for which full and accurate information is available. • Contemporary, completed transactions of similar real estate for which full data may not be available but for which enough reliable data can be obtained to use as evidence. • Similar real estate being marketed where offers may have been made but a binding contract has not been completed. • Asking prices.
Category B: general market data	Data that offers guidance, rather than direct indication of value. It includes: • Information from published sources or commercial databases. • Other indirect evidence (e.g. indices). • Historic evidence. • Demand/supply data for rent, owner occupation, or investment.
Category C: other sources	Data that might provide broad indications of value, including: • Transactional evidence from other real estate types and locations. • Other background data (e.g. interest rates, stock market movements, and returns, which can give an indication of real estate yields).

Source: Adapted from RICS (2019, p. 14 and 15).

be used in markets where transactional data is inexistent, incomplete, or insufficient. If it is not possible to find category A information, category B can be used. Category B includes indirect insights through published sources, indices, historical evidence, and supply-demand data. Finally, category C, the lowest tier in the hierarchy, encompasses broad value indicators from different property types or economic data, like interest rates and stock trends.

According to the RICS guidance note, 'Comparable evidence in real estate valuation', from 2019, comparable evidence should be:

- **Comprehensive.** There should be several comparables rather than just one. Typically, valuation reports include at least five comparables in markets with ample data, ensuring a robust basis for comparison.
- **Very similar or, if possible, identical to the subject property.** The comparables should closely resemble or, if possible, be identical to the subject property. This similarity enhances the accuracy of the valuation by ensuring that the comparables reflect the same characteristics as the subject property.
- **Recent.** That is, representative of the market on the date of valuation. The comparables should be representative of the market as of the valuation date. Recent transactions provide the most relevant market conditions and price levels. As a general guideline, in markets with ample data, comparable evidence should not be older than two years.
- **The result of arm's length transaction in the market.** The comparables should result from transactions between willing buyers and sellers with no relationship to each other, ensuring the prices reflect true market conditions without any bias or external influence.
- **Verifiable.** The details of the comparables should be verifiable – meaning the transaction data can be confirmed through reliable sources.
- **Consistent with local market practices.** The comparables should align with the norms and practices of the local market, ensuring they reflect the market's specific dynamics and conditions.
- **The result of underlying demand.** The prices of the comparables should reflect genuine market demand, indicating they were not artificially influenced by temporary or atypical factors.

Applications

Whilst the comparison method is a standalone approach for property valuation, its principles and the data it utilises play a foundational role in multiple other valuation techniques. In the investment method, for example, the comparison method is used to evaluate market rent and to estimate all-risk yields and discount rates, aligning these inputs with current market conditions. Similarly, in the residual method, it may be one way of determining the gross development value (GDV), by providing market-based evidence for valuing the completed project, as will be discussed in Chapter 10. In the cost method, the comparison method is applied to assess land value.

Overall, the comparison method has two main applications: as a stand-alone method to estimate market value or market rent through direct comparison and to derive inputs, such as market rent and the all-risks yield, for other valuation methods, including investment and residual methods.

Direct capital comparison

Direct capital comparison is the most straightforward application of the comparison method. It estimates a property's market value by comparing it to similar properties that have recently sold. This approach is commonly used for real estate types with frequent transactions, such as residential properties, because it relies on actual market data to determine value. By analysing recent sales, this method provides an insight into what buyers are willing to pay for comparable properties, which is crucial for reflecting true market conditions.

The process begins by identifying properties comparable to the subject property in terms of location, size, type, and other key characteristics. The valuer then gathers sales data from these properties, ideally from the past year or two, to ensure that the information is relevant and reflects current market trends. Once comparable properties are selected, the valuer adjusts for differences, since no two properties are exactly alike. For example, if the subject property has an additional bedroom, a larger garden, or unique amenities, these factors will be adjusted in the sales data to create a more accurate comparison (California Department of Real Estate, 2010). Adjustments are necessary to account for features that either add to or detract from the value, allowing the valuer to tailor the final estimate to the subject property's specific characteristics.

The next step is to analyse the adjusted comparable to estimate the market value of the subject property. If several comparables yield similar adjusted values, the valuer can have confidence in the robustness of the estimate. Since this method is grounded in real market transactions, it provides a transparent and straightforward approach, which makes it easy to explain to clients and stakeholders. It directly reflects the current market environment and the behaviour of buyers and sellers, offering a reliable basis for valuation.

Despite its strengths, direct capital comparison has some limitations. The method requires enough comparable data, making it less effective in areas with few recent transactions or unique properties, in which finding direct comparables can be challenging. Additionally, the adjustments made for differences between properties can introduce subjectivity, as these adjustments rely on the valuer's professional judgement. While these challenges exist, the method remains one of the most widely accepted approaches for valuing residential properties and is often supplemented by other valuation techniques in more complex cases or markets with limited transaction data.

Overall, direct capital comparison is particularly well-suited to active markets, where comparable sales data is readily available (McAllister and Loizou, 2009; IVSC, 2024). Its simplicity and transparency make it a preferred method for valuing residential properties, as it provides a realistic estimate based on actual buyer and seller behaviour. In markets with consistent transaction data, this method can be a highly effective way to assess property value, though it may need to be complemented by additional methods in less active or more specialised markets.

Increasingly, the availability, credibility, and transparency of comparable data are improving with the sharing of data from actual transactions from legal sources, such as the HM Land Registry in the UK, and in the UAE the Dubai Land Department. In many countries, third-party companies gain access to the data, analyse it, and present it for valuation and other professionals. For example, the Property Monitor system in the UAE, originally developed by Cavendish Maxwell, even provides real-time price data in a ticker format similar to Bloomberg for many residential areas in Dubai. As it states on their website (https://propertymonitor.com) under the heading Comparable Data, they are 'the UAE's largest source of transferred sales, pending sales, memorandum of understandings (MOU), rental contracts, mortgage transactions, valuations, and property listing data provided in near real-time the same day the transactions occurs'.

In some cases, similar data is indirectly available through free-to-use online services, such as Zoopla (www.zoopla.co.uk) in the UK, which uses historic data from the HM Land Registry and AI algorithms to provide price estimates and instant valuations. Many countries have opened land registry data to the public, but its availability and commercialisation, such as the examples given earlier, vary from country to country.

Example 7.1

Estimate the market value of a house with a gross internal floor area of 620 m². In the same vicinity, a comparable property of 550 m² recently sold for £300,000. Another similar property, measuring 650 m², was sold two months ago for £400,000.

Solution

Comparable A = £300,000/550 = **£545.45/m²**
Comparable B = £400,000/650 = **£615.38/m²**

By calculating a price per square metre, you remove the size factor from the equation, allowing you to more easily compare properties of different sizes on a like-for-like basis. Since the subject property is between the sizes of Comparable A and Comparable B, we can interpolate to estimate the value.

Linear interpolation formula:

$$y = y_1 + \frac{(y_2 - y_1)}{(x_2 - x_1)} \times (x - x_1)$$

Where:

- y_1 and y_2 are the known values at x_1 and x_2.
- x is the value at which you want to interpolate.
- y is the interpolated value at x.

$$y = 545.45 + \frac{(615.38 - 545.45)}{(650 - 550)} \times (620 - 550)$$

y= **£594.40/m²**

Using this interpolated rate, we can estimate the market value for the subject property:

$$\textbf{\textit{Market value}} = 594.40 \times 620 = £368,529$$

This example demonstrates a quantitative adjustment, where only the differences in property size were considered, with all other factors (such as quality, condition, and amenities) assumed to be identical. However, qualitative adjustments can also be applied. For example, if one of the houses has air conditioning and this feature typically attracts a 1% price premium in the market, an adjustment can be made to reflect this. It is important to emphasise that quantitative adjustments should be based on a careful analysis of market evidence.

Market rent

Where the *income approach* is to be used, whether in relation to tenant-occupied property or owner-occupied property, an initial step is to estimate the market rent or current rental value of that property. The assessment requires analysis of current rents being paid, rents being quoted, and the vacancy rate of comparable properties.

Detailed analysis of the letting market must precede analysis of a specific letting. Thus, if it can be demonstrated within the market for office space that users will pay a higher rent for space on the higher floors than for space on the ground floor, then this may be reflected in the way in which a letting of a whole office building is analysed in terms of rent per square foot. Continued analysis will indicate whether or not bids are influenced by, for example, the quality and location of entrances and common areas (providing toilet, vending, and catering facilities); the provision of air conditioning, car parking, security, accessibility (in some cases to facilitate 24/7 working practices), lifts, communal facilities, recreation, leisure, food, banks, cash points, first aid, design and layout, floor plate and structural grid layout, and flexibility; and other matters such as Internet access and speeds, energy-efficiency, and increasingly, the provision of advanced amenities, including co-working and health and well-being facilities, including gyms and relaxation areas.

Reliable estimates of current rents can only flow from analysis of rents actually being paid, and under no circumstances should a valuation be based on an estimate of rent derived from analysis of an investment sale where there is no current rack rent passing. This is because rent analysed in this manner depends upon the assumption of a capitalisation rate. For example, if the only information available on a transaction is that a building of 5,000 ft^2 has just been sold for $500,000 with vacant possession, the only possible analysis is of sale price per square foot.

In every valuation, the principal terms and conditions of the lease must be noted. The following key terms can affect the level of negotiated rent:

- Total length, or lease term
- Rent reviews and frequency
- Payment frequency, for example, monthly or quarterly in advance
- Incentives, for example, rent-free periods, capital inducement
- Responsibility and liability for repairs, insurance, and management
- Service charge and shared facility costs and management operations
- Restrictions on use and opening hours
- Alienation (the right to assign or sublet the whole or part)

Market rental evidence is gathered from various sources, as outlined in Table 7.1, and the comparables, along with key factors to consider, can be summarised in a rent comparison schedule, such as the one shown in Table 7.3. Most valuation reports include a comparison schedule to transparently inform clients about the comparables used, the rationale for their selection, and the sources of information.

Table 7.3 presents an example of a market rent comparable schedule. The subject property in this fictitious example is a vacant ground-floor office unit with 15,000 ft^2, located in submarket Y, and the valuation date is 16 November 2024. The building is three years old and has achieved the highest Platinum rating in the US Green Building Council rating system, LEED. Table 7.3 meets the requirement for thorough comparables by including at least five examples from the same submarket (Y).

The 'Address with picture' column is useful to provide the reader with the location of the comparables, along with some hints about the quality of the building. Although images are not included here, they are common in most valuation reports. The 'Date of transaction' column ensures the evidence is current (i.e. two years from the valuation date), while other columns, such as 'Tenant', 'Lease information', 'Size', and 'Comments', allow valuers to demonstrate

Table 7.3 Example of a rent comparison schedule

Address with picture	*Date of transaction*	*Tenant*	*Lease length/break option/ repairing terms/incentives*	*Size (ft²)*	*Rent ($)*	*Rent per ft² ($/ft²)*	*Comment*
105 Capitol Road, **submarket Y**	11 November 2024	Tenant A	10 years, with a mutual break in year 3; full repairing and insuring; 1 year rent-free	14,000	450,000	32.14	Located on the 5th floor of a 5-year-old building, rated LEED **Platinum**
45 Violet Street, **submarket Y**	16 September 2023	Tenant B	8 years, with a break option in year 4; internal repairing; 6 months rent-free	12,500	397,000	31.76	Situated on the ground floor of a 10-year-old building, rated LEED **Gold**
32 Wall Lane, **submarket Y**	23 June 2023	Tenant C	5 years, with no break option; full repairing and insuring; no incentives	11,000	395,000	35.91	Located on the 3rd floor of an 8-year-old building, unrated
67 Thistle Street, **submarket Y**	7 April 2023	Tenant D	7 years, with a break option in year 5; full repairing and insuring; 3 months rent-free	13,000	350,000	26.92	Positioned on the 2nd floor of a recently refurbished 13-year-old building, rated LEED **Platinum**
88 Anemone Plaza, **submarket Y**	28 December 2022	Tenant E	10 years, with a mutual break in year 3; full repairing and insuring; 9 months rent-free	16,500	411,000	24.91	Located on the ground floor of an 8-year-old building, Unrated

similarities and highlight any differences between the comparables and the subject property that may impact valuation. These details are crucial for assessing the appropriateness of the comparables, ensuring the valuation is both accurate and well-supported.

The comparable schedule is typically followed by an in-depth analysis comparing the selected evidence to the subject property. This analysis involves examining both the quantitative data (e.g. rent per square foot, lease duration, floor area) and qualitative aspects (e.g. building quality, amenities, location, and tenant profile). The goal is to identify how each comparable property aligns with or diverges from the subject property, considering specific characteristics that might justify differences in rent levels. The estimated market rent for the subject property is then derived based on the evidence from these comparable properties and market analysis. The analysis, along with the estimated market rent, is typically presented as a rationale within the valuation report, detailing the relevant market context and the comparable evidence considered.

Example 7.2

Your client wants to let a warehouse with two docks built in 1960 which has a NIA of 10,000 ft^2 and is currently vacant. Give your opinion as to the warehouse's market rent, considering the following comparables, all located in close proximity to the subject property:

Comparable	*Information*
A	12,000 ft^2 NIA, 2 docks, modern building, recently let at **\$9.50/ft^2**
B	9,000 ft^2 NIA, 1 dock, built in the 1970s, let a year ago at **\$7/ft^2**
C	8,000 ft^2 NIA, built before the 1940s, with no docks, let two years ago at **\$5/ft^2**
D	Recently refurbished warehouse with 3 docks, with a NIA of 15,000 ft^2, let six months ago at **\$8.50/ft^2**
E	A warehouse built in the 1960s with a NIA of 11,000 ft^2 and two docks, let one year ago at **\$7.50/ft^2**

Indicative answer

Step 1: Analyse the differences and similarities between each comparable and the subject property.

Comparable	*Comment*
A **\$9/ft^2**	This property shares the same number of docks as the subject property. Despite being 2,000 ft^2 larger, potentially leading to a rent discount due to economies of scale, it is worth noting that the comparable is a modern building, which tends to command higher rents.
B **\$7/ft^2**	This property has a similar age as the subject property. While the smaller size might trigger a higher rent per square foot, the property has been let a year ago and has fewer docks than the subject property; thus, expected is a higher rent per square foot for the latter.

(Continued)

(Continued)

Comparable	*Comment*
C **$5/ft²**	Comparable C is considerably older and lacks docks, which make it less desirable compared to the subject property. Despite being smaller, the lack of docks and the much older age suggest that the subject property should command a higher rent per square foot.
D **$8.5/ft²**	Comparable D is larger than the subject property, with one extra dock, and has recently been refurbished, indicating a better condition. The fact that it was let six months ago is relevant as it reflects current market conditions. Given that the subject property is not refurbished and is smaller, it is expected to achieve a rent slightly below this comparable.
E **$7.5/ft²**	This is the most directly comparable property as it is of a similar age and size and has the same number of docks as the subject property. The slight difference in size suggests that the subject property might achieve a rent per square foot close to this, but possibly slightly higher, as the subject property is smaller, and the lease for Comparable E is from one year ago.

Based on the analysis of the comparables, a reasonable market rent for the subject property would likely be in the range of \$7.50 to \$8.00/ft². Given that Comparable E is the most similar, and considering the subject property's slightly smaller size, along with recent market conditions, a rate of \$8.00/ft² is deemed appropriate.

Consequently, the market rent for the subject property is calculated at

$8/ft² × 10,000 ft² = $80,000

Headline rents versus net effective rents

Generally, rents will be analysed based on both headline rents – that is, the contracted gross rent – and net effective rents – that is, the contracted gross rent receivable, considering any rent-free periods or other tenant incentives (e.g. capital contribution). To move from headline to net effective, the RICS guidance note suggests four methods:

1. Straight line
2. Calculation reflecting time value of cash flow utilising a money discount rate
3. Comparison by reference to the effect on investment value
4. Discounted cash flow calculation reflecting time value of cash flow utilising equated yields (see following text) and explicit rental growth rates

In practice, most valuers use the straight-line method. This is because the length of lease, or the period over which the incentive acts as an incentive to sign up and take occupation, is frequently for five years or less, which means that there is minimal variation in respect of each basis of analysis, it is easier to use, and it reflects the view of most tenants as to the effect the incentive has on the rent they have agreed to pay. Many lettings will have a three-month fitting-out period as standard. If this is a typical market allowance, then it will be excluded in the net effective rent calculation.

Example 7.3

Using the straight-line method, assess the net effective rent for a five-year term at £300,000 pa, with a one-year rent-free period, which includes a three-month fitting-out period.

Basically, the total rent payable over the five years is calculated and then spread over the actual rent-paying period, and so:

Headline rent	£300,000
Payment period	4.25 years (i.e. 5 years less 9 months [0.75 of a year] as 3 months [0.25] is rent-free)
Total rent payable	£1,275,000
Spread over 5 years	÷ 5
Net effective rent	**£255,000**

An important issue with all such inducements is assessing the period of time to which the inducement applies. The authors' opinion is that it should apply only to the period up to the first full rent review, and for accuracy, a discounting approach should be adopted to reflect the time value of money. But for now, the straight-line market approach will be used.

In the analysis of both headline and net effective rent, valuers work on a net basis, that is, net of operating costs. So the first step is to analyse the lease terms to determine whether the rent is simply rent or whether it includes an element of operating or running costs.

Landlord's expenses

Where investment property is already subject to a contracted lease rent and/or where it is customary to quote rents for a specific type of property on gross terms, it is necessary to deduct a landlord's outgoings (operating expenses) before capitalising the net operating income (net income for valuation purposes). The valuation must be based on net income, that is, net of all the normal expenses that the owner of an interest in that property is required to meet out of the rents received from ownership of that interest in that property other than tax.

Expenses may be imposed upon the landlord by legislation, or they may be contractual, as in an existing lease. An inspection of the property may suggest that, though neither party is statutorily or contractually liable, there are annual expenses that will have to be met by the owner of that interest in the property. The valuer must identify all such liabilities and make full allowance for them in the valuation. To do this, reference must be made to all existing leases in respect of the property.

The principal items of expenditure can be broadly classified under the headings of Insurance, Management, Taxes, Running Expenses, and Repairs. Of these, only the cost of complying with repairing obligations should cause any real difficulty in accurate assessment.

Insurance

The valuer, armed with plans and detailed measurements of the building, will be able to estimate or obtain an accurate quotation for all insurances, particularly fire insurance. Fire insurance is an extremely involved subject, complicated by the variation in insurance policies offered. A valuer should therefore be acquainted with the terms and conditions of policies offered by two or three leading insurance companies and should always assess the insurance premium in accordance with those policies. If reinstatement value (Chapter 11) is required, this can be referred to a building surveyor or, preferably, a quantity surveyor or may be based on adjusted reinstatement cost figures extracted from current architects' and builders' prices, or in some cases, use can be made of the RICS Building Cost Information Service (BCIS). Due to the 'averaging provisions' of most policies, it is better to be over-insured than under-insured and, hence, to overestimate rather than underestimate this item. Insurance cover for all third-party liabilities, such as the potential accident or injury to visitors on or about the property or injury to the general public which might be caused by falling masonry or roof tiles, is usually the occupier's responsibility, but again, cover is essential, and this may have to be arranged by a landlord and recovered from the tenants in a multi-let property. Quotations for other insurances necessary on boilers, lifts, etc. can always be obtained from a broker or insurance company.

A deduction for insurance is only necessary if the lease has no provision for the recovery of all insurance charges from the tenant in addition to the rent. For instance, in the UK, the wording of a 'full repairing and insuring lease' leaves the responsibility for insurance with the tenant, but a 'full repairing lease' will leave the responsibility with the landlord, but with provision for the premium to be recovered as rent from the tenant, and hence, payment is enforceable in the same way as rent. In the larger multi-let property, the landlord or owner retains the responsibility for insurance but recovers the cost on a proportionate basis from the tenants; this will include any residual insurance for common or shared parts.

Management

Management refers to the property owner's supervising costs equivalent to the fee that would be due to a management agent for rent collection, inspections of the premises, and instructions to builders to carry out repairs. It is frequently considered to be too small to allow for in the case of premises let to first-class tenants on full repairing and insuring terms. The issue of licences to assign or sublet and to carry out alterations is subject to additional fees payable by the tenant. Valuers must use their experience and follow market norms, but it is suggested that if the valuer's firm would charge a fee for acting as managing agents for the property being valued, then any owner of the property will incur a similar cost, and a deduction for management should be made.

The fee may be based on a percentage of rents and service charges collected or on a negotiated annual fee. The valuer will be aware of the appropriate market adjustments to make, based on current fees charged by managing agents. Where VAT is payable on rents, it will impose an additional management cost, which will be reflected in the fees negotiated by managing agents. Where the management fee is recoverable as part of a full service charge, no deduction need be made in a valuation of the landlord's interest.

Taxes

The Uniform Business Rate (UBR) is paid by the occupier of business premises in the UK; similar taxes based on the annual value of a property are to be found in most countries or states. Inspection of the lease will indicate who has contracted to meet the 'rates' demand. If premises are let at an inclusive rental (i.e. inclusive of UBR), then rates should be deducted. If the property is let exclusive of the UBR, then no deduction need be made from the rent. If the letting is inclusive and there is no 'excess rates clause', then the deduction must represent the average annual figure expected up until the end of the lease or next review, not a figure representing current rates. If an excess rates clause is included, then the sum to be deducted is the amount of rates due in the first year of the lease, or the lowest sum demanded during the current lease. This is because such a clause allows for the recovery from the tenant of any increase in rates over and above the amount due in the first year of the lease.

Other 'community and government-based outgoings' may include water consumption, drainage/sewage charges, and charges for environmental and other purposes, most of which are payable by the tenant in most situations, either directly or through a service charge or management agreement in a multi-tenanted situation.

It has always been the custom in the UK to value before deduction of tax, on the grounds that income and corporation tax are related to the individual or company and not to the property. However, if *market value* implies the most probable selling price, there is automatically an implication of the most probable purchaser. The valuer without this knowledge cannot be held to be assessing market value; thus, it is held by some that the tax liability of the most probable purchaser should be reflected in the valuation. Market valuations are based on before-tax income, and so no allowance for tax will be made in this book unless specifically stated.

Running expenses

Where the owner of an interest is responsible for the day-to-day running of a property, such as a block of flats, a serviced office building, or a modern shopping centre, then a deduction from rent may be necessary to cover the cost of items such as heating, lighting, cleaning, security, and general staffing. Current UK practice is to include provision for a separate service charge to be levied to cover the full cost of most running expenses.

An inspection of the lease will indicate the extent to which such expenses are recoverable. Older leases tend to include partial service charges, in which case the total income from the property should be assessed, and the total cost of services deducted.

All the items falling under the general head of 'running expenses' are capable of accurate assessment by reference to current accounts for the subject property; by comparison with other properties; by enquiry of electricity, gas, and oil suppliers; by enquiry of staff agencies; and so on. The valuer operating within a firm with a large management department is at a distinct advantage as that department should be able to provide fairly accurate estimates based on detailed analyses of comparable managed property.

Where a full service charge is payable by the tenant, this is usually adjustable in arrears. In other words, the annual service charge is based on last year's expenses. During periods of rapidly rising costs, the owner will have to meet the difference between service cost and service charge. This will lead to a reduction in the owner's cash flow, and this amount may need to be considered in a valuation.

Increasingly, these cost issues are being met by more complex service charge clauses and schedules which provide for interim increases, for example, to cover the uncertain energy element in running expenses. (The statutory requirements relating to service charges and management charges in residential property are complex. Readers involved or interested in the residential sector must refer to all current housing and landlord and tenant legislation.) Allowances for running costs will frequently play an important part in the construction of a cash flow for DCF valuation or analysis. But the critical first step is identifying liability for these and determining whether they are recoverable in full or in part.

Repairs

The lease(s) will indicate those items of repair that have to be met by the landlord out of rental income. The sum to be deducted from an annual income before capitalisation must be an annual averaged figure. Thus, liability to redecorate a building every five years should be estimated and averaged over the five years. If the cost of redecoration is recoverable by a direct proportionate charge to tenants, then no deduction need be made.

A check should be made against double accounting – if the cost of repairs to boilers, lifts, etc. is covered by a service charge, then no allowance should appear under repairs for such items. But it may be necessary to include an annual deduction by way of a sinking fund or capital provision fund to build up reserves to cover the cost of, say, the replacement of the air conditioning plant after ten years. Where an allowance must be made, every effort is required to estimate the amount as accurately as possible. An excess allowance will lead to an undervaluation, and an inadequate allowance to an overvaluation.

Advice, if needed, should be obtained from builders and building surveyors as well as by comparison with other known repair costs for comparable managed property.

Essential works

Any obvious immediate renewals or repairs as at the date of the valuation should be allowed for by a deduction from the estimate of capital value to reflect the cost. If on a valuation inspection serious structural or other major defects are suspected, further investigation, including structural surveys, may be needed to assess the true impact of the problem(s) detected.

Averages and percentages

Valuers should use averages with care. An example of this is the use of average heating costs per square metre. The valuer is required to value a specific property, not an average property. The requirement is to estimate the average annual cost of heating that specific property. Some valuers, texts, and correspondence courses suggest that it is a reasonable approach to base insurance premiums and repair costs on a percentage of market rent. This approach is not recommended unless the valuer can prove that the figure adopted is correct. A percentage allowance may be widely inaccurate; a high street shop could be let at £100,000 pa, and a factory could be let at £100,000 pa, but the latter will almost certainly cost more to keep in repair. Two properties may let at identical rents, but one may be constructed of maintenance-free materials, the other more cheaply constructed with short-life material. A thousand square metres of space in Oxford Street, London, and in Exeter, Devon, may cost the same to maintain but could have very different market rents. A large old building may let at the same rent as a small modern building, but the repairs will differ significantly.

Similar points may be made in respect of the use of percentages to estimate insurance premiums. Students may be permitted to use percentages in class examples if the data needed for accurate assessment is not provided by their tutors or readily determinable. For insurance, this might be

0.5–1% of market rent; for external and internal repairs, a figure of 10% might be adequate, possibly split 6% and 4% between external and internal. Management fees might be taken at 1–5% of market rent, but a percentage should never be taken of the rent collected. A figure of 10–15% could be used as the fee for managing a full service charge for, say, a shopping centre. Students must reflect on the reality of the figures they calculate when using percentages. For example, inadvertently, by using a percentage, the management cost of collecting and accounting for a multi-million-pound rent and nothing else could cover the annual costs of a property management department. How many staff would a 5% management fee on a £10,000,000 rent roll cover? For this reason, in practice, all allowances should be based on current costs and fee rates and not on percentages.

The rent, whether headline or net effective, having been adjusted for owners' (landlords') expenses, can now be expressed as a rent per square foot/metre. The area of the property used for this stage is the area required to be used under IPMS and/or local measurement practices or legal requirements. The basis must be consistent within a market to allow for comparison, which is one of the main forces behind the need for an international measurement standard.

Net operating rent and units of comparison

As previously discussed, net operating rent (also known as net rent, or NOI) is calculated by deducting all the landlord's expenses from the gross rent. Once the net rent or NOI for the property is determined, it can be expressed, for comparison purposes, in terms of a suitable unit of comparison. Thus:

- Agricultural land rent per hectare or rent per acre
- Offices rent per m^2/ft^2 NIA/NIA or per m^2/ft^2 IPMS 3.2
- Factory rent and warehouses per m^2/ft^2 GIA, or per m^2/ft^2 IMPS 2
- Shops rent per m^2/ft^2 overall or per m^2/ft^2 in terms of Zone A (ITZA) (see following text)

With the introduction of IPMS, it is essential to make specific reference to the basis of the area measurement used, for example, IPMS 2 or IPMS 3 for office. For some time to come, areas of offices might be given in terms of IPMS, but rental analysis might continue in the UK to be analysed in terms of NIA. Other local practices may continue until the IPMS is fully adopted, and whilst a valuer should be aspiring to use IPMS, they should also be aware of continuing local practice.

Analysis of rents to arrive at a comparable rent per square metre per square foot is a relatively simple exercise of dividing the total rent by the appropriate total area.

Retail rent

In the retail market in the UK and some other countries or states, the property industry has developed the concept of analysing the rent in terms of zones of activity. Three alternative approaches have been developed for analysing shop rent:

- Overall analysis
- Arithmetic zoning
- Natural zoning

Overall analysis

The rent for the retail space is divided by the lettable rental space to obtain an overall rent per square metre per square foot. This is a simple approach but is complicated by the retail leasing practice of letting the retail space together with space on upper floors used for storage, sales, rest rooms, offices, or residential accommodation at a single rental figure. In these circumstances, it is desirable to isolate the rent for the retail space.

Some valuers suggest there is a relationship between ground-floor space and space on the upper floors, but this is only likely to be the case where the use is the same or ancillary (e.g. storage). Here, custom or thorough analysis will indicate the relationship, if any, between ground-floor and upper-floor rental values. Where the use is different, the rent of the upper floor should be assessed by comparison with similar space elsewhere and deducted from the total rent before analysis. Thus, if the upper floors comprise flats, then the rent for these should be assessed by comparison within the residential market and then deducted from the total rent.

Overall analysis tends to be used for shops in small parades and for large space users. Dealing with standard shop units, for example, within a shopping mall, the unit of comparison can often be left as the 'shop unit'. Rent will be a factor of location/position and not size. In the case of large space users, it is reasonable to argue that tenants of such premises will pay a pro rata rent for every additional square meter per square foot up to a given maximum. The problem here is that what one retailer might consider is a desirable maximum could be excessive for all other retailers. Such a point should be reflected as a risk accounted for by the valuer in the consideration of a capitalisation rate or discount rate.

Arithmetic zoning

This approach is preferred in many cases to an overall rent analysis because, in retailing, an assumption is made, namely, that the space used to attract the customer into the premises is the most valuable (i.e. the frontage to the street or mall). Again, the rent for the retail space should be isolated before analysis.

Where arithmetic zoning is used, it is custom and practice to use zones categorised from **A** (the front, most valuable area) to **D** (the back of the shop), each with a depth of 6 or 6.1 metres. For units deeper than 24 or 24.3 m, any remaining area is classified as **R** (remainder). To give greater weight to the space at the front of the shop, the area in Zone **A** remains constant, while each subsequent zone is halved in size. This zoning process is used to calculate the area in terms of ITZA.

Calculating the ITZA allows for a 'Zone A' rent to be a standardised metric for comparison. The zone fractions are as follows: **A (1), B (1/2), C (1/4), D (1/8), and R (1/16).**

Upper and lower floors, as well as non-sales areas, are not zoned but assigned different values based on their function. Common ratios include:

- First-floor sales: 1/6, 1/8, or 1/10
- Basement or lower ground sales: 1/10
- Ground floor ancillary or storage areas: 1/40

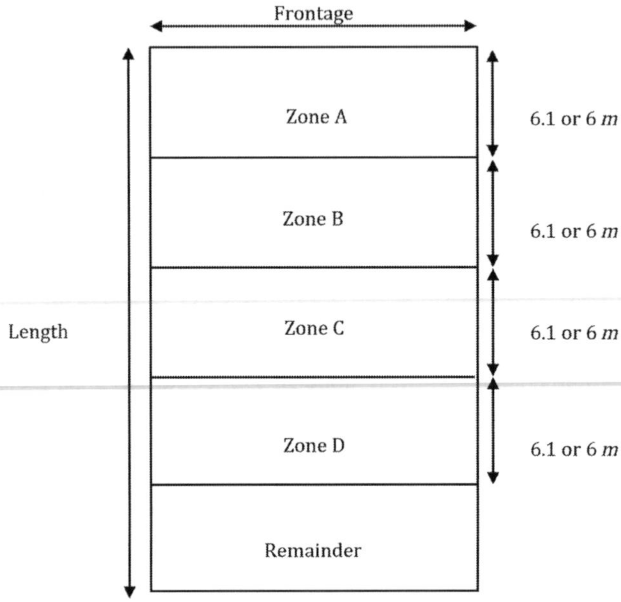

Figure 7.1 Graphic explanation of the zoning.

Example 7.4

Analyse the rent of £50,000 being paid for shop premises with a frontage of 6 m and a depth of 21 m.

Overall:

£50,000 ÷ 126 (6 × 21) = £396.82/m^2

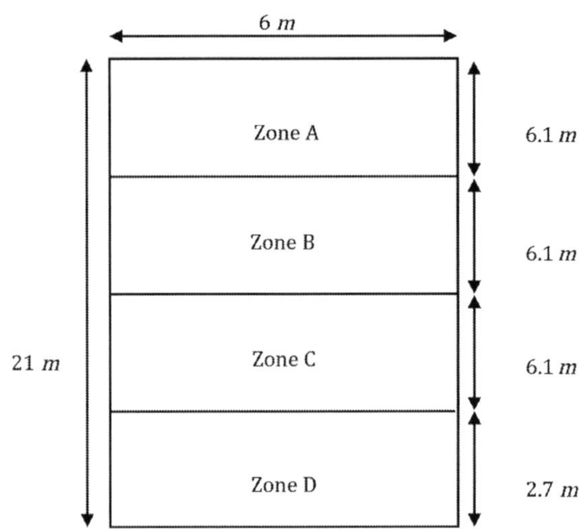

Figure 7.2 Zoning analysis for example 7.4.

Zoning: Assume zones of 6.1 m depth and £x per square meter rent for zone A.

Zone A = 6 × 6.1 = 36.60 m^2/1	= 36.60 m^2 ITZA
Zone B = 6 × 6.1 = 36.60 m^2/2	= 18.30 m^2 ITZA
Zone C = 6 × 6.1 = 36.60 m^2/4	= 9.15 m^2 ITZA
Zone D = 6 × 2.7 = 16.2 m^2/8	= 2.03 m^2 ITZA
Total	66.08 m^2 ITZA

£50,000 ÷ 66.08 m^2 ITZA = **£756.72/m^2 ITZA**

Example 7.5

A retail unit was recently leased for €250,000 per annum. The unit has a rectangular shape with a clear internal floor area and is a single-storey structure. Its frontage measures 10 m, with a floor depth of 27 m, plus an additional 7 m of storage space at the rear.

Using this market evidence, calculate the market rent value (MR) for the adjacent shop, which shares similar characteristics except for its size. The adjacent shop has a frontage of 8 m, a floor depth of 20 m, and an additional 5 m of storage at the rear.

Solution

Step 1: Analyse the comparable evidence.

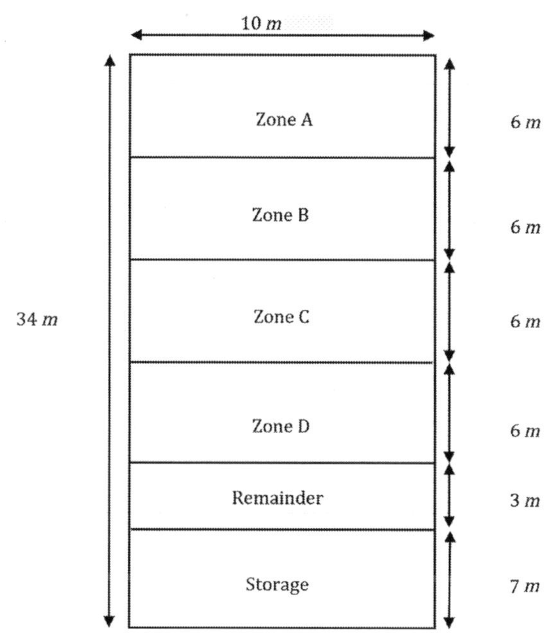

Figure 7.3 Zoning analysis of the comparable evidence.

Zone	*Frontage length*	*Depth*	*Area (m^2)*	*Area ITZA (m^2)*
A	10	6	60	60
B	10	6	60	30
C	10	6	60	15
D	10	6	60	7.5
R	10	3	30	1.875
Storage	10	7	70	1.75
Total			**340**	**116.125**
Rental value: €250,000/116.125 = **€2,152.85/m² ITZA**				

Step 2: Analyse the subject property.

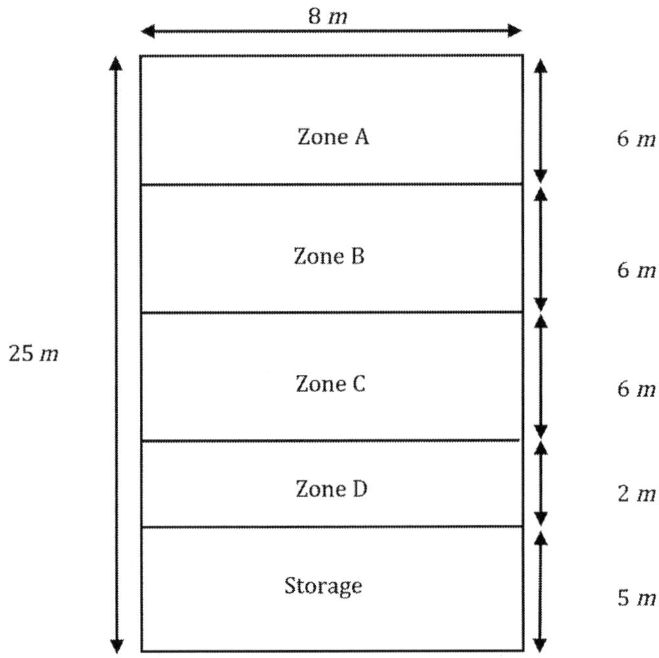

Figure 7.4 Zoning analysis of the subject property.

Zone	*Frontage length*	*Depth*	*Area (m^2)*	*Area ITZA (m^2)*
A	8	6	48	48
B	8	6	48	24
C	8	6	48	12
D	8	2	16	2
Storage	8	5	40	1
Total			**200**	**87**

Step 3: Calculate the market rent, considering the rent per square meter ITZA from Step 1 and the area, in square meters, ITZA from Step 2.
Market rent for subject property: €2,152.85 × 87 = **€187,297.95**

Analysis of market rents will reveal a pattern, which should reflect the footfall of shoppers. There is likely to be a high spot, with rents diminishing the further one moves away from the highest footfall.

The retail space could be divided into any number of zones. Different regions and different retail centres display different trading patterns, and whilst the convention is to use zones from A to D, other practices will be found, and thus it is important to check the local market convention (Scarrett, 2008). Custom and practice do not necessarily reflect market behaviour, and adapting different zones and zone depths for analysis can produce considerable variations in opinion as to the ITZA rent for a given location. Currently, the Valuation Office Agency uses Zone A rents for rateable value calculations; in so doing, they converted the historical use of 20 ft zones to the nearest centimetre and rounded to 6.1 m for zone A and zone B; 6.1 m has therefore become the industry standard.

In business, retailers do not see the premises divided into rigid zones. Every rental estimate must be looked at in light of the current market and common sense – Who would be the most probable tenant for premises 7 m wide by 100 m deep? Is there any user who operates in such space, or is the last 50 m waste or valueless space for most retailers in that locality?

Upper floor and basement sales space is included in the analysis. The relationship with Zone A must be determined by considering the level of sales or density of shoppers to be found on the first-floor and basement areas compared to the ground floor; this might be 1/6 or 1/8 of Zone A. In the case of corner properties where there is a double frontage, zoning will be based on the most valuable frontage, and a market addition will be made should the corner frontage add any value. For example, a corner shop unit on the corner of two busy shopping malls will have added value, but in a secondary position where the side frontage is onto a residential cul-de-sac, there will be little or no added value.

Having established market rents in terms of Zone A, the valuer can now assess the market rent for other shops in the same locality. Market experience will help the valuer assess the prime position in a mall or retail street and to determine prime Zone A and the rate at which prime Zone A diminishes the further the unit under consideration is from the prime spot. Prime is often where there is the highest footfall or movement of shoppers.

A note of caution here that retailing is constantly evolving, and in many cases, the use of storage within the unit is diminishing, with a potential impact on the methodology for zoning. This is because many retailers adopt a 'brick and click' approach, with the ability to order online to the store or to home. In the case of a large shopping mall, such as Meadowhall in the UK, stock can be ordered from a nearby warehouse within 20–30 minutes where an item is not available in store. Equally with the adoption of more experiential retailing where the purpose of the store is evolving, the traditional zoning methodology may not be fully applicable anymore.

Example 7.6

Estimate the market rent for a shop in a high street location of a market town. Zone A rents on ten-year full repairing and insuring (FRI) leases, with a single five-year rent review, at **£500/m² ITZA**. The shop trades on the ground and first floor and measures 8 m × 24 m. The first floor has a retail area of 150 m². First-floor retail space, by comparison, is 1/8 of Zone A. This market rent calculation could be set out as:

Zone	*Area*	*Area ITZA*
A	8 × 6.1 = 48.8	48.8
B	8 × 6.1 = 48.8	48.8/2 = 24.4
C	8 × 6.1 = 48.8	48.8/4 = 12.2
D	8 × 5.7 = 45.6	45.6/8 = 5.7
First floor	150	150/8 = 18.75

Total = 109.85 m² ITZA.
Market rent = £500/m² ITZA × 109.85 m² **ITZA** = **£54,925.**

Natural zoning

This method is rarely, if ever, used and can only effectively be used to analyse rents within a shopping street or centre where information is available on a number of units, as it requires comparison between units. As previously explained, the rents for retail space must be isolated from the rent for the premises as a whole. Natural zoning assumes a base rent is payable for a given area of space, and that the extra rent paid for larger areas represents the additional rent for the extra floor area. It is difficult to use as an ongoing basis of analysis. This technique was originally detailed in Emeny and Wilks (1982), as it is not in common use and only of academic interest; the reader seeking further information is referred to the fifth edition of *The Income Approach.*

In terms of the rental analysis of retail space, warehouse accommodation, and factory space, the valuer needs to have a sound knowledge of the specific requirements of different retailers and of different manufacturers.

An alternative technique, although not yet found to be any more satisfactory than those listed, is multivariate analysis. Here the dependent variable rent is considered against several independent variables which could include size, location, distance from car parks, bus station, etc.

Estimating rental comparables in practice

It would be ideal if all rental estimates could be based on true comparables, that is, those of the same size, design, facilities, location, etc. This is rarely the case in practice, however; with experience, adjustments can be made for some of these variations, but wherever possible, estimates of rent should be based on close comparables.

As far as office records are concerned, strict procedures should be adhered to so that the format of data is consistent. In this respect, it is recommended that all rents be analysed net of landlord's outgoings (see earlier), and that apart from obvious factors, such as the address of the property, the record should contain details of facilities included (e.g. central heating, air conditioning) and the lease terms.

The valuer must be in possession of all the facts of a given letting before any rental figure can be analysed. In the case of residential and commercial property, statute intervenes to protect the tenant in several ways. Thus, a tenant of business premises who carries out improvements would not expect to pay an increase in rent for these improvements on the renewal of the lease. The valuer must therefore know the basis upon which a given rent was agreed upon before he or she can make proper use of the information. An earlier lease might have been surrendered, or the tenant might have undertaken to modernise the premises. Either of these could have resulted in a rental lower than a market rental agreed upon between the parties. Of increasing concern is the lease definition of rent, where there is provision for a rent review – there is almost certainly a difference between 'full market rent' and a 'reasonable rent', the latter inferring something less than the maximum rent achievable if offered in the open market. The RICS has advised that where rental evidence is obtained from other valuers, it should be provided on a consistent basis.

It is arguable that the rent agreed upon for a 20-year lease without review will differ from that agreed upon for the same lease with five-year reviews. Whether these rents would be higher or lower will depend upon whether the rental market is rising or falling. For preference, comparability should entail estimating rents from analyses of lettings with similar review patterns. Where this is not possible, the valuer will have to exercise professional judgement in making adjustments from the comparable evidence before using the evidence to estimate the current market rent of the subject property.

The need in a weak letting market for landlords to offer inducements, such as fitting out costs and rent-free periods, has been considered earlier and simply re-emphasises the need for valuers to be in full possession of all the facts before reaching any conclusion on market rents. Sometimes, intricate deals to secure a tenant with incentives or create a situation to lengthen the lease, thereby impacting the total average lease length of a multi-let property, may be subject to a non-disclosure agreement, and care must be taken in establishing whether published headline rents are what they seem.

A concern in some markets is that market evidence is based on quarterly in advance rent payments, but in an increasing number of cases, tenants are paying rent on a monthly in advance basis. There are two issues here: The first is simply the added cost of accounting on a monthly basis, that is, 12 times a year rather than 4 times a year, and the second is the question of equivalence between quarterly payments and monthly payments. Again, this can be resolved using discounting techniques. These points of principle are considered in more detail in Chapter 3.

As noted earlier, rents agreed will also be affected by a number of other lease clauses, including the ability to transfer the lease/license, user, repair, and service charges. Particular attention should be paid to those dealing with alienation, that is, the ability to transfer a lease or licence agreement to another party. In some parts of the world, this transfer is called an assignment, where the tenancy/licence is completely transferred, or a subletting, where a sub-lease or licence is created below the original parties. Although in countries such as the UK, legislation such as the Landlord and Tenant Act 1988 ensures that landlords do not unduly delay the granting of consents where permission is conditional upon granting of approval, the position in relation to an absolute prohibition on alienation remains unchanged.

Where there is an absolute prohibition of transfer, the risk to the tenant is higher as it gives no flexibility in options to exit other than at the end of the lease. Therefore, this may have an adverse effect on rents, as tenants will price in this extra risk. User restrictions in shopping centres and multiple-tenanted commercial properties are common, as in a retail environment the landlord wants to control and optimise the tenant mix to maximise the total rental return.

The tenant may also take comfort from some restriction on competition which flows from a well-balanced tenant mix. In this case, it can have a beneficial effect on rents.

In different parts of the world, when a lease is fully transferred or assigned, the original party may not fully relinquish their responsibilities. Privity of contract exists in some countries, including the UK; it is a common-law doctrine that prevents a person who is not party to a contract from enforcing a term of that contract, even where the contract was made for the purpose of conferring a benefit on the third party. In the UK, The Landlord and Tenant (Covenants) Act 1995 has amended the law on privity of contract. It allows landlords to define in a lease the qualifications of acceptable assignees, and if a transfer is made to a party deemed acceptable, the continuing legal liability of the original lessee is removed. Whilst 'reasonableness' remains a factor, some leases will be seen to be more or less onerous than others depending on how an acceptable assignee is defined.

The valuer has now established the current net income and/or the market rent net of landlord's expenses; the next step is to assess all risks yield.

All risks yield (ARY)

The ARY reflects the return required by investors to compensate for all risks associated with a property and is derived by analysing the net initial yields (NIYs) of comparable property transactions. The NIY is calculated as the ratio of a property's net income to its purchase price and is typically found in real estate sales data.

To assess the appropriate ARY, the valuer must know what return on capital investors are looking for when buying comparable property. The following summary list is not comprehensive but is indicative of the factors that a valuer must consider when assessing an appropriate yield:

- General level of interest rates
- State of the national and local economies
- Government financial and fiscal policies
- Location of the property in terms of labour, transport, and general and special accessibility
- Legal factors, such as planning, highways, landlord and tenant legislation, health and safety, disability requirements, fire, working conditions
- Rental growth prospects
- Capital growth prospects
- Security of rental income, tenants' covenant, unexpired lease term, rent reviews
- Taxation of income and capital, including, for example, in the UK, VAT; Stamp Duty Land Tax (SDLT), which in other countries may also be called deed taxes; and real estate transfer taxes, which range, as of 2024, from 2% in Poland to 12.25% in Belgium
- Liquidity issues, to include not only the cost of disposal but, more importantly, also the time it might take to realise the value of the property
- Volatility of the market
- Management costs

Most valuation reports include a comparison schedule that clearly outlines the comparables used, the reasoning behind their selection, and the sources of information. Table 7.4 presents such a schedule for ARY estimation using the same subject property as in the market rent example.

The 'Date of transaction' column ensures the evidence is current (i.e. two years from the valuation date), while other columns, such as 'Tenant', 'Lease information', 'Size', and 'Comments',

Table 7.4 All risks yield comparison schedule

Address (usually provided with an image)	*Date of transaction*	*Tenant(s) and covenant strength*	*Lease length/ break option/ repairing terms/ WAULT/ULT*	*Size (sf)*	*Price ($)*	*Price per square foot ($/ft²)*	*Net initial yield (%)*	*Comment*
10 Shamrock Rd, **submarket Y**	11 December 2023	Tenant A (strong covenant).	Lease expires in 2033. Mutual break in year 2026, FRI.	14,000	4,500,000	321.43	6.5	Located on the ground floor of a 5-year-old building with a LEED **Platinum** rating.
45 Peony St, **submarket Y**	16 September 2024	Tenant B (moderate covenant) occupies 50% of the building, and Tenant C (moderate covenant) occupies 50% of the building.	WAULT is 5.8 years. Break options were not exercised, and there are none left. Both tenants are in FRI leases.	12,500	3,970,000	317.60	7.0	Ground floor of a 10-year-old building rated LEED **Gold**.
47 Oaktree Rd, **submarket Y**	15 June 2024	Tenant D (poor covenant)	Lease expires in January 2025.	13,000	3,500,000	269.23	7.2	Ground floor of a 13-year-old building; good state of repair; LEED **Gold** rating.
80 Hibiscus Plaza, **submarket Y**	07 June 2024	Vacant	–	16,000	2,900,000	181.25	8.3	Ground floor of a 7-year-old building; good state of repair; **NOT LEED-RATED**.
100 Sunflower Rd, **submarket Y**	10 October 2024	Vacant	–	15,500	2,700,000	174.19	8.5	Ground floor of a 10-year-old building; poor state of repair; **NOT LEED-RATED**.

allow valuers to demonstrate similarities and highlight any differences between the comparables and the subject property that may impact valuation. These details are crucial for assessing the appropriateness of the comparables, ensuring the valuation is both accurate and well-supported.

As with the comparable schedule for market rent, the ARY comparison schedule is typically followed by an analysis comparing the selected evidence to the subject property.

Adjusting comparables

Once all relevant data is collected and collated, the valuer with experience will need to adjust the comparable evidence, be it of rent or capitalisation rate or whatever, to arrive at the subject property's market rent, net operating income, and capitalisation rate.

In support of any adjustments that might be needed to the market evidence, the valuer will consider the property and its specific risks in terms of five measures of quality:

- Quality of the legal title
- Quality of the lease(s)
- Quality of the location
- Quality of the building
- Quality of the tenant(s) in occupation or those that can be assumed in the case of a new or owner-occupied property

While these five dimensions are the focus of the assessment, the IVSC (2024) emphasises the need to analyse and adjust for any significant difference between the comparable evidence and the subject property. Beyond these five measures of quality, IVSC (2024) also identifies historical and expected growth and profitability or profit-making capability of the property as further sources of variation. Dunn (2004) further points to the nature of financing, the motivation behind the sale, capital expenditures made immediately after purchase, and prevailing market conditions at the time of the transaction as important areas of comparison. While a broad range of factors may influence the assessment, the following discussion focuses on the five measures of quality, as they are often the most apparent sources of difference between the comparable evidence and the subject property.

Quality of the legal title

Appropriate checks must always be made regarding the quality of the legal title. In the UK, Canada, and India, for example, in most cases, the title will be either a freehold or a leasehold, but other titles, such as commonhold (in the UK), condominium ownership (in Canada), and co-operative housing societies (in India), are also possible. In many European countries, dominant is the concept of a usufruct, which is a legal right accorded to a person or party that confers the temporary right to use and derive income or benefit from someone else's property. This limited right can be found in many mixed and civil law jurisdictions.

Checks must be made (for instance, on the UK Land Registry site) to see if there are any restrictive covenants which might restrict the use of the property, for example, land to be used for educational purposes only, or the density of development, which could be restricted to protect an adjoining property, rights of way, rights of light, and where let, all leases should be read and any risks evaluated.

Quality of the lease(s)

Investors would prefer long leases of 15–20 years, with regular rent reviews or indexed rents and which have a medium to long unexpired term. The trend in the market has been for a shorter lease of 10 years or less. In fact, in many parts of the world, this has reduced further, with 5- and 3-year leases being quite common. Table 7.5 illustrates how, even pre and post the COVID-19 pandemic, average lease length continues to reduce.

Table 7.5 Average office lease length **(months)** comparison

Quarter	*United Kingdom*	*Australia*	*New Zealand*
2019 Q1	52	41	43
2023 Q1	34	29	27
Percentage change	**−34%**	**−28%**	**−37%**

Source: Re-leased (2024); Market Lens (2023).

The question of unexpired term has evolved significantly over the past 20 years as the property market has evolved, and in office markets, the COVID-19 pandemic accelerated the growth of flexibility and hybrid working and reduced lease lengths further. Taxation changes brought about by International Accounting Standard (IAS) 17 also fuelled the drive to flexibility, co-working, and ever-shortened leases.

This has driven a fundamental shift in the office market from passive management driven by long leases with little or no contact between landlord and tenant (unless something went wrong) to highly engaged and amenitised spaces with short leases and multiple revenue options from additional services. Or as Steven Skinner, CEO of HB Reavis, was quoted as saying, 'there has been a pivot from what I would call the lazy wholesaling of office space to the retailing of office space'. Putting that into practice, HB Reavis developments feature co-working and many other amenities to promote tenant retention. All this changes the focus from long stable cash flows to more volatility, which developers are managing by focusing on what tenants require and providing the amenities and added value services that will foster tenant retention. This has gone so far as the asset manager of the iconic ICD Brookfield Place in Dubai stating that what they do is 'curating experiences for [their] tenants'.

Nevertheless, in the case of single-tenanted buildings, it is important to assess the unexpired lease term (ULT), that is, for how long this single tenant will remain in the property. In the case of multi-tenanted buildings, it will be important to check both the unexpired lengths and the quantity of space occupied by each tenant. So for example, if there are four tenants and one occupies 50% of the space and the other three share the balance, then there will be a significantly different assessment of tenant security if the single tenant has a lease of, say, seven years to run than if that tenant only had two years. For this reason, valuers consider very carefully the WAULT and weighted average unexpired lease term certain (WAULTC) as well as the actual occupation terms.

A significant risk of managing and owning property is vacancy risk. No tenant means no income and, in most cases, increased landlord or owners' holding costs. This risk is represented by the relationship between a property's (or portfolio's) unexpired contracted rental payment in proportion to the percentage of the total rent roll. WAULT is a calculation representing the remaining contractual fixed lease payments of a building (or portfolio) divided by the total contracted annual rent of the building (or portfolio) at a specific point in time; for a valuation, this is as at the date of valuation.

A simple and popular way to calculate WAULT is to:

- Calculate each tenant's current contracted rent as a percentage of the total passing rent.
- For each tenant, multiply this factor by the tenant's unexpired lease term in years.
- Add the figures for all the tenants together to find the WAULT.

Interpretation of WAULT varies from investor to investor; as a general rule, a WAULT of less than four years would be considered to be 'short-term income', four to eight years as 'medium-term income', and over eight years as long-term income. So given a matching pair of office buildings in the same location, one with a WAULT of 3.25 and the other with a WAULT of 9.2, investors would prefer the latter and would bid slightly more for it. Hence, when adjusting market yields for valuation purposes, knowledge of the respective WAULT figures can be important. If the comparable yield was 6% from the 3.25 WAULT, then, all other things being the same, the yield or capitalisation rate for a valuation of the 9.2 WAULT building might need to be lower, at, say, 5.8%.

A similar metric is to weight each tenant's income stream within a building based upon total rentable area. For example, a building has the following profile:

Tenant 1: Occupies 5% of rentable area, and the lease expires in 1.5 years.
Tenant 2: Occupies 80% of rentable area, and the lease expires in 9 years.
Tenant 3: Occupies 15% of rentable area, and the lease expires in 1.25 years.

The weighted average lease to expiry is:

$(0.05 \times 1.5) + (0.8 \times 9) + (0.15 \times 1.25)$, which gives **7.46 years**.

These measures must be viewed with all the other information. Here, the WAULT is 7.46 years, but two leases (20%) expire within 1.5 years.

In the US and APAC markets, WAULT is better known as weighted average lease expiry (WALE), and where leases have break clauses, WAULTC is used. The calculation is the same as WAULT/WALE, but the unexpired term in the UK is taken to the tenant's break option.

The longer the unexpired term, the better investment the property will be seen as, all other issues remaining the same. But if the property was one suitable for refurbishment, conversely, a shorter WAULT might be preferred.

The terms and conditions in respect of repairs, insurance, and management should be such as to leave the rent payable as a net income to the landlord. For an individual building, a lease on FRI terms is preferred. In multi-tenanted properties, a service and maintenance charge should be included to provide full recovery of all expenses, including the associated costs of management. The rent review should be upward only, index-linked or stepped. The user clause should not be overly restricted, as this can impact on market rents. Assignment or subletting should only be with the landlord's consent, which should not be unreasonably withheld. Alterations and improvements should only be with the landlord's consent. Properties that are let on lesser terms will be less desirable even though in the valuation the actual cost of any impediments can be calculated and adjustments in value made.

Quality of the location

Similar properties can have very different values due simply to location. In office markets, this may include agglomeration factors, where some companies want to be close to competitors, for example, banks in Canary Wharf in London. In other cases, retailers will want to be located away from direct competition, unless, like Aldi in the UK, they want to be placed close to, say, Sainsbury's to generate comparative and complementary shopping. Location is also subject to changing preferences and change; in recent decades, large financial and professional firms have favoured prime locations, such as Wall Street in New York, the City in London, or La Défense in Paris. But with changing employment, younger generations of employees, and hybrid working, evidence from Urbanite suggests that areas that are more vibrant, especially in the evening, such as Shoreditch in London and Soho in New York, are sometimes being favoured over traditional locations. Valuers need to be aware of these changing preferences in demand.

Environment checks must be made to check for flood risks and water run-off, contamination risks, crime levels, access, parking, and any other factor which might impact on the suitability of the location for the permitted use.

Quality of the building

The valuer must be able to assess the quality of the building in terms of its construction and suitability for tenants seeking buildings for that permitted use. New buildings are likely, due to recent changes to building regulations and carbon footprint requirements, to be built to a higher occupation standard and to be greener. Older buildings may have poor energy ratings, may be costly to maintain, and may not be suitable for today's market due to floor-to-ceiling heights or room areas. Space requirements change over time, particularly for retailers; typically, storerooms are no longer required, and the space needs to be sufficient for all stocks to be on display. As part of the building inspection, notice will have been taken as to whether the building meets with the requirements of building regulations, health and safety, and disability compliance. A particular concern to valuers is the need to identify the possibility of the property containing deleterious materials and/or of the site or buildings being contaminated. Note will also be taken as to whether all required certificates are current, such as fire certificates, energy certificates, air conditioning equipment or boilers, lift inspection certification, etc. A photographic record of the building and its surroundings is actively encouraged. The valuer's market knowledge is essential in being able to assess the property's fitness for purpose. Tenants' demands change over time in respect of space requirements and space performance. Valuers will classify different properties according to the extent to which they meet current best build standards from an occupier's perspective.

As we explore in Chapter 13, published credentials, especially environmental standards, such as the UK-originated BREEAM, the US-based LEED, or the Australian Green Star, but also WELL Building Standard Certification and even a building's WiredScore, are increasingly impactful on a tenant's decision-making to occupy a building and therefore have implications for value. In many countries, public occupiers are required to occupy buildings above a stated rating – meaning a huge market is deselected from lower-rated buildings, seriously impacting value. Many private organisations also have benchmarks and aspirations to make sure they occupy buildings above a certain rating, again impacting the available tenants for a property and, hence, value.

Quality of the tenant(s)

Where tenanted, the valuer will be checking the tenant's covenant strength (effectively, the risk of default) or credit rating. It is important to judge whether the tenant(s) are capable of paying the rent and of meeting their other obligations under the lease. Checks can be made through Dun and Bradstreet, Experian, banks, and other suppliers. Other checks will include, in the UK, Company House to examine annual reports and accounts and, in the case of quoted companies, share value and share movement. Tenants can be public limited companies (PLCs), limited liability companies (LLCs), partnerships, individuals, government departments, or charities, and within each sector and sub-sector, there will be tenants deemed to be of sound covenant and others where there will be added risk of non-payment. Many countries have similar, often online facilities to examine company accounts and the status of both PLCs and LLCs, such as the Bundesanzeiger Verlag in Germany and the Australian Securities & Investments Commission (ASIC). The 2007/2008 credit crunch produced a strong warning not to assume anything about a tenant without very careful analysis.

The factors which make a building and location good for a tenant will also make it attractive to an investor. RICS Global Valuation Practice Guidance – Applications (VPGA) 8 specifies a minimum of matters to be considered by a valuer in the process of the valuer's investigation. VPGA 8 provides some idea of the mass of information that needs to be collected and assessed in order for a valuer to arrive at an objective opinion of the property's marketable qualities, which, in turn, will affect the valuer's judgement as to the income-generating capabilities of the property and its suitability as an investment. As previously detailed, the inspection will include accurate measurement in accordance with IPMS or other country or practice standards.

Not all the answers come from the inspection, although good observation during an inspection will alert the valuer to many potential issues which will need to be reviewed in formulating an opinion of value. As part of the process of inspection and investigation, the valuer will undertake desk searches of historic uses, searches of the environmental agency's sites, and will check planning and highway matters with the relevant authorities. The actual use or uses within a property may not always comply with current planning consents.

In terms of the economy, the valuer will be considering the national and local economic picture. Looking at the economic base of the area, employment and unemployment data, wages, population structure, growth prospects, stock of similar property, vacancy rates, rent levels, and investor interests all will be considered within the context of the property market and the alternative competing investment markets.

All this data is collected for analysis if considered significant in terms of value. The valuer needs to know what other properties are on the market; whether they are better in terms of location, etc.; who is in the market as potential buyers and as potential tenants; and whether the market for the subject property is active.

To collect the level of information required entails either very sound local knowledge or the need to make enquiries of the local planning authorities, rate collecting authorities, highway authorities, transport companies, local census statistics, and a range of other sources, including issues such as crime rates.

The valuer will look at the capitalisation rates or initial yields achieved on sales of comparable properties, then make brief notes as to the quality of the title, building, location, tenant. Lease terms will be noted, and the best comparables, if not known to the valuer, may need to be viewed, albeit from the outside only. The valuer is looking for close matches and so will reject evidence which needs substantial adjustment on every comparator, but provided they are basically similar, an adjustment, based on experience, will be made up. These adjustments will be in the range of 0–0.5%. It is suggested that if a larger adjustment is needed, then the valuer may need to look for other comparables. If none are available, then the report will need additional commentary on the lack of comparable data.

Voids

Where there is an over-supply of space within a given area, the probability of voids occurring when leases terminate is increased. If a single investment building is let in suites to a number of tenants, voids may occur sufficiently regularly for a valuer to conclude that the average occupancy is only, say, 80%. In such cases, having estimated the market rent, the figure should be reduced to the level of the most probable annual amount. Voids are less likely to affect operating expenses, so a pro rata allowance on operating costs should not be made. Indeed, in some exceptional cases, it may have to be increased to allow for empty rates, non-recoverable service costs, and additional security.

In the case of a building let to a single tenant, a void is only likely to occur at the end of the current lease or if there is a break clause. If this is a reasonable expectation, then the income pattern must be assumed to be broken at the renewal date for the length of time considered necessary to allow for finding a new tenant. Additional allowances may have to be made for negative cash flows if empty rates and other running expenses must be met by the owner during this period.

The time taken to renew leases to current tenants and to negotiate rent reviews can represent a significant loss to a landlord. New leases will contain 'interest' clauses to cover interest lost on rent arrears and delayed settlement of reviews and renewals.

In other cases, the valuer may wish to adjust rental income down to reflect the fact that agreement is unlikely to coincide with the rent review or lease renewal date.

Purchase expenses

It has long been valuation practice when giving investment advice to include in the total purchase cost an allowance to cover stamp duty, solicitors' fees, and any other expenses to the purchaser occasioned by the transaction, including VAT. This in total is currently assumed to come to 6.80% of the purchase price, but for many investors with internal teams of lawyers and surveyors, it might be a lot less.

It should be remembered that analysis of a transaction might also take these expenses into account to reveal the buyer's yield on total outlay rather than the yield realised on the contract price alone.

It is current practice in investment valuation work to express value net of these sums, but market value is normally expressed without deduction for selling expenses.

The key role of the comparative method in other valuation methodology methods

It is important before moving on to the other main methods of valuation to pause and reflect on the significant contribution that the comparable method makes to many forms of valuation.

As previously stated in this chapter, the comparison method involves comparing the subject property with similar, recently sold properties and then adjusting for differences between the properties. This principle of making adjustments based on empirical data is used, implicitly or explicitly, in many other valuation contexts. Therefore, the comparison method becomes a foundational tool (***cornerstone***) in property valuation. Whilst primarily involving the juxtaposing of the subject property against similar, recently sold properties to derive value, its influence extends well beyond this simple application.

The comparison method not only stands as a distinct approach but also weaves its principles seamlessly into various other valuation methods, enhancing their accuracy and relevance. This involvement highlights the method's significance, showcasing its versatility and integral role in the broader spectrum of property valuation. In the following discussion, we delve into the pivotal role the comparative method plays in complementing and reinforcing other prevalent valuation techniques.

1. *Income capitalisation approach.* Even when using the income approach, one might need to refer to comparative data to estimate market rent, especially if the subject property or its units are vacant or if lease agreements are considered below or above market rates. It is also used to estimate yields, growth, percentage of expenses, and discount rates.
2. *Cost approach.* Whilst this method is based on the cost of constructing a replica of the subject property, the value of the land on which the property has been constructed is often determined using the comparative method by looking at recent sales of similar vacant land plots. Of course, the estimated replacement also involves comparison.
3. *Residual and development appraisal methods.* When determining the GDV, essentially the value of a developed property, the comparative method should be used. Comparing the potential property with recently sold developed properties can help in estimating the GDV. Absorption rates, discount rates, developer's profit percentage, and development costs are also the outcome of comparison.
4. *Investment method.* When assessing the market value of an investment property, the comparative method can be used to gauge the current market sentiment, with the key parameters being market rents, yields, vacancy rates, cap rates, and discount rates, even if the primary valuation is based on cash flow projections.
5. *Profits method.* Similar to the investment method, with key parameters depending on the type of business under valuation, such as average daily rates, occupancy rates, etc.

Last but not least, the comparison method can be used for validation and for establishing base rates.

Validation. Even if primary valuation is conducted using another method, valuers often cross-check their results using the comparative method as a sanity check, ensuring that their valuation is in line with recent market transactions.

Establishing base rates. For methods that rely on rates such as capitalisation rates or discount rates, the comparative method can assist in establishing a baseline by analysing transaction data and extracting implicit rates from market comps.

The importance of data in property valuations

Professional valuers are constantly confronted with changes in the way they work. For example:

- What are the big changes today?
- How are trends impacting markets?

- How does a geopolitical shock impact the property sector?
- How does a disruptive new developer shake up an existing submarket?

This is all about big data

Big data is a broad term that encompasses the ongoing upheaval in analytics. In the past, the challenge was that there was a lack of good data, and it had to be presented with interpretation of patterns. Today, on the contrary, the challenge is how to sort, analyse, and draw conclusions from the mass of readily available but not always credible data. This new situation requires a rethinking of data and data analysis.

As a key takeaway of this chapter, it must be clearly understood that ***data*** plays the most significant role in the valuation process, especially, but not exclusively, in respect of the comparative method, and will impact valuations in the following ways:

1. *Accuracy and reliability.* At the heart of the valuation process is the pursuit of accurate and reliable results. The accuracy of a valuation is heavily dependent on the **quality and relevance of the data used**. For instance, when using the comparable sales method (a common approach in property valuation), it is essential to have access to recent, local, and relevant sales data to make accurate comparisons.
2. *Standardisation and consistency.* Institutions such as RICS, IVS, TEGoVA, and the Appraisal Institute in the United States emphasise the importance of consistency and standardisation in the valuation process. To achieve this, valuers require **a consistent set of data** which is processed in a standardised manner. This ensures that different valuers or firms can come to relatively similar valuation conclusions given similar data.
3. *Market trends and analysis.* The property market, like any other, is subject to fluctuations and trends. **Historical and current data provides a lens** into the market's direction, enabling valuers to adjust their assessments based on market conditions. For instance, in a market showing a consistent upward trend in property values, a valuation would be adjusted accordingly.
4. *Risk management.* **Good data allows valuers to understand the risks associated** with a property. Factors like location, historical price volatility, and local market conditions can provide insight into the potential risks and rewards of a property investment.
5. *Compliance and regulation.* As professionals who follow and adopt certain standards and ethical considerations, valuers need to ensure their assessments are defensible and compliant with relevant guidelines. The IVS, for instance, sets out how valuations should be conducted on an international scale. Using **comprehensive and reliable data ensures that a valuation adheres to these standards**.
6. *Informed decision-making.* Beyond the valuers, the end users of valuations (like investors, lenders, and property owners) rely on these assessments to make informed decisions. **Accurate and comprehensive data ensures that these stakeholders can trust the valuation as a foundation for their decisions**.
7. *Technological evolution.* With advancements in technology, particularly in the realms of **big data and artificial intelligence**, the valuation process is becoming more sophisticated. This new wave of tech-driven valuation tools requires a **robust dataset** to operate effectively.
8. *Diversity of data sources.* The property market is affected by a wide array of factors, from macroeconomic indicators to local infrastructure developments. A valuer with a broad background understands the importance of **sourcing data from a diverse range of channels** to get a comprehensive view of the property's value.

9. *Transparency and trust.* In today's information-driven world, stakeholders often demand transparency in the valuation process. By **showcasing the data sources and methodologies** used, valuers can enhance trust in their assessments.
10. *Adaptability.* The property market is always evolving. **New data allows valuers to remain adaptable** and to update their methodologies and perspectives in light of changing circumstances.

> The paradox here lies in the dual role of quality data. While robust datasets empower valuers to enhance the precision and consistency of their assessments, they simultaneously pave the way for automated valuation models (AVMs) that can perform portions of their routine tasks. Conversely, when the data is subpar, not only do valuers struggle to produce reliable valuations (however, daily work increases due to the high demand for valuations), but AVMs also become inoperative.

Data acts as the backbone of the property valuation process. It is not just about numbers; it is about understanding market dynamics, managing risks, and ensuring that valuations are both accurate and reliable. An expert valuer should always rely on quality data to uphold the highest professional and ethical standards in their practice.

How data is utilised in each primary valuation method

Method	*Data needed*	*How data affects it*
Comparative sales method (or sales comparison approach)	Recent sale prices of comparable properties (comps), features of comps (e.g. square metres, number of rooms, location), date of sale, and terms and conditions of sales.	The value of a property is directly compared to recent sales of similar properties. The quality, relevance, and how recent the comps are can significantly affect the valuation.
Income capitalisation or the income approach (investment)	Potential rental income, operating expenses, vacancy rates, yields, rental growth, and capitalisation rate. Expected future cash flows and a discount rate to calculate the present value.	This approach determines the value of a property based on the income it can potentially generate. Accurate data on potential rent, expenses, and prevailing cap rates ensures the valuation is reflective of the property's income potential. Accurate cash flow projections and appropriate discount rates ensure a reliable valuation.
Cost (contractor's) approach	Cost to build a replica of the subject property, current value of the land, and depreciation since the construction.	This method values a property based on how much it would cost to build an identical or equivalent property. Up-to-date construction cost data, land values, and accurate depreciation rates are crucial.

(Continued)

(Continued)

Method	*Data needed*	*How data affects it*
Residual (development appraisal) method	Value of the completed development, costs of development, and profit margin expected by developers, absorption rates, professional fees, contingencies, and others.	Used mainly for properties with development potential, this method determines the value of a property (land) based on the end value minus costs and required profit. Accurate projections and cost estimates are pivotal.
Profit method	Gross income from the business, operating expenses, capitalisation rate, and analyses of similar business for proper rating and adjustments.	This approach is used primarily for properties like hotels or pubs, where the property and business are intertwined. Reliable income statements and operating cost data are essential.

To summarise the preceding list, someone could say that *having the most recent and relevant data is crucial for ensuring that valuations reflect current market conditions*. Consistent and comprehensive data sources boost the reliability of valuations, enabling stakeholders to make informed decisions based on the assessment.

In a professional arena that requires trust, showcasing the data and methodologies used can enhance credibility. Transparent data sources and clear methodologies can reduce disputes and disagreements. International and local valuation standards often require valuers to base their judgments on concrete evidence. Proper data ensures compliance with these standards.

Last but not the least, with consistent data over time, valuers can identify trends in the property market. This can be essential for forecasting or understanding market cycles as well.

Using regression analysis to interpolate data to estimate comparable pricing

The following step-by-step example will guide you through the process of using a simple regression model in Excel (or other applications) to model prices based on collected data.

Example 7.7 Using regression analysis

The scenario

You are a real estate analyst tasked with developing a model to estimate apartment prices in a seaside resort in Cyprus in a specific neighbourhood. You have collected data on 20 recently sold apartments (see following), including their internal area, covered veranda area, uncovered veranda area, floor level, age, and number of parking spaces.

The following data has been collected from completed transactions using online access to the national land registry database.

Flat	*Internal area (m^2)*	*Covered verandas (m^2)*	*Uncovered verandas (m^2)*	*Floor*	*Age (years)*	*Parking spaces*	*Price (€1,000)*
1	80	10	5	2	5	1	200
2	90	12	8	3	2	2	250
3	75	8	3	1	10	1	180
4	100	15	10	4	1	2	320
5	85	10	5	2	7	1	220
6	95	12	8	3	4	2	280
7	70	8	3	1	12	0	160
8	110	18	12	5	3	2	350
9	90	12	8	3	6	1	240
10	85	10	5	2	9	2	230
11	75	8	3	1	15	1	170
12	105	16	11	4	2	2	330
13	95	12	8	3	8	1	260
14	80	10	5	2	11	0	190
15	115	20	15	6	1	3	380
16	90	12	8	3	5	2	270
17	85	10	5	2	14	1	210
18	70	8	3	1	18	0	150
19	100	15	10	4	4	2	300
20	95	12	8	3	7	1	250

The task

1. *Perform a regression analysis.* Using a statistical software package like Excel, R, or Python, perform a multiple linear regression analysis, with 'Price' as the dependent variable and the other characteristics as independent variables.
2. *Interpret the results.* Examine the output and identify:
   - The coefficients for each independent variable
   - The intercept
   - The R-squared value
   - Discussion of the statistical significance of the coefficients
3. *Develop a valuation model.* Construct a regression equation using the coefficients and intercept.
4. Estimate a property value: Use the model to estimate the value of an apartment with the following characteristics:

Internal area	**85 m^2**
Covered verandas	**12 m^2**
Uncovered verandas	**0 m^2**
Floor	**3**
Age	**9 years**
Parking spaces	**1**

The solution

Step 1: Prepare the data

- *Open Excel.* Start with a fresh Excel spreadsheet.
- *Enter the data.* Input the data for the 20 apartments. Make sure each feature (internal area, covered verandas, etc.) has its own column with a clear header.

Step 2: Activate the data analysis tool ToolPak

- Go to **Tools** > **Excel Add-Ins** and you should have **Analysis ToolPak** available as an option.
- Check the box next to **Analysis ToolPak** and click **OK**.

Step 3: Perform the regression analysis

- Go to the **Data** tab and click **Data Analysis**.
- Select **Regression** from the list and click '**OK**'.
- **Input Y range.** Select the entire column of 'Price (€)' data (including the header).
- **Input X range.** Select all the columns with the apartment features (Internal area, Covered verandas, etc), including the headers.
- Check the '**Labels**' box since we have included the headers.
- **Choose an output option.** You can select '**New Worksheet Ply**' to have the results displayed on a new sheet.
- Click **OK**.

Step 4: Generate the output.
Excel will generate a table with the regression results.

Step 5: Interpret the output.

- *R square.* This tells you how well the model fits the data. A higher value (closer to 1) means a better fit.
- *Coefficients.*
- *Intercept.* This is the estimated price when all other features are zero. It might not have a practical interpretation in this context. The intercept is −94.49684839. As mentioned earlier, this might not have a practical interpretation in this context since it represents the predicted price when all features are zero.
- *The coefficients for each feature.* These tell you how much the price changes for a one-unit increase in that feature, holding other features constant.
- *Internal area* (m^2). The coefficient is 2.799261814; this means that for each additional square metre of internal area, the price of the apartment is expected to increase by approximately €2.80, holding all other features constant.
- *Covered verandas* (m^2). The coefficient is 9.722348283; this indicates that each additional square metre of covered veranda space adds about €9.72 to the price.

SUMMARY OUTPUT

Regression Statistics	
Multiple R	0,994923072
R Square	0,989871919
Adjusted R Square	0,983963872
Standard Error	8,238629926
Observations	20

ANOVA

	df	*SS*	*MS*	*F*	*Significance F*
Regression	7	79605,49972	11372,21	167,5464	5,265E-11
Residual	12	814,5002766	67,87502		
Total	19	80420			

	Coefficients	*Standard Error*	*t Stat*	*P-value*	*Lower 95%*	*Upper 95%*
Intercept	-94,49684839	54,07474896	-1,7475226	0,1060622	-212,3156052	23,32190841
Flat	0,054441631	0,564397355	0,0964598	0,9247477	-1,175274567	1,284157829
Internal Area (sq.m)	2,799261814	0,764160667	3,6631849	0,0032472	1,13429875	4,464224878
Covered Verandahs (sq.m)	9,722348283	4,834289061	2,0111227	0,0673274	-0,810662746	20,25535931
Uncovered Verandahs (sq.m)	2,137338734	6,569408807	0,3253472	0,7505209	-12,17617346	16,45085092
Floor	-17,1033622	18,09552582	-0,9451708	0,3632148	-56,53012603	22,32340162
Age (years)	-1,273893987	1,251993065	-1,0174928	0,3289953	-4,00175254	1,453964566
Parking Spaces	11,57010751	4,323189012	2,67629	0,0201748	2,150687821	20,98952719

RESIDUAL OUTPUT

Observation	*redicted Price (*	*Residuals*	*dard Residuals*
1	208,402628	-8,402628049	-1,28335
2	260,5948279	-10,59482785	-1,61817
3	181,5297205	-1,529720473	-0,23364
4	306,3085834	13,69141664	2,091123
5	220,0689157	-0,068915671	-0,01053
6	272,2611155	7,738884523	1,181978
7	153,6332824	6,366717553	0,972404
8	348,3095402	1,690459831	0,258188
9	244,3102358	-4,310235819	-0,65831
10	229,3634434	0,636556641	0,097223
11	175,5957836	-5,595783588	-0,85466
12	331,3262185	-1,326218512	-0,20256
13	255,9765234	4,023476559	0,614515
14	189,8968978	0,103102174	0,015747
15	385,5581867	-5,558186704	-0,84892
16	257,5353287	12,46467127	1,903759
17	211,8049573	-1,804957336	-0,27568
18	146,5887765	3,411223532	0,521004
19	303,3035259	-3,303525868	-0,50456
20	257,6315088	-7,631508847	-1,16558

PROBABILITY OUTPUT

Percentile	*Price (€)*
2,5	150
7,5	160
12,5	170
17,5	180
22,5	190
27,5	200
32,5	210
37,5	220
42,5	230
47,5	240
52,5	250
57,5	250
62,5	260
67,5	270
72,5	280
77,5	300
82,5	320
87,5	330
92,5	350
97,5	380

Figure 7.5 Excel output for example 7.7.

- **Uncovered verandas** (m^2). The coefficient is 2.137338734, suggesting that uncovered verandas also contribute positively to the price, but to a lesser extent than covered verandas.

- *Floor.* The coefficient is −17.1033622; this negative value indicates that, in this dataset, apartments on higher floors tend to have lower prices. This might be due to specific preferences in this market or other factors not captured in our data.
- *Age* (years). The coefficient is −1.273893987. As expected, older apartments tend to be less expensive. Each additional year of age decreases the price by about €1.27.
- *Parking spaces.* The coefficient is 11.57010751. This indicates that having more parking spaces increases the price, with each additional space adding approximately €11.57.
- *P-values.* Most of the features have p-values less than 0.05, indicating that they are statistically significant predictors of price. This means their impact on price is likely not due to random chance. The exceptions are 'flat' and 'uncovered verandas', which have p-values greater than 0.05, suggesting they might not be as important in determining price in this specific dataset.

Step 6: Building the price prediction equation.
Based on the coefficients, the equation for predicting apartment prices is:

Price = −94.49684839 + (2.799261814 × Internal Area) + (9.722348283 × Covered Verandas) + (2.137338734 × Uncovered verandas) + (−17.1033622 × Floor) + (−1.273893987 × Age) + (11.57010751 × Parking spaces)

Step 7: Testing the model.
Now we must test the model with the same apartment we used before:

Internal area	**85 m²**
Covered verandas	**12 m²**
Uncovered verandas	**0 m²**
Floor	**3**
Age	**9 years**
Parking spaces	**1**

Entering these values into the equation:

Price = −94.49684839 + (2.799261814 × 85) + (9.722348283 × 12) + (2.137338734 × 0) + (−17.1033622 × 3) + (−1.273893987 × 9) + (11.57010751 × 1)

This gives us a predicted price of approximately €208,402.63.

Discussion

How well does the model explain the variation in apartment prices?

Quite well, actually! The R-squared value of 0.99 (or 99%) indicates that our model explains 99% of the variation in apartment prices. This is a very high value, suggesting that the features we included (internal area, covered verandas, etc.) are strong predictors of price in this dataset.

What are the limitations of this model?

Even with such a high R-squared, it is essential to acknowledge the limitations:

- *Limited dataset.* Our model is based on a relatively small dataset of only 20 apartments. A larger and more diverse dataset would provide more robust and generalisable results.
- *Omitted variables.* Many other factors that we have not included in this model can influence apartment prices. These could include the following:
  - *Location.* Proximity to amenities, schools, parks, and transportation.
  - *Condition.* The overall state of repair, recent renovations, and quality of construction.
  - *View.* Apartments with desirable views might command higher prices.
  - *Market conditions.* Economic factors, interest rates, and supply and demand dynamics in the local market.
  - *Specific features.* Unique characteristics, like balcony size, floor within a building (e.g. penthouse vs ground floor), and the presence of a fireplace or other amenities.
  - *Linear assumption.* Linear regression assumes a linear relationship between the features and the price. In reality, these relationships might be more complex.
  - *Multicollinearity.* There might be some correlation between the independent variables (e.g. larger apartments might also have larger verandas). This can sometimes affect the stability and interpretation of the coefficients.

What other factors might influence apartment prices in this neighbourhood?

In addition to the omitted variables mentioned earlier, consider these factors that could be specific to this neighbourhood:

- *Neighbourhood characteristics.* Safety, noise levels, access to green spaces, and the overall desirability of the neighbourhood.
- *Building amenities.* Presence of a gym, swimming pool, concierge service, or other shared amenities in the building.
- *Homeowners association or condominium fees.* The monthly fees charged by the homeowners association can affect the overall cost of ownership.
- *Local regulations.* Zoning laws, building codes, and other regulations can influence property values.

How could the model be improved?

Here are some ways to improve the model:

- *Gather more data.* Increase the sample size and include a wider range of apartments.
- *Incorporate more variables.* Add relevant features, like those mentioned earlier (location, condition, view, etc.), to capture more of the factors that influence price.

- *Feature engineering.* Create new variables by combining or transforming existing ones. For example, you could create a variable for 'total veranda area' by adding covered and uncovered verandas.
- *Consider non-linear relationships.* Explore non-linear regression techniques or transformations of variables to capture more complex relationships.
- *Address multicollinearity.* If there is significant correlation between independent variables, consider techniques like principal component analysis (PCA) or removing one of the correlated variables.
- *Validate the model.* Split the data into training and testing sets to evaluate the model's performance on unseen data.

By addressing these limitations and incorporating improvements, you can build a more robust and accurate model for predicting apartment prices. We recommend you try to evolve the model using the aforementioned recommendations if necessary, creating additional dummy data to see how it impacts the results.

Summary

- The comparison method is central to real estate valuation, based on comparing the subject property with similar recently transacted properties.
- The method relies on accurate, reliable data, which may be challenging in less transparent markets.
- The comparison method requires a valuer to be aware of both 'big picture' trends which are influencing change in property markets and tenants' behaviours and the detailed examination of property markets and submarkets and specific sectors.
- The comparison method has two main applications: as a stand-alone method to estimate market value or market rent through direct comparison and to derive inputs – such as costs, market rent yields, discount rates, etc. – for other valuation methods.

Questions

1. A residential property has a gross internal floor area of 400 m^2. Comparable A, measuring 390 m^2, sold for €250,000, while Comparable B, at 420 m^2, sold for €280,000.

   a. Adjust Comparable A's price for an additional bedroom in the subject property, which adds 10% to its value.
   b. Adjust Comparable B's price for a smaller garden, reducing its value by 5%.
   c. Calculate the price per square metre for each comparable.
   d. Estimate the market value for the subject property.

2. Your client is considering selling a detached house built in 2010 with a total external area of 300 m^2. This two-storey property includes three en-suite bedrooms, a kitchen, a living room, an office space, and two parking spaces and features outdoor amenities, such as a garden and a pool. Using the following evidence of comparable properties recently sold in the area, provide your opinion on the likely market value of this property.

House	*Information*
A	Built in 2008, **280 m²**, 3 bedrooms (2 en-suites), garden only, 1 parking space, sold for **£950,000**
B	Built in 2012, **320 m²**, 4 bedrooms (3 en-suites), garden and pool, 2 parking spaces, sold for **£1,200,000**
C	Built in 2015, **290 m²**, 3 bedrooms (2 en-suites), garden, 1 parking space, sold for **£1,050,000**
D	Built in 2020, **310 m²**, 3 bedrooms (3 en-suites), garden, pool, 2 parking spaces, sold for **£1,150,000**
E	Built in 2008, **290 m²**, 3 bedrooms (3 en-suites), garden, 2 parking spaces, sold for **£1,000,000**

3. Using the straight-line method, assess the net effective rent for a five-year term at $180,000 pa, with a one-year rent-free period.
4. A retail unit was recently leased for €100,000 per annum. The unit has a rectangular shape with a clear internal floor area and is a single-storey structure. Its frontage measures 12 m, with a floor depth of 24 m, plus an additional 6 m of storage space at the rear. Using this market evidence, calculate the market rent for the adjacent shop, which shares similar characteristics except for its size. The adjacent shop has a frontage of 10 m, a floor depth of 20 m, and an additional 5 m of storage at the rear. *Use 6 m deep zones for the calculation.*
5. A property has three tenancies, as follows:

Tenant	*Area (ft²)*	*Lease expiry (months)*	*Rent ($)*
A	10,000	12	500,000
B	50,000	36	1,500,000
C	40,000	48	1,400,000

Calculate the WAULT weighted by the area and weighted by the rent. Critically evaluate how each weighting affects the perceived security of income. How would these differences influence your estimation of the ARY?

References

California Department of Real Estate (2010). *Reference Book – A Real Estate Guide*. Available at: www.dre.ca.gov/publications/ReferenceBook.html

Dunn, C. (2004). *The Appraisal of Real Estate*. 2nd Canadian Edition. Vancouver: The University of British Columbia.

International Valuation Standards Council (IVSC) (2024). *International Valuation Standards*. London: IVSC.

JLL (2024). *Global Real Estate Transparency Index 2024 Rankings*. Available at: www.jll.co.uk/en/trends-and-insights/research/global-real-estate-transparency-index/greti-global-rankings-and-methodology

McAllister, P., Loizou, P. (2009). The Appraisal of Data Centres: Deconstructing the Cash Flow. *Journal of Property Investment and Finance*, 27(1), pp. 65–80.

Re-Leased (2024). Available at: www.re-leased.com/blog/july-market-lens-a-close-look-at-the-changing-landscape-of-office-lease-lengths

RICS (2019). *Comparable Evidence in Real Estate Valuation*. London: Royal Institution of Chartered Surveyors.

Scarrett, D. (2008). *Property Valuation: The 5 Methods*. 2nd Edition. London: Routledge.

Chapter 8

The investment method

Introduction

Investment in real estate is a multifaceted and dynamic field that plays a crucial role in the global economy. As a significant asset class, real estate offers diverse opportunities for individuals and institutions to build wealth and generate income. Understanding the principles and techniques of property investment is essential not only for investors but also for valuers seeking to provide accurate and reliable assessments of property worth.

The investment method is one of the most widely used techniques in the *income approach* to property valuation. It is grounded in the principle of anticipating future benefits, focusing on the income-generating potential of real estate assets. By evaluating expected rental income streams and translating them into a present value, this method provides a clear and rational framework for valuing a wide variety of properties.

From retail malls and office buildings to multifamily residences and industrial warehouses, the investment method applies universally to income-producing real estate. It is particularly favoured by investors and stakeholders who rely on quantitative metrics to make informed decisions.

This chapter explores the theoretical foundation of the investment method and its practical applications. Beginning with an explanation of yield and its various forms, the text delves into direct capitalisation, traditional valuation techniques, and advanced discounted DCF models. To provide a comprehensive understanding, key concepts, such as the discount rate, net present value (NPV), and IRR, are discussed in depth.

Real-world examples, worked calculations, and case studies illustrate how the investment method adapts to different property types and market conditions. The chapter concludes with a look at future trends, including the integration of technology, the growing emphasis on sustainability, and the evolving demands of modern investors.

By the end of this chapter, readers will have a thorough understanding of how to apply the investment method effectively, making it an indispensable tool in the field of property valuation.

What yield is, and the types of yields and rates

As discussed in earlier chapters, *yield* is a critical metric in property valuation, representing the relationship between a property's income and its value, expressed as a percentage. It quantifies the return on investment, allowing investors to compare the performance of different assets.

DOI: 10.1201/9781003490357-10

What is a yield?

- It is a rate of interest.
- It indicates the level of earnings of a particular investment.
- It also indicates the level of risk associated with a particular investment.
- The lower the yield, the lower the risk associated with it.

Types of yields

There are several types of yield mentioned in valuation literature.

- *Net initial yield (NIY)*. Predominantly used in the UK, the NIY is calculated as the ratio of the property's net income to its purchase price, including acquisition costs. It reflects the immediate return an investor can expect at the time of purchase. For example, if a property generates €50,000 in annual net income and is purchased for €1,000,000 and the purchaser's costs is 6.8%, then the net initial yield is:

$$\text{Net Initial Yield} = \frac{\text{Net operating income}}{\text{Property price} + \text{Acquisition cost}}$$

$$\text{Net Initial Yield} = \frac{50{,}000{,}000}{1{,}000{,}0000 + 0.068 \times 1{,}000{,}000} = 4.68\%$$

- *ARY.* A comprehensive measure, ARY reflects the return required by investors to compensate for all risks associated with a property. It is derived by comparing NIYs of similar properties in the market and includes considerations for location, lease terms, tenant quality, and market volatility, as seen in Chapter 7.
- *Reversionary yield.* This represents the anticipated yield once the property reaches its full income potential, often after lease renewals or rent adjustments. For instance, a property currently rented at below-market rates may have a higher reversionary yield once rents align with market levels.
- *Equivalent yield.* Equivalent yield, or the IRR of a property assuming no changes in rental value or exit yields, balances initial and reversionary yields. It is derived through trial and error to find the rate that equates the present value of future rental income to the purchase price. This yield is especially relevant for properties with mixed income profiles, balancing between current contracted rent and anticipated market rent.

Example 8.1

Consider a property purchased for €1,000,000 with an initial annual rent of €60,000 and an anticipated market rent of €80,000 after five years. The equivalent yield reflects the blended income stream by balancing initial and future incomes. Here is how we calculate it:

1. *Calculate the present value (PV) of initial rent (€60,000 per year for five years).* Using a discount rate (trial equivalent yield), assume 6.5%:

$$PV = PMT \times \frac{\left[1 - \frac{1}{(1+r)^n}\right]}{r}$$

$$PV = 60,000 \times \frac{\left[1 - \frac{1}{(1+0.065)^5}\right]}{0.065}$$

$PV = €249,340.77$

2. *Calculate the PV of the reversionary rent (€80,000 starting from year 6 onwards).* Calculate the PV in perpetuity of €80,000 starting from year 6:

$$PV = PMT \times \frac{1}{r}$$

$$PV = 80,000 \times \frac{1}{0.065}$$

$PV = €1,230,769.23$

Then, discount this value back to present (year 0):

$$PV = \frac{FV}{(1+r)^n}$$

$$PV = \frac{1,230,769.23}{(1+0.065)^5}$$

$PV = €898,314.88$

3. *Combine the two components.* Add the present value of the initial rent and the reversionary rent:

   Total PV = €249,340.77 + €898,314.88 = €1,147,655.64

4. *Iterate to find the equivalent yield.* Adjust the discount rate iteratively until the total present value equals the purchase price (€1,000,000). After trial and error, the equivalent yield is found to be **7.4%**.

   Note: This can be done in Excel by using the Goal Seek technique:

- *Set cell.* Select the cell containing the total PV €1,147,655.64.
- *To value.* €1,000,000.
- *By changing cell.* Select the cell containing the yield.

This calculation demonstrates the blended return of current and anticipated future income, accounting for the time value of money.

- *Equated yield.* The equated yield takes into account anticipated rental growth over time and inflation adjustments, providing a realistic return that reflects future income potential. It differs from the equivalent yield by explicitly including growth assumptions. For example, a property with rents that increase by 2% annually would have an equated yield that adjusts the return to reflect this projected growth. When tied to the risk-free rate, the equated yield includes a baseline return (e.g. government bond yields) plus a growth component and risk premiums for property-specific factors. This makes it particularly useful for analysing investments where rental growth plays a significant role in long-term returns. Additionally, while equivalent yield balances current and future incomes without assuming growth, equated yield explicitly incorporates expected income growth. For example, if a property's rent grows annually by 2% and its current net income is €40,000, equated yield adjusts the return to reflect this growth, offering a dynamic perspective compared to static measures like ARY.

While the aforementioned yields are widely referenced in UK property valuation literature, the American approach primarily emphasises the capitalisation rate and discount rate. The focus on these two metrics reflects their simplicity and direct application in comparing investment opportunities and modelling future cash flows. Capitalisation rates provide a snapshot of current income performance relative to market value, while discount rates allow for detailed cash flow modelling, accounting for risk and return expectations over time.

- *Capitalisation rate (cap rate).* Often used interchangeably with NIY, the cap rate is similar in calculation; however, it does not account for acquisition costs, making it a less reliable measure of profitability in real estate markets where transaction costs are high.

$$Cap\ Rate = \frac{Net\ operating\ income}{Property\ price}$$

For example, if a property generates €50,000 in annual net income and is purchased for €1,000,000 and the purchaser's costs is 6.8%, then the cap rate is:

$$Cap\ Rate = \frac{50,000}{1,000,000} = 5\%$$

- *Discount rate.* The discount rate, often used in conjunction with DCF models, reflects the required rate of return for an investor, incorporating both the risk-free rate and property-specific risk premiums. For example, if the risk-free rate is 2% and the risk premium is 5%, the discount rate would be 7%. Unlike the capitalisation rate, which provides a snapshot of income performance relative to property value, the discount rate is used to discount future cash flows to their present value, making it a critical tool for detailed investment analysis. The focus on these two metrics reflects their simplicity and direct application in comparing investment opportunities and modelling future cash flows.

Implicit vs explicit methods

The investment method is based on the concept of determining the present value of future net rents, and this can be done using implicit or explicit methods.

Implicit methods, such as direct capitalisation, term and reversion, and hardcore and top slice, use the ARY derived from comparable evidence to discount future rents. Any rental growth, depreciation, or risk is *implied*, rather than explicitly shown, in one single ARY. Implicit methods are useful for properties with stable, predictable income streams but offer limited transparency.

In contrast, and as the name suggests, explicit methods, such as DCF, make all the cash flow assumptions explicit. Explicit methods require detailed assumptions about rental growth, inflation, expenses, and terminal value. For instance, a DCF analysis may project a property's €70,000 annual income over five years, with an 8% discount rate applied to future cash flows to determine present value. Explicit methods provide greater transparency and flexibility, making them more suitable for properties with variable income or in dynamic markets.

Implicit methods encapsulate all factors in a single rate, while explicit methods break down components such as growth, risk, and timing. Implicit methods are quicker to use but less detailed, whereas explicit methods demand more data and modelling. Both methods have their place in property valuation. Implicit methods are ideal for quick assessments in stable markets, while explicit methods provide a robust framework for understanding complex income patterns and future uncertainties.

Direct capitalisation

Direct capitalisation is the simplest form of investment method used to determine the value of an income-producing property by applying a market-derived capitalisation rate to its NOI. This approach is commonly used for properties with stable and predictable income streams, as it provides a straightforward means to estimate the property's current market value. If the income is fixed for a period much in excess of 60 years, or if a property is let at its market rent (rack rented) and there is market evidence of NIYs, then it can be treated as a level annuity in perpetuity.

One of the primary advantages of direct capitalisation is its simplicity. The method requires minimal data and is easy to apply, particularly when reliable market comparables are available. Additionally, direct capitalisation is efficient as it provides a quick estimate of value for

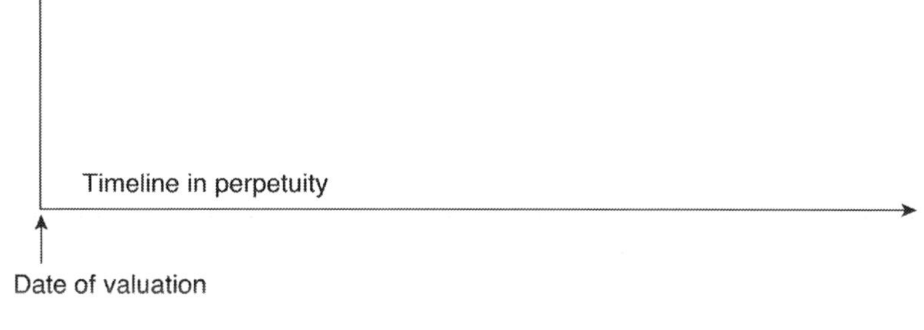

Figure 8.1 A level annuity in perpetuity.

properties with stable income streams. Another key benefit is that it reflects current investor expectations and market sentiment. Because the capitalisation rate is derived from market transactions, the valuation aligns with prevailing investment trends, ensuring that the estimated value is grounded in real-world market conditions.

Despite its advantages, direct capitalisation has notable limitations. A significant drawback is its assumption of income stability. The method treats NOI as constant, which may not adequately account for future income growth or decline. This can lead to inaccuracies when valuing properties with variable revenue streams, such as those in emerging or transitional markets. Another limitation is its heavy reliance on market data. The accuracy of the valuation depends on the availability and quality of comparable transactions, which may be difficult to obtain in less active or opaque markets. Furthermore, direct capitalisation lacks flexibility, making it less suitable for properties with uncertain or fluctuating income, such as assets undergoing repositioning, redevelopment, or lease renegotiations.

Direct capitalisation involves the following steps:

Step 1: Assess the NOI.
NOI represents the annual income (rent) generated by the property after deducting operating expenses, such as maintenance, taxes, and property management fees, and it does not include financing costs or depreciation, as it reflects the property's operational performance.

$$NOI = Gross\ Income - Operating\ Expenses$$

Example 8.2

If a property generates a gross income of €100,000 annually and operating expenses are €30,000, the NOI is:

$$NOI = Gross\ Income - Operating\ Expenses = 100{,}000 - 30{,}000 = 70{,}000$$

Step 2: Estimate the ARY.
As seen in Chapter 7, the ARY derives from market comparables and reflects investor expectations for similar properties. Factors such as location, property type, risk profile, and market conditions influence this rate. If comparable properties exhibit NIYs of around 6%, this indicates the expected annual return for an investor. When the subject property carries a higher risk than the comparable evidence, a valuer may apply a slightly higher rate, such as 6.25% or 6.5%. Conversely, for a lower-risk property, the rate may be adjusted downward to 5.75% or 5.5%.
Step 3: Apply the direct capitalisation formula.
With the NOI and ARY determined, the MV of the subject property can be calculated using the direct capitalisation formula.

$$MV = \frac{NOI}{ARY}$$

or

$$MV = NOI \times \frac{1}{ARY} = NOI \times YPp$$

As evident from the preceding formula, the NOI is treated as an ordinary annuity with no fixed term, effectively a perpetuity.

Example 8.3

A retail property generates an annual NOI of €80,000, and the market data suggests an ARY of 6% for similar properties, with the difference that the property under study is leased to a public government authority. Public government authorities are considered grade A tenants, which implies safe return. In that case, the valuer is adjusting the cap rate to a lower figure, say, 5.5%. Therefore, the property value is calculated as:

$$MV = \frac{80,000}{0.055} = 1,454,545 \ \textit{say}\ €1,540,000$$

This valuation assumes the property will continue generating the same NOI over time without significant changes in income or expenses.

In the UK, a conventional valuation would appear as:

Net income	*€80,000*
YPp at 5.5%	1/0.055 = 18.1818
MV	€1,454,545

Example 8.4

A retail park was developed on the basis of a ground lease from a landowner who wished to retain ownership of the freehold. The ground lease was for 125 years at a fixed ground rent of £100,000 pa – that is, there were no break clauses or rent reviews. The lease has 80 years to run.

- Fixed income of £100,000
- Unexpired term 80 years as at valuation date
- Capitalisation rate 10% based on market comparables

Net income	*£100,000*
YPp at 10%	10
MV	£1,000,000

A perpetuity approach is used because the years purchase in perpetuity (YPp) factor for 80 years is considered equivalent to that in perpetuity (9.9951 cf 10.00). But in some circumstances, it might be necessary to use the YPp for 80 years and add in a reversion to site value in 80 years' time. This could be a preferred approach where today's highest and best use of the site is a more valuable use than the current permitted use.

There may be market comparables to support the capitalisation rate of 10%. However, in the absence of evidence, the rate can be compared to that achieved on irredeemable government stocks, with an adjustment for any extra risks or security attaching to the property investment. Investments of this nature are very secure in money terms, but not in real terms. Money security is provided by the occupation rents exceeding the ground rent and by the fact that the ground lease would be forfeited in English law if, for any reason, the rent of £100,000 was not paid.

Example 8.5

Value a freehold retail property in a prime trading position let at its market rent of £120,000 pa on FRI terms – meaning that the rent is all net investment income to the owner, that is, NOI. The lease is for ten years, with a rent review after five years, should market rents increase over the next five years. No break clauses. Similar properties are selling on the basis of ARY or capitalisation rate of 5.5%.

Rent	£120,000
YPp at 5.5%	18.1818
MV (gross of fees)	£2,181,818

Note:

1. Where the lease is for ten years but it can be assumed that it would continue to let readily and that market rents will always be receivable, then it can be treated as a perpetual income. Such an assumption could not be made if the trading position was under threat from planned development, such as an out-of-town centre, or where the tenant's financial position is not secure.
2. Although the rent is reviewable every five years, there is no need to provide for this explicitly in a capitalisation approach if at each reversion the rent would also be equivalent to today's market rent of £120,000. The key distinction between Examples 8.4 and 8.5 is that the former is actually fixed for 80 years, whereas the latter is inflation-proofed by the rent reviews.
3. All properties incur a management cost, but customarily, when property is let on FRI terms, no deduction is made for management.
4. For some valuation purposes, market value is the capitalised sum adjusted for the costs of acquisition (survey fees, legal fees, and Stamp Duty Land Tax (SDLT)). Where the contract price (market value) plus fees add up to £2,181,818, then £120,000 will represent a return on total outlay of 5.5%. To ensure the initial return of 5.5% is achieved,

the before-costs sum has to be calculated. Currently, in some areas of the market, an approximate figure of 6.8% is used to represent legal and agent/surveyor fees, inclusive of VAT and 5% for SDLT, but a more correct figure can be found as follows:

£2,181,818 + £12,500 − £2,000 ÷ 1.068 = £2,052,732.21 (contract price, two decimal places used to show details of calculations).

At a contract price of £2,052,732.21, a buyer incurring standard costs taken in the UK market to be 1.8% (1.5% agent's fee + 0.5% legal fee + 20% VAT), or £36,949.18, plus SDLT on £2,052,732.21, namely, £2,000, being 2% on the first tranche of £100,000 above the £150,000 SDLT exemption, plus 5% of the excess above £250,000, namely, £90,136.61 (5% of £1,802,732.21), giving £2,052,732.21 + £36,949.18 + £2,000 + £ 90,136.61, a total expenditure including fees and SDLT of £2,181,818.

Example 8.6

An office is rented for €1,500 per month (net). The lease agreement was signed for five years with a fixed rent and started three years ago. Similar properties are yielding 6%. However, due to significant demand over the last three years, the market rent is now estimated at €2,000 per month (net).

The scenario described illustrates a common challenge in valuation methodology when there is a significant disparity between market rent and passing rent. Each approach has its strengths and weaknesses, highlighting the need to tailor the valuation methodology to the specific circumstances of the property and market. Following is an analysis of three approaches in order to understand the necessity of the traditional investment approaches such as term and reversion and hardcore and layer.

Approach 1: Capitalisation of passing rent

The valuer uses the passing rent and capitalises it in perpetuity. In this case, MV_1 is calculated as:

$MV_1 = 1{,}500 \times 12/6\% = €300{,}000.$

Issue: This approach does not take into account the difference between the passing rent and the MR.

Approach 2: Capitalisation of MR

The valuer ignores the passing rent, considering that the lease contract expires in two years, and instead capitalises the market rent in perpetuity. In this case, MV_2 is calculated as:

$MV_2 = 2{,}000 \times 12/6\% = €400{,}000.$

Issue: This approach overlooks the two years during which the property will generate a lower income under the existing lease terms.

Approach 3: Capitalisation with yield adjustments

The valuer uses the passing rent but reduces the ARY to account for the lower risk associated with the lease agreement, given that the passing rent is significantly below the MR. For example, the ARY might be adjusted to 5.5%. In this case, MV_3 is calculated as:

$$MV_3 = 1{,}500 \times 12/5.5\% = €327{,}273.$$

Issue: This approach relies on an implicit and subjective adjustment of the ARY, which may lack transparency.

All three approaches have limitations when there is a significant discrepancy between the passing rent and the market rent. Each method fails to fully capture the combined effect of current income and future potential in a balanced and transparent manner.

Direct capitalisation is widely used for its practicality and ease of use, especially for properties like office buildings, retail spaces, or warehouses with predictable cash flows. However, valuers must recognise its limitations and consider alternative methods. When the current rent is at market level, direct capitalisation can work well as it provides a simple and effective way to estimate property value. However, most of the time, the lease signed is not fresh, and the current rent may be higher or lower than the market rent. In such cases, traditional methods like term and reversion, hardcore and layer, or even better, the DCF become necessary. These approaches account for discrepancies between current and market rent by separating income into components, such as term (current rent) and reversion (market rent), or by modelling income changes over time.

Term and reversion

Term and reversion is a valuation technique often used when there is a difference between the current passing rent (under an existing lease) and the expected market rent. Term and reversion accounts for both the current passing rent during the lease period (term) and the expected (market) rent after the lease expires (reversion). The concept of the method is that it separates the valuation into two components (vertical division):

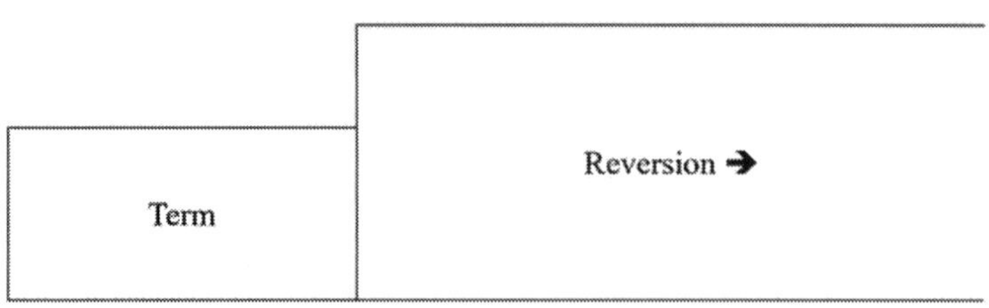

Figure 8.2 Term and reversion.

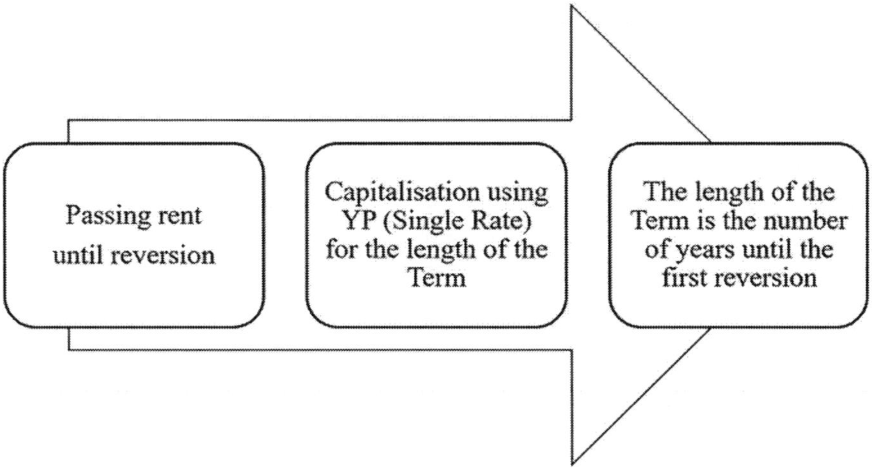

Figure 8.3 Calculating the term.

The term

The *term* refers to the period during which the property is let at the current passing rent. This component focuses on the income the property generates while the lease is in place. The income for the remaining lease term is capitalised at a yield that reflects the secure nature of this income (typically the ARY derived from comparable evidence). Since the income is limited to a specific period, the valuation is calculated as the present value of the passing rent over the remaining lease period.

The formula for the term value is as follows:

$$Term\ Value = Passing\ Rent\ (yearly) \times \frac{1 - \left(\frac{1}{(1+r)^n}\right)}{r}$$

or

$$Term\ Value = Passing\ Rent\ (yearly) \times PV\ £1\ pa\ for\ n\ years\ @\ r$$

The reversion

The *reversion* refers to the property's value after the lease expires, when it is expected to produce income at the prevailing market rent. This component reflects the potential long-term value of the property.

The MR is capitalised in perpetuity, as the property is assumed to produce this level of income indefinitely after the lease ends. The resulting value is then discounted back to the present value, as this income will not begin until the lease expires.

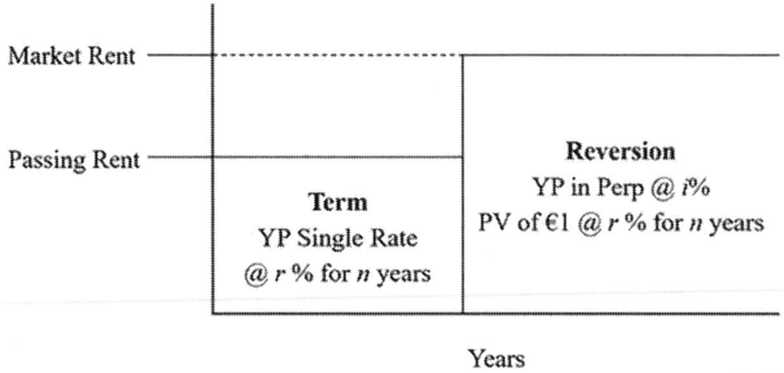

Figure 8.4 Illustration of term and reversion.

The formula for the reversion value is:

$$Reversion\ Value = Market\ Rent\ (yearly) \times \frac{1}{r} \times \frac{1}{(1+r)^n}$$

or

$$Reversion\ Value = Market\ Rent\ (yearly) \times YP\ in\ perp\ @\ r \times PV\ of\ £1\ for\ n\ years$$

The market value

The MV of the subject property will be the sum of the term value and the reversion value.

$$MV = Term\ Value + Reversion\ Value$$

Example 8.7

An office is rented for €1,500 per month (net). The lease agreement was signed for five years with a fixed rent and started three years ago. Similar properties are yielding 6%. However, due to significant demand over the last three years, the market rent is now estimated at €2,000 per month (net). Estimate the MV of the commercial property.

Term

Net Income = €1,500 × 12 = €18,000 pa

$$\times\ YP\ 3\ yrs\ @\ 6\% = \left(1 - \frac{1}{(1+6\%)^3}\right) / 6\% \times \underline{2.673}$$

€48,114.22 [1]

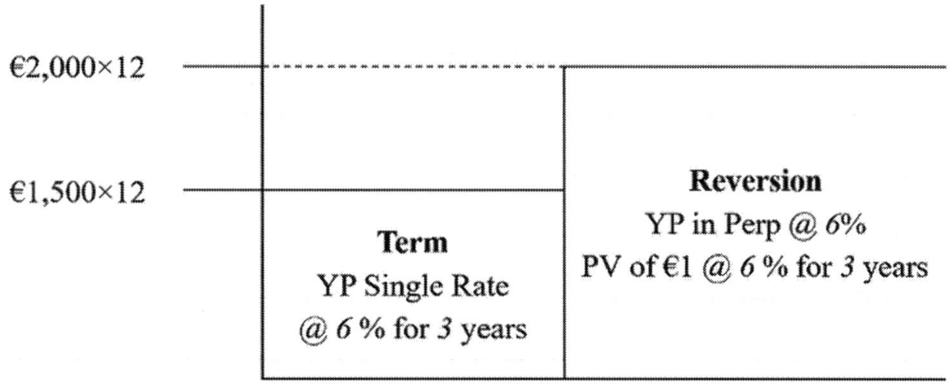

Figure 8.5 Graphic representation of Example 8.7.

Reversion

Net Income = €2,000 × 12 = €24,000 pa

$\times$ YP in perp deferred 3 yrs @ 6% $= \frac{1}{6\%} \times \frac{1}{(1+6\%)^3} = 16.667 \times 0.84 = \times \underline{13.99}$

€335,847.7 [2]
Capital Value [1] + [2] = *€383,961.9*
Say, **€384,000**

Example 8.8

Value a freehold retail property in a prime trading position let at £100,000 pa on FRI terms with two years to run to a review to the market rent of £120,000 pa on FRI terms. The lease is for ten years, with no break clauses. Similar properties are selling on the basis of ARY or capitalisation rate of 5.5%.

Term income	*£100,000*	
PV £1 pa 2 years at 5.5%	*1.8463*	*£184,630*
Reversion income	*£120,000*	
YPp at 5.5%	*18.1818*	
× PV £1 in 2 years at 5.5%	*0.8984*	*£1,960,260*
Value		***£2,144,890***

Note: The reversion is today's MR even though the actual reversion is two years away; this is the conventional approach.

The term and reversion method is particularly useful for properties that are under-rented (reversionary) – meaning, the passing rent is below MR, and its MV is known. Here is how it works:

1. *Term value.* This represents the income generated during the period in which the property is rented at the below-market rent. The net income is discounted using an appropriate yield or capitalisation rate to calculate its present value.
2. *Reversion value.* This is the estimated value of the property at the end of the lease term, when it reverts to market conditions. It assumes that the property will generate market rent thereafter. The reversion value is also discounted to its present value using a suitable yield.
3. *Market value calculation.* The total market value of the property is derived by adding the term value (income during the lease term) and the reversion value (future income post-lease term). This method is ideal for situations when the current rent is below market levels due to long-term lease agreements, controlled rents, or other contractual obligations. It separates the income streams and discounts them based on their individual risk and timing.

Example 8.9

Since the ARY is not provided, we can estimate it based on the given data as follows:

€1,400/€22,000 = 0.0636 = 6.36%

Similar to the previous example, we need to separate the income stream into two components: the *term*, which reflects the current tenant's rental situation, and the *reversion*, which accounts for the property's income potential once the lease expires and reverts to market conditions.

To simplify, the property's market value, based on comparable evidence, is calculated as:

$100 \text{ m}^2 \times €22{,}000/\text{m}^2 = €2{,}200{,}000$

The issue in this case is that the property is not vacant but is leased at below-market rent for the next four years and until the lease expires. The proposed solution is:

Term

Net Income $= €900/\text{m}^2 \times 100 = €90{,}000$ pa

$$\times \text{ YP 4 yrs @ } 6.36\% = \frac{1-\left(\dfrac{1}{(1+6.36\%)^3}\right)}{6.36\%} = \times \underline{3.437}$$

€309,305.8 [1]

Since the current market value and the annual growth rate are known, we can estimate the market value of the office in four years as follows:

MV today = 100 m^2 × €22,000/m^2 = €2,200,000

However, since property prices are expected to grow at 3% annually, we apply this growth over the four-year period:

$FMV\left(future\ market\ value\right) = Current\ Market\ Value \times \left(1 + growth\ rate\right)^{\wedge} n$
Substituting the values: €2,200,000 × $(1 + 3\%)^4$ = €2,476,119

Next, to estimate the value of the property today while accounting for this growth, we discount the future market value back to the present using an appropriate discount rate. In that case, and having in mind that the opportunity cost of money should be higher than the ARY, we adopt a rational discount rate, say, 10%.

Reversion

MV in 4 years: €2,476,119
PV for 4 yrs @ 10% = $(1 + 10\%)^{-4}$ × 0.683
€1,691,223.6 [2]
Capital Value [1] + [2] = €2,000,528.4
Say, **€2,000,000**

An amount that is rational due to the lower rent that the property will produce for the next four years.

In some cases, a whole leasehold agreement is provided with more than one review happening during the remaining period. This is illustrated in the following example:

Example 8.10

A shop was rented 3 years ago on a 20-year lease with 5-year rent reviews. The passing rent is €50,000 FRI (full repairing and insuring) per year, while the MR is €58,000 FRI, and ARY is estimated at 6%. Calculate the MV of the property.

Step 1: *Draw a timeline.* Create a straight line representing the total duration of the analysis, divided into equal intervals (e.g. five years per interval).
Step 2: *Mark review patterns.* Indicate the points on the timeline where reviews occur. If the reviews happen every five years, you would place markers at the 0-year, 5-year, 10-year, etc. points. Use visual indicators like dots, lines, or annotations to clearly represent review events.
Step 3: *Label key points, and add labels to clarify:*

- The current year or starting point
- The years until the next review
- The points of review along the timeline

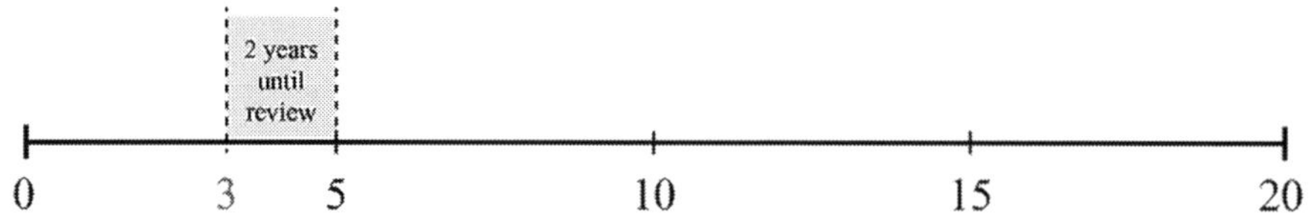

Figure 8.6 Graphic representation of timeline.

From the preceding, we can see that the next review will occur in two years. Following this, the situation can be addressed in a manner similar to the previous example, separating the income stream into term (until the next review) and reversion (post-review adjustments).

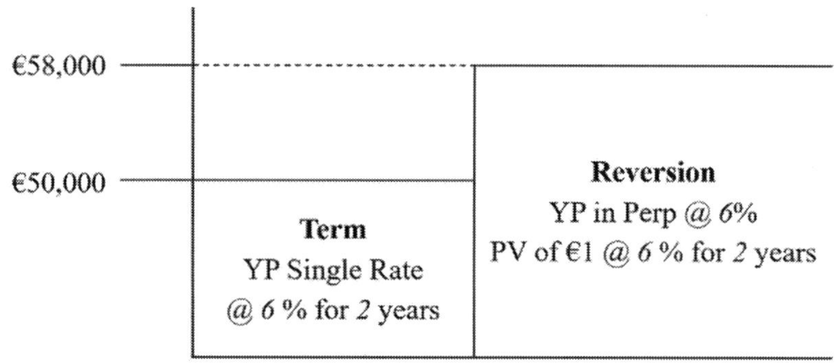

Figure 8.7 Graphic representation of Example 8.10.

Term

Net Income = €50,000 pa

$\times$ YP 2 yrs @ 6% $= \left(1 - \dfrac{1}{(1+6\%)^2}\right) / 6\% = \times\ \underline{1.833}$

€91,669.63 [1]

Reversion

Net Income = €58,000 pa

$\times$ YP in perp deferred 2 yrs @ 6% $= \dfrac{1}{6\%} \times (1+6\%)^\wedge -2 = 16.667 \times 0.889 = \times\ \underline{14.8336}$

€860,347.1 [2]
Capital Value [1] + [2] = *€952,016.7*
Say, **€950,000**

Despite the fact that multiple reviews were scheduled, the 'term and reversion' approach considered only the first review in the analysis. This simplification allows for focusing on the immediate impact of the next review rather than accounting for all subsequent reviews, which could complicate the valuation process.

Example 8.11

An office property is let at £100,000 pa on FRI terms, that is, net. The MR on the same terms is £120,000. The lease is due to end in three years and will take one year to re-let, and a rent-free period of six months will have to be offered as an incentive. There is evidence of market capitalisation rates of 7% for similar properties with five years or longer to run.

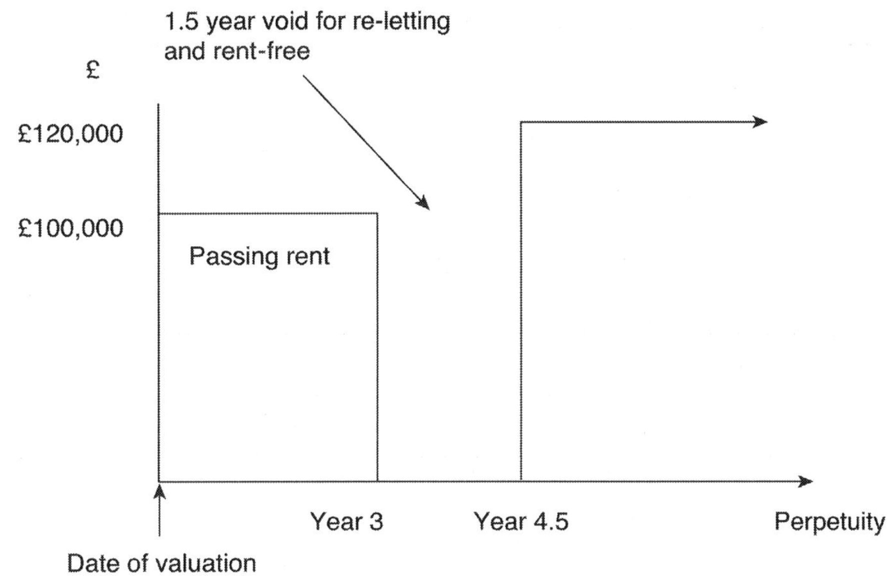

Figure 8.8 Allowing for a break in income.

Term		
Current rent	£100,000	
PV £1 pa for 3 years at 7%	2.6243	£262,430
Reversion		
MR	£120,000	
PV £1 pa in perp. at 7%	14.285	
	£1,714,286	
PV £1 in 4.5 years at 7%	0.737518	£1,264,317
MV gross of fees		**£1,526,747**

Where commercial property is empty, a liability for property taxes in some countries (such as empty 'rates' in the UK) can occur. This would mean that there is not only a loss of rent but also a liability for rates producing a negative income for the landlord. To this has to be added the cost of insurance, to be borne by the landlord whilst the premises are vacant, and in some cases, there will be non-recoverable service charge costs.

Example 8.12

Value the ***leasehold*** investment in a retail outlet in a popular shopping street. The lease has 20 years to run at a fixed head rent of £50,000. The current sublease has five years to run to the lease renewal at a rent of £80,000, and the market rent is £100,000. The appropriate leasehold capitalisation rate is 8% based on market comparables.

Term		
Rent receivable	£80,000	
Rent payable	£50,000	
Profit rent	£30,000	
PV £1 pa for 5 years at 8%	3.9927	£119,781
Reversion		
Rent receivable	£100,000	
Rent payable	£50,000	
Profit rent	£50,000	
PV £1 pa for 15 years at 8%	8.5595	
	£427,975	
PV £ in 5 years at 8%	0.68058	£291,271
Value		**£411,052**

Note: So far, all the examples in this chapter dealt with freeholds with occupation leases. This example refers to a long leasehold, as seen in Chapter 3.

The following key facts about leaseholds should be noted:

1. A leasehold interest has a finite duration.
2. In investors' terms, a lease can only have a market value if it produces an income, and if the lease is assignable, the leasehold income is called the profit rent.
3. A profit rent arises when a freeholder grants a lease to a tenant. At some time over the life of the lease (the lease term), the market rent or rent received from a sublessee under a sublease is greater than the rent reserved and payable under the head lease. Normally, this is when the rent actually received or potentially receivable is less than the rent paid to the superior landlord or head lessor after adjusting for any irrecoverable operating costs or outgoings. However, even when the rent receivable and the rent payable are the same, the lease may still have an investment value if in the foreseeable future the rent receivable will exceed the rent payable.

The main issue facing valuers who are undertaking genuine leasehold investment valuations is the lack of market comparable evidence. Every leasehold interest in property is unique in terms of lease, and so where market analysis is available, valuers will find they have to make adjustments to overcome differences between the comparable and the subject property. The other issues are the same and require a thorough inspection of the property and documents, and clearly, the valuation can be based on an in-arrears assumption or quarterly in advance. The valuer may even need to deal with situations where the head rent is quarterly in advance and the sublease rents are monthly in advance. Careful application of basic principles will resolve all these problems.

Hardcore and layer method (top slice)

Following the term and reversion approach, another commonly used method for property valuation is the hardcore and layer technique. This method is particularly useful when analysing rental income that includes an element of over-rent or under-rent compared to the market rent.

The concept of the method is that it separates the valuation into two components (horizontal division), unlike with term and reversion, which is a vertical division of income streams.

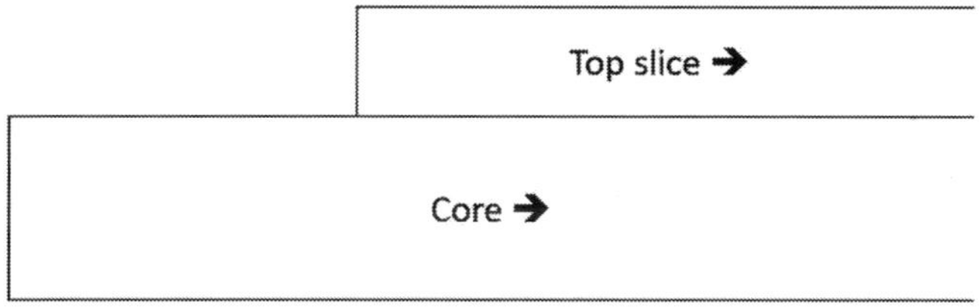

Figure 8.9 Hardcore and layer illustration.

Hardcore

The *hardcore* (or core) refers to the portion of a property's income that is considered secure, stable, and sustainable over the long term. It is typically based on the current passing rent – the rent that the tenant is currently paying under the terms of the lease agreement. This income is viewed as having low risk because it is already established and contractually guaranteed, making it less susceptible to market fluctuations. As a result, the hardcore income is capitalised in perpetuity at a lower yield (reflecting its lower risk), producing a higher capital value for this portion of the income. Essentially, the hardcore represents the foundation or base level of income, upon which additional, less certain income layers (such as over-rent or future market rent adjustments) are analysed.

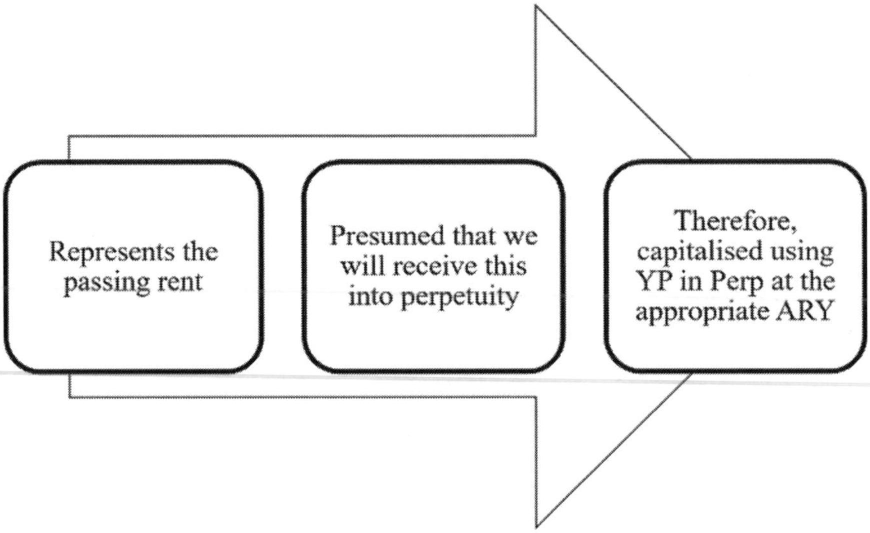

Figure 8.10 Calculating the core.

Layer (top slice)

The *layer* represents the portion of income that exceeds the hardcore (i.e. the passing rent) and reflects the potential for future adjustments to the property's rent, such as over-rent or the expectation of reaching market rent at the next review. This income is considered less secure and carries a higher risk, as it is contingent upon future events, such as market growth, lease renewals, or changes in tenant agreements. Consequently, the layer income is capitalised in perpetuity at a higher yield, reflecting the increased risk and uncertainty associated with it and deferred back. The layer provides a detailed understanding of the property's additional income potential, which might not yet be realised but has implications for the overall valuation. This approach ensures a more comprehensive breakdown of income streams, allowing for a nuanced assessment of risk and value.

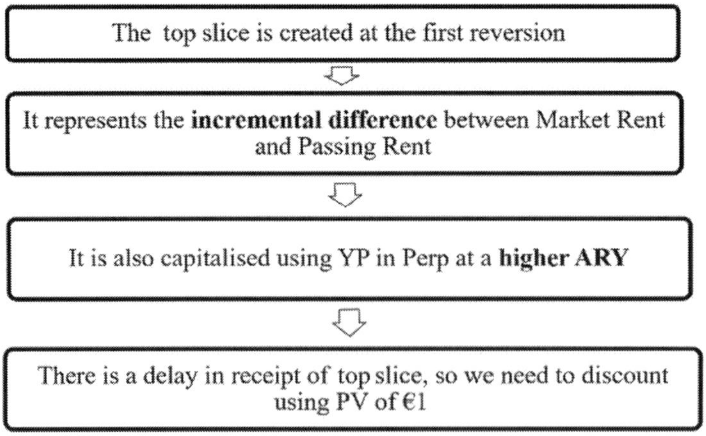

Figure 8.11 Calculating the top slice (layer).

The completed method is illustrated in the following diagram:

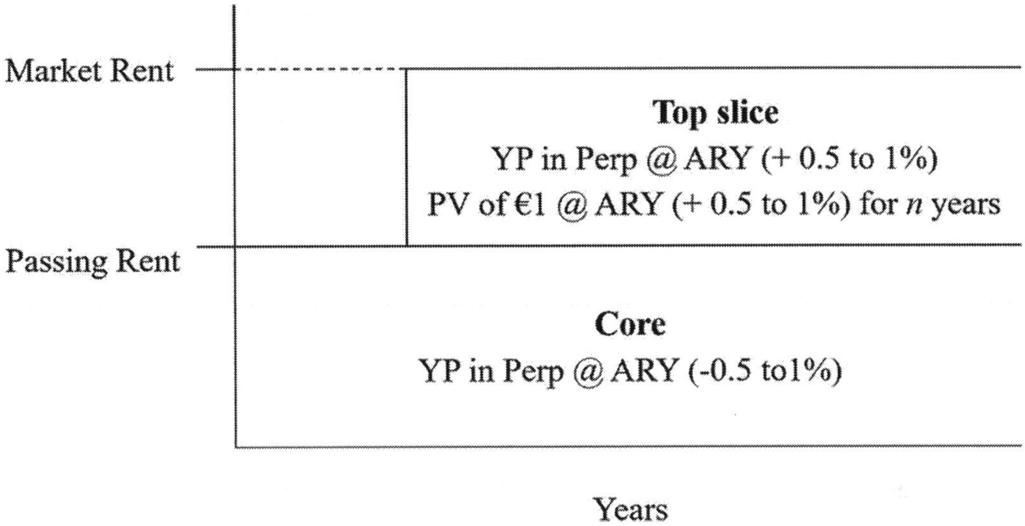

Figure 8.12 Graphic illustration of hardcore and layer.

Example 8.13

1. What is the **top slice** if the current rent for a building is €58,200 pa and MR is estimated at €69,750 per annum?
    - €11,550 pa
2. What would the core income be for the same building?
    - €58,200 pa
3. What formula would you use to capitalise both incomes?
    - YP in perp at the appropriate ARY
4. How do we reflect the additional risk associated with the top slice?
    - Increase the ARY by 0.5–1%

For better understanding, and having applied the term and reversion method (vertical division of income stream) to assess the investment value of the subject properties, we now turn to the hardcore layer method (horizontal division/layers), where the 'hardcore' layer represents the secure base rent, which is capitalised at a lower yield, and a 'top slice' captures any rental uplift beyond the initial rent, capitalised at a higher yield to reflect its increased risk. The same numerical examples will be illustrated.

Example 8.14

An office is rented for €1,500 per month (net). The lease agreement was signed for five years with a fixed rent and started three years ago. Similar properties are yielding 6%. However, due to significant demand over the last three years, the MR is now estimated at €2,000 per month (net). Estimate the MV of the commercial property.

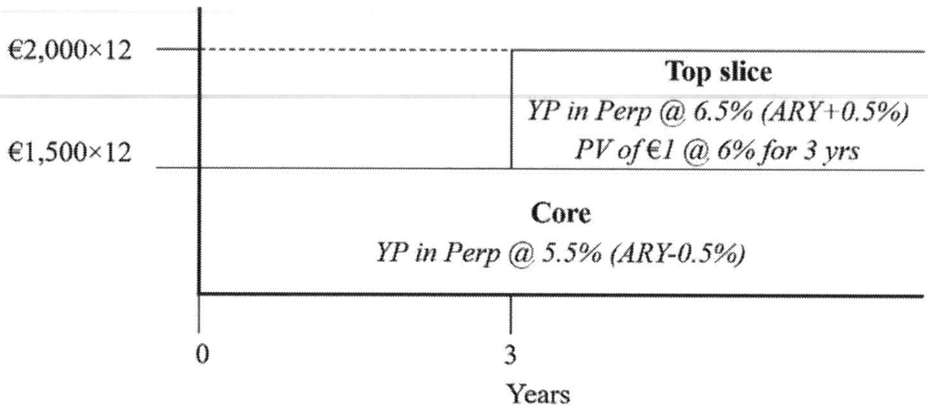

Figure 8.13 Graphic representation of Example 8.14.

Top slice = MR − Passing Rent = €2,000 × 12 − €1,500 × 12 = €6,000

Core

Passing rent pa = €1,500 × 12 = €18,000 pa
× YP in perp @ 5.5% = 1/5.5% = × 18.18
€327,272.7 [1]

Top slice

Overage (incremental income) €6,000

× YP in perp @ 6.5% deferred 3 yrs @ 6% = $\frac{1}{6.5\%} \times \frac{1}{(1+6\%)^3}$ = 15.38 × 0.84 = × 12.91

€77,503.3 [2]
Capital Value [1] + [2] = *€404,776*
say, **€405,000**

Compared with Example 8.7, the hardcore approach indicated a 5.2% higher value than the term and reversion. It is not something unexpected and has to do with the yield adjustments.

Unlike the term and reversion method, which is used commonly for under-rented properties, the hardcore and layer method is applicable to both under-rented and over-rented properties. This flexibility makes it particularly useful in assessing investment value where rental income consists of a stable core portion and a fluctuating or higher-risk element. By applying the hardcore and layer method, we can better reflect the distinct risk profiles of different income streams, ensuring a more precise valuation.

See the following example:

Example 8.15

A small shop in Prague is rented for €30,000 per year (net). The lease agreement was signed for eight years with a fixed rent and started six years ago. Similar properties are yielding 6%. However, due to a crisis in the rent over the last years, the MR is now estimated at €20,000 per year (net). Estimate the MV of the commercial property.

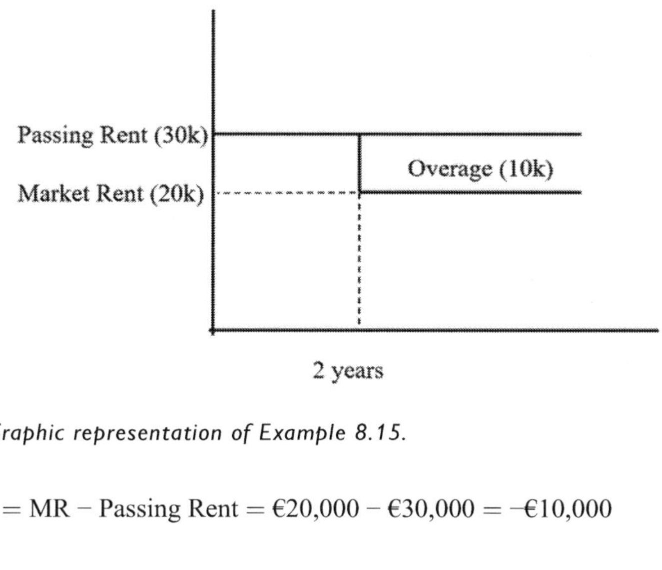

Figure 8.14 Graphic representation of Example 8.15.

Top Slice = MR − Passing Rent = €20,000 − €30,000 = −€10,000

Core

Passing rent per annum = €30,000 pa
× YP in perp @ ARY = 1/6% = × <u>16.67</u>
€500,000 [1]

Top slice

Overage (decremental income) €10,000

× YP in perp @ 6% deferred 2 yrs @ 6% = $\frac{1}{6\%} \times \frac{1}{(1+6\%)^2}$ = 16.667 × 0.889 = × <u>14.833</u>

<u>€148,332</u> [2]

Capital Value [1] − [2] = *€351,667.3*
say, **€350,000**

In cases where the passing rent exceeds the market rent, we account for the overage by deducting the excess portion. To validate the rationale behind the result, we can consider the property's capital value assuming vacant possession, which, when capitalised in perpetuity, amounts to €333,000. However, since the shop will generate a higher income for the next two years until the lease expires, this additional income is reflected as an extra €17,000 after applying the appropriate discounting factor.

Similar to Example 8.11, in what follows we deal with the same problem with rent review intervals, but solved with the hardcore and layer approach.

Example 8.16

A shop was rented 3 years ago on a 20-year lease with 5-year rent reviews. The passing rent is €50,000 FRI (full repairing and insuring) per year, the MR is €58,000 FRI, whilst ARY is estimated at 6%. Calculate the MV of the property.

(Similar to Example 8.11, but with the hardcore and layer approach.)

Step 1: *Draw a timeline.* Create a straight line representing the total duration of the analysis, divided into equal intervals (e.g. five years per interval).

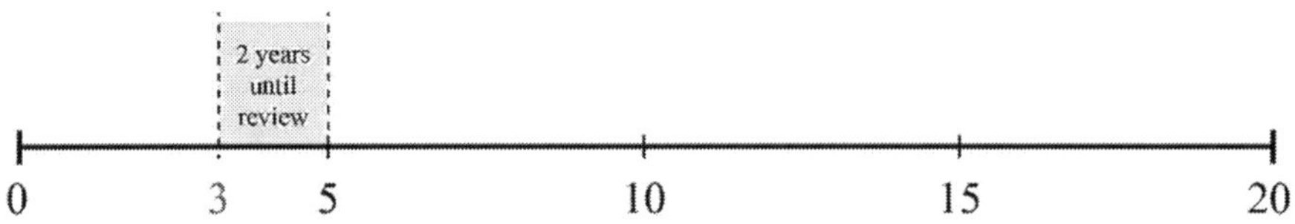

Figure 8.15 Graphic representation of timeline.

Step 2: *Mark the review patterns.* Indicate the points on the timeline where reviews occur. If the reviews happen every five years, you will place markers at the 0-year, 5-year, 10-year, etc. points. Use visual indicators like dots, lines, or annotations to clearly represent review events.

Step 3: *Label key points and add labels, to clarify:*

- The current year or starting point
- The years until the next review
- The points of review along the timeline

From the preceding, we can see that the next review will occur in two years. Following this, the situation can be addressed in a manner similar to the term and reversion example, but separating the income streams horizontally.

Top Slice = €58,000 − €50,000 = €8,000

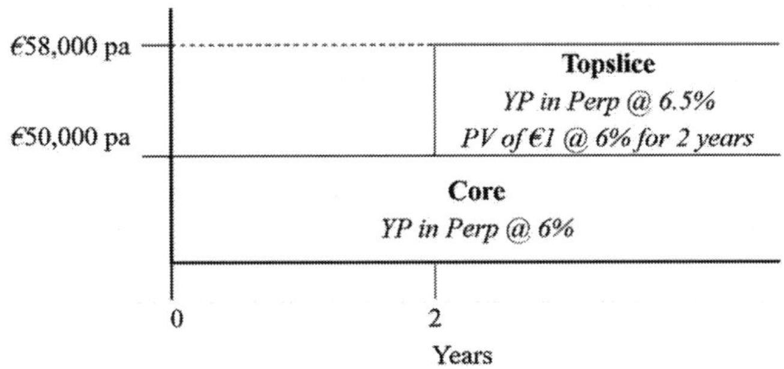

Figure 8.16 Graphic representation of Example 8.16. (This is similar to Example 8.11.)

Core

Passing rent pa = €50,000 pa
× YP in perp @ ARY = 1/6% = × 16.67
€833,500 [1]

Top slice

Top slice is capitalised in perpetuity, but with higher ARY due to the additional risk:

Overage (incremental income) €8,000

× YP in perp @ 6.5% deferred 2 yrs @ 6% = $\frac{1}{6.5\%} \times \frac{1}{(1+6\%)^2}$ = 15.385 × 0.889 =

× 14.833
€118,666.2 [2]
Capital Value [1] + [2] = *€952,166.2*
say, **€952,000**

In Examples 8.7 and 8.11, the results are identical.

Example 8.17

An office property is let at £100,000 pa on FRI terms, that is, net. The MR on the same terms is £120,000. The lease is due to end in three years and will take one year to re-let, and a rent-free period of six months will have to be offered as an incentive. There is evidence of market capitalisation rates of 7% for similar properties with five years or longer to run. Voids and empty rates are estimated at £40,000 per year, based on current assessments, and an insurance premium of £11,750 is expected for the one-year marketing period before securing a new tenant. Certain software applying a hardcore approach follows the methodology outlined in what follows.

Hardcore	£100,000	
PV £1 pa in perp at 7%	14.2857	£1,428,572
Top slice	£20,000	
PV £1 pa in perp at 7% × PV £1 for 3 years at 7% (14.2857 × 0.8162979)	11.66145	£233,229
Value assuming no void costs or loss of rent		£1,661,801
Adjustments		
Insurance payable at the end of year 3, start of year 4	(£11,750)	
PV £1 for 3 years at 7%	0.8162979	(£9,592)
Empty rates payable after 3 months of £30,000	(£30,000)	
PV £1 for 3.25 years at 7%	0.802606	(£24,078)
Adjust for two years' rent of £120,000 not received due to voids	(£120,000)	
PV £1 pa for 2 years at 7% × PV £1 for 3 years at 7% (1.8080 × 0.8162979)	1.47586	(£177,104)
Plus 6 months' rent on in arrears assumptions to be paid at the end of year 5	£60,000	
PV £1 for 5 years at 7%	0.7129862	£42,779
Value allowing for voids and costs		**£1,493,805**

Note: Some approaches also deduct letting agent and legal fees at the end of the valuation.

Term and reversion or hardcore and layer?

Term and reversion	*Hardcore and layer*
• When the current rent is different from the market rent (under-rented properties usually) • When a lease is in place and will revert to a market rent at expiry • When valuing properties with a mix of older leases and expected rental growth upon reversion • Simple to apply in cases with clear reversion expectations • Recommended when the valuer feels more comfortable with the market rent at the review	• When the current rent is different from the market rent (under-rented and over-rented properties) • When the property generates a stable income stream but includes an over-/under-rented portion • When valuing properties with different income layers (e.g. stepped rents or rent reviews) • Can handle over-rented situations better • Recommended when the valuer feels the passing rent (core) is safer

Introducing DCF and discount rates

This section starts with DCF in blocks of income as a natural continuation of the traditional approaches, while at next sessions, discount rates and DCF will be discussed more analytically.

What makes discount rate better than the ARY and DCF preferable than traditional approaches?

While ARY-based methods can be useful for quick assessments, they lack transparency, fail to account for future rental performance comprehensively, and do not explicitly consider risk-adjusted returns. The DCF method using a discount rate provides a more structured, accurate, and investor-friendly approach to valuation. The discount rate is considered superior to the ARY in several key ways, particularly in investment valuations that involve changing cash flows over time. Here is why:

Transparency and explicit risk assessment

The ARY is often derived from market transactions, making it a blended yield that implicitly accounts for growth, risk, and expenses. However, it lacks transparency in how these components interact.

The discount rate, on the other hand, is explicitly chosen based on risk factors, such as property type, location, tenant strength, and macroeconomic conditions. This makes risk assumptions more transparent and justifiable.

Captures full lease cash flow, not just initial yield

Traditional methods like term and reversion and hardcore and layer focus mainly on the initial lease period and the first rent review, often oversimplifying future rental growth and risks. The DCF method using the discount rate accounts for the entire lease term, incorporating rental uplifts, void periods, and exit value, leading to a more accurate and complete valuation.

Time value of money consideration

ARY-based methods do not explicitly account for the timing of cash flows, which can be misleading, especially in long-term investments. The discount rate in DCF properly discounts future cash flows to present value, recognising that €1 today is worth more than €1 in the future.

Adaptability to changing market conditions and flexibility in valuation scenarios

The ARY is backward-looking, often derived from past transactions, and can fail to adjust quickly to changes in the economic climate, rental growth expectations, or interest rates. The discount rate is forward-looking, allowing for adjustments in market risk premiums and economic forecasts.

More suitable for institutional and international investors

Many institutional and international investors favour DCF models because they align with standard financial valuation techniques (e.g. those used in corporate finance and stock valuation). The discount rate is more universally understood, making it easier to compare property investments across different markets.

Investors typically assess property investments using IRR and NPV, both of which rely on discounted cash flow techniques rather than simple yield calculations. Using a discount rate aligns real estate valuation with broader financial industry standards, making it easier to compare property investments with alternative assets, such as bonds or equities.

DCF in blocks of income

While traditional investment methods, such as term and reversion and hardcore and layer, rely on the ARY, which can be challenging to determine due to market variations and subjectivity, a more transparent and structured approach is introduced in the following paragraphs. This is where the DCF in blocks becomes relevant. By applying a discount rate instead of an assumed implicit yield, ARY-DCF explicitly accounts for the time value of money, offering a clearer and more objective assessment of a property's worth. DCF explicitly accounts for the entire lease term, incorporating all rental uplifts and potential risks over time. This forward-looking approach provides a clearer and more objective evaluation by breaking down cash flows into distinct periods, ensuring a comprehensive assessment of income streams and their present value.

Constructing a DCF (in blocks)

Building a DCF model requires a structured approach that accurately reflects a property's future income potential while incorporating risk and time value considerations. The process begins with forecasting future cash flows, typically over the length of a lease or investment horizon. These cash flows include rental income, expected rent reviews, operating expenses, void periods, and potential capital expenditures. Each projected cash flow is then discounted to its present value using an appropriate discount rate, which reflects the investor's required return based on risk factors, such as market volatility, tenant strength, and economic conditions. The sum of these discounted cash flows determines the NPV of the property. Additionally, a terminal (exit) value is often calculated to account for the resale value at the end of the investment period, discounted back to today's value. By structuring the valuation in blocks, each period's cash flow is separately assessed, ensuring a transparent and adaptable methodology that captures the full income potential of the asset.

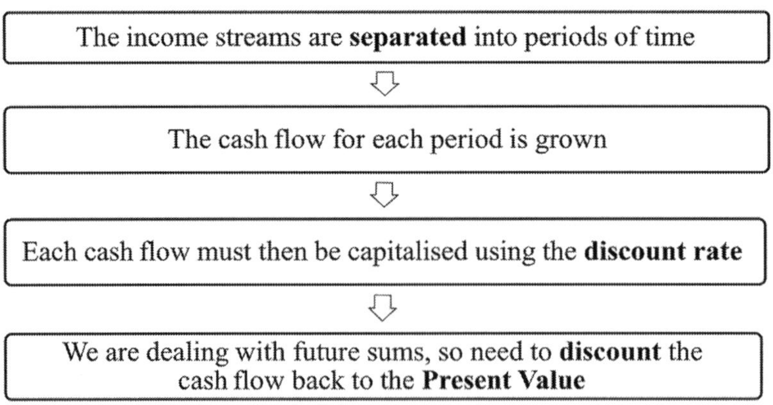

Figure 8.17 Steps in constructing a DCF in blocks of income.

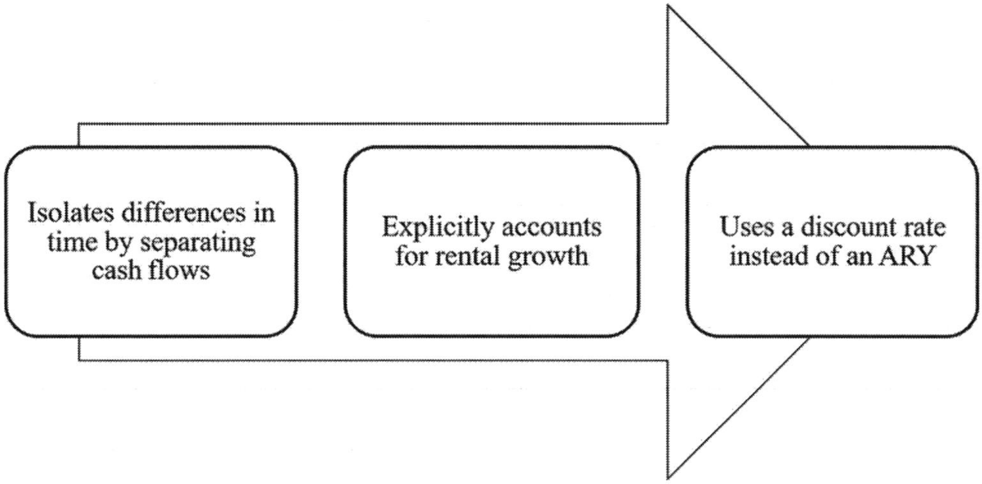

Figure 8.18 DCF mechanics.

In the following example, we build a simple cash flow.

Example 8.18

A shop was rented 3 years ago on a 20-year lease with 5-year rent reviews. The passing rent is €50,000 FRI (full repairing and insuring) per year, the MR is €58,000 FRI, whilst the ARY is estimated at 6%, and the discount rate is 8%. Calculate the MV of the property assuming that the annual rental growth is 2%. (Similar to Examples 8.11 and 8.16, but with the DCF in blocks approach, including explicitly a rental growth.)

Step 1: *Draw a timeline.* Create a straight line representing the total duration of the analysis, divided into equal intervals (e.g. five years per interval).

Step 2: *Mark review patterns.* Indicate the points on the timeline where reviews occur. If the reviews happen every five years, you would place markers at the 0-year, 5-year, 10-year, etc. points. Use visual indicators like dots, lines, or annotations to clearly represent review events.

Step 3: *Label key points.* Add labels to clarify the current year or starting point:

- The years until the next review
- The points of review along the timeline

Step 4: *Income for every period.* Calculate the income stream for every period. Use the YP formula with the discount rate to bring the income stream at the beginning of every period and to defer back to today with the PV formula. When the contract expires, calculate the reversion (capitalise the last income stream in perpetuity at ARY and defer back to today with PV).

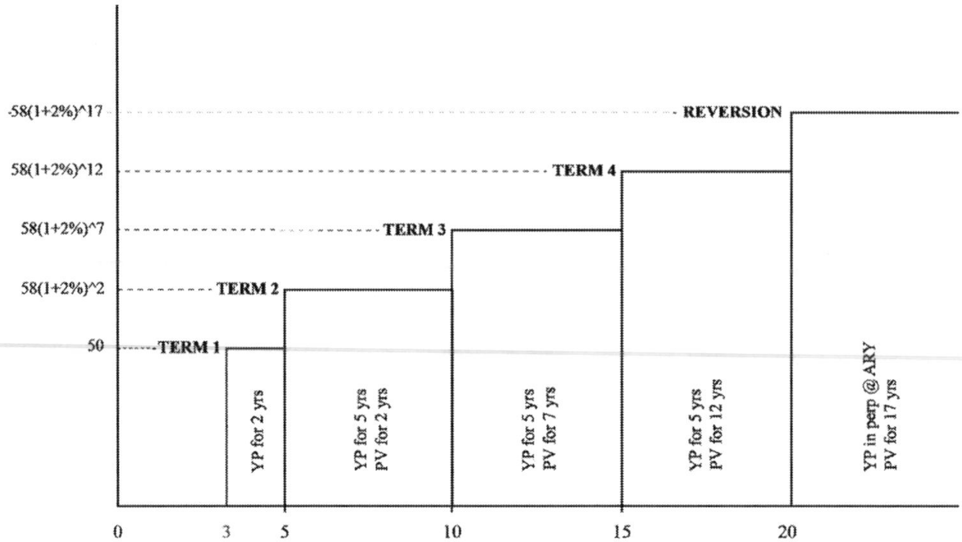

Figure 8.19 Graphic representation of Example 8.18. (This is similar to Examples 8.11 and 8.16.)

From the preceding, we can see that the first review will occur in two years. Following this, the situation can be addressed in a manner similar to the term and reversion example, but with more terms.

Term 1 (years 4–5) – 2 years
Income: €50,000

$$\text{YP for 2 years} = \left(1 - \frac{1}{(1+8\%)^2}\right) / 8\% = \underline{1{,}783}$$

€89,163 [Term 1]

Term 2 (years 6–10) – 5 years
Today's MR is estimated at €58,000, and a 2% growth is expected in the next review (2 years).
Income: $€58{,}000 \times (1+2\%)^2 = €60{,}343$

$$\text{YP for 5 years} = \left(1 - \frac{1}{(1+8\%)^5}\right) / 8\% = 3{,}993$$

$$\text{PV for 2 years} = \frac{1}{(1+8\%)^2} = \underline{0{,}857}$$

€206,494 [Term 2]

Term 3 (years 11–15) – 5 years
MR at year 11 is estimated at €58,000, with a 2% growth for 7 years.
Income: $€58,000 \times (1+2\%)^7 = €66,624$

$$\text{YP for 5 years} = \left(1 - \frac{1}{(1+8\%)^5}\right) / 8\% = 3,993$$

$$\text{PV for 7 years} = \frac{1}{(1+8\%)^7} = \underline{0,583}$$

€155,214 [Term 3]
Term 4 (years 16–20) – 5 years
MR at year 16 is estimated at €58,000, with a 2% growth for 12 years.
Income: $€58,000 \times (1+2\%)^{12} = €73,558$

$$\text{YP for 5 years} = \left(1 - \frac{1}{(1+8\%)^5}\right) / 8\% = 3,993$$

$$\text{PV for 12 years} = \frac{1}{(1+8\%)^{12}} = \underline{0,397}$$

€116,631 [Term 4]
Reversion (years 21 to perp)
MR at year 21 is estimated at €58,000, with a 2% growth for 17 years.
Income: $€58,000 \times (1+2\%)^{17} = €81,214$
YP in perp @ ARY $= 1/6\% = 16,667$

$$\text{PV for 17 years} = \frac{1}{(1+8\%)^{17}} = \underline{0.27}$$

€365,827 [Reversion]
MV = Term 1 + Term 2 + Term 3 + Term 4 + Reversion = €933,396.

Alternatively, this example is presented in the following data table format and in a full spreadsheet format.

Period	*Income*	*YP*	*PV*	*NCF*
Years 3–5	€50,000	1.783	1	€89,163
Years 5–10	€60,343	3.993	0.857	€206,561
Years 10–15	€66624	3.993	0.583	€155,214
Years 15–20	€73558	3.993	0.397	€116,631
Exit	€81214	16.667	0.270	€365,827
MV				**€933,396**

	A	B	C	D	E	F	G	H
1	Pasing rent	50000	(years)	period	Income	YP	PV	NCF
2	Market Rent	58000	2	years 3-5	=B1	=(1-(1+$B$5)^-C2)/$B$5	1	=E2*F2*G2
3	growth	0.02	5	years 5-10	=B2*(1+B3)^2	=(1-(1+$B$5)^-C3)/$B$5	=(1+$B$5)^-2	=E3*F3*G3
4	ARY	0.06	5	years 10-15	=B2*(1+B3)^7	=(1-(1+$B$5)^-C4)/$B$5	=(1+$B$5)^-7	=E4*F4*G4
5	Discount Rate	0.08	5	years 15-20	=B2*(1+B3)^12	=(1-(1+$B$5)^-C5)/$B$5	=(1+$B$5)^-12	=E5*F5*G5
6				exit	=B2*(1+B3)^17	=1/B4	=(1+$B$5)^-17	=E6*F6*G6
7							MV	=SUM(H2:H6)

Figure 8.20 Illustration of spreadsheet for Example 8.18.

The DCF in blocks of income is used in property valuation and investment analysis when income is received in distinct time periods rather than as a continuous stream. This method is commonly applied in cases where lease structures involve different terms, rent escalations, or expiration dates, making it necessary to model income changes accurately over time. It is also used in phased developments or refurbishments, where income may commence only after specific stages of completion. Additionally, DCF in blocks of income is crucial when dealing with rental reversions, as it accounts for the transition from current contracted rents to expected future market rents.

Furthermore, properties with planned lease breaks or renewals benefit from this approach, as it allows for the segmentation of income based on different occupancy scenarios. The method is also widely used in valuing mixed-use properties, where diverse income sources, such as residential, retail, and office spaces, generate cash flows at different rates and times.

Within property valuation, the investment method often employs DCF in blocks of income to reflect anticipated cash flows over a given period, discounting each block at an appropriate rate to determine present value. The structured approach of DCF in blocks of income provides a more precise valuation by considering variations in rental income, lease durations, and potential void periods, ultimately ensuring a more accurate representation of a property's financial performance and investment potential.

DCF (yearly) and discount rate

The DCF technique allows valuers to determine the present worth of future cash flows from a property investment by applying a discount rate. This technique aligns property valuation with financial investment analysis, providing a more explicit and structured approach to estimating value than traditional methods.

DCF has gained prominence due to its ability to account for time value of money, anticipated rental growth, and investment risk. While it was initially met with scepticism in property markets outside the United States, its adoption has been facilitated by advances in computing and financial modelling. This chapter will explore the concept of the discount rate, the mechanics of DCF calculations, and the practical application of this approach to property valuation.

Understanding the discount rate and its influencing factors

The *discount rate* is the required rate of return that an investor expects to achieve from a property investment. It reflects the opportunity cost of capital, risk profile, and anticipated future market conditions. The choice of an appropriate discount rate is critical, as it significantly affects the valuation outcome.

Various factors influence the selection of an appropriate discount rate, ranging from broad economic conditions to specific property characteristics. These factors help investors and valuers assess the potential return on investment while accounting for inflation, liquidity, market trends, and regulatory impacts. Understanding these influences allows for a more accurate and market-aligned valuation, ensuring that the applied discount rate appropriately reflects the risk-return profile of the asset in question. The main factors are discussed in what follows.

Risk profile

Higher-risk investments typically require higher discount rates to compensate for uncertainty in cash flows and market fluctuations.

Market conditions

Economic trends, inflation expectations, and interest rates significantly impact the discount rate. A low-interest-rate environment typically leads to lower discount rates, while periods of economic instability increase required returns.

Comparability with alternative investments

Investors assess property investments relative to bonds, equities, and other asset classes to determine if the property market provides an attractive risk-adjusted return.

Inflation expectations

A higher expected inflation rate often leads to a higher discount rate to ensure real returns remain competitive.

Liquidity and marketability

Less-liquid assets may warrant a higher discount rate due to difficulties in exiting the investment without significant price reductions.

Regulatory and tax considerations

Government regulations, tax incentives, and policy changes can impact required returns, influencing the discount rate applied in valuations.

Lease structures and investment horizons

The stability and structure of lease agreements, break clauses, and tenant creditworthiness influence the discount rate. Shorter leases with uncertain renewals typically require a higher discount rate than long-term, secure leases.

Capital market trends

The overall performance of financial markets, investor sentiment, and economic cycles play a significant role in setting discount rates. For example, during economic downturns, required returns tend to rise due to increased risk perception.

Property-specific risks

The condition, location, and usage of a property influence its risk profile. A high-risk investment, such as an underperforming retail centre in a declining market, would require a higher discount rate than a well-located, fully occupied office building.

Investment strategy and exit considerations

The intended holding period of an investment and expected exit strategies also affect the discount rate. Short-term investments with a high reliance on resale value may demand higher discount rates compared to long-term holdings focused on stable income generation.

Methods for determining the discount rate

Selecting an appropriate discount rate is a critical step in property valuation, as it directly affects the present value of future cash flows and ultimately determines investment feasibility. Various methods exist to establish a discount rate, each tailored to different valuation contexts, risk assessments, and market conditions. Some approaches rely on theoretical financial models, such as the capital asset pricing model (CAPM) and the weighted average cost of capital (WACC), which factor in systematic risk and capital structure. Others, like the market extraction method and survey-based approaches, derive discount rates from observed investor behaviour and market data. Additionally, historical benchmarking and the build-up method allow for practical estimations based on comparable property returns and specific risk premiums. The choice of method depends on the nature of the asset, the available market data, and the level of precision required in the valuation process.

Capital Asset Pricing Model (CAPM)

Determines the discount rate using the formula

$$r = R_f + \beta \times \left(R_m - R_f\right)$$

Where:

R_f = risk-free rate (e.g. government bond yield)
β = property beta (systematic risk measure relative to the market)
R_m = expected market return

CAPM is particularly useful in institutional property investments, where systematic risk factors influence returns.

Example 8.19

Suppose an investor is evaluating a commercial real estate property and wants to determine the appropriate discount rate using the CAPM, given the following data:

- **Risk-free rate (R_f) = 3%** (based on a ten-year government bond yield)
- **Market return (R_m) = 8%** (average return of the real estate market index)
- **Property beta (β) = 1.2** (measure of property's sensitivity to market movements)

Using the CAPM formula, $r = Rf + \beta(Rm - Rf) = 3\% + 1.2 \times (8\% - 3\%) = 3\% + 6\% = 9\%$.

Thus, the required discount rate for this property investment, based on CAPM, is 9%.

This means the investor would expect at least a 9% return per year to justify investing in this property, given its market risk exposure.

But how is beta (β) calculated?

Beta (β) measures the sensitivity of a property's or an asset's returns compared to the overall market returns. It quantifies systematic risk, indicating how much an asset's price moves in relation to market fluctuations. A beta of 1.0 means the asset moves in line with the market, while a beta greater than 1 suggests higher volatility than the market, and a beta less than 1 suggests lower volatility.

Formula for beta (β) calculation:

$$b = \frac{\operatorname{cov}(R_i, R_m)}{\operatorname{var}(R_m)}$$

Where:

Ri = returns of the individual property or asset
R_m = returns of the overall market (e.g. real estate index)
$cov(R_i, Rm)$ = covariance between the asset's returns and the market's returns
$var(R_m)$ = variance of the market's returns

Steps to calculate beta for real estate investments

1. *Collect data.* Gather historical return data for the property (or real estate sector) and a relevant market benchmark (e.g. real estate investment index, stock market index).
2. *Calculate returns.* Compute periodic percentage changes for both the asset and the market.
3. *Determine covariance.* Measure how the asset's returns move relative to the market.
4. *Calculate market variance.* Compute the variance of the market returns.
5. *Estimate beta.* Divide the covariance by the market variance.

Example 8.20

Estimating the beta of an individual property is challenging due to the lack of historical return data. However, beta can be assessed at the property sector level or from REITs. For example, in the US real estate market, betas can be derived from the annual total returns by property sector for REITs.

Example: beta estimation for real estate sectors from REITs

The following table displays the historical total returns for various REIT sectors – office, industrial, retail, residential, etc. – alongside the overall REIT market from 1994 to 2022, as published by Nareit (www.reit.com/). To estimate the beta for each sector, we calculate the covariance between each sector's returns (R_i) and the overall market (R_m), then divide by the market variance (R_m).

Year	*All REITs*	*Office*	*Industrial*	*Retail*	*Residential*	*Healthcare*	*Lodging/ resorts*	*Self-storage*
1994	0.8%	2.9%	18.7%	3.0%	2.3%	4.1%	−8.9%	8.9%
1995	18.3%	38.8%	16.2%	5.1%	12.0%	24.9%	30.8%	34.4%
1996	35.8%	51.8%	37.2%	34.6%	29.5%	20.4%	49.2%	42.8%
1997	18.9%	29.0%	19.0%	16.9%	16.3%	15.8%	30.1%	3.4%
1998	−18.8%	−17.3%	−11.7%	−4.7%	−8.1%	−17.4%	−52.8%	−7.2%
1999	−6.5%	4.3%	3.9%	−11.8%	9.5%	−24.8%	−16.1%	−8.0%
2000	25.9%	35.5%	28.6%	18.0%	34.3%	25.8%	45.8%	14.7%
2001	15.5%	6.6%	7.4%	30.4%	9.0%	51.9%	−8.6%	43.2%
2002	5.2%	−6.3%	17.3%	21.1%	−6.0%	4.8%	−1.5%	0.6%
2003	38.5%	34.0%	33.1%	46.8%	25.9%	53.6%	31.7%	38.1%
2004	30.4%	23.3%	34.1%	40.2%	32.7%	21.0%	32.7%	29.7%
2005	8.3%	13.1%	15.4%	11.8%	13.7%	1.8%	9.8%	26.5%
2006	34.4%	45.2%	28.9%	29.0%	38.9%	44.5%	28.2%	40.9%
2007	−17.8%	−19.0%	0.4%	−15.8%	−25.2%	2.1%	−22.4%	−24.8%
2008	−37.3%	−41.1%	−67.5%	−48.4%	−24.9%	−12.0%	−59.7%	5.0%
2009	27.4%	35.5%	12.2%	27.2%	30.8%	24.6%	67.2%	8.4%
2010	27.6%	18.4%	18.9%	33.4%	46.0%	19.2%	42.8%	29.3%
2011	7.3%	−0.8%	−5.2%	12.2%	15.4%	13.6%	−14.3%	35.2%
2012	20.1%	14.2%	31.3%	26.7%	6.9%	20.4%	12.5%	19.9%
2013	3.2%	5.6%	7.4%	1.9%	−5.4%	−7.1%	27.2%	9.5%
2014	27.1%	25.9%	21.0%	27.6%	40.0%	33.3%	32.5%	31.4%
2015	2.3%	0.3%	2.6%	4.6%	17.1%	−7.2%	−24.4%	40.6%
2016	9.3%	13.2%	30.7%	0.9%	4.5%	6.4%	24.3%	−8.1%
2017	9.3%	5.2%	20.6%	−4.8%	6.6%	0.9%	7.2%	3.7%
2018	−4.1%	−14.5%	−2.5%	−5.0%	3.1%	7.6%	−12.8%	2.9%
2019	28.1%	31.4%	48.7%	10.7%	30.9%	21.2%	15.7%	13.7%
2020	−5.9%	−18.4%	12.2%	−25.2%	−10.7%	−9.9%	−23.6%	12.9%
2021	39.9%	22.0%	62.0%	51.9%	58.3%	16.3%	18.2%	79.4%
2022	−25.1%	−37.6%	−28.6%	−13.3%	−31.3%	−22.2%	−15.3%	−26.7%

Beta calculation results:

Sector	*Covariance (R_i, R_m)*	*Variance (R_m)*	*Beta (β)*
Lodging/resorts	0.0495	0.0384	1.2876
Office	0.0420	0.0384	1.0927
Industrial	0.0400	0.0384	1.0401
Retail	0.0393	0.0384	1.0233
Residential	0.0380	0.0384	0.9890
Self-storage	0.0310	0.0384	0.8082
Healthcare	0.0308	0.0384	0.8030

From the preceding estimates, we observe that lodging/resorts REITs have the highest beta ($\beta = 1.2876$), indicating they are the most volatile compared to the overall REIT market. Healthcare REITs have the lowest beta ($\beta = 0.8030$), suggesting they experience less systematic risk relative to the market.

WACC

Utilised for projects financed through a mix of debt and equity:

$$WACC = \left(VE \times R_e\right) + \left(VD \times R_d \times \left(1 - T_c\right)\right)$$

Where:

E = market value of equity
D = market value of debt
V = total value of financing (E + D)
R_e = cost of equity
R_d = cost of debt
T_c = corporate tax rate

WACC is commonly used in corporate real estate portfolios where different capital structures influence the required return.

Market extraction method

This approach derives the discount rate by analysing recent property transactions, identifying discount rates applied by investors in similar assets. Market-based discount rates provide practical and real-time valuation benchmarks.

Survey-based approaches

Institutional investors, real estate funds, and valuation firms conduct surveys to estimate prevailing discount rates. These are particularly useful when market transactions are scarce.

Historical data and benchmarking

Analysing historical returns of comparable property investments helps determine discount rates for similar assets, accounting for risk, growth expectations, and economic conditions.

Build-up method

A discount rate is constructed by adding a risk-free rate to different risk premiums associated with the property's characteristics, market conditions, and capital costs. This method is frequently applied in real estate development valuations.

The relationship between discount rate and capitalisation rate

The discount rate and capitalisation rate (cap rate) are interrelated but distinct concepts in valuation.

Discount rate

Applies to future cash flows and includes both income and capital appreciation components.

Cap rate

A simplified measure based on a single period's income, calculated as:

$$Cap\ rate = \frac{Net\ Operating\ Income}{Property\ Value}$$

While the discount rate considers long-term investment performance, the cap rate is often used for quick comparisons of property investments. The cap rate can be derived from the discount rate using the following approximation:

$$Cap\ Rate = Discount\ Rate - Expected\ Growth\ Rate$$

This relationship demonstrates how higher expected income growth allows for lower cap rates while maintaining investor return expectations. Understanding this interplay helps valuers differentiate between implicit and explicit income growth assumptions in different valuation approaches.

By accurately determining the discount rate, valuers and investors can make informed decisions that align with market trends, risk tolerance, and investment objectives.

DCF steps

Step 1: *Project future cash flows.* Identify expected income streams, including rental income and other revenues, along with anticipated expenses, such as maintenance, management, and taxes.

Step 2: *Determine the discount rate.* Select an appropriate rate based on investment risk and market conditions.

Step 3: *Discount future cash flows.* Apply the PV formula discount future cash flows to their present value:

$$PV = \frac{FV}{(1+r)^n}$$

Where:

FV = future value
PV = present value
r = discount rate
n = number of periods

Step 4: *Calculate the terminal value.* Estimate the property's value at the end of the holding period. The most common method is the direct capitalisation approach:

$$TV = \frac{NOI_{n+1}}{r-g}$$

Where:

TV = terminal value
NOI_{n+1} = NOI in the year after the holding period
r = discount rate
g = expected growth rate of NOI

The terminal value is then discounted back to the present value using the discount rate.

Step 5: *Sum the present values.* Add the discounted cash flows and the terminal value to derive the total present value of the investment.

By following these steps, investors and valuers can accurately estimate the intrinsic value of a property and make informed investment decisions.

Example 8.21

A real estate investor is evaluating the purchase of a commercial office building that generates an initial NOI of €500,000 per year. The investor expects the NOI to grow annually at 2% due to rental escalations and market appreciation. The planned holding period is ten years, after which the investor intends to sell the property.

At the end of the tenth year, the investor estimates the resale price (also known as terminal value) by applying a capitalisation rate of 7% to the NOI in year eleven.

The investor's required discount rate is 9%, reflecting the opportunity cost of capital and risk associated with the investment. The objective is to determine the present value of the projected cash flows and estimate the fair value of the investment.

Step 1: *Forecast the NOI for the holding period.* Since the NOI grows at **2% per year**, we calculate the NOI for each year using the following formula:

$$NOI_t = NOI_0 \times (1+g)^t$$

Where:

NOI_t = the net operating income in year t
NOI_0 = 500,000 (initial income)
g = 2%, or 0.02 (growth rate)

The NOI projections for the ten-year period are calculated.

Step 2: *Estimate the TV.* The TV is the expected sale price at the end of year 10. It is determined by capitalising the NOI in year 11 using the formula:

$$TV = \frac{NO_{11}}{Cap\ Rate}$$

Where:

$$NOI_{11} = NOI_{10} \times (1+g)$$

$Cap\ Rate$ = 7%, or 0.07

Step 3: *Discount each cash flow to PV.* Each year's NOI and the terminal value must be discounted back to the present value using the formula:

$$PV = \frac{NOI_t}{(1+r)^t}$$

Where:

r = 9% (discount rate).

The TV is also discounted: $PV_{TV} = \dfrac{TV}{(1+r)^{10}}$

Step 4: *Sum the present values.* The sum of all discounted NOI values and the discounted TV gives the total present value (property value).

Solution calculation

Inputs

NOI	**€500,000**
Discount rate	**9%**
Cap rate	**7%**
Growth	**2%**

Year	*NOI*	*TV*	*PV factor*	*DCF*
1	€510,000		0.9174	€467,890
2	€520,200		0.8417	€437,842
3	€530,604		0.7722	€409,724
4	€541,216		0.7084	€383,411
5	€552,040		0.6499	€358,788
6	€563,081		0.5963	€335,747
7	€574,343		0.547	€314,185
8	€585,830		0.5019	€294,008
9	€597,546		0.4604	€275,127
10	€609,497	€8,881,245	0.4224	€4,008,992
11	€621,687			
Total present value (estimated MV)				**€7,285,714**

Showing how you would execute the calculation in a spreadsheet format with formulas:

	A	B	C	D	E	F	G
1	NOI	500000	YEAR	NOI	TV	PV	CF
2	discount rate	0.09	1	=$B$1*(1+$B$4)^C2		=(1+$B$2)^-C2	=(D2+E2)*F2
3	Cap rate	0.07	2	=$B$1*(1+$B$4)^C3		=(1+$B$2)^-C3	=(D3+E3)*F3
4	Growth	0.02	3	=$B$1*(1+$B$4)^C4		=(1+$B$2)^-C4	=(D4+E4)*F4
5			4	=$B$1*(1+$B$4)^C5		=(1+$B$2)^-C5	=(D5+E5)*F5
6			5	=$B$1*(1+$B$4)^C6		=(1+$B$2)^-C6	=(D6+E6)*F6
7			6	=$B$1*(1+$B$4)^C7		=(1+$B$2)^-C7	=(D7+E7)*F7
8			7	=$B$1*(1+$B$4)^C8		=(1+$B$2)^-C8	=(D8+E8)*F8
9			8	=$B$1*(1+$B$4)^C9		=(1+$B$2)^-C9	=(D9+E9)*F9
10			9	=$B$1*(1+$B$4)^C10		=(1+$B$2)^-C10	=(D10+E10)*F10
11			10	=$B$1*(1+$B$4)^C11	=D12/B3	=(1+$B$2)^-C11	=(D11+E11)*F11
12			11	=$B$1*(1+$B$4)^C12			
13				Total Present Value (estimated MV)		PV	**=SUM(G2:G11)**

Figure 8.21 Illustration of spreadsheet for Example 8.21.

Example 8.22

A multinational real estate developer purchased commercial real estate in Riga, Latvia, and constructed a mixed-use structure with offices on the higher levels and stores on the first floor. The project is totally rented and just completed. While offices are rented for €350,000 per year, with contracts offering a steady increase of 3% annually and an occupancy rate of 80%, shops are rented for €200,000 per year under a five-year contract with fixed rent. Additionally, an electronic billboard that generates an average of €1,500 each month was built by the developer. It is anticipated that management expenses will account for 10% of overall revenue. The developer intends to keep his investment for five years before selling. It is expected that the marketing period after the end of the five-year period will be 12 months until sale, and the amount of €60,000 is the estimated cost for a small renovation (to be used beginning of year six). In year six, the only income will be from the billboard.

Market data

- The current expected ARY for offices is 5.75%, while for stores it is 6.25%.
- At the time of valuation, the risk-free rate on five-year government bonds was 4.1%.
- According to a recent estimate based on data from a well-known Latvian property index, commercial stores will rise by 8% overall over the next five years.
- It is projected that taxes and selling costs will account for 7% of the sale price.
- All payments are in arrears unless stated otherwise.

It appears that this scenario is more complicated, so the proposed solution will be presented in a spreadsheet.

Step 1: Structure your Excel file

Before starting any calculations, it is essential to set up your Excel spreadsheet logically. **Use separate sections** for inputs, calculations, and final valuation.

Inputs and assumptions

- Clearly define **key input variables**, such as rental income, growth rates, expenses, discount rate, and cap rates.
- Label each input **clearly** to avoid confusion.
- Use **named ranges** (e.g. 'Shop_Rent', 'Office_Rent', 'Growth_Rate') to make formulas easier to read.

Income and expense forecasts

- Create a **yearly projection table** (years 1–5) for each income stream:
  - Retail shops
  - Office spaces
  - Billboard advertising
- Include **growth adjustments** for offices and inflation impacts if applicable.
- Calculate **total revenue** by summing all income sources.

NOI calculation

- Deduct **management costs** (as a percentage of revenue).
- Compute **NOI** for each year.
- Check your formulas to ensure there are no **logical errors**.

DCF and TV calculations

- Apply **discounting formulas** to bring future cash flows to **PV**.
- Use =**PV(discount rate, year, cash flow)**, or manually discount each cash flow.
- Calculate the **TV** at the end of year 5.
- Capitalise **year 6 NOI** using cap rates.
- Deduct **selling costs and taxes** from the estimated sale price.
- Discount the **net TV** back to the present.

Summary and valuation conclusion

- Sum all discounted NOI and discounted TV to get the total present value of the property.
- Compare this with potential purchase price to evaluate if the investment is viable.

Key tips for students

1. Use Excel formulas effectively.
    - Avoid hard-coding values; instead, link cells to input data.
    - Use formulas like =**FV()**, =**PV()**, and =**NPV()** for discounting cash flows.
2. Check for logical consistency.
    - Ensure office rent grows correctly (compounding 3% per year).
    - Verify total income calculations each year.
3. Make your spreadsheet readable.
    - Use **cell formatting**, colour-coding (e.g. blue for inputs, black for calculated values), and comments to clarify formulas.
    - Apply **data validation** to prevent incorrect entries.
4. Scenario/sensitivity analysis (optional but recommended).
    - Add a sensitivity table to test different cap rates, rent growth rates, or discount rates.
    - Use **data table or Goal Seek** to explore various outcomes.

Key inputs/assumptions	
Offices rent	€350,000
Offices occupancy	80%
Offices yearly increase	3%
Shops rent	€200,000
Shops occupancy	100%
Billboard (monthly avg.)	€1,500
Management expenses	10%
Offices cap rate	5.75%
Shops cap rate	6.25%

(Continued)

(Continued)

Key inputs/assumptions	
Selling costs/taxes	7%
Five-year bond (RF)	4.10%
Total growth for five years (shops)	8%
Renovation cost	€60,000
Discount rate (RF + premiums)	10%

With the use of the *build-up model*, we adopt 10% as the appropriate discount rate.

	year	*1*	*2*	*3*	*4*	*5*	*6*
Income	Offices Rent	€ 280,000	€ 288,400	€ 297,052	€ 305,964	€ 315,142	
	Shops Rent	€ 200,000	€ 200,000	€ 200,000	€ 200,000	€ 200,000	
	Billboard	€ 18,000	€ 18,000	€ 18,000	€ 18,000	€ 18,000	€ 18,000
	Sale Offices						€ 5,645,161
	Sale Shops						€ 3,456,000
	Total Income	€ 498,000	€ 506,400	€ 515,052	€ 523,964	€ 533,142	€ 9,119,161
Expenses	Mngmnt Expenses	€ (49,800)	€ (50,640)	€ (51,505)	€ (52,396)	€ (53,314)	
	Selling costs						€ (637,081)
	renovation					€ (60,000)	
	Total Expenses	€ (49,800)	€ (50,640)	€ (51,505)	€ (52,396)	€ (113,314)	€ (637,081)
	FREE CASH FLOW	€ 448,200	€ 455,760	€ 463,547	€ 471,567	€ 419,828	€ 8,482,079
					Market Value (NPV)		€ 6,503,065

Figure 8.22 Calculations for Example 8.22.

And showing the calculations and formulae in full spreadsheet format:

	A	B	C	D	E	F	G	H	I	J	K
1	KEY INPUTS / ASSUMPTIONS								year		
2	Offices Rent	350000			*year*	*1*	*2*	*3*	*4*	*5*	*6*
3	Offices Occupancy	0.8		*Income*	=A2	=$B$2*$B$3	=F3*(1+$B$4)	=G3*(1+$B$4)	=H3*(1+$B$4)	=I3*(1+$B$4)	
4	Offices yearly increase	0.03			=A5	=$B$5	=$B$5	=$B$5	=$B$5	=$B$5	
5	Shops Rent	200000			Billboard	=$B$7*12	=$B$7*12	=$B$7*12	=$B$7*12	=$B$7*12	=$B$7*12
6	Shops Occupancy	1			Sale Offices						=J3*(1+B4)/B9
7	Billboard (monthly avg)	1500			Sale Shops						=J4*(1+B13)/B10
8	Mngmnt Expenses	0.1			Total Income	=SUM(F3:F7)	=SUM(G3:G7)	=SUM(H3:H7)	=SUM(I3:I7)	=SUM(J3:J7)	=SUM(K3:K7)
9	Offices Cap Rate	0.0575		*Expenses*	=A8	=-F8*$B$8	=-G8*$B$8	=-H8*$B$8	=-I8*$B$8	=-J8*$B$8	
10	Shops Cap Rate	0.0625			Selling costs						=-B11*(K6+K7)
11	Selling costs / Taxes	0.07			renovation					=-B14	
12	5-year Bond (RF)	0.041			Total Expenses	=SUM(F9:F11)	=SUM(G9:G11)	=SUM(H9:H11)	=SUM(I9:I11)	=SUM(J9:J11)	=SUM(K9:K11)
13	Total Growth for 5yrs (shops)	0.08			FREE CASH FLOW	=F8+F12	=G8+G12	=H8+H12	=I8+I12	=J8+J12	=K8+K12
14	Renovation cost	60000									
15	Discount Rate (RF + premiums)	0.1							Market Value (NPV)		=NPV(B15,F13:K13)

Figure 8.23 Illustration of spreadsheet for Example 8.22.

Introducing NPV and IRR

DCF is used predominantly in the analysis of all forms of investment opportunities, but it is the preferred method in some countries or states for undertaking an income approach valuation. In its general form for investment analysis, it has two standard products, NPV and IRR. It is worth considering these first before looking at DCF as a valuation method where the method is used to determine the PV of a cash flow.

NPV is the sum of the present values of all future cash flows associated with an investment, discounted at a specified rate (the investor's required rate of return or cost of capital). It determines whether an investment will add value or result in a net loss.

The NPV formula is:

$$NPV = \sum_{t=0}^{n} \frac{C_t}{(1+r)^t} - C_0$$

Where:

C_t = net cash inflow at time t
r = discount rate (required rate of return)
t = time period
C_0 = initial investment

If NPV > 0, the investment is profitable and should be accepted.
If NPV = 0, the investment breaks even but does not create additional value.
If NPV < 0, the investment results in a net loss and should be rejected.

Future net benefits receivable from the investment to be analysed are discounted at a given *target rate*. The sum of the discounted benefits is found, and the initial cost of the investment deducted from this sum, to leave what is termed the NPV of the investment, which may be positive or negative. A positive NPV implies that a rate of return greater than the target rate is being yielded by the investment; a negative NPV implies that the yield is lower than the target rate. The target rate is the minimum rate the investor requires in order to make the investment worthwhile, taking into account the risks involved and all other relevant factors. It will be governed, to some extent, by the investor's cost of capital. The target rate may be called the hurdle rate or discount rate. Excel and other software programmes can complete NPV calculations for known cash flows with constant cash flows once the user has provided the target rate, the constant payment per period, and the number of payments due.

Investments may require the initial outlay of a large capital sum, and investors will often borrow such sums of money to complete their purchases. The interest to be paid on that loan will need to be covered by the investment income with a sufficient additional amount to provide an adequate return to the investor on the balance of capital provided by them. The investor's target rate will need to reflect both the interest on capital borrowed and an adequate return on the equity provided by the investor.

An alternative way of looking at this is that investors will always have alternative opportunities for the investment of their capital. Money may be lent quite easily to earn a rate of interest based on the cost of capital. The return from any investment should therefore compare favourably with the opportunity cost of the funds employed, and this will usually be related to the cost of capital. For these reasons, the target rate should compare well with the cost of capital. From this basis, a positive or negative NPV will be the result of the analysis, and upon this result, an investment decision can be made.

Example 8.23

Find the NPV of the following project using a target rate of 13%.

Outlay £10,000

	Income £s	*PV £1 at 13%*	*Discounted sum £s*
Returns in year 1	5,000	0.8849	4,425
Returns in year 2	4,000	0.7831	3,133
Returns in year 3	6,000	0.6930	4,158
			£11,716
		Less outlay	£10,000
		NPV	**£1,716**

The investment yields a return of 13% and, in addition, a positive NPV of £1,716. In the absence of other choices, this investment may be accepted because the return exceeds the target rate. However, it is more usual for the investment decision to be one which requires a choice to be made between two or more investments.

Example 8.24

An investor has £1,400 to invest and has a choice between investments A and B. The following returns are anticipated:

Year	*Income flow (£s)*	
	A	*B*
1	600	200
2	400	400
3	200	400
4	400	600
5	400	600

The investor's target rate for both investments is 10%. The investments are mutually exclusive (only one of the two can be undertaken). Which investment should be chosen?

Income flow A			
Year	*Income £s*	*PV £1 at 10%*	*Discounted sum £s*
1	600	0.9091	545
2	400	0.8264	330
3	200	0.7513	150
4	400	0.6830	273
5	400	0.6209	248
			1,546
		Less outlay	1,400
		NPV	146

Income flow B			
Year	*Income £s*	*PV £1 at 10%*	*Discounted sum £s*
1	200	0.9091	182
2	400	0.8264	330
3	400	0.7513	300
4	600	0.6830	409
5	600	0.6209	373
			1,594
		Less outlay	**1,400**
		NPV	**194**

From this information, Investment B should be chosen. Each investment gives a return of 10% plus a positive NPV. B produces the greater NPV of £194 compared to A at £146.

If either investment had been considered to be subject to more risk, that factor would have been reflected in the choice of target rate.

In Example 8.24 both investments involved the same amount of capital or initial outlay. However, this will not always be the case, and when outlays on mutually exclusive investments differ, the investment decision will not be so simple. An NPV of £200 from an investment costing £250 is considerably more attractive than a similar NPV produced from a £25,000 outlay. How can this be reflected in an analysis? A possible approach is to express the NPV as a percentage of the outlay.

Which of these mutually exclusive investments should be undertaken when the investor's target rate is 10%?

Year	*Investment A*	*Investment B*
	Outlay £5,000	*Outlay £7,000*
1	£3,000	£4,000
2	£2,000	£3,000
3	£1,500	£2,000

(Continued)

(Continued)

Income flow A

Year	*Income (£s)*	*PV £1 at 10%*	*Discounted sum (£s)*
1	3,000	0.9091	2,727
2	2,000	0.8264	1,653
3	1,500	0.7513	1,127
			5,507
		Less outlay	5,000
		NPV	**507**

Income flow B

Year	*Income (£s)*	*PV £1 at 10%*	*Discounted sum (£s)*
1	4,000	0.9091	3,636
2	3,000	0.8264	2,479
3	2,000	0.7513	1,503
			7,618
		Less outlay	7,000
		NPV	**618**

At first sight, B might appear to be more profitable, but if the NPV is expressed as a percentage of outlay, the picture changes.

$$A = \frac{£507}{£5,000} \times 100 = 10.14\%$$

$$B = \frac{£618}{£7,000} \times 100 = 8.33\%$$

On this basis, A, and not B, should be chosen.

The NPV method is a satisfactory aid in the great majority of investment problems but suffers from one particular disadvantage. The return provided by an investment is expressed in two parts, a rate of return and a cash sum in addition, which represents an extra return. These two parts are expressed in different units, which may make certain investments difficult to compare.

This problem is not present in the following method of expressing the results of a DCF analysis.

Internal rate of return

IRR is the discount rate that makes the NPV of an investment equal to zero. It represents the rate at which future cash inflows equal the initial investment outlay.

The IRR formula is:

$$0 = \sum_{t=0}^{n} \frac{C_t}{(1+IRR)^t} - C_0$$

Where:

C_t = net cash inflow at time t
r = discount rate (required rate of return)
t = time period
C_0 = initial investment

Solving for IRR requires numerical methods, such as trial and error, or using financial software.

If IRR > required rate of return, the investment is attractive.
If IRR = required rate of return, the investment is acceptable but not superior to other opportunities.
If IRR < required rate of return, the investment should be rejected.

It produces an NPV of zero and may be found by the use of various trial discount rates. It may also be defined as the discount rate at which the NPV is zero and so represents an annualised rate of return throughout the life of the cash flow period.

Example 8.26

Find the IRR of the following investment.

Outlay £6,000

	Returns	Year 1	£1,024
		Year 2	£4,000
		Year 3	£3,000

Trying 10%

Year	*£s*	*PV £1 at 10%*	*Discounted sum (£s)*
1	1,024	0.9091	931
2	4,000	0.8264	3,306
3	3,000	0.7513	2,253
			6,490
		Less outlay	6,000
		NPV	**£490**

At a trial rate of 10%, a positive NPV results. £490 is too high – an NPV of 0 is required. The trial rate must be too low, as the future receipts need to be discounted to a greater extent.

Trying 16%			
Year	*£s*	*PV £1 at 16%*	*Discounted sum (£s)*
1	1,024	0.8621	886
2	4,000	0.7432	2,972
3	3,000	0.6407	1,923
			5,778
		Less outlay	6,000
		NPV	**−222**

This time the receipts have been discounted too much. A negative NPV is the result, so the trial rate is too high. The IRR must be between 10% and 16%.

Trying 14%			
Year	*£s*	*PV £1 at 14%*	*Discounted sum (£s)*
1	1,024	0.8772	899
2	4,000	0.7695	3,076
3	3,000	0.6750	2,025
			6,000
		Less outlay	6,000
		NPV	**0,000**

As the NPV is £0,000, the IRR must be 14%.

Calculation of the IRR by the use of trial rates will be difficult when the IRR does not happen to coincide with a round figure, as in the following illustration.

Outlay £4,925			
Year	*£s*	*PV £1 at 11%*	*Discounted sum (£s)*
1	2,000	0.9009	1,802
2	2,000	0.8116	1,623
3	2,000	0.7312	1,462
			4,887
		Less outlay	4,925
		NPV	**−38**

Using 11%, the trial rate is too high.

Trying 10%			
Year	*£s*	*PV £1 at 10%*	*Discounted sum (£s)*
1	2,000	0.9091	1,818
2	2,000	0.8264	1,652
3	2,000	0.7513	1,503
			4,973
		Less outlay	4,925
		NPV	**48**

The IRR is therefore between 10% and 11%.

To continue using trial rates to make a more accurate estimate of the IRR would require the analyst to work from, say, 10.001% in 0.001% steps until the NPV is zero. This is impracticable. The analysis has shown that the IRR lies between 10% and 11%, but the accuracy of such an analysis is limited.

In practice, IRR calculations are run on software such as Excel (see p. 224 (Figure 8.24)). It is possible to use linear interpolation, as set out in the following, to obtain a better result, but again, this cannot be precise. This material has been retained in this edition for academic completeness.

Linear interpolation can be carried out by the use of the following formula, where the lower trial rate = *ltr* and the higher trial rate = *htr*. The formula simply assumes a straight-line relationship between the trial rates and the resulting NPVs:

$$\text{IRR} = ltr + \frac{NPVltr}{NPVltr + NPVhtr} \times (htr - ltr)$$

In this case, the lower trial rate is 10%, the higher trial rate is 11%, the NPV at the lower trial rate is £48, and the NPV at the higher trial rate is −£38, a total of £86 ignoring the signs.

The difference in trial rates is 1%.

$$IRR = 10\% + \left(\frac{48}{48+38}\right) \times (11-10)$$

$$\text{IRR} = 10\% + \left(\frac{48}{36} \times 1\right)$$

$$\text{IRR} = 10\% + 0.558$$
$$\text{IRR} = 10.558\%$$

The equation is spreading £86 out equally along a line representing 1%. At the point on the line where it reaches £0, the basis points (1/10 or 1/100) are noted and added to the lower figure of 10% to arrive at 10.588%, say, 10.56%.

Over a difference of 1%, linear interpolation to two decimal places will be sufficiently accurate for most purposes. However, it must be borne in mind that the mathematical relationship here is geometric, not arithmetic; it is not a straight line. Because of this, linear interpolation is inaccurate to a certain extent, and so results should not be expressed to too many decimal places, that is, should not infer accuracy.

As mentioned earlier, IRR calculations are more readily undertaken using an investment calculator or 'Goal Seek' on Microsoft Excel, which provides a full solution, which in this example is 10.560884981651%.

Excel repeats the IRR calculations very rapidly until it finds the percentage rate which discounts the cash flows less the cash outlay to equal exactly £0.

	A	B	C	D	E
1					
2			Target Rate	11%	
3			Outlay	£4,925	
4		Period	Cash Flow	PV £1 at D2	Discounted sum
5		1	£2,000	0.904479	£1,808.96
6		2	£2,000	0.8180823	£1,636.16
7		3	£2,000	0.7399383	£1,479.88
8		Present value			£4,925.00
9		Less outlay			£4,925
10		NPV			-£0.00

Figure 8.24 Goal Seek to the nearest whole percent.

This Excel table shows a Goal Seek solution for Example 8.26, where 10.650% is shown rounded to 11%.

A normal DCF with a target rate is set up for the problem on Excel. The procedure is as follows:

1. Click Data Tab on the toolbar.
2. Click What-If Analysis.
3. Click Goal Seek in the drop-down menu.
4. Set NPV cell, in this case, E10, to the value of £0.
5. Insert target rate, here cell D3, in 'By changing cell'.
6. Click OK.
7. Move cursor to the target rate cell and the full IRR is displayed in the equation bar; alternatively, cell D2 can now be expanded to the required number of decimal places.

NPV and IRR as investment metrics

NPV and IRR provide insights into whether an investment will generate acceptable returns given a required discount rate. While NPV measures absolute profitability in monetary terms, IRR expresses returns as a percentage rate of return. These concepts are critical in real estate investment, corporate finance, and infrastructure projects.

Feature	*NPV*	*IRR*
Output	Monetary value (£, $, €, etc.)	Percentage (%)
Decision rule	Accept if NPV > 0.	Accept if IRR > required return.
Indicates	Absolute value creation	Efficiency of investment
Multiple projects	Works well for different scales	Can be misleading with different scales

Advantages of NPV	Limitations of NPV
• Accounts for TVM • Provides a direct profitability measure • Works well for projects with different lifespans and scales	• Requires an appropriate discount rate • Does not express return as a percentage
Advantages of IRR	*Limitations of IRR*
• Easy to interpret as a percentage return • Useful when comparing different investments	• Can give multiple values in non-conventional cash flows • Does not work well for mutually exclusive projects

NPV and IRR are essential tools in investment analysis, particularly in real estate, project finance, and corporate decision-making. NPV is preferred when selecting projects based on absolute profitability, whereas IRR is more intuitive when comparing investments based on percentage returns, and understanding the differences and applications of these methods ensures sound financial decision-making and helps maximise investment returns.

Example 8.27

A real estate investor is evaluating the purchase of a commercial office building that generates an initial NOI of €500,000 per year. The investor expects the NOI to grow annually at 2% due to rental escalations and market appreciation. The planned holding period is ten years, after which the investor intends to sell the property.

At the end of the 10th year, the investor estimates the resale price (also known as TV) by applying a capitalisation rate of 7% to the NOI in year 11.

The investor's required discount rate is 9%, reflecting the opportunity cost of capital and risk associated with the investment. The objective is to determine the present value of the projected cash flows and estimate the fair value of the investment.

Step 1: *Forecast the NOI for the holding period.* Since the NOI grows at **2% per year**, we calculate the NOI for each year using the formula:

$$NOI_t = NOI_0 \times (1+g)^t$$

Where:

NOI_t = the net operating income in year t
NOI_0 = 500,000 (initial income)
g = 2%, or 0.02 (growth rate)

The NOI projections for the ten-year period are calculated.

Step 2: *Estimate the TV.* The *TV* is the expected sale price at the end of year 10. It is determined by capitalising the NOI in year 11 using the formula:

$$TV = \frac{NO_{11}}{Cap\ Rate}$$

Where:

$$NOI_{11} = NOI_{10} \times (1+g)$$

Cap Rate = 7% or 0.07

Step 3: *Discount cash flows to PV.* Each year's NOI and the TV must be discounted back to present value using the formula:

$$PV = \frac{NOI_t}{(1+r)^t}$$

Where:

r = 9% (discount rate)

The TV is also discounted: $PV_{TV} = \frac{TV}{(1+r)^{10}}$

Step 4: *Sum the present values.* The sum of all discounted NOI values and the discounted TV gives the total present value (property value).

Solution calculation

Year	*NOI*	*TV*	*PV factor*	*DCF*
1	€510,000		0.917	€467,890
2	€520,200		0.842	€437,842
3	€530,604		0.772	€409,724
4	€541,216		0.708	€383,411
5	€552,040		0.65	€358,788
6	€563,081		0.596	€335,747
7	€574,343		0.547	€314,185
8	€585,830		0.502	€294,008
9	€597,546		0.46	€275,127
10	€609,497	€8,881,245	0.422	€4,008,992
11	€621,687			
Total present value (estimated MV)				**€7,285,714**

Showing how you would execute the calculation in a spreadsheet formula:

	A	B	C	D	E	F	G
1	NOI	500000	YEAR	NOI	TV	PV	CF
2	discount rate	0.09	1	=$B$1*(1+$B$4)^C2		=(1+$B$2)^-C2	=(D2+E2)*F2
3	Cap rate	0.07	2	=$B$1*(1+$B$4)^C3		=(1+$B$2)^-C3	=(D3+E3)*F3
4	Growth	0.02	3	=$B$1*(1+$B$4)^C4		=(1+$B$2)^-C4	=(D4+E4)*F4
5			4	=$B$1*(1+$B$4)^C5		=(1+$B$2)^-C5	=(D5+E5)*F5
6			5	=$B$1*(1+$B$4)^C6		=(1+$B$2)^-C6	=(D6+E6)*F6
7			6	=$B$1*(1+$B$4)^C7		=(1+$B$2)^-C7	=(D7+E7)*F7
8			7	=$B$1*(1+$B$4)^C8		=(1+$B$2)^-C8	=(D8+E8)*F8
9			8	=$B$1*(1+$B$4)^C9		=(1+$B$2)^-C9	=(D9+E9)*F9
10			9	=$B$1*(1+$B$4)^C10		=(1+$B$2)^-C10	=(D10+E10)*F10
11			10	=$B$1*(1+$B$4)^C11	=D12/B3	=(1+$B$2)^-C11	=(D11+E11)*F11
12			11	=$B$1*(1+$B$4)^C12			
13				Total Present Value (estimated MV)		PV	**=SUM(G2:G11)**

Figure 8.25 Illustration of spreadsheet for Example 8.27.

Given that the purchase price of the investment is €7,000,000, calculate the NPV and IRR of the investment. To do so, include year 0 and the purchase amount in the calculation.

YEAR	Purchase	NOI	TV	CASHFLOW	PV	CF
0	-€ 7,000,000			-€ 7,000,000	1	-€ 7,000,000
1		€ 510,000		€ 510,000	0.9174	€ 467,890
2		€ 520,200		€ 520,200	0.8417	€ 437,842
3		€ 530,604		€ 530,604	0.7722	€ 409,724
4		€ 541,216		€ 541,216	0.7084	€ 383,411
5		€ 552,040		€ 552,040	0.6499	€ 358,788
6		€ 563,081		€ 563,081	0.5963	€ 335,747
7		€ 574,343		€ 574,343	0.5470	€ 314,185
8		€ 585,830		€ 585,830	0.5019	€ 294,008
9		€ 597,546		€ 597,546	0.4604	€ 275,127
10		€ 609,497	€ 8,881,245	€ 9,490,742	0.4224	€ 4,008,992
11		€ 621,687				
					NPV	**€ 285,714**
					IRR	**9.58%**

Figure 8.26 Data table for Example 8.27.

Showing how you would execute the calculation in a spreadsheet with formulae:

	A	B	C	D	E	F	G	H	I
1	NOI	500000	YEAR	Purchase	NOI	TV	CASHFLOW	PV	CF
2	discount rate	0.09	0	=-B5			=SUM(D2:F2)	1	=G2*H2
3	Cap rate	0.07	1		=$B$1*(1+$B$4)^C3		=SUM(D3:F3)	=(1+$B$2)^-C3	=G3*H3
4	Growth	0.02	2		=$B$1*(1+$B$4)^C4		=SUM(D4:F4)	=(1+$B$2)^-C4	=G4*H4
5	Purchase price	7000000	3		=$B$1*(1+$B$4)^C5		=SUM(D5:F5)	=(1+$B$2)^-C5	=G5*H5
6			4		=$B$1*(1+$B$4)^C6		=SUM(D6:F6)	=(1+$B$2)^-C6	=G6*H6
7			5		=$B$1*(1+$B$4)^C7		=SUM(D7:F7)	=(1+$B$2)^-C7	=G7*H7
8			6		=$B$1*(1+$B$4)^C8		=SUM(D8:F8)	=(1+$B$2)^-C8	=G8*H8
9			7		=$B$1*(1+$B$4)^C9		=SUM(D9:F9)	=(1+$B$2)^-C9	=G9*H9
10			8		=$B$1*(1+$B$4)^C10		=SUM(D10:F10)	=(1+$B$2)^-C10	=G10*H10
11			9		=$B$1*(1+$B$4)^C11		=SUM(D11:F11)	=(1+$B$2)^-C11	=G11*H11
12			10		=$B$1*(1+$B$4)^C12	=E13/B3	=SUM(D12:F12)	=(1+$B$2)^-C12	=G12*H12
13			11		=$B$1*(1+$B$4)^C13				
14								**NPV**	**=SUM(I2:I12)**
15	or with excel formula	**=G2+NPV(B2,G3:G12)**						**IRR**	**=IRR(G2:G12)**

Figure 8.27 Illustration of spreadsheet for Example 8.27.

Instead of calculating yearly PVs, you can use the integrated Excel formula for NPV.

Figure 8.28 Excel formula for NPV.

In this situation, a different result for NPV will be obtained, specifically, NPV = €262,123.20. This occurs because Excel automatically considers year 0 as year 1. Therefore, to use the formula directly, it should be written as NPV = G2 + NPV(B2, G3:G12), which would produce the same result with the PVs per year.

The IRR formula is:

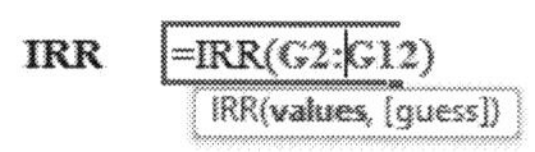

Figure 8.29 Excel formula for IRR.

Example 8.28

An investment fund has engaged a property investment analyst as an advisor to evaluate two potential property investments. The options available are Investment A, a leasehold shop, and Investment B, a small freehold office building. Both properties are being offered for sale at a price of €1,500,000, with the following details provided for analysis.

Investment A

The property consists of a single shop unit arranged over two floors, with a total net floor area of 3,000 m^2. The current leaseholder holds the property on a net lease for a term of 35 years, which will expire five years after the anticipated completion of the purchase, at a fixed rent of €30,500 per annum. The entire property is let on a net lease to the current occupier for a term expiring two days before the head lease, at a rent of €470,000 per annum, subject to a review to MR three years before the lease expiration. The current market rental value is €530,000 per annum. Acquisition fees are estimated at 4% of the purchase price. Rent review fees are estimated at 5% of the revised rent. Management costs are 12% of rent collected per annum. Rents are currently growing at 5% per annum, with little change expected in the short term.

Investment B

The property B is a small freehold office building with a total area of 2,500 m^2 let with five years of the current lease to run. Last year's rent was €150,000, and around 27% of this rent was lost in outgoings, including management fees. Acquisition and sale fees are estimated at 5% of price. Rents are annually reviewable in line with RPI (the retail price index, Germany). Current capitalisation rates are between 7% and 9%.

General market data

RPI increased by 3.5% last year. German government fixed-interest securities medium-dated currently yield around 5% if held to maturity. Rents from each property can be assumed to be received annually in arrears. All fees can be regarded as reliable cost estimates.

Solution

First, we need to establish the discount rate. Considering that government fixed-interest securities with medium-term maturities currently yield approximately 5% if held to maturity, we have decided to ***set the discount rate at 11%*** for both investments by applying one of the standard methods for estimating the discount rate or based on investor demands.

NPV and IRR estimation for **Investment A**:

Type	*Leasehold*
Purchase price	€1,500,000
Head rent	€30,500
Passing rent	€470,000
Market rent	€530,000
Acquisition fees	4%
Rent rev. fees on rev. rent	5%
Management costs (on coll. rent)	12%
Annual rental growth	5%
Discount rate	11%

year	0	1	2	3	4	5
purchase	-€ 1,500,000					
Acq. Fees	-€ 60,000					
Head Rent		-€ 30,500	-€ 30,500	-€ 30,500	-€ 30,500	-€ 30,500
Rent In		€ 470,000	€ 470,000	€ 584,325	€ 584,325	€ 584,325
Rent Rev Fees				-€ 29,216		
Mgm. Costs		-€ 56,400	-€ 56,400	-€ 70,119	-€ 70,119	-€ 70,119
Cashflow	-€ 1,560,000	€ 383,100	€ 383,100	€ 454,490	€ 483,706	€ 483,706
NPV	€ 34,074.77					
IRR	11.82%					

Figure 8.30 Data table for Investment A.

Showing how you would execute the calculation in a spreadsheet with formulae:

	A	B	C	D	E	F	G	H	I	J
1	Type:	LEASEHOLD								
2	Purchase price	1500000		year	0	1	2	3	4	5
3	Head Rent	30500		purchase	=-B2					
4	Passing Rent	470000		Acq. Fees	=E3*B6					
5	Market Rent	530000		Head Rent		=-$B$3	=-$B$3	=-$B$3	=-$B$3	=-$B$3
6	Acq. Fees	0.04		Rent In		=$B$4	=$B$4	=$B$5*(1+$B$9)^2	=$B$5*(1+$B$9)^2	=$B$5*(1+$B$9)^2
7	Rent Rev. Fees on Rev. Rent	0.05		Rent Rev Fees				=-H6*B7		
8	Mngment Costs (on coll. Rent)	0.12		Mgm. Costs		=-$B$8*F6	=-$B$8*G6	=-$B$8*H6	=-$B$8*I6	=-$B$8*J6
9	Annual Rental Growth	0.05		Cashflow	=SUM(E3:E8)	=SUM(F3:F8)	=SUM(G3:G8)	=SUM(H3:H8)	=SUM(I3:I8)	=SUM(J3:J8)
10	Discount rate	0.11								
11				NPV	=NPV(B10,F9:J9)+E9					
12				IRR	=IRR(E9:J9)					

Figure 8.31 Illustrative spreadsheet for Investment A.

We observe that Investment A returns a positive NPV and, as expected, IRR higher than the discount rate.

NPV and IRR estimation for **Investment B:**

Type	*Freehold*
Purchase price	€1,500,000
Passing rent	€150,000
Acquisition and sale fees	5%
Outgoings	27%
RPI	3.5%
Discount rate	11%
Capitalisation rate	8%

year	0	1	2	3	4	5
Purchase	-€ 1,500,000					
Purchase Fee	-€ 75,000					
Rent In		€ 155,250	€ 160,684	€ 166,308	€ 172,128	€ 178,153
Outgoings		-€ 41,918	-€ 43,385	-€ 44,903	-€ 46,475	-€ 48,101
Sale						€ 2,304,854
Sale Fee						-€ 115,243
CashFlow	-€ 1,575,000	€ 113,333	€ 117,299	€ 121,405	€ 125,654	€ 2,319,663
NPV	€ 170,453					
IRR	13.6%					

Figure 8.32 Data table for Investment B.

And showing the calculations in full spreadsheet format:

	A	B	C	D	E	F	G	H	I	J
1	Type:	FREEHOLD								
2	Purchase price	1500000		year	0	1	2	3	4	5
3	Passing Rent	150000		Purchase	=-B2					
4	Acq. And Sale Fees	0.05		Purchase Fee	=B4*E3					
5	Outgoings	0.27		Rent In		=$B$3*(1+$B$6)^F2	=$B$3*(1+$B$6)^G2	=$B$3*(1+$B$6)^H2	=$B$3*(1+$B$6)^I2	=$B$3*(1+$B$6)^J2
6	RPI index	0.035		Outgoings		=-$B$5*F5	=-$B$5*G5	=-$B$5*H5	=-$B$5*I5	=-$B$5*J5
7	Discount rate	0.11		Sale						=J5*(1+B6)/B8
8	Cap. Rate	0.08		Sale Fee						=-J7*B4
9				CashFlow	=SUM(E3:E8)	=SUM(F3:F8)	=SUM(G3:G8)	=SUM(H3:H8)	=SUM(I3:I8)	=SUM(J3:J8)
10										
11				NPV	=NPV(B7,F9:J9)+E9					
12				IRR	=IRR(E9:J9)					

Figure 8.33 Illustrative spreadsheet for Investment B.

Comparing the two investments:

	Investment A	*Investment B*
NPV	€34,074	€170,453
IRR	11.82%	13.6%

Both investments exhibit positive NPVs and IRRs exceeding the required discount rate, thus meeting the fund's criteria. Nevertheless, Investment B appears to be more favourable. To confirm this assessment, however, it is advisable to conduct a sensitivity analysis.

Introducing sensitivity and scenario analysis in DCF

One of the most critical steps in applying the DCF method is carrying out a thorough sensitivity analysis. While DCF techniques allow a valuer or investor to derive a present value for a property (or a portfolio of properties) by projecting future net incomes and discounting them at an appropriate rate, these inputs are always subject to uncertainty and market fluctuations. Sensitivity analysis addresses that uncertainty by systematically testing the impact of varying each input within realistic ranges. By doing so, it clarifies how the valuation result (the net present value) or the project's return (internal rate of return) changes when assumptions deviate from initial expectations.

Identifying the Key Inputs

A DCF relies on several assumptions that interact to determine the present value of the property's future benefits. The most influential typically include:

- *Discount rate or target rate of return.* Represents the opportunity cost of capital and the perceived risk of the asset. Small adjustments in this rate can produce substantial differences in the final valuation, especially for projects with longer horizons.
- *Rental growth rate or exit yield.* Growth in future rental income, or the sale price at the end of the forecast, can have a proportionately large effect on final estimates of worth.
- *Void periods and re-letting assumptions.* The length of any downtime between tenants, the costs associated with securing a new tenant, and the reversion to a new rent level can all significantly affect the income stream.
- *Operating costs and capital expenditures.* Property taxes, insurance, maintenance, management fees, and periodic refurbishments can fluctuate with inflation, changes in market demand, or shifts in local regulations.

Measuring the impact of uncertainty

Once the valuer has identified the key variables, small percentage changes, both up and down, are applied to each in turn. This process helps demonstrate how sensitive the valuation or rate of return is to realistic modifications in future assumptions. For example, by modelling a 0.5% increase or decrease in the discount rate, one can observe the size of the swing in net present value. Likewise, adjusting the growth rate of rents by one percentage point may reveal significant upside or downside potential.

Such sensitivity testing can be presented in a tabular form, 'tornado' diagram, or multi-factor scenario approach, enabling the valuer (and any client or stakeholder) to see at a glance which factors have the greatest impact.

Stress testing and scenario analysis

In addition to single-variable sensitivity checks, scenario analysis or stress testing is often applied. Rather than adjusting each input in isolation, scenarios combine several shifting parameters to explore coherent future states of the market, for instance, a 'bear' scenario (weaker tenant demand, longer voids, slightly higher yield) versus a 'bull' scenario (strong economic growth, higher rental growth, stable yields). By considering multiple scenarios, the valuer illustrates best-, worst-, and middle-case outcomes, providing stakeholders with an understanding of not only a single 'expected' value but also the potential range of outcomes.

Supporting risk management and decision-making

The primary benefit of sensitivity analysis is improved decision-making. Lenders, equity investors, and other stakeholders can see precisely which assumptions drive the investment's performance and, therefore, where risk-mitigation efforts are most needed. If, for instance, the DCF model proves especially sensitive to even moderate increases in operating costs or to small changes in the exit yield, an investor might:

- Revisit lease structures or negotiate different maintenance obligations with tenants
- Set aside higher capital reserves for contingencies
- Explore hedging strategies or diversified funding structures

Hence, by laying bare the effects of underlying assumptions, sensitivity analysis reduces over-reliance on a single 'headline' valuation figure and ensures that an appraisal or feasibility study realistically reflects possible deviations from the initial projection.

Key takeaways

1. *Uncertainty is inevitable.* No matter how carefully a DCF projection is assembled, future outcomes can diverge from assumptions.
2. *Small changes can have large effects.* Even modest deviations in discount rates, growth rates, or re-letting periods can significantly change returns.
3. *Scenario analysis enriches insight.* Scenario-based sensitivity testing shows the interplay between variables, not just single-factor changes.
4. *Better decisions, reduced risk.* By highlighting which drivers have the most impact, sensitivity analysis guides risk management and strategic planning, making the DCF output more robust for investors, lenders, and other real estate stakeholders.

By incorporating sensitivity and scenario analyses, we can better understand the impact of various assumptions on our DCF model. This will allow us to pinpoint which parameters are most influential and how they interact under different scenarios. With this approach, we can enhance the accuracy and reliability of our projections, ultimately leading to more informed decision-making for all stakeholders involved.

Having that in mind, let us revisit Example 8.28, but this time we will add the sensitivity and scenario analyses.

Example 8.28b

When preparing the DCF model on Example 8.28, certain assumptions were made. These assumptions will be referred to as ***base scenarios*** in both cases. Upon revisiting the investments, it is evident that for Investment A, the sensitive parameters are management costs and annual rental growth. For Investment B, the sensitive parameters include outgoings (such as management fees), rental growth based on the RPI, and the cap rate. The following table synopsises the two investments and the sensitive variables.

	Sensitive variable	*Base scenario*	*Range*
Investment A	Management cost	12%	9–15%
	Annual rental growth	5%	3– 7%
Investment B	Outgoings	27%	20– 35%
	Growth by index	3.50%	0–7%
	Cap rate (exit yield)	8%	6–10%

Scenario analysis for Investment A

Base scenario: Management costs 12%, rental growth 5%
Optimistic scenario: Management costs 9%, rental growth 7%
Pessimistic scenario: Management costs 15%, rental growth 3%

Investment A	*Base scenario @ 50%*	*Optimistic scenario @ 25%*	*Pessimistic scenario @ 25%*
NPV	€34,074	€132,727	€61,204
IRR	11.82%	14%	9.49%
Weighted NPV	€34,918		
Weighted IRR	11.82%		

We can observe that with the assumptions of the pessimistic scenario, NPV becomes negative and target rate (discount rate) was not achieved.

Scenario analysis for Investment B

Base scenario: Outgoings 27%, index growth 3.5%, cap rate 8%
Optimistic scenario: Outgoings 20%, index growth 7%, cap rate 6%
Pessimistic scenario: Outgoings 35%, index growth 0%, cap rate 10%

Investment B	*Base scenario @ 50%*	*Optimistic scenario @ 25%*	*Pessimistic scenario @ 25%*
NPV	€170,453	€1,078,368	€368,982
IRR	13.60%	24.6%	4.45%
Weighted NPV	€262,573		
Weighted IRR	14.07%		

On Investment B, while the optimistic scenario gives great NPV, on the other hand, the pessimistic scenario shows very low IRR and negative NPV, which should problematise the analyst.

Alternatively, sensitivity analysis can be made with the use of a **spreadsheet**.

A two-variable data table helps analyse how two inputs (e.g., management costs and rental growth) affect a result for Investment A.

Set up a table (two-variable data table):

- List different management costs in a column.
- List different rent growth rates in a row.
- In the top-left corner, place the formula that calculates the valuation (NPV in our case).
- Select the entire range (including discount rates and growth rates).
- Go to *Data → What-If Analysis → Data Table.*
- In the *Row Input Cell*, select the cell where rent growth is referenced.
- In the *Column Input Cell*, select the cell where the discount rate is referenced.
- Click *OK* to generate results.

Investment A data table

Sensitivity analysis on rental growth (row) with management costs (column):

€34,075	*3%*	*4%*	*5%*	*6%*	*7%*
9%	€54,002	€73,402	€92,989	€112,764	€132,727
10%	€34,801	€53,983	€73,351	€92,904	€112,643
11%	€15,600	€34,565	€53,713	€73,044	€92,559
12%	€(3,601)	€15,146	***€34,075***	€53,184	€72,475
13%	€(22,802)	€(4,272)	€14,437	€33,324	€52,391
14%	€(42,003)	€(23,691)	€(5,202)	€13,464	€32,307
15%	€(61,204)	€(43,109)	€(24,840)	€(6,396)	€12,223

Investment B data tables

For Investment B, with three variables instead of two, three additional tables are needed to analyse the interrelations of assumptions.

Sensitivity analysis with RPI growth (row) and outgoings (column):

€170,453	*0.00%*	*1.75%*	*3.50%*	*5.25%*	*7.00%*
20.0%	€(74,407)	€63,709	€213,222	€374,905	€549,569
23.5%	€(93,811)	€43,337	€191,838	€352,463	€526,024
27.0%	€(113,214)	€22,964	***€170,453***	€330,022	€502,479
30.5%	€(132,618)	€2,591	€149,068	€307,580	€478,934
34.0%	€(152,021)	€(17,781)	€127,683	€285,138	€455,389

Sensitivity analysis with RPI growth (row) and cap rate (column):

€170,453	0.00%	1.75%	3.50%	5.25%	7.00%
6.0%	€239,148	€413,981	€603,595	€809,006	€1,031,279
7.0%	€37,798	€190,543	€356,085	€535,301	€729,107
8.0%	€(113,214)	€22,964	***€170,453***	€330,022	€502,479
9.0%	€(230,668)	€(107,375)	€26,072	€170,360	€326,212
10.0%	€(324,631)	€(211,646)	€(89,433)	€42,631	€185,199

Sensitivity analysis with outgoings (row) and cap rate (column):

€170,453	20.00%	23.50%	27.00%	30.50%	34.00%
6.0%	€646,365	€624,980	€603,595	€582,211	€560,826
7.0%	€398,855	€377,470	€356,085	€334,701	€313,316
8.0%	€213,222	€191,838	***€170,453***	€149,068	€127,683
9.0%	€68,842	€47,457	€26,072	€4,687	€(16,697)
10.0%	€(46,663)	€(68,048)	€(89,433)	€(110,817)	€(132,202)

Conclusion

Despite Investment B having a higher IRR and NPV than Investment A, it is riskier due to reliance on key assumptions, like outgoings, growth by index, and cap rate.

One might argue that Investment A is safer because it is a leasehold and is not impacted by capitalisation at the end of the leasing period.

Sensitivity analysis is a must-use tool for all investment appraisals.

Monte Carlo simulation for sensitivity analysis in property valuation

Monte Carlo simulation is a powerful statistical technique used in property valuation and investment analysis to model uncertainty and risk by running thousands of scenarios with varying input assumptions. It is particularly useful in sensitivity analysis, allowing valuers and investors to assess how changes in key variables, such as rental growth rates, discount rates, or vacancy rates, affect property value projections.

Why use a Monte Carlo simulation?

Traditional deterministic sensitivity analysis, such as data tables in Excel, evaluates only a fixed set of input values. However, Monte Carlo simulation goes beyond this by considering probability distributions for inputs, providing a more realistic and probabilistic range of outcomes.

This method is especially valuable when analysing:

- Uncertainty in future rental growth rates
- Variability in yield or discount rates
- Impact of vacancy fluctuations
- Sensitivity of capital expenditure projections

Steps to implement a Monte Carlo simulation in Excel

To perform a Monte Carlo simulation in Excel, follow these steps:

1. Define key input variables and probability distributions.

Instead of using single fixed values, define key assumptions as probability distributions:

- *Rental growth rate.* Normal distribution (mean = 3%, standard deviation = 1%)
- *Discount rate (yield).* Triangular distribution (min = 5%, most likely = 6%, max = 7%)
- *Vacancy rate.* Uniform distribution (between 2% and 8%)

These distributions reflect the **real-world variability** of inputs.

2. Generate random inputs.

Use **Excel functions** to generate random values for these inputs:

=NORM.INV(RAND(), 3%, 1%) for rental growth rate
=TRIINV(RAND(), 5%, 6%, 7%) for discount rate
=RAND()*(8%-2%)+2% for vacancy rate

These formulas generate a random value for each variable based on its probability distribution.

3. Calculate property value for each scenario.

Use **NPV or DCF models** to calculate the property's value for each randomly generated scenario:

$$NPV = \sum \frac{Cash\ Flow_t}{(1 + Discount\ Rate)^\wedge t}$$

Where $CashFlow_t$ is influenced by rental growth and vacancy rate.

4. Run thousands of simulations.

- Copy these calculations across 10,000 rows to simulate different possible outcomes.
- Each row represents one randomised scenario with different assumptions.
- Use Excel's data table or VBA macro to automate the iterations.

5. *Analyse results with descriptive statistics.*

- Use =AVERAGE(range) to find the mean projected property value.
- Use =STDEV(range) to determine risk/volatility.
- Create histograms and probability distributions using Excel's data analysis ToolPak.

Interpreting Monte Carlo simulation results

The simulation produces a distribution of possible property values rather than a single-point estimate. The following insights can be derived:

- *Expected property value.* The average of all simulated valuations provides a probable estimate.
- *Risk assessment.* The standard deviation indicates how much property values fluctuate under different conditions.
- *Probability of undervaluation/overvaluation.* Investors can determine the likelihood that a property's value falls below a critical threshold (e.g. probability of value being under a lender's required amount).

Example 8.29

A real estate investment firm is evaluating the potential purchase of a commercial office building. The investor wants to understand how **rental growth rates**, **discount rates**, and **vacancy rates** impact the valuation. Since these factors are uncertain, they decide to apply a **Monte Carlo simulation** to model different scenarios and estimate the most probable property value.

Assumptions

- The property currently generates €80,000 in NOI annually.
- The rental growth rate follows a normal distribution with thus:
  - Mean = 3% (0.03)
  - Standard deviation = 1% (0.01)
- The discount rate (yield) follows a triangular distribution:
  - Minimum = 5% (0.05)
  - Most likely = 6% (0.06)
  - Maximum = 7% (0.07)
- The vacancy rate follows a uniform distribution:
  - Minimum = 2% (0.02)
  - Maximum = 8% (0.08)

The investor uses the income capitalisation approach to estimate property value:

Where:

$$Property\ Value = \frac{Net\ Operating\ Income(NOI)}{Discount\ Rate}$$

$$NOI = Rental\ Income \times (1 - Vacancy\ rate)$$

The firm runs 10,000 simulations to determine the expected property value.

Solution

To run the Monte Carlo simulation:

Using **Excel** or **Python**, students can simulate different property values by generating **random values** for rental growth, discount rates, and vacancy rates. What follows is the suggested methodology for Excel users:

Step 1: *Generate random inputs.*

- Rental growth rate: =NORM.INV(RAND(), 3%, 1%)
- Discount rate: TRIINV(RAND(), 5%, 6%, 7%)
- Vacancy rate: RAND()*(8%-2%)+2%

Step 2: *Calculate NOI for each scenario.*

$$\text{NOI} = €80{,}000 \times (1 + \text{Rental Growth}) \times (1 - \text{Vacancy Rate})$$

Step 3: *Calculate property value for each scenario.*

$$Property\ Value = \frac{NOI}{Discount\ Rate}$$

Step 4: *Repeat 10,000 times.* Use Excel's data table or Python's NumPy to generate and store 10,000 property values.

Step 5: *Analyse results.*

- Compute the mean and standard deviation of the simulated property values.
- Plot a histogram to visualise the distribution of values.

Interpreting the results

After running **10,000 simulations**, the firm finds:

- **Expected property value (mean): €1,250,000**
- **90% confidence interval: €1,100,000 to €1,400,000**
- **Probability of property value below €1,100,000: 15%**
- **Probability of property value above €1,400,000: 10%**

Investment decision

- If the seller is asking **€1,350,000**, the investor may hesitate, as **only 25% of simulated values exceed this amount**.
- If the asking price is **€1,100,000**, this might be a **good deal**, as **85% of the simulations suggest a higher value**.

Conclusion

Monte Carlo simulation transforms traditional property valuation by integrating uncertainty and risk into the analysis. Instead of relying on fixed inputs, it provides a probability-driven approach, making valuation more realistic and data-driven. This helps investors make better-informed decisions by quantifying potential risks and opportunities under different market conditions.

- Monte Carlo simulation helps **quantify risk and uncertainty** in property valuation.
- Instead of a single estimate, students learn how **distributions of possible outcomes** provide a better decision-making framework.
- Investors should use **probability-driven insights** rather than relying on **fixed-input values**

Questions

1. What are the key principles behind the investment method of property valuation?
2. How does the *income approach* differ from other property valuation methods, such as the comparison and cost approaches?
3. What is the significance of yield in the investment method, and how does it influence property valuation?
4. Explain the different types of yields used in property valuation, such as initial yield, reversionary yield, equivalent yield, and ARY.
5. What are the advantages and limitations of using direct capitalisation in investment property valuation?
6. How is NOI calculated, and why is it a crucial factor in the investment method?
7. What is the formula for direct capitalisation, and how is it applied in property valuation?
8. How does the DCF method differ from direct capitalisation in valuing income-generating properties?
9. In what situations would a valuer use the term and reversion method instead of the hardcore and layer method?
10. A property generates an annual net income of €50,000, and the market yield is 6%. Calculate the capital value using the direct capitalisation method.
11. What factors influence the selection of an appropriate capitalisation rate for a property investment?
12. How do market fluctuations, interest rates, and economic conditions impact the valuation of investment properties?
13. What role does risk assessment play in property valuation using the investment method?
14. How do the length of a lease and tenant strength affect the capitalisation rate applied to an investment property?
15. What is the difference between implicit and explicit valuation methods, and how do they impact the results in the investment method?
16. The property under study is a retail shop (7 m frontage, 12 m depth, 40 m^2 mezzanine) at Nowy Swiat Street, Warsaw. The current tenant is a clothing retailer. The tenancy agreement was signed two years ago. The period of the lease is ten years (reviewed every five years). The rent was agreed at €3,000 per month. Estimate the MV of the property by considering the following comparable evidence (all shops are of similar condition and equivalent location). The yearly increase is agreed on inflation rate +1%.

Comparable	*Frontage (m)*	*Depth (m)*	*Mezzanine (m^2)*	*Monthly market rent (€)*
A	8	10	40	3,800
B	9	10	45	4,050
C	11	10	55	4,180
D	8	9	0	3,500

ARY is estimated at 5.5%. Risk-free rate (10-year government bonds) is currently at 3.5%. Inflation is expected to be 4%.

17. A client wants to purchase a shop as part of an investment portfolio. The shop was let on a 25-year FRI lease 8 years ago at the rent of €100,000 per annum and is subject to five yearly, upwards-only rent reviews. The investor's target (equated yield) is 8%. The probable average future growth has been assessed to be 1.5% per annum compound. Estimate the freehold value of the property assuming an ARY of 5%. You are asked to use the DCF approach (in blocks).
18. A high street retail unit was let in March 2009 on a 15-year lease with 3-year rent reviews. The current rent is €20,000 per annum, and the MR is estimated to be €24,000. Assuming an ARY of 5%, calculate the MV of the unit using a traditional approach. If needed, assume risk-free rate of 3%.
19. The property under study is a block of apartments on the Mediterranean island of Santorini that is rented short-term but on a monthly basis. The building started operating 2 years ago on a 20-year FRI (full repairing and insuring) lease with 5-year rent reviews. The passing rent of the property is €180,000, and the MR is estimated today at €190,000 per year. ARY is estimated to be 6% based on a local index. Assume 0% rental growth the next 6 years, 3% for the subsequent 5 years, and 1% rental growth for the remaining period. Estimate the value of the freehold using Excel with:
    - Traditional hardcore/layer method
    - DCF in blocks
    - Shortcut DCF

    Critically assess the approaches used and explain analytically the calculations.
20. You have been appointed as an advisor to a tax-exempt investment fund that has to make a choice between three alternative property investments in Cyprus which they have been offered. Investment A is a freehold warehouse in an industrial area in Limassol, Investment B is a freehold residential building comprised of apartments in Larnaca, and Investment C is a freehold office in Paphos. Each property is for sale at €1,500,000, and the required target rate of return in every case is 10%. You are allowed to use data from estate agents and make other logical assumptions.
21. **Investment A.** The property is a freehold warehouse unit in Limassol, Cyprus, 4,500 m^2 and 450 m^2 of office space. It is rented on a 20 years' lease, and 6 years remain until the end of the lease. Last year's rent was €14,500 per month. Approximately 5% of the rent is lost on outgoings (including management fees). Acquisition and sale fees are estimated at 3% of price. Rents are annually reviewed and, according to the agreement, are following CPI plus 1%. Investment is expected to be sold at the end of the lease.

    Investment B. This property is a 15-year-old block of 12 flats in Larnaca, each renting for €1,200 per month. Management fees are 14% of income. Rent is increasing by 5% annually. Major renovations will cost €120,000 after four years, and the property must be sold within 8 years. Acquisition and sale fees are 5% of the price. Calculate the investment worth.

Investment C. This property is a freehold grade A, 1,500 m^2 office in Paphos's CBD. A foreign company that was occupying the property was paying €12/month per square metre, but they have just moved out. Another company from Germany is interested to rent the office, and they are offering a five-year contract (FRI lease) but will move not earlier than six months and no later than one year. They are willing to pay the amount of €30,000 as a reserve deposit, and €14/m^2 per month (with a 2% yearly increase) as soon as they move in. Acquisition and sale fees are estimated at 5% of the price.

- Calculate NPV and IRR of every investment.
- Identify sensitive variables.
- Prepare scenarios.
- Prepare sensitivity analysis tables.
- Comment on the results.

Further Reading

Baum, A., & Crosby, N. (2020) *Property Investment Appraisal*, 4th edn. Wiley-Blackwell.

Baum, A., Mackmin, D., & Nunnington, N. (2017) *The Income Approach to Property Valuation*, 7th edn. Routledge.

Blackledge, M. (2019) *Introducing Property Valuation*, 2nd edn. Routledge.

Brown, G.R., & Matysiak, G. (2020) *Real Estate Investment: A Capital Market Approach*, 3rd edn. Routledge.

Fomina, O., Moshkovska, O., Prokopova, O., Nikolenko, N., & Slomchynska, S. (2018) 'Methodological approaches to investment property valuation', *Investment Management and Financial Innovations*, 15(4), pp. 367–381.

Geltner, D., Miller, N.G., Clayton, J., & Eichholtz, P. (2021) *Commercial Real Estate Analysis and Investments*, 3rd edn. OnCourse Learning.

Kucharska-Stasiak, E. (2019) 'Valuation schools and the evolution of the income approach: An evaluation of change trends', *Real Estate Management and Valuation*, 27(2), pp. 66–76.

Sayce, S., Smith, J., Cooper, R., & Venmore-Rowland, P. (2021) *Real Estate Appraisal: From Value to Worth*, 3rd edn. Routledge.

Trojanek, M. (2010) 'The application of income approach in property valuation in Poland', *Economics & Sociology*, 3(2), pp. 35–42.

Chapter 9

The profits method

Introduction

With most types of commercial property, it is possible to arrive at a rental value by considering evidence of similar properties on a rent per square metre/foot basis, interpreting that evidence, and then applying a rent per square metre/foot to the subject property. This is not possible with some classes of property, as the rental value will be driven by the trading potential rather than purely by the location, size, and specification of the demise. Increasingly, these valuations are derived from the performance of the business, not the real estate components, to derive a rental and/or capital value.

Properties which are valued with regard to trading potential include licensed premises, such as public houses, bars, clubs, restaurants, and hotels. They also include theatres, cinemas, care homes, caravan parks, and petrol filling stations.

The RICS Valuation Standards, which are compliant with IVS, enshrine the methodology for profits-based valuations. Within the RICS Valuation Global Standards 2021 is VPGA 4, the valuation of individual trade-related property. There is also the RICS guidance note (GN 67/2010), on the capital and rental valuation of public houses, bars, restaurants, and nightclubs in England and Wales, providing more detailed guidance on licensed premises. These should be compulsory reading for both valuers and students who wish to undertake profits method valuations.

Overview of the profits method

The profits method is based on the principle that the market value or market rent of trade-related property is determined by its trading potential, or fair maintainable turnover (FMT), assuming it is operated by a reasonably efficient operator (REO).

A REO is a concept where the valuer assumes that the market participants are competent operators acting in an efficient manner. The idea of a REO is one of the most important principles that make sure the value is reliable, objective, and relevant to the market. According to the REO assumption, the property's business is run by a made-up, skilled person who acts in a way that is reasonable from a business point of view. This made-up operator is neither incredibly skilled nor notably inefficient. Instead, their actions are about what you would expect from a normal operator in the market. The assumption makes sure that the valuation takes into account long-term income generation and managing operational costs without any distortions that could come from the specific circumstances of the real operator.

DOI: 10.1201/9781003490357-11

The concept of FMT is that the landlord should not be punished for having an under-performing tenant, and the tenant should not be penalised for performing above the level that might be expected for a REO.

The profits method involves the following steps:

Step 1: Estimate the FMT that could be generated at the property by a REO.
Step 2: Estimate the gross profit by deducting purchase costs from the FMT.
Step 3: Estimate the net profit or fair maintainable operating profit (FMOP), also called 'divisible balance', by deducting running costs from the gross profit.
Step 4: Either:

a. Capitalise FMOP at an appropriate yield to assess capital value from an owner-occupier's perspective.
b. Split the FMOP – one part will go to the landlord, taking a percentage as rent, and one part will go to the tenant as profit. A 50/50 split is common, although valuers can debate a lower or higher percentage of the 'divisible balance' as rent depending on the supply and demand for a specific property and other factors, such as the lease terms. The rent portion may be capitalised at an appropriate yield to arrive at a capital value from an investor's perspective.

The profits method connects property value directly to business performance, making it particularly suitable for unique properties where comparable sales data is unavailable. Additionally, it captures the impact of operational efficiency on value. The method, however, relies heavily on accurate financial records, and its use of assumptions and projections can introduce a degree of subjectivity. Furthermore, it is sensitive to external factors, such as economic fluctuations, which can affect the reliability of valuations.

Regarding capital values, freehold sales with vacant possession will often be as a fully equipped trading entity or 'unit'. With vacant possession, this includes the inherent goodwill that runs with the property. The personal goodwill of the seller is lost when the seller leaves, unless it has become inherent.

Where freehold trading entities are sold either freehold or leasehold, a multiple of net profit can be used to arrive at capital value.

Earnings before interest, taxation, depreciation, and amortisation (EBITDA) and *earnings before interest, taxation, depreciation, amortisation, and rent* (EBITDAR) are essentially accounting terms relating to actual performance and can be derived from actual accounts. Valuers adopting the profits method of valuation will, however, apply FMOP. To arrive at a figure for FMOP on freeholds, the valuer needs to adjust EBITDAR, and on leaseholds, the valuer needs to adjust EBITDA. In both cases, this is to arrive at a figure that a purchaser is likely to adopt, rather than the actual performance of the incumbent operator.

The liabilities taken on by a purchaser will include compliance with relevant legislation and regulations, which vary according to the property type. For example, transfer of undertakings (protection of employment) (TUPE) regulations in the UK may mean that the rights of existing staff have to be dealt with as part of a sale.

The amount of detail required to undertake a profits method valuation will, in practice, be driven by the proportionate use of time, which is an issue even when acting as an expert witness. Initial assessments can be undertaken using a shorthand method, whereas a valuer making a case to an arbitrator or tribunal will normally adopt a more detailed approach.

It is possible to check the calculations with reference to elements found in each property type, such as rent per cinema seat, rent per restaurant cover, rent per barrel of beer sold, but the valuer has to endeavour to analyse comparables consistently in what are generally imperfect submarkets. That is a challenging task.

What follows examines the four steps in more detail.

Step 1: Assess the FMT

A main driver of the method is the assessment of FMT. This is the annual turnover, net of VAT, that a REO is capable of achieving without over-trading or under-trading. It involves estimating the trading potential of a business in the premises rather than adopting the actual level of trade under the existing ownership. It excludes personal goodwill, which is any goodwill above market expectations which would be lost if that operator moved on.

In assessing the FMT, the target audience has to be considered. The current trading style might be wrong for the area, and consequently, the unit is under-performing. In assessing the rental value, the valuer is entitled to assume that the unit is trading to its FMT potential.

The exterior appearance of an outlet may be poor and could be affecting trade. Repainting, new signage, and planting need not be expensive and can improve trade. Landscaping of outside areas (e.g. decking in trade gardens of pubs) is more expensive but can dramatically increase trade. The valuer may assume that work not requiring the landlord's consent as 'tenant's improvements' has been undertaken and make an allowance for the capital expenditure.

The internal layout is something that often cannot be changed. There may, however, be non-structural alterations which can be undertaken to increase trade, and if the REO would do it, then it should be assumed that the current tenant would do so. Leases will often say that it is the tenant's responsibility to comply with statutory obligations. Any structural alterations, however, are likely to be tenant's improvements and should be disregarded unless a capital contribution was obtained from the landlord or a rent-free period was granted in lieu of a tenant's capital contribution. Cosmetic improvements are easier to deal with, and it is appropriate to consider if trade is being affected by the appearance of the interior.

Whilst consideration of these factors and experience will lead the valuer to an FMT figure, it is useful, as with all valuations, to consider comparable evidence. The turnover of comparables can be used to check the assessment of FMT. The comparables can also be analysed to produce some other key indicators.

Once a figure for FMT has been adopted, the remainder of a shadow profit and loss sheet can be compiled, which ends with an FMOP. This is net profit before rent, otherwise known as the divisible balance.

Step 2: Assess the gross profit

Between establishing the FMT and arriving at an FMOP, calculations have to be made to determine gross profitability, from which purchasers' costs are subtracted. Purchasers' costs include all the raw materials used in the trading-related property by the REO. For example, in the case of a bar, these would be the alcoholic and soft drinks purchased; for a restaurant, it would also include the raw ingredients used for cooking; and for a petrol station, it would be the fuel. It could also include stock, where appropriate.

Gross profitability will depend on the products being offered and the wholesale supply line, which could be tied under the terms of a lease or franchise, and the sustainable retail prices of the products in that location.

Gross Profit = FMT − Purchases costs

Step 3: Assess the FMOP

Once the total likely gross profit figure is calculated, likely running costs have to be deducted to arrive at the FMOP.

The costs are:

- *Operational costs*, including energy costs, rates, insurance, repairs, accountancy, sundries.
- *Staff wage costs*, including holiday pay and national insurance.
- *The tenant's capital*, including working capital and fixtures and fittings. Here, the aggregate capital cost is calculated, and an interest rate is then applied to produce an annualised deduction. This is based on a reasonable rate of return and/or the cost of money.

Benchmarking costs for restaurants and bars can be sourced from a variety of global industry reports and trade associations. For instance, in the United States, the National Restaurant Association publishes the Restaurant Operations Data Abstract, which is based on financial and operating data from nearly a thousand restaurant operations (National Restaurant Association, 2024). In the UK, the British Beer and Pub Association (BBPA) and the Association of Licensed Multiple Retailers (AMLR) can be used as a check on the level of costs for particular styles of outlets, either operator-run or managed. The valuer needs to decide whether the style of outlet would be managed, that is, run by a multiple operator with a manager or large/busy enough to warrant an owner occupier employing a manager, before allowing for a manager's salary.

Once the total gross profitability from FMT and the likely costs have been calculated, the following calculation can be made:

FMOP = Gross Profit − Running Costs

Step 4: Assess the capital valuation

Where the purchaser is an owner-occupier, valuers will normally calculate value based on an FMOP (adjusted EBITDAR, that is, not actual performance) that is capitalised at an appropriate rate of return, reflecting the risk and rewards of the property and its trading potential. An individual purchaser may bid for a site on the basis of their own perception of potential (which will reflect their acquisition costs, including finance costs and their own target return on investment), but the valuer must disregard special purchasers, unless making a special assumption.

Purchasers will need to fund their acquisition, and lending institutions will normally require the accounts of the current operator to compare with the purchaser's own forecasts. Where lending becomes driven by existing accounts, the amount a purchaser will be able to pay may be linked to the actual accounts or the purchaser's concept of future trading potential, which may not be in line with the valuer's opinion of FMT and FMOP.

$$\textbf{Capital Value} = FMOP \times \frac{1}{rate\ of\ return}$$

Example 9.1

A restaurant's financial details are provided as follows:

Turnover	$800,000
Purchases	$50,000
Operational costs	$50,000
Staff costs	$200,000

Calculate the value of the property considering a rate of return of 10%.

Solution

Step 1: The FMT = **$800,000**
Step 2: The gross profit = $800,000 – $50,000 = **$750,000**
Step 3: The FMOP = $750,000 – $50,000 – $200,000 = **$500,000**

Step 4: Value $= \$500,000 \times \frac{1}{10\%} = \$5,000,000$

Example 9.2

Value a 200-bedroom student accommodation located in close proximity to a university in Portugal.

The property is 90% occupied during term time (33 weeks), and the average rent per week is €250, with bills included. During the off-peak season, the occupancy rate drops to 60%, and the rent drops to an average of €180 per week. With regard to outgoings, there is €3,000 per week, and general expenses, including staff, energy, and marketing costs, are €5,000 per week. From analysis of other trade-related properties, an all risks yield of 8% is appropriate.

Solution

Step 1: Calculate the FMT.

(0.90 × 200 *rooms* × 33 *weeks* × €250 *per week*) + (06 × 200 *rooms* × 19 *weeks* × €180 *per week*) = €1,895,40

Step 2: Calculate the gross profit.

€1,895,40 – €3,000 *per week* × 52 *weeks* = €1,739,400

Step 3: Calculate the FMOP.

= €1,739,400 – €5,000 *per week* × 52 *weeks* = €1,479,400

Step 4: Calculate the value.

$$= €1,479,400 \times \frac{1}{8\%} €18,492,500$$

Step 4b: Assess the rental valuation

On properties occupied by tenants, the FMOP is available to be split between the property landlord as rent and the tenant as net profit after rent. Hence, the other name for FMOP is the *divisible balance*.

From the available FMOP, or divisible balance, the landlord would like to *maximise* the rent, and the tenant would like to *maximise* the 'net profit after rent'. If demand is strong for a unit, there is an argument that the tenant's bid would be higher than 50%, and if the lease is onerous and/or on a weak site, or where management is difficult, there is an argument that the tenant should have more than 50% to reflect the risk (i.e. the tenant's percentage rental bid is lower).

The percentage split of the divisible balance depends on the supply and demand at the time of the type of property in question and the risk/reward relationship to the figures.

For freeholds, where it is an investment valuation, an investment yield can be applied in the normal way to a rental stream, although estimated rental value can be calculated first to establish whether the property is ***under***- or ***over*-rented**.

There may be an overbid from a multiple operator who may have economies of scale and/or may wish to tie the site for a wholesale income stream in addition to rent. In the latter case, a higher or lower yield than the rent may be applied to the wholesale income from the tied products to reflect the state of the market and the possible volatility of product sales.

Example 9.3

Value a town centre car park occupying 1,200 m^2 with 100 parking spaces. The car park operator pays 45% of the FMOP as 'rent'. Accounts show an average occupancy of 70% over the past few years, with each space generating an average daily income of £8.50. With regard to outgoings, purchases average £500 per week, and general expenses, including energy, and maintenance, are £1,000 per week.

Assume an ARY of 6%.

Solution

Step1: $\text{FMT} = 0.7 \times 100\,\textit{spaces} \times 365\,\textit{days} \times £8.50\,\textit{per parking space per day} = £217,175$

Step 2: $\text{Gross profit} = £217,175 - (52\,\textit{weeks} \times £500\,\textit{per week}) = £191,175$

Step 3: $\text{FMOP} = £191,175 - (52\,\textit{weeks} \times £1,000\,\textit{per week}) = £139,175$

Step 4: $\text{Value} = £139,175 \times 0.45 \times \frac{1}{6\%} = £1,043,812$

Commentary on use of actual accounts

The shadow profit and loss calculations will follow a similar format to the actual profit and loss sheet of the type of business in question.

There is a distinction, however, between the actual accounts of the business and the valuer's calculations, which in theory should reflect what the REO will expect to achieve at the date of valuation. Although the history of the site will be a strong factor indicating future trading potential, the prospective purchaser, and therefore the valuer, will look forward, not back.

As stated previously, the adoption of the trading potential of the REO is used so that the property landlord is not penalised for the under-trading of a poor tenant in the form of a low rent and the tenant is not penalised for over-trading in the form of a higher rent.

The valuer, when undertaking a valuation of an owner-operated entity, or when acting for a tenant, should have access to the actual accounts. An assessment as to whether the actual operator is trading above or below the level of a REO has to be made. The actual trading level is always useful in assessing FMT, but only if the valuer stands back, looks at the relevant factors, and considers whether, given all the circumstances, the tenant is over-trading or under-trading. The valuer can do this based on experience and trading figures from other trading entities in the same market or submarket. The level of competition and the prices that are sustainable to customers are important factors.

Commentary on the use of audited accounts in the profits method

The profits method predominantly depends on financial data to ascertain the value of income-generating properties, with audited accounts regarded as a vital source of this information. These accounts furnish comprehensive summaries of a business's financial performance, encompassing income, expenses, and profitability. Nonetheless, although audited accounts are theoretically dependable and standardised, their primary objective – often focused on tax minimisation – can pose considerable challenges in property valuation. Audited accounts are anticipated to provide an impartial perspective of a company's financial condition, forming the basis for calculating net operating profit, which is pivotal to the profits method. For valuers, these accounts facilitate the determination of gross income, operating expenses, and the net profit generated by a property. They ensure uniformity in calculations and align valuations with the operational realities of a business.

Challenges with audited accounts

The profits method primarily relies on financial data to determine the value of income-generating properties, with audited accounts being considered a critical source of this information. These accounts provide detailed, comprehensive summaries of the financial performance of a business, including income, expenses, and profitability. However, despite the theoretical reliability and standardisation of audited accounts, their primary objective, which is frequently centred on tax minimisation, can present significant obstacles to property valuation. It is anticipated that audited accounts will offer an unbiased assessment of the financial status of a business, serving as the foundation for the calculation of net operating profit, which is essential to the profits method. These accounts simplify the process of calculating the net profit, operating expenses, and gross income of a property for valuers. They guarantee that valuations are consistent with the operational realities of a business and that calculations are consistent.

Addressing the challenges

To mitigate the limitations of audited accounts, valuers must normalise financial data by:

- Excluding personal expenses
- Estimating undeclared income
- Aligning salaries to market rates
- Removing extraordinary costs

Comparing performance metrics to industry benchmarks helps identify discrepancies, while interviews with owners or managers provide insights into unreported income or unique circumstances. Additionally, cross-verifying financial data with market research and third-party reports enhances accuracy. Transparency in reporting adjustments and data limitations ensures credibility and manages client expectations. By combining these measures, valuers can counteract the challenges posed by audited accounts and deliver reliable property valuations.

The profits method has both possibilities and risks because it depends on audited accounts. Even though they give us an organised way to value things, their flaws mean we need to be very careful and well-informed. Valuers need to use both audited accounts and their own professional judgement. They need to make sure that any changes or assumptions they make are based on real-world data and knowledge of the business. Valuers can make property assessments more accurate and trustworthy by dealing with these problems.

The importance of the REO assumption in the application of the profits method

The REO assumption plays a pivotal role in the profits method of valuation. It ensures that the valuation reflects the income potential of the property as if it were operated by a skilled, experienced, and efficient operator, rather than being influenced by the unique capabilities, inefficiencies, or financial circumstances of the current owner or operator. This assumption standardises the valuation process, providing an objective measure of the property's intrinsic ability to generate revenue under typical market conditions. By doing so, it aligns the valuation with market expectations and removes biases related to individual performance or operational anomalies, thereby enhancing the reliability and comparability of the results.

Ensuring realistic revenue and cost projections

Estimates of how much money the business will make and how much it will cost to run are necessary for accurate value. Using the REO assumption ensures that the projections reflect what is achievable under typical market conditions. By adopting this standardised perspective, the method avoids overestimating income or underestimating costs due to unique factors, such as poor management or exceptionally skilled management, associated with specific operations.

Aligning with market behaviour

Property values in income-driven sectors are shaped by the expectations of market participants. The REO assumption ensures that valuations align with what a typical buyer or investor would anticipate from the property's trading performance.

Producing fair and objective valuations

Valuations must be based on the property's true value, not on how efficiently or inefficiently the operators run the business. For instance, a business performing poorly due to ineffective management should not be undervalued if the property has the potential to generate higher income under competent management. Similarly, exceptional performance resulting from outstanding managerial skills should not artificially inflate the property's value.

Standardising the valuation process

The application of a consistent framework facilitates the comparability of valuations across properties through the REO assumption. Standardised benchmarks for income, expenses, and profit margins guarantee that values are justifiable and consistent with industry standards, which is especially crucial in institutional or regulatory environments. STR and CoStar provide a comprehensive set of benchmarks for clients including GIS-based tools that can examine many of the key hotel performance indicators by area, hotel chain, star rating, etc. and can support hotel operators, investors, and developers. Readers may want to explore their services at https://str.com/benchmarking. Other benchmarking operators include CBRE and the Benchmarking Alliance.

Mitigating subjectivity and bias

The REO assumption minimises the impact of subjective evaluations related to the specific circumstances of the actual operators. Without this assumption, factors such as the operator's financial situation or management style could distort the valuation.

Practical application of the profits method

The profits method involves several key steps where the rational operator assumption is integral:

Step 1: *Estimation of the gross income.* Income is projected based on what a REO could achieve, using market data and industry benchmarks (as discussed earlier), rather than the current operator's specific performance.

Step 2: *Deduction of the operating costs.* Expenses are estimated assuming efficient management practices and reasonable control over costs, reflecting industry norms rather than the idiosyncrasies of the current operator.

Step 3: *Calculation of the net profit.* The net profit figure is derived as the difference between gross income and operating costs, representing the sustainable earning potential of the property under competent management.

Step 4: *Capital value derivation.* A market-derived multiplier is applied to the net profit to estimate the capital value, ensuring the valuation reflects the income-generating potential of the property as perceived by the market.

The implications of the REO assumption for valuation accuracy

The REO assumption serves as a safeguard against the risks of over-reliance on the current operator's specific performance. By focusing on sustainable income generation, it ensures that the valuation reflects the true earning potential of the property under normal market conditions. Furthermore, it enhances the defensibility of the valuation by aligning with market expectations and standard practices.

The REO assumption is crucial in the proper application of the profits method. It provides a framework for generating valuations that are fair, objective, and reflective of market realities. By assuming a competent operator, the profits method ensures that valuations are insulated from the unique circumstances of individual operators, producing results that accurately capture the property's intrinsic value as an income-generating asset.

Freehold valuation and sales

Once figures for FMT and FMOP are calculated, it is possible to calculate rental or freehold values. For freeholds, where it is an investment valuation, an investment yield can be applied in the normal way to a rental stream, although estimated rental value can be calculated first to establish whether the property is under- or over-rented.

There may be an overbid from a multiple operator who may have economies of scale and/or may wish to tie the site for a wholesale income stream in addition to rent. In the latter case, a higher or lower yield than the rent may be applied to the wholesale income from the tied products to reflect the state of the market and the possible volatility of product sales.

Where the purchaser is an owner-occupier, valuers will normally calculate value based on a multiple of FMOP (adjusted EBITDAR, i.e., not actual performance). An individual purchaser may bid for a site on the basis of their own perception of potential (which will reflect their acquisition costs, including finance costs, and their own target return on investment), but the valuer must disregard special purchasers, unless making a special assumption.

Purchasers will need to fund their purchase, and lending institutions will normally require the accounts of the current operator to compare with the purchaser's own forecasts. Where lending becomes driven by existing accounts, the amount a purchaser will be able to pay may be linked to the actual accounts or the purchaser's concept of future trading potential, which may not be in line with the valuer's opinion of FMT and FMOP.

Going concern valuation

When an entire business is sold, the concept of 'going concern value' often arises. ASA, Chartered Business Valuers (CBV) Institute, RICS, and Taqeem (2022) define *going concern value* as a premise of value that reflects an assumption about the property's operational context. This premise assumes the business is a viable, ongoing commercial enterprise with a realistic expectation of sustained future earnings. The IVS highlights that financial statements are normally produced on the assumption that the entire entity is a going concern. This implies that, in a hypothetical transaction, it is reasonable to assume that any existing contracts would be transferred to the buyer.

The *going concern* term, though commonly used, is more suitable for business valuations than for property valuation. In such cases, the total value includes the land and buildings, the trade fixtures, fittings, furniture and equipment, any transferable licences, and the company goodwill, an intangible asset.

Goodwill can be challenging to assess, as it involves elements that are either specific to the current business setup or advantageous to the site's operation. The former, sometimes referred to as personal goodwill, is tied to the reputation or customer loyalty associated with the business owner and will likely dissipate if they leave. On the other hand, commercial or inherent goodwill, that is, the advantages tied to the property's characteristics for a specific type of business, might be more durable. Valuers should include only those aspects of goodwill that add tangible, saleable value based on sustainable profit potential, excluding personal goodwill, which will not transfer with the property.

The most common approach in valuing goodwill is to take the purchase price of a company and subtract the fair market value of the assets and liabilities. In the UK, another popular approach is to apply a multiplier to the FMOP of a business. The range of multiples usually varies between 1 and 5. Typically, higher multipliers are used for businesses with increasing profits. In contrast, for businesses with stable or declining earnings, lower multipliers may apply.

Applications of the profits method

Licensed property

Licensed property, that is, a property that is occupied by someone who has a license to do so, rather than being the legal owner, is perhaps the largest sector where the profits test method of valuation is applied. This approach is particularly useful for valuing a diverse range of licensed properties, including bars, pubs, restaurants, and nightclubs. There are a wide variety of types of licensed property, with different markets and submarkets.

The subjective nature of the profits valuation methodology has brought it under scrutiny from time to time. Where possible, it may be avoided, such as town and city bars let as a shell, where there is sufficient evidence of similar space let on a floor area basis. Where this is not possible, however, no one has yet successfully promoted a better method of assessing properties which are sold or let based on their trading potential.

Other valuation approaches may be applicable to either complement or replace the FMT/profits method. In strong city/town centre drinking circuits, 'superpubs' can be analysed on the basis of capacity and average spend. In areas with flexible use clauses, properties might also be compared using methods like weighted or zoning approaches.

The regulatory burden on licensed properties is significant and varies globally. For example, local laws often address public safety, nuisance control, and operating hours. The introduction of smoking bans in public spaces, as seen in many countries, has influenced profitability by restricting indoor smoking to outdoor areas. Additionally, noise complaints and public disorder can lead to license reviews, affecting business continuity and profitability.

Operational costs, such as wages, health and safety compliance, and environmental regulations, further impact the profitability of licensed properties. Valuers must also consider the broader economic environment, which affects entry costs for new operators and the financial viability of properties. For example, during economic downturns, the costs associated with long leases may deter new entrants, allowing established operators to dominate prime locations.

In terms of leasing structures, models vary. In the UK, tied leases are common, where tenants must buy stock, such as beer, cider, and other drinks, from the landlord. This model often offers lower rents to offset the reduced purchasing flexibility, but it has faced criticism for limiting competitive opportunities. Tied leases also exist in other countries, though the specifics vary widely depending on local regulations.

Beyond pubs, other types of licensed properties, like nightclubs and restaurants, have unique valuation considerations. Nightclubs often face high costs for security and entertainment, and their revenues may be impacted by regulatory changes, such as those governing 24-hour licenses. Restaurants are typically valued by per square metre or cover basis, and some markets also recognise distinctions between restaurant and takeaway use, each with its own demand and valuation implications.

Finally, in the hotel sector, valuation approaches often include income capitalisation and DCF models, particularly for higher-end properties. Metrics like Revenue per Available Room (RevPAR) are commonly used to gauge performance. Hotel valuations are highly sensitive to external factors, such as tourism trends, economic conditions, and local competition. The valuation approach for licensed hotels may involve multiple income streams from rooms, food, and on-site services, all contributing to overall profitability.

Example 9.4

Set out in the following is an example of a profits method rental calculation of a free-of-tie leased public house.

A free-of-tie lease is a type of commercial lease that allows a tenant to operate their business independently of the landlord or owner of the property. This means that the tenant is not bound by any obligations to purchase goods or services from the landlord or to use a particular supplier; in the case of public houses or bars, this means without a tie or restrictions from a particular brewery or operator.

The public house is in a good suburban location in the south east of the UK, with a solid clientele. It has an assessed FMT barrelage of 300 barrels and a healthy bar snack food trade. There is one amusement machine with prizes, a quiz machine, and a pool table. Entertainment is provided once a week. There are two bed-and-breakfast rooms that are full Monday to Friday at £30 per night. Note that VAT is deducted at 17.5%, and VAT is subject to change.

Solution

	A	B	C	D	E	F	G
1							
2	Wet Trade Based Upon:	300	Barrels				
3							
4	WET TRADE	Pints/litres	Avg. price exc VAT	Comments	Turnover	Gross profit %	Gross Profit
5	Beer and Cider	84000	£3.75	Pint	£315,000	65.0%	£204,750
6	Packaged Beer & Cider	1000	£3.75	Bottle	£13,650	65.0%	£8,873
7	Wines	2500	£4.00	250 ml	£40,000	70.0%	£28,000
8	Spirits	300	£3.00	25ml	£36,000	68.0%	£24,480
9	PPS	300	£3.00	bottle	£3,276	65.0%	£2,129
10	Minerals	2000	£1.30	various	£17,316	70.0%	£12,121
11	SUB TOTAL WET				£425,242	65.93%	£280,353
12							
13	TOTAL	Numbers per week	Avg. spend exc VAT	Comments	Turnover	Gross Profit %	Gross Profit
14	DRY TRADE	200	£6.00		£62,400	65%	£40,560
15	NET MACHINE INCOME	200	£1.00		£10,400	100%	£10,400
16	ACCOMM	10	£41.66		£21,663	85%	£18,414
17	SUB TOTAL				£94,463	73.44%	£69,374
18							
19	TOTAL FMT				£519,705	67.29%	£349,727
20							
21	LESS COSTS			% of FMT			Costs
22	ESTABLISHMENT			17.00%			£88,350
23	WAGES			21.00%			£109,138
24	BANDS & ENTERTAINMENT		£100				£5,200
25	SECURITY DOOR STAFF		£0				£0
26				INVENTORY	£30,000		
27				2 WEEKS STOCK	£6,538		
28				WKNG CAPITAL	£15,000		
29	FINANCE COSTS				£51,538	10%	£5,154
30	TOTAL COSTS						£207,842
31							
32	TENANTS' IMPROVEMENTS	Description:			£0	10%	£0
33							
34	DIVISIBLE BALANCE						£141,885
35	TENANT'S BID				50%	Say=	£70,943

Figure 9.1 A profits-based valuation of a leased public house.

Cinema and theatres

Cinemas can follow a similar format to the licensed leisure sector in terms of working out FMT and then deducting costs, which include film hire, to arrive at an FMOP. The income streams are divided between admissions and confectionary sales, with additions for screen advertising and car parking if appropriate. Where there is sufficient comparable evidence (e.g. in the mature multiplex market), calculations can be based on rent per square metre/foot. Alternatively, calculations based on rent per annum per seat or per screen can be used, but there has to be awareness that the different style of auditoria can result in widely differing operating costs. For example, an 'art' cinema may fall into a different submarket. The multiplex market has matured, and older-style cinemas are becoming rarer, with many being converted into bingo halls and superpubs or mixed-use venues.

Live theatres have their revenue sources from ticket sales, concessions (food and drinks), and merchandise. Factors affecting the profitability of live theatres include the popularity of the performance, location and size of the venue, and the production costs. In major cities, or with hit productions that draw large audiences, live theatres can be profitable. Small or independent theatres may find profitability more challenging without the external support of sponsors or government grants. The appropriate valuation approach may be DRC if there is no commercial market for the building.

Care or nursing homes

Care homes are also valued with reference to trading potential. Capacity, occupancy rate, and average fee rates can be used to calculate FMT. The valuer has to be aware of the different regulations covering different care situations. Like hotels, calculations based upon rent per room and DCF calculations can be used.

The size and quality of the accommodation are important, with single-storey care homes providing single-occupancy en-suite rooms, meeting modern minimum requirements, being more readily saleable than older homes with shared occupancies and/or sub-standard facilities. A growing number of countries are implementing quality indicators in care homes, which can provide valuable insights for the valuer. For instance, in Australia, the Aged Care Quality and Safety Commission oversees the National Aged Care Mandatory Quality Indicator Program, in which residential care providers must submit data every three months of a range of indicators, such as consumer experience, physical restraint, and activities of daily living (Australian Government, 2024). In the UK, the Care Quality Commission (CQC) monitors care homes through inspections and ratings based on safety, effectiveness, and resident experience (Care Quality Commission, 2024). In the United States, the Centres for Medicare and Medicaid Services (CMS) use the Nursing Home Compare programme, offering ratings based on health inspections, staffing levels, and quality measures (Centers for Medicare & Medicaid Services, 2024).

The location of a care home will determine the level of fees, and higher-than-average fees will be expected in areas with a higher proportion of privately funded residents. Specialist care homes with the need for more acute care can result in higher revenues, but the costs of providing care will inevitably be higher.

Petrol filling/gas stations

Petrol filling stations are assessed based on throughput per litre/gallon. The key is establishing 'genuine' sustainable trade. Ties to petrol companies and franchise agreements are common, but value will be driven by whether or not operators will be interested in the site, with major oil companies having a minimum throughput threshold. Petrol forecourts at superstores have affected the viability of older roadside stations.

Gas station sales have become very significant over the past 20 years, with partnerships developed between petrol retailers and supermarket operators. Care needs to be taken to identify income streams and the rights and liabilities that run with the property interest being valued. Many petrol filling stations also act as convenience stores, and some have introduced coffee and snacks, sometimes in partnership with national or international operators, which provide a separate income stream to be considered.

In the future, electric vehicle charging will have to be considered based on kilowatt-hour.

Golf courses

Golf courses can be assessed based on the quality of the course, membership numbers, revenue generated by the rounds of golf played, and ancillary income. *Ancillary income* refers to additional revenue streams not directly tied to playing golf but related to services and facilities, such as pro shops, equipment rentals, event hosting, and clubhouse.

Other important aspects to consider when valuing golf courses include, but are not limited to, the location and catchment population; the difficulty, the history, and the length of the golf course; the age and state of repair of the irrigation system; the frequency of closures during winter or poor weather; the number of public rights of way and their impact on golf rounds; and the competition in the area.

In some markets, like the UK, since the golf development boom of the late 1980s and early 1990s, the bottom end of the market has become oversupplied in a lot of areas. Top-end courses, however, can still command good demand from investors and operators.

Established member-owned clubs built many decades ago are rarely sold and can command premium values. These are run for the benefit of members, with the target being a surplus each year to cover major expenditure in the future. They compete with commercially run courses, although they are run on a different basis.

Racecourses, racetracks, and stadia

Racecourses, racetracks, stadia, and other outdoor venues can be evaluated by considering the status of the venue within the sector, the sustainability of events, and income streams.

A particular venue may have a unique value which is difficult to quantify, but the valuer needs to comment that a premium value would be likely should the property ever be sold. Stadia are generally attached to a team. Whilst the ownership can often be separated from the ownership of the team whose home venue it is, the base value is driven by sustainable sporting revenue (which can be volatile, depending on team performance over time). The stadium can, however, be seen as a trophy in its own right, and ancillary accommodation, such as conferencing facilities, can add value over and above a capital amount for the pitch, seating, and concourses.

Incorporating the DCF model in the profits method

The DCF model offers a dynamic and comprehensive technique of evaluating property value based on future cash flows. It can be integrated into the profits method to further enhance it and manage the complexity of many business-related situations.

DCF methodology in profits valuation

The DCF model involves projecting the net cash flows the property is expected to generate over a specific time period and discounting these cash flows to their present value, using a discount rate.

What follow are the major steps in applying DCF to the profits method:

Step 1: *Forecast the future income and expenses.*

- Project the gross income and operating expenses over a specified time frame (e.g. five to ten years).
- Incorporate anticipated growth rates for revenue and expenses in accordance with market trends.

Step 2: *Calculate the net cash flows.*

- To find the annual net cash flow, take the expected income and subtract the expected costs for each year (this is different than FMOP).

Step 3: *Select an appropriate discount rate.*

- Pick a discount rate that takes into account how risky both the business and the property are.
- The rate often includes a risk-free rate plus a risk premium for the property type.

Step 4: *Calculate the exit (terminal) value.*

- Use an exit capitalisation rate applied to last year's net cash flow to get an idea of how much the property will be worth at the end of the estimate period.

$$Exit\,value = \frac{CF\ final\ year}{Exit\ yield}$$

Step 5: *Discount the cash flows to present values.*

- Use the chosen discount rate to calculate the present value of each year's net cash flow and the terminal value.

Step 6: *Sum the present values.*

- Calculate how much the property is worth; this is calculated by adding the present value of the yearly cash flows and the present value of the property's end value using the formula:

$$Property\ Value = \sum_{t=1}^{n} \frac{CF_t}{(1+r)^n} + \frac{TV}{(1+r)^n}$$

Where:

CF_t = the net cash flow in year t
r = the discount rate
TV = the terminal value (property value at the end of the projected cash flow period)
n = the number of projected years in the cash flow

Turnover rents

Turnover rents, also known as percentage leases, are a lease structure where the rental payment is partially or entirely based on the tenant's revenue or sales generated from the leased property. The approach originated in the United States during the Great Depression of the 1930s as a way to accommodate struggling businesses, and by the 1970s, it had spread to the UK and Australia (McAllister, 1996; Blackledge, 2009). Traditionally, turnover rents were mainly used in specialist retail spaces, such as discount outlet villages and airport units. However, the COVID-19 pandemic accelerated the adoption of turnover rents across a broader range of retail properties, as both landlords and tenants sought ways to share financial risks and adapt to an unpredictable market.

During the pandemic, with retailers facing significant declines in revenue and landlords concerned about vacancies, turnover rents became a widespread solution. This structure allowed tenants to pay lower rents during periods of reduced sales, thereby helping them remain in business, whilst landlords could benefit from occupation and potential revenue increases when sales improved. While initially perceived as a temporary concession to fill vacant units or as part of restructuring efforts like company voluntary arrangements (CVAs), turnover rents appear to signal a lasting shift in the market. According to recent research by IPF (2021), turnover-based leases now make up around 8% of all retail leases and approximately 15% within UK shopping centres. Among stakeholders, turnover-based leases are increasingly standard for shopping centres and specific retail sectors, such as fashion.

Turnover rents ensure that rent increases only when business turnover increases, providing landlords with an incentive to support their tenants' success (Blackledge, 2009). Conversely, tenants benefit from the assurance that rent remains affordable, as it adjusts in line with business performance. However, this dynamic can also pose risks to landlords; if turnover declines sharply, rental income may drop, reducing the property's value as an investment. To mitigate this risk, turnover rental agreements often include a minimum base rent, typically set at 75–80% of the estimated market rent (Blackledge, 2009). This provides the landlord with a guaranteed minimum income, while still allowing the possibility of additional income linked to the tenant's performance. Turnover rents can be structured in various ways, depending on the agreement between the landlord and the tenant (McAllister, 1996). The main structures include the following:

- *Pure turnover rent.* The rent is solely based on a percentage of the tenant's gross sales or revenue. This type of rent structure is relatively uncommon, as it places all the risk on the landlord. However, it became more prevalent during the peak of the COVID-19 pandemic. For example, in late 2020 and early 2021 in the UK, retailers initiated CVAs that involved switching stores to a turnover-based rent model, often with zero base rent terms.
- *Base rent plus turnover component.* A more common structure, this model includes a fixed base rent that is supplemented by a turnover rent component. The tenant pays a guaranteed minimum base rent, ensuring the landlord a steady income, and then pays an additional amount based on a percentage of their sales above a specified threshold. This structure balances risk between the landlord and the tenant.
- *Threshold-based turnover rent.* In this structure, turnover rent is only paid once the tenant's sales exceed a predefined threshold. For example, a retailer might pay a base rent up to a certain level of sales, and any sales above this threshold incur an additional rent based on a percentage of those sales. This allows the tenant to achieve a certain level of performance before sharing profits with the landlord.

There is a link between the base rent and the turnover percentage, and clearly, an agreement can vary from 100% based on turnover to 100% on market rent and anything in between. Where a valuation is required of property let on such terms, full details of total rents actually collected and checked against audited accounts for a minimum of three years should be used as the basis for determining income cash flows for valuation purposes. Increasingly, technology is promoting greater transparency, with some large international malls having access to a tenant's real-time electronic point of sale system (EPOS).

One of the primary advantages of turnover rents is the alignment of interests between landlords and tenants. Both parties are incentivised to boost the tenant's sales, as increased revenue benefits them both. This mutual interest often encourages landlords to take a more active role in property management and marketing efforts, as they share directly in the tenant's success. Additionally, turnover rents facilitate risk sharing. Tenants pay less during periods of lower sales, which can help them sustain operations during economic downturns and maintain occupancy levels within the property. This flexibility can be particularly beneficial in challenging market conditions, where traditional fixed rents might strain tenants' finances.

For landlords, turnover rents also offer the potential for increased income, especially during periods of high sales or seasonal peaks. When sales are strong, the landlord's rental income can rise accordingly, allowing them to benefit directly from the tenant's performance. This potential for revenue growth can be attractive in dynamic retail environments, where tenant sales may fluctuate significantly over time.

Despite these advantages, turnover rents also come with several drawbacks. One significant challenge is revenue volatility. Because rent is tied to the tenant's sales, landlords may face fluctuating income, which can complicate cash flow management and long-term financial planning. This uncertainty may deter some landlords who prefer the stability of fixed rental agreements. Another disadvantage is the administrative burden associated with turnover rents. These leases require detailed sales reporting and often necessitate audits to ensure accurate revenue tracking. This process can increase administrative costs and add complexity for both landlords and tenants. Additionally, turnover rents can lead to potential conflicts between the parties, as they rely on transparent sales reporting. If a landlord suspects that a tenant is underreporting sales to reduce rent payments, it can lead to disputes and strain the landlord-tenant relationship. As a result, turnover rents require a high level of trust and clear communication to be effective.

Despite the somewhat widespread use of turnover rents in the retail sector, this rental structure is also employed in management agreements for flexible workspace providers within the office market, as well as in the hotel sector (RICS, 2019). In these agreements, a percentage of the operator's income is used as rent, aligning the interests of the landlord and operator while sharing financial risks and benefits based on occupancy and revenue performance. This approach is gaining traction in the office market, where flexibility and adaptability have become increasingly important for both tenants and landlords. In the flexible workspace market, turnover-based rent arrangements provide landlords with a share of income from services such as private office rentals, hot desk rentals, meeting rooms, and other amenities, allowing landlords to benefit from the space's popularity and utilisation. The hotel sector can similarly leverage turnover rents, where rental payments are tied to room occupancy rates and additional revenue streams, such as food and beverage (F&B) sales or event hosting. This model helps hotel operators manage rent payments in line with business cycles, while landlords benefit from higher rental income during peak seasons.

Leases with turnover rents share similarities with the profits method of valuation in that they both tie the property's value to the profitability of the tenant's business. However, turnover rents are not valued by the profits method itself; instead, these leases are often better suited to valuation using a DCF model. The DCF model is generally considered more appropriate because it offers a detailed, forward-looking approach that can robustly capture the projected cash flows associated with turnover-based income, which an implicit model may not fully address.

The main advantage of using a DCF model lies in its ability to account for the variability and timing of income, which is crucial when dealing with turnover rents. This model allows valuers to project different scenarios for the tenant's revenue growth, seasonality, and market fluctuations, offering a clearer picture of potential income streams. Unlike the implicit model, which uses a single capitalisation rate to reflect all risks and growth expectations, the DCF model incorporates these elements directly into the cash flow projections. By discounting these cash flows at an appropriate rate, the DCF model provides a nuanced valuation that reflects the risk and growth potential associated with turnover rents.

Example 9.5

You are the landlord of a newly built class A office building in a central location in a provincial town in the United States. However, due to the shift toward remote work following the pandemic, finding tenants has become increasingly challenging, and your building currently has a concerning 0% occupancy rate. You recently read an article suggesting that a flexible co-working space provider can add vibrancy to office buildings and potentially attract other tenants. Coincidentally, a flexible co-working space operator contacts you, expressing interest in occupying the ground floor of your building (8,000 ft^2).

The FW provider offers you two leasing options:

Option A: A ten-year lease at a rent of **\$30/ft^2**, which is below the current market rent.

Option B: A ten-year lease under a 3 + 7 partnership structure. For the first three years, the provider pays no base rent but instead pays 60% of the operating income. From years 4 to 10, the provider pays a fixed rent of **\$55/ft^2**, plus an incentive rent of 20% on any revenue exceeding \$1.5 million. The provider has presented you with its projection:

Year	*Total income*	*Total operating costs*	*Operating income*
Year 1	500,000	−600,000	−100,000
Year 2	700,000	−550,000	150,000
Year 3	800,000	−550,000	250,000
Year 4	1,000,000	500,000	500,000
Year 5	1,100,000	500,000	600,000
Year 6	1,200,000	500,000	700,000
Year 7	1,300,000	500,000	800,000
Year 8	1,400,000	500,000	900,000
Year 9	1,500,000	500,000	1,000,000
Year 10	1,600,000	500,000	1,100,000

In both options, the FW provider also asks for \$1,000,000 as the capital requirement for fitouts and furniture to be paid before occupation. Which option is more advantageous for you as the landlord, assuming the FW projections are accurate and your required rate of return is 12%?

Comment

This example is a simplified version of a real-life proposal. In practice, the landlord would be interested in understanding how the FW provider derives its income, the projected occupancy rates, desk rates, provider's reputation, whether it is an established operator, its membership base, and other factors. Nonetheless, the example illustrates the landlord's decision-making process. While **Option A** provides stable income, it is below market rent levels.

In **Option B**, two distinct turnover models are apparent: During years 1 to 3, rent is based purely on turnover (a pure turnover rent). By year 10, a threshold-based turnover rent – termed 'incentive rent' by the FW provider – comes into play, allowing the landlord to earn a 20% premium on revenue exceeding $1.5 million. Although this only appears in Year 10, it presents a potential for higher income.

Solution

Option A

Year	*Cash flow A*
Year 0	−1,000,000
Year 1	240,000
Year 2	240,000
Year 3	240,000
Year 4	240,000
Year 5	240,000
Year 6	240,000
Year 7	240,000
Year 8	240,000
Year 9	240,000
Year 10	240,000
NPV = $614,246	
IRR = 21.43%	

Option B

Year	*Total income*	*Total cost*	*Operating income*	*Rent*	*Incentive rent*	*Cash flow B*
Year 0	–	–	–	–	–	−1,000,000
Year 1	500,000	−600,000	−100,000	–	–	–
Year 2	700,000	−550,000	150,000	90,000	–	90,000
Year 3	800,000	−550,000	250,000	150,000	–	150,000
Year 4	1,000,000	−500,000	500,000	440,000	–	440,000
Year 5	1,100,000	−500,000	600,000	440,000	–	440,000
Year 6	1,200,000	−500,000	700,000	440,000	–	440,000
Year 7	1,300,000	−500,000	800,000	440,000	–	440,000
Year 8	1,400,000	−500,000	900,000	440,000	–	440,000
Year 9	1,500,000	−500,000	1,000,000	440,000	–	440,000
Year 10	1,600,000	−500,000	1,100,000	440,000	20,000	460,000
NPV = $614,246						
IRR = 21.43%						

Option B offers higher NPV ($614,246) and IRR (21.43%) compared to Option A's NPV of $356,054 and IRR of 20.18%, making it more financially advantageous under the given assumptions and required rate of return. The profit-sharing element and incentive rent structure in Option B provide opportunities for potentially greater returns as the FW provider's revenue grows.

However, the profit-based rent in the first three years means that income is closely tied to the financial performance of the FW provider, which may be volatile, especially in the current post-pandemic office market.

By contrast, **Option A** offers a stable and predictable income stream at a fixed rate, making it less vulnerable to market fluctuations and the operator's revenue levels. While Option B has the potential for higher returns, it also carries greater income variability and depends on the provider's ability to achieve profitability and revenue growth. Therefore, whilst Option B may yield better financial returns, it requires a higher tolerance for risk and income uncertainty.

To complete this chapter, we provide an extended case study using a DCF model to determine a hotel's present value, considering factors such as revenue growth, operating expenses, and capital expenditures, seeking a comprehensive valuation and financial analysis.

Extended case study: a profits method valuation of a 75-bedroom hotel (Example 9.6)

Introduction

Hotel valuation is a complex process that requires an in-depth understanding of financial performance, market trends, and operational factors. This case study applies the DCF model to determine the present value of a hypothetical hospitality establishment in the Mediterranean region. By analysing key financial metrics, such as revenue growth, operating expenses, and capital expenditures, this method provides a comprehensive valuation framework.

Data collection and key metrics

To conduct a thorough valuation, the valuer must collect financial and operational data, including the following:

- Audited financial statements from the past five years (*historical data*)
- Operational (*current*) data, such as:
  - Monthly/annual occupancy rates
  - Average daily room rate (ADR)
  - Number of rooms and room types
  - Days of operation per year

Sample historical data

Set out in Table 9.1 are illustrative examples of the historical evidence to be provided.

Table 9.1 Examples of the historical data to be collected

	Historical evidence				
Year	*−4*	*−3*	*−2*	*−1*	*0*
Days of operation	365	365	365	365	365
Rooms	75	75	75	75	75
Occupancy rate	82%	76%	80%	85%	87%
ADR (€)	109.2	99.0	117.9	107.1	115.8
Room revenue (€)	2,450,832	2,059,063	2,582,661	2,492,193	2,758,533
F&B (€)	1,770,715	1,845,659	1,545,750	1,848,027	2,045,406
Other (€)	329,089	*413,426*	371,805	327,908	335,889
Total revenue (€)	4,550,635	4,318,148	4,500,216	4,668,128	5,139,827
Cost of sales (€)	861,453	894,746	772,288	920,087	995,843
Payroll (€)	1,467,975	1,420,258	1,290,558	1,424,385	1,614,397
General expenses (€)	1,087,193	876,193	954,556	920,770	967,894

Table 9.2 Analysis of historical data by percentage

	Historical evidence				
Year	*−4*	*−3*	*−2*	*−1*	*0*
Days of operation	365	365	365	365	365
rooms	75	75	75	75	75
Occupancy rate	**82%**	**76%**	**80%**	**85%**	**87%**
ADR (€)	109	99	118	107	116
Room revenue (€)	2,450,832	2,059,063	2,582,661	2,492,193	2,758,533
%	**53.9%**	**47.7%**	**57.4%**	**53.4%**	**53.7%**
F&B (€)	1,770,715	1,845,659	1,545,750	1,848,027	2,045,406
%	**38.9%**	**42.7%**	**34.3%**	**39.6%**	**39.8%**
Other (€)	329,089	413,426	371,805	327,908	335,889
%	**7.2%**	**9.6%**	**8.3%**	**7.0%**	**6.5%**
Total revenue (€)	4,550,635	4,318,148	4,500,216	4,668,128	5,139,827
Cost of sales (€)	861,453	894,746	772,288	920,087	995,843
%	**18.9%**	**20.7%**	**17.2%**	**19.7%**	**19.4%**
Payroll (€)	1,467,975	1,420,258	1,290,558	1,424,385	1,614,397
%	**32.3%**	**32.9%**	**28.7%**	**30.5%**	**31.4%**
General expenses (€)	1,087,193	876,193	954,556	920,770	967,894
%	**23.9%**	**20.3%**	**21.2%**	**19.7%**	**18.8%**

Forecasting the revenue and expenses

To develop an accurate cash flow forecast, we must establish a set of well-reasoned assumptions based on historical data and industry benchmarks. The first step is to express all income sources and expenses as a percentage of **room revenue**, as this provides a standardised basis for projection.

Historical performance data

Table 9.2 presents historical financial and operational data for the hotel under analysis.

Table 9.3 The weighting structure to be used for forecasting revenues and sales

Weights allocated (out of 100%)	*5%*	*10%*	*15%*	*30%*	*40%*	*Projected year 1 (weighted average)*
Year	*−4*	*−3*	*−2*	*−1*	*0*	*Year 1 forward*
Occupancy rate	82%	76%	80%	85%	87%	**84.00%**
Room revenue (€)	53.9%	47.7%	57.4%	53.4%	53.7%	**53.55%**
F&B (€)	38.9%	42.7%	34.3%	39.6%	39.8%	**39.17%**
Other (€)	7.2%	9.6%	8.3%	7.0%	6.5%	**7.28%**
Total revenue (€)	100.0%	100.0%	100.0%	100.0%	100.0%	
Cost of sales (€)	18.9%	20.7%	17.2%	19.7%	19.4%	**19.26%**
Payroll (€)	32.3%	32.9%	28.7%	30.5%	31.4%	**30.92%**
General expenses (€)	23.9%	20.3%	21.2%	19.7%	18.8%	**19.86%**

Developing a revenue forecast

To create a reliable forecast, historical performance is weighted, with more emphasis placed on recent years as they are more indicative of future trends. This approach mitigates the impact of outdated data and outliers.

Estimating future revenue

Step 1: Projecting room revenue

To project room revenue, we rely on **comparable market data** and the historical ADR of the subject hotel. The ADR is determined using either:

1. The **historical ADR** trend of the hotel itself
2. **Market ADR** from comparable hotels of a similar category, location, and positioning

Once the ADR is decided, we have a set of assumptions for **year 1** that looks like this:

Occupancy rate	84.00%
ADR	€115
Room revenue	53.55%
Food and beverage	39.17%
Other	7.28%
Cost of sales	19.26%
Payroll	30.92%
General expenses	19.86%

The revenue forecast follows this formula:

Room Revenue = Days of Operation × ADR × Occupancy Rate × Total Rooms

For the following years, we can apply an ADR and occupancy growth (or reduction) depending on the performance of the hotel and the economy. In addition, we can determine a *stabilisation* year.

The stabilisation year is usually set between years 3 and 6 of the cash flow, depending on the performance of the hotel. Growth cannot continue in perpetuity; therefore, it is capped where appropriate. Set out in what follows are the growth assumptions to be applied.

Annual occupancy rate growth *(OR)*	2.00%
Stabilisation year	5
ADR	€115
Annual revenue growth	2.00%

With the preceding information, you can now predict the room revenue of the hotel for year 1 and years to come:

Year 1: Room revenue = 365 days × ADR × occupancy rate × number of rooms
= 365 × 115 × 84% × 75 = **€2,644,425**

Year 2: Room revenue = 365 days × (ADR × ((1 + 2%)^1) × (occupancy rate × ((1 + 2%)^1)
× number of rooms
= 365 × (115 × (1.02^1) × (84%) × (1.02^1) × 75
= 365 × 117.3 × 85.68% × 75 = **€2,749,654**

Step 2: Calculate other revenue streams and expenses

Once **total revenue** is determined, all other revenue streams and expenses are calculated as a percentage of **total revenue**, based on historical weighted averages. Operating costs are also expressed as a percentage of **total revenue**, ensuring expenses scale with revenue growth, and are set out in Table 9.4.

Key takeaways

- *Total revenue is a function of room revenue*, with historical revenue distribution percentages applied to estimate F&B and other revenue.

Table 9.4 Projected years 1 and 2 revenues and costs

Assumptions	*Projected year 1* (weighted average)	*Projected year 2* (weighted average)
Days of operation	365	365
Rooms	75	75
Occupancy rate	84.00%	86.00%
ADR	115	117.3
Room revenue (€)	2,644,425	2,761,535
F&B (€)	1,934,307	2,019,969
Other (€)	359,504	375,424
Total revenue (€)	4,938,235	5,156,929
Cost of sales (€)	950,894	993,005
Payroll (€)	1,526,969	1,594,592
General expenses (€)	980,498	1,023,920

- *Expenses are projected proportionally*, ensuring cost estimates reflect realistic operating conditions.
- *Future growth factors (e.g. ADR and occupancy rate increases)* drive overall revenue expansion, while cost percentages remain stable, unless operational efficiencies or cost pressures dictate otherwise.

Step 3: Building the hotel's cash flow model

To assess the financial performance of a hotel, all relevant data must be structured into annual financial blocks categorising income and expenses, to ultimately determine the hotel's Gross Operating Profit (GOP) and NOI/EBITDA.

Revenue and expense classification

Hotels generate revenue from multiple streams, while operating expenses encompass a broad range of costs necessary to sustain operations. The primary categories include the following.

Main revenue sources

1. *Room revenue.* The core revenue driver, influenced by occupancy rates and ADR.
2. *F&B revenue.* Income from hotel restaurants, bars, banquets, and catering services.
3. *Meetings, incentives, conferences, and exhibitions (MICE).* Revenue from hosting corporate and social events.
4. *Spa and wellness revenue.* Income from spa treatments, wellness services, and fitness facilities.
5. *Rental income.* Revenue from leasing retail spaces, office areas, or other facilities within the hotel to third parties.

Main operating expenses

1. *Administrative and general.* Office expenses, accounting, legal fees, IT, and other administrative costs.
2. *Payroll and staff expenses.* Salaries, wages, benefits, and training for employees.
3. *Sales and marketing.* Advertising, promotions, digital marketing, and booking platform fees.
4. *Property operations, and maintenance (POM).* Repairs, maintenance, security, and groundskeeping.
5. *Utilities.* Electricity, water, gas, heating, and internet services.
6. *management fees.* Base and incentive fees paid to hotel management companies.
7. *Fixed costs.* Insurance, property taxes, lease payments, and contributions to the Furniture, Fixtures, and Equipment (FF&E) reserve.

Step 4: Structuring the cash flow model

The valuer can pick a layout of headings to build the cash flow model based on the hotel accounts. An example layout is set out in Table 9.5:

Table 9.5 Inputs to the cash flow model

Hotel details
Number of rooms
Days of operation
Number of available room nights
Number of bookings
Occupancy rate
Average room daily rate (ADR)
Revenue per available room (RevPar)
Revenues
Rooms
F&B
Other
Minor operated departments
Total revenue
Costs of sales
F&B
Other
Minor operated departments
Total cost of sales
Operating income
Expenses
Administration and general
Payroll
Staff expenses
Marketing
Repairs and maintenance
Utilities
Base management fee
Franchise fee
Total undistributed expenses
Operating profit
Fixed charges
Incentive management fee
Property tax
Property insurance
Rental and lease payments
FF&E reserve
NET OPERATING INCOME (EBITDA)

Whilst most of the preceding headings are self-explanatory, it is worth making reference to a number for the purpose of clarity.

ADR. Hotel rates fluctuate daily and depend on a range of drivers, including night of the week, seasonality, local events, etc. The ADR is the average rate achieved (or forecast) across the entire year. This is net of VAT and other local taxes.

RevPAR. This is the combination of occupancy and ADR and is a very important figure for a hotel. Occupancy can be 'sacrificed' for higher rate or, alternatively, may potentially be driven higher with rate cutting. The skill of the hotelier is to deliver the most profitable combination of rate and occupancy, and the valuer must make an assessment of what the market would expect.

FF&E reserve. This is a 'notional' allowance (sinking fund) to build up a reserve over time to repair and renew the fixtures, fittings, and equipment. Such cash reserve, if it exists, may be transferred to a purchaser, and a valuer needs to be aware of this.

EBITDA. This is the level of the profit which is valued and excludes tax and other matters which are personal to the owning entity; incorporation of such items would prevent like-for-like comparison across different properties. This is often also referred to as adjusted net operating income.

Step 5: Exit value and DCF valuation

Once the annual **EBITDA** has been projected for the 10-year cash flow period, the **exit value** (or **terminal value**) must be determined. The exit value represents the hypothetical **sale value** of the hotel after 10 years based on its **stabilised financial performance**.

The exit value is equal to the EBITDA of the 11th year multiplied by the capitalisation rate. The cost of sales is deducted from this value; it is then added to the EBIDTA of the 10th and final year of the cash flow.

A capitalisation rate for a hotel usually ranges between 7 and 10%, depending on the type of asset and market factors. It is intended to determine the rate of return on invested capital demanded by the entities financing a theoretical hotel acquisition. It reflects various factors, such as perceived risk, liquidity, inflation, etc., as previously discussed. The cost of sales includes agent's fees, legal fees, and any other fees which may arise. In the case of our hotel example, a capitalisation rate of 8% and a cost of sales of 3% were applied. The 3% is taken off the 'sale value of the hotel'.

Therefore, in the case of our hotel:

EBITDA year 10 = **€1,798,473**
EBITDA year 11 = **€1,852,427**

The exit value is calculated as:

€1,852,427 × 8% = **€23,155,342**
€23,155,342 × 3% = **€694,600**
Cash flow of year 10: €1,798,473 + €23,155,342 − €694,600 = **€24,259,155**

The cash flows for the projected 10-year period represent future values; therefore, they then need to be discounted in order to represent current-day values. The discount rate for a hotel is determined in a similar way to any other discount rate, but factors such as the performance of the tourism industry and the risk associated with the specific hotel are also taken into account.

In the case of our hotel example, a discount rate of 10% was applied, as it was considered to be a relatively low-risk hotel investment, and the discounted cash flow is set out in Table 9.6, which follows.

The addition of the preceding results in an NPV of **€17,954,617**.

Table 9.6 The discounted cash flow

Year	*Cash flow*	*Discount rate at 10%*	*Discounted cash flow*
Year 1	€1,377,540	0.909	€1,252,309
Year 2	€1,299,255	0.826	€1,073,765
Year 3	€1,378,380	0.751	€1,035,597
Year 4	€1,462,323	0.683	€998,786
Year 5	€1,551,379	0.621	€963,284
Year 6	€1,597,920	0.564	€901,984
Year 7	€1,645,858	0.513	€844,585
Year 8	€1,695,233	0.467	€790,839
Year 9	€1,746,090	0.424	€740,513
Year 10	€24,259,155	0.386	€9,352,954

Step 6: Interpreting the valuation results

The **NPV of €17,954,617** represents the estimated **present-day value of the hotel**, based on the 10-year projected cash flows and the applied assumptions.

If the **purchase price** of the hotel is **below** the NPV, the investment is considered **financially viable**.

If the **purchase price** is **above** the NPV, the investment may be **overpriced**, unless justifiable by strategic benefits.

Key takeaways

- **Exit value** is derived from EBITDA of year 11 and the capitalisation rate, adjusted for transaction costs.
- The **DCF methodology discounts future cash flows** to present value, ensuring an accurate risk-adjusted valuation.
- **NPV determines the hotel's intrinsic value**, assisting investors in making informed acquisition or disposition decisions.

Summary

- The profits method excels for properties like cinemas, care homes, or niche retail outlets where comparable sales data is limited or unavailable. Instead of relying on market trends, the profits method evaluates a property's specific trading potential, allowing valuers to derive its value based on business income and operational conditions. This flexibility makes it invaluable for unique or specialised property types.
- The profits technique eliminates anomalies resulting from inefficient or excessively efficient operators by postulating a hypothetical REO. This presumption standardises income and expenditures, matching valuations with pragmatic and sustainable operational circumstances. It guarantees that the assessment reflects the property's genuine potential, rather than the peculiarities of an individual business owner. A poorly managed restaurant's low profitability would not unjustly diminish its valuation if a capable operator could generate significantly higher revenue.

- The assumption of an REO is essential for the effective implementation of the approach. An experienced appraiser or decision-maker is needed to employ logical, systematic reasoning in assessing property values, utilising objective evidence and market data. This process entails the analysis of quantitative data (e.g. income, expenses, market trends) alongside qualitative factors (e.g. property condition, location). This basis is crucial for attaining transparency, precision, and confidence in valuation procedures.
- Given the possible combinations of trading style, lease, and use, the profits method provides a way of assessing the rental value of each outlet on its potential merits and has much to offer – but its effectiveness depends significantly on the valuer's expertise and skill in adapting the method to each unique scenario.
- The valuer must acquire the knowledge and experience of the market or submarket they wish to work in when adopting the profits method of valuation that knowledge and experience has to be applied to make relatively subjective assessments.
- It is more difficult to apply objective modelling to valuations as every trading establishment is individual, with different pricing and business potential. The valuer can adopt key market ratios with assistance from benchmarking sources.
- While comparables remain important across all property sectors, the valuer must judiciously interpret and adjust data from these sources. The profits method adds complexity, as the dynamics between landlords and tenants, including the presence or absence of supply ties, influence the valuation process.

Questions

1. Estimate the value of an owner-occupied bar in Barcelona with the following financial details. The rate of return is 8%.

Turnover	**€800,000**
Purchases	**€100,000**
Operational costs	**€70,000**
Staff costs	**€150,000**

2. Value a 100-bedroom hotel located in a capital European city knowing that its average daily rate is €115. The occupancy rate of the hotel varies throughout the year. During the first quarter, the average occupancy rate is 65%; in the second quarter, the occupancy rate is 72%; in the third quarter, the average occupancy rate increases to 78%; and in the final quarter, it drops to 70%. The hotel also has a restaurant and a spa. Considering past financial accounts, you estimate that the turnover of the restaurant is €190,000, and the FMT of the spa is €50,000 per year. The purchases of raw materials are €500,000 per year. Staff, energy, and marketing costs average €670,000 per year. From analysis of other trade-related properties, an all risks yield of 7% is appropriate.
3. A cinema complex with ten screens is located in a busy urban area in the UK. The average ticket price is £12, and the cinema has an average annual attendance of 100,000. In addition to ticket sales, the cinema generates £450,000, on average, from concessions involving F&B. Operating costs, including staff wages, utilities, and maintenance, amount to £1,200,000 per year. Estimate the value of the cinema, given that comparable properties suggest an ARY of 8%.
4. Value a self-storage facility with a floor area of 40,000 ft^2 located in a suburban area of a city. The average annual rate per square foot is €35, and the facility has an occupancy rate

of 73% on average. Running costs of the self-storage operator, including wages, average €2,500 per week. The split between the landlord and the self-storage operator is 50/50. Other self-storage facilities suggest a 6% ARY is appropriate.

5. A struggling but promising retailer has recently agreed with its landlords to pay turnover rents. The turnover rent structure will consist of a base rent of £10,000 per annum, plus 65% of the retailer's turnover for a period of ten years. To value the retail unit, assume the retailer's turnover is zero in the first year, £50,000 in the second year, with an expected annual growth rate of 2% thereafter. Comparable evidence indicates a 7.5% ARY is appropriate.
6. A care home generates an annual net operating profit of €150,000, which is expected to grow at 3% annually for the next five years. The discount rate is 8%, and the terminal value is calculated using a capitalisation rate of 7%. Calculate:

    a. The present value of cash flows for the next five years
    b. The terminal value and its present value
    c. The total property value

7. A restaurant generates a reported gross income of €500,000 and operating expenses of €300,000. During an audit, it was found that €50,000 in personal expenses were recorded as business expenses. Calculate:

    a. The adjusted net operating profit
    b. The property value if the capitalisation rate is 10%

8. Why is it difficult to apply objective modelling to unique trading establishments, and how can benchmarking help?
9. A retail outlet pays a base rent of €50,000 annually and an additional turnover rent of 10% on gross income exceeding €300,000. If the gross income for the current year is €500,000:

    a. Calculate the total rent paid by the tenant.
    b. Determine the property value if the total rent is capitalised at 7%.

10. What is the importance of comparables in the profits method, and how should they be adjusted for unique trading styles?
11. A hotel reports an occupancy rate of 60% and an average room rate of €90. However, industry benchmarks for similar hotels indicate an average occupancy rate of 70% and a room rate of €100. Assuming 100 rooms and 200 operational days:

    a. Calculate the reported gross income.
    b. Adjust the gross income to industry benchmarks.
    c. Calculate the property value if the capitalisation rate is 8% and the discount rate is 10% (use annual indexation for three to four years until it gets to market standards and five years DCF).

12. What are some common challenges associated with using audited accounts in the profits method?

References

ASA, CBV Institute, RICS, Taqeem (2022). *International valuation glossary- business valuation*. Available at: www.rics.org/content/dam/ricsglobal/documents/standards/feb_24_2022_international_business_valuation_glossary_revision_english.pdf

Australian Government (2024). *Quality indicators*. Available at: www.agedcarequality.gov.au/providers/assessment-monitoring/quality-indicators

Blackledge, M. (2009). *Introducing property valuation*. Routledge: London.

Care Quality Commission (2024). *Care homes and supported living services: evidence categories*. Available at: www.cqc.org.uk/guidance-regulation/providers/assessment/care-homes-and-supported-living-services-evidence-categories/safe

Centers for Medicare & Medicaid Services (2024). *Five-star quality rating system*. Available at: www.cms.gov/medicare/health-safety-standards/certification-compliance/five-star-quality-rating-system

IPF (2021). *Turnover-based leases: issues, drivers and challenges – a survey of current practice*. Available at: www.ipf.org.uk/asset/B083A038-725C-496F-B81AC929168DCB44/

McAllister, P. (1996). Turnover rents: comparative valuation issues. *Journal of Property Valuation and Investment*, 14(2), pp. 6–23. Available at: www.emerald.com/insight/content/doi/10.1108/14635789610112637/full/html

National Restaurant Association (2024). *New report provides operational data on restaurants*. Available at: https://restaurant.org/research-and-media/research/economists-notebook/analysis-commentary/new-report-provides-operational-data-on-restaurants/

RICS (2019). *Insights into the valuation of flexible workspace*. Available at: www.rics.org/news-insights/research-and-insights/valuation-of-flexible-workspace

Chapter 10

The residual method and development appraisal

Introduction

Development projects are, by their nature, forward-looking, often involving large mixed-use schemes which may not be fully completed until many years into the future. This means that the traditional approach of adopting recent comparables to underpin valuation assumptions, rents and capitalisation rates, etc. may not be valid. There is likely to be a significant period of time between the available comparable information and when it would be applied to the rents and/or sales prices to be achieved in the future.

In recent years, changes in markets and huge shocks, such as COVID-19, have not just fundamentally changed but actually disrupted the office market. With many occupiers only requiring 50% of their office portfolios, but of a much higher quality to attract staff back to the office, the fundamentals of office plate sizes, standard of finish, and amenities have been altered dramatically. From 2026 onwards, more generational change, with generation alpha both entering employment and seeking residential accommodation, is likely to further dramatically shift the demand for many real estate products. These shifts impact upon many aspects of the valuation, such as the size and shape of apartments, the need for home offices, and the corresponding adjustments in both cost and price points. All these 'big picture' changes may need to be anticipated within an appraisal of a multi-phase development project extending over, say, a five- to ten-year period. Whilst challenging, this is what makes development appraisals interesting, and we hope in this chapter to not only demonstrate the basic theory but, through an extensive case study, also illustrate the practical considerations of project viability.

The changing nature and complexity of development projects

To add to the complexity, development projects in recent years have become much more sophisticated, with developers focusing on 'products', amenities, and services, not just the provision of a basic building. In the residential market, more sophisticated approaches are being applied which create differentiated, specially designed, and sometimes branded products (e.g. Yoo, designed by Philippe Starck) to meet the needs of target customers. The development process therefore has an increasing marketing bias, using strategic marketing techniques to match the products more closely to target customer groups than in the past. Whilst this is outside the scope of this text, it is worth noting that development-based valuations for large schemes have to evolve to embrace this sophistication. For example, in the office market, successful schemes require a wide range of amenities calibrated to the end user; these include wellness facilities, cafés, restaurants, co-working spaces, and more. Understanding the end user is essential to

DOI: 10.1201/9781003490357-12

make the right decisions as to what amenity products are included within the project. This, in turn, impacts the allocation of space and significantly impacts rental and capitalisation. In many ways, the market and end user analysis, in what we call a 'highest and best use' study, cannot be separated from the appraisal process, or if it is, then it is, in our opinion, at their peril. Whilst for teaching purposes, we have to simplify the process to start with, we end the chapter with a more complex real application so learners can appreciate the complexity and connections needed to successfully explore viability through an appraisal.

Residual valuations and development appraisals

Historically, the valuation methodologies for development appraisal are clearly split into two: the residual valuation and the development appraisal. In essence, the methodologies are the same in that they both seek to assess the surplus that is left after estimating the completed value of a development project and deducting all the costs. In a residual valuation, the surplus is the land value, and in a development appraisal, the surplus is the developer's profit, having considered the actual or projected value of the site. Taking a more contemporary approach, it is easier to say that the process is the same, but the outputs and the contexts of application are different.

The appraisal of development projects requires the valuer to assess income from a property not yet built, potentially on a site not yet purchased, and therefore likely to be subject to many decisions and stakeholder requirements, from planning permissions to planning gain requirements, shifts in market fundamentals, and consumer behaviour. Therefore, the valuer is not only making assumptions about the macro- and micro-economic context, market behaviour, and target customer preferences some time in the future but is also doing so in an environment of increasing uncertainty and risk.

Both approaches require the valuer to estimate the future value of the development, with reference to income from rents and capital values upon completion of the project, and take away all the costs of the project, including finance. Some professional guidelines require the rents and values to be based on current values (or at the valuation date), whereas in practice and in some situations, explicit projections of costs and revenues are recommended.

Historically, this has led to a sustained criticism of the approach. However, we see through the more sophisticated use of computer modelling and forecasting techniques the need to embrace the 'problem' of uncertainty as the nature of the beast and not to cling to looking backwards to comparability, which for all but the simplest of schemes is likely to be too far removed from the development products value once completed.

When a developer is considering purchasing a site to develop a project, the residual valuation will be used to determine the price to be paid for the site. The valuer will estimate the project's value, based upon what could be built under planning and other regulations, and deduct all costs associated with the project, including a target profit set by the client.

When a developer is undertaking a development appraisal, where the land has been acquired or the price set, the same process as earlier will be applied. However, in this case, the land value will be included as a cost and not in the profit. The residual amount calculated will now equal any profit (or loss, if negative) and will be used to test the viability of a scheme or alternative schemes.

So to summarise, the valuation of development projects is historically based upon two approaches using the same fundamentals:

The value of the completed development	*less*	**all the costs associated with the development**	*equals*	a **SURPLUS** (*a profit*) or a **DEFICIT** (*a loss*)

> Where a valuer is tasked to determine what should be paid for a site, the surplus will be the land value, and a target profit is included as a cost; this is called a ***residual valuation***.
>
> Where a valuer is tasked to determine the profitability of a scheme, the surplus will be the developer's profit, and land value is included as a cost; this is called a ***development appraisal***.

The need for development appraisals or residual valuations occurs in a number of different contexts. For example:

- An untouched area of land (known as a greenfield site) or underdeveloped site, where planning permission for development has been, or is likely to be, obtained
- A derelict or abandoned site (known as a brownfield site) where planning permission for development has been, or is likely to be, obtained
- An existing building for refurbishment or where planning permission has been, or is likely to be, obtained for a refurbishment and/or change of use
- An existing building where planning permission has been, or is likely to be, granted for its demolition and replacement

In practice, the methodology adapts to different contexts, usually with increasing complexity and certainty as a scheme progresses. When used to calculate the price to pay for a development site, the input variables to the appraisal may be based on broad assumptions and generalised data, for example, building costs based on published indices and historic datasets. Later on, when the site has been purchased and construction is ongoing, detailed cost information on a monthly cash flow basis may be provided by a building cost consultant/quantity surveyor and used in the appraisal to monitor project viability, thereby increasing its accuracy and complexity. The problems and criticisms of the initial appraisals are discussed later in this chapter.

Development costs are likely to be made on a monthly or even shorter cycle, and so a detailed cash flow is required to ensure an accurate picture of costs and revenues, especially on complex schemes, which may have a long period of phased development. The *income approach* requires the capitalisation of the potential future income of a scheme as part of a risk and return analysis, which is often undertaken at the site acquisition stage. The direct comparison method is preferred where possible, but as we have indicated earlier, this may be problematic, given the time difference between comparable evidence and when the development products are available in the market. At the very least, however, it is recommended that market-based comparable evidence be used as a 'reality check', especially when assessing residual site values. However, it is in this very market for property with development potential that the most difficulty in finding good comparable evidence is experienced.

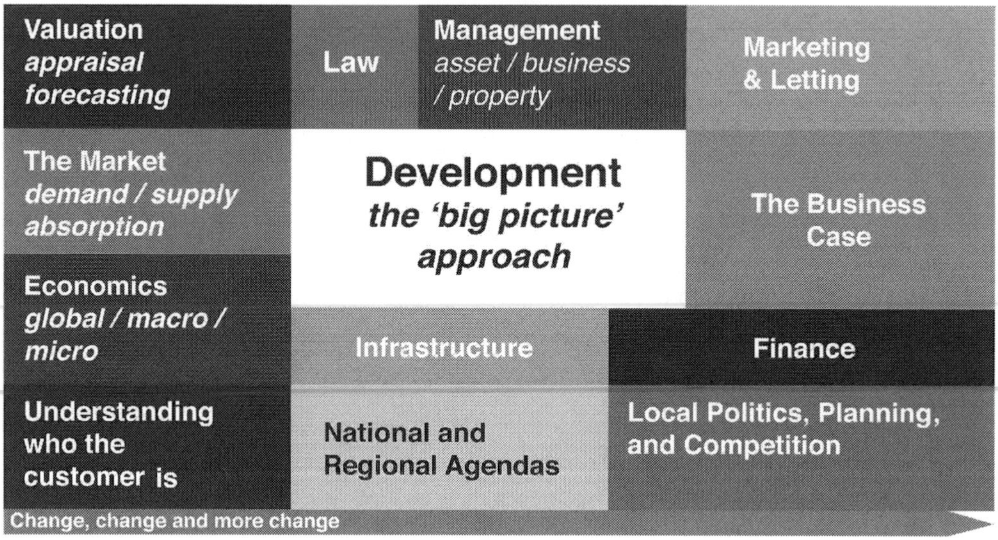

Figure 10.1 The 'big picture' approach to development viability analysis.

Incorporating a 'big picture' approach

Development valuations, especially on major projects, require the valuer to incorporate a significant and disparate number of inputs – some direct, others more intangible and indirect – and the authors suggest that a 'big picture' methodology be incorporated to ensure that all aspects of the project and their direct or indirect impacts are considered. This is diagrammatically shown in what follows as a dynamic jigsaw puzzle where all elements of the development link together in terms of their impact upon the development viability analysis.

This concept is brought to life and applied in our case study at the end of the chapter.

Elements of the 'big picture' approach

Understanding the customer (end user)

This has become absolutely essential and requires a thorough analysis of the *shape* of demand, that is, what the various sub-sectors of the markets identified require from a development. As we stated earlier, property development is becoming more target market–based, and therefore a thorough understanding of submarkets and target group preferences and behaviour is becoming part of the development valuer's considerations. Companies such as JLL now produce reports which set out specific characteristics and behaviours of customers for, say, co-living, who may vary from digital nomads to postgraduate students or what could be called transient executives. Whilst all these groups want flexible, communal-focused living, they have specific individual preferences.

Consideration of customer preferences is required in respect of:

- Appropriate finishes and appliances in residential property.
- Wellness and sports amenities, such as pools, gyms, and saunas.

- Lifestyle amenities, including cinema rooms, rooftop BBQ areas, and co-working spaces in the lobby of residential units.
- The provision of specialist services in residential buildings designed for older people.
- The ability to create contemporary office layouts and space plans in office premises suitable for hybrid working practices.
- Increasingly, the demand for sustainable development by companies who wish to align their ESG status with the premises they occupy should be fully understood.

All these factors impact directly on the appraisal, for example, the shift to including flexible office and co-working spaces within a traditional office tower is now normal, but it means a significant shift in the amount of income security from, say, three- to five-year leases to shorter term or even membership-based income. But without this amenity, the office tower may not lease out effectively, which may be a greater risk. All this complexity emphasises the dramatic changes in the industry from the early editions of this book, when real estate markets acted as wholesalers, with little direct contact with the end user. Tenants were simply there to be controlled, pay their rent, and occupy under the terms of their leases. Now, especially in office markets, developers are retailers of space and have sustained and direct contact with customers, rather than just tenants. To illustrate this point, the asset managers of ICD Brookfield Place in Dubai describe themselves as 'curators of a workplace experience', providing a wide variety of wellness and food and beverage amenities for their customers. This direct contact, greater management responsibilities, and more diverse income flows have a direct impact on the complexity and depth of analysis required for a contemporary development appraisal.

National and regional agendas

This includes planning frameworks, sustainability targets, infrastructure provision, transport policy, regeneration, and other initiatives which may provide both costs and financial support for projects. For example, major rail links, light railways, and other transport policies will have a dramatic impact on sites in close proximity, especially with a growing sustainability focus on not using individual cars for commuting. Many countries now regulate that public occupiers can only occupy buildings which meet a minimum sustainability criteria, say, LEED Gold. This directly impacts the shape of office developments, with any building not meeting these standards carrying a huge risk of not being able to satisfy the requirements of any public occupier.

Local politics, planning, and competition

In addition to, and not necessarily consistent with this, there will be local planning and other political agendas which may impact upon development proposals. In some cases, a change of political control in a local authority within the time frame of a scheme may impose significant change. In others, an authority may introduce inward investment incentives to attract businesses, which may impact upon the demand for commercial premises. It is also essential to understand the potential competition for the anticipated development in terms of schemes both currently under construction and in the planning approval pipeline.

Macro- and micro-economics

This should increasingly be considered at global, national, and local levels of analysis, as markets are intertwined and may impact upon development in many ways, including the availability and cost of finance. In a capital city, occupational demand may be international, and finance for major schemes may come from international investors, which are prone to political and other risks. For example, the unexpected outcome of the Brexit vote in the UK has had major consequences for the location of financial institutions in the city of London. The ill-advised policies of the Liz Truss Government in the UK brought chaos and misery to both major institutions and individual mortgage holders in a very short space of time. Geopolitical shocks are sadly not uncommon in this decade, from interruption of gas and oil supplies to significant restrictions in movement and immigration. Finally, the volatility in markets caused by the unpredictable global tariffs of the Trump administration cannot go without a mention here.

Finance

There is the need to recognise the source and cost of finance. Increasingly, this is not on a purely debt basis, and if equity is provided, there may be complex arrangements for the equity provider to share in future profits. In major schemes, a complex consortium of investors may be involved, each with their own profile of inputs and outputs from the scheme. Historically, in many cases, development is 'forward-funded', using long-term finance from an investor, such as a pension fund or a consortium of private and/or corporate investors. Developers may also require short-term funding to finance the costs of construction before the project is sold to an investor. Some contemporary developments, for example, co-working spaces, are turning to *crowdfunding* solutions to build a large network of small investors with a bias towards a more social form of funding rather than the historic institutional funding of development projects. In rare cases, such as ICD Brookfield Place, very high-quality products may be individually rated by an agency such as Moody's. Evaluating the quality and certainty of income, the amenities, and the end user attractiveness, amongst many other things, allows for a specific rating to be applied, which may open up access to more attractive bond-based funding.

Infrastructure

Public transport, in particular, and the provision of local services are very important and, if not available immediately, may impact upon the letability or saleability of a scheme. Infrastructure may have to be provided as part of the planning requirement, and therefore, the cost and timing of appropriate infrastructure to support the development are essential. Airports and their capacity and reach are a good example of a driver of demand, especially for the hotel market. The best example is probably Dubai, coupled with the strategic expansion of Emirates and Dubai International Airport and free stopover programmes.

The relevant property market and submarkets

A very detailed analysis of market conditions will be required. This should include an analysis, for each development submarket, of the potential demand, the vacancy rates, and the absorption rate. The *absorption rate* is how many apartments or square meters of commercial space are being let or sold over a given period of time. Demand drivers should also be recognised; for example, are there existing and planned activities and projects which will act as drivers of potential users to a site and its products? Understanding what is happening in and around the site is essential, mapping both vehicular and pedestrian transport and proximity.

The business case

This has two dimensions; first, for investors, it is the case for investing and will be dependent upon the risk and return profile of both the investor and the investment. Second, where the development is, for example, a leisure facility, such as a golf course, the valuation must be based upon future projections of income from the business and will involve specialist consideration of projected incomes and costs. Hotel developments, which are typically retained by a developer/investor and let to a branded hotel operator, may require a forecast of a profit and loss account to be made to determine the viability of the scheme and to forecast the operator rent, which will equate to the income from the hotel.

Marketing and letting

This is becoming increasingly sophisticated and is used at different times of the development to raise awareness through to closing deals. It is also becoming a very significant cost to be included in the development appraisal with the use of drones, augmented reality (AR), and virtual reality (VR) to capture *off-plan* sales to support an improved cash flow position.

It may also include television and radio advertising, promotional videos and interactive brochures, open days (both physical and virtual), and events and launches. Many mixed-use and residential developments require pre-sales of apartments to help the cash flow of the whole scheme. For a purchaser to buy ‘off-plan’, much more marketing effort is required to reassure the buyer of the quality of the finished product. For example, the creation of a show suite in another building or high-quality virtual reality presentations.

Social media platforms are increasingly driving customers to projects and generating traffic to websites through the creative use of content and the capturing of leads. Promotion also has to adapt to changing generations and how they interact with media. As an example of how things have changed, we recommend you look at the Instagram site for ICD Brookfield Place. It really is all about the user experience and the amenities, not the fabric of the building or the offices themselves, and illustrates both the changing priorities and the promotional content that is relevant for a modern office development.

Law

The legal position is an important constraint and may impact significantly upon the development process. For example, rights of way, easements, and boundary issues may directly affect the scale, massing, and layout of the design and restrict value. Existing tenants may also have legal rights, and legal impositions may flow from planning permissions and funding agreements.

Management

It is absolutely critical to consider the future asset management of a scheme. Creating a scheme with onerous management issues and complexity may impact upon an investor’s view of a scheme, as they will take over potential problems. Costs in use and life cycle analysis of materials and components are increasingly common, and with the application of information management systems (IMS) and building information modelling (BIM) systems, the complex ecosystem of a large project can be mapped, collated, and recorded in detail, providing invaluable information for the future asset management of a development.

Finally, it is the appraisal or valuation that has to bring all these elements together and quantify input variables, which are heavily influenced by all the other elements earlier discussed. All these sit in a context of rapid and constant change, risks, and disruption.

Garbage in, garbage out

In the past, development appraisals and, especially, residual valuations have been notoriously unreliable and have been frequently criticised by the Lands Tribunal in the UK and in court judgements.

The adage 'garbage in, garbage out' is particularly pertinent to these appraisals for the following reasons:

- They involve an element of forecasting, as the calculations are undertaken some time before the development is finished and lettings or sales are completed. In the case of complicated development projects, completion may be some years into the future due to the need to obtain planning consents, complete infrastructure works, and undertake building operations.
- They are complex and involve many variables which are highly sensitive to changes in both macro-economic conditions and local micro-economic factors.

Where the residual method is used to assess the value of bare sites for property taxation or statutory purposes, it is often divorced from the realities of actual development. It has been heavily criticised in legal cases, especially in the UK. The UK has a distinct legal entity for the Lands Tribunal which resolves a range of disputes about the value of land and buildings, and about their occupation, use, or development. In one case, its members denounced the method as 'far from a certain guide to values' (*Cuckmere Brick Co Ltd and Fawke v Mutual Finance Ltd* [1971] 218 EG 1571) and have suggested that 'once valuers are let loose upon residual valuations, however honest the valuers and however reasoned their arguments, they can prove almost anything' (*First Garden City Ltd v Letchworth Garden City Corporation* [1986] 200 EG 123).

The Lands Tribunal reconfirmed its distrust of the residual approach in a case involving the compulsory purchase of a development site. In *Ridgeland Properties Ltd (Claimant) v Bristol City Council (Acquiring Authority)* (2009) UKUT 102 (LC), the following extract from the ruling restates the Lands Tribunal's concerns with this methodology:

> Finally we would draw the parties' attention, as we did several times during the hearing, to the need to consider pragmatically and sensibly how much information a developer would expect and require in order to formulate an open market bid at the valuation date using the residual method of valuation. This Tribunal has repeatedly stressed its reluctance to use this valuation method. Its enforced use in this reference does not mean that its faults are any the less; it remains a valuation method of last resort, which is inherently very sensitive to even small changes in the input variables.

This case highlighted a number of areas where different assumptions had made huge differences between the valuations of each party. For example, a difference of £4.5 million between the parties regarding the cost of superstructure works, due mainly to the difference in the experts' opinions about the cost of curtain wall cladding.

However, as will be explained later in this chapter, when the technique is used to appraise projects where land has been purchased for a known sum and a building cost consultant/quantity surveyor has provided accurate cost information, the technique can become more robust.

We could provide many more case examples with similar stories, and more recently, disputes centred on the calculation of planning gain to be provided in major development schemes have also been the subject of criticisms of developers inflating their cost assumptions to reduce the surplus available for planning gain, such as social housing. The debate rumbles on and has generated legal disputes not only in the UK but also around the globe.

However, we hope that with the increased use of modelling, forecasting, and DCF-based appraisal techniques, the criticisms become less valid, and the appraisals become more robust.

As in the previous edition, we believe that the profession is moving forward in embracing the fact that most development projects do have extended time periods, which may divorce them from current valuation realities, including current comparable evidence. Whilst comparability is strongly advocated as a 'reality check' in the appraisal process, we feel that more effort should be placed on using more sophisticated techniques to forecast variables with greater accuracy, to undertake detailed sensitivity analysis (which we cover later in this chapter), and to model different scenarios based on variations in the input variables and different potential economic out-turns. Indeed, we propose a question: 'Should development appraisals adopt the principle of worth?' It is our contention that, in many ways, major development projects mirror the principles of investment worth. When considering the viability of a development scheme, there can be very significant differences between the positions of developers and/or investors in a development project, particularly in respect of the finance and funding of the project. Different developers may also have very different views as to what constitutes the highest and best use for the site, in particular the types of 'products' to be provided. Examples of this include the number of different-sized apartments; the number of premium apartments, such as penthouses; the brand; the star rating of a hotel; and the sustainability rating of an office tower. Therefore, five developers contemplating bidding for a site could undertake appraisals and produce entirely different results based upon their approach, assumptions, attitude to risk, and access to funding. It does not mean that any of them are necessarily 'wrong' or contravene professional guidelines, but they reflect the worth of the site based upon their assumptions, the products they believe to be most suitable in establishing the highest and best use, and perhaps most significantly, their access to funding and the cost of that funding. Indeed, the latest professional guidance note discusses these special assumptions and how to manage them. This is explored in more detail in our detailed practice case study at the end of the chapter.

Special assumptions: a case study illustration

The following simplified case study, used by one of the authors as a graduate development field studies project, illustrates both the big picture and the influence of the developer's position and access to capital in a real-life example.

The case study site is large, located in a European capital city and representing a whole city block. It has planning permission for a mixed-use scheme of offices, residential areas, a hotel, retail, and food and beverage uses. This example reflects the reality of major development projects in many cities around the world.

The site is owned by the local authority. It has detailed planning requirements which include fixed proportions of each use that can be built, and restrictions on the height and density (indicated by a maximum gross floor area) of those uses. It requires a specific amount of the site to be kept as green areas for public amenity and a number of other conditions. As is common in many countries, the permission also requires the developer to provide an element of social housing either within the site or nearby.

In this, as in many schemes, the uses of the site may be subject to interpretation and some negotiation dependent upon the planning gain or benefits a developer may provide in addition to paying for the site. For example, a developer may seek additional density or variations depending upon the planning gain, in this case, the amount of social housing provision and whether or not it is provided on the site or elsewhere in the city. This represents the first significant planning-oriented variable.

Developers may then have very different opinions as to what product is appropriate for the market; in this case, some developers proposed traditional offices, for which rents and yields were determined from current market comparables. However, other developers had, through extensive research, identified a significant demand for a new style of office concept, for which a considerable premium could be achieved and where there was no competition. Comparables do not help in this situation, as there is nothing locally to compare this new concept to.

In the residential component, one developer wanted to introduce branded residences, working in partnership with a five-star hotel developer. Again, there were no comparable schemes to help provide evidence of relevant price points. Branding has been proven to bring premium pricing, but this was yet to be established in this location. These variations, together with different assumptions about the type, distribution, and quality of the apartments within the residential part of the scheme, add further variation to each bid.

Finally, and perhaps most significantly, different developers had wildly different assumptions about financing and risk. Some local developers were heavily reliant upon debt funding from within the country concerned and took a highly conservative approach to risk, which was reflected in the phasing of the development to maximise early cash flows from pre-sales. One developer, from the Middle East, had access to significant equity and required virtually no debt funding; their exposure to risk therefore reflected a very different approach to cash flow management, phasing of the site, and which uses were constructed first.

It is easy to see that mixing all these assumptions between different developers interested in the site is likely to produce significantly different calculations of the surplus available for the required profit and, therefore, the funds available to bid for the land. This is why a movement to development appraisal embracing this reality in a form of worth calculation is more appropriate and validates the more sophisticated appraisal approaches being used in practice, which reflect the individual circumstances and assumptions of a developer.

Indeed, as will be explored later, the RICS, in its Professional Standard, refers to this concept and states:

> Notwithstanding the disregard of special value, where the price offered by prospective buyers generally in the market would reflect an expectation of a change in the circumstances of the asset in the future, the impact of that expectation is reflected in market value.

Professional frameworks and methodologies

A full consideration of the development appraisal process is outside the scope of this book. However, before considering the *income approach* in relation to development opportunities, we feel it wise to discuss preferred methodologies and professional frameworks to development appraisal which help reduce uncertainty and address some of its critics' concerns. The first is a 'sifting' methodology that filters potential uses and is a precursor to the financial analysis of a development scheme for particular use or combination of uses, embodied in the North American USPAP Regulations. The second is the RICS approach, incorporated in Valuation Information Paper 12. VIP

12 was subject to criticism, and incorporating aspects of the 2012 RICS guidance note on the Financial Viability in Planning (FVIP), it has evolved since the last edition into the *RICS Global Professional Standard: Valuation of Development Property*, which is effective from 1 February 2020.

The revisions made in the 2012 guidance note and incorporated into the 2020 Valuation Standard included a strong recommendation as to the use of advanced analysis, including sensitivity analysis; this is also recommended in the IVS 2017 and onward and is therefore likely to be incorporated in all future guidance notes and standards. Later in this chapter, we discuss in detail the issue of sensitivity to changes in variables and how such techniques can be used to explore these sensitivities and make them explicit.

The move to a professional standard brings about a shift to more rigorous and enforceable expectations and a greater global emphasis on consistency. More information about the detailed expectations and guidance is set out later in the chapter.

The highest and best use methodology

A development appraisal can be required in a number of different circumstances, and depending on these, the emphasis given to various aspects of the solution will vary. The most common development scenarios are where a developer owns or has an option on a site (which may be greenfield, brownfield, or cleared or contain existing buildings), and we wish to determine the optimum use for it. Conversely, we may have a development scheme in mind for which we know there is a proven market demand (e.g. a development of upmarket apartments or a mixed-use scheme), but the developer needs to find the best location. Although differing in certain aspects, these are two sides of the same coin, and both involve a good deal of similar work. Crucial to both sets of circumstances is the recognition of the need to find the optimum use for specific sites or locations. In North American appraisal practice, this concept of 'optimum' or *highest and best* use is the litmus test against which all proposals are judged. Essentially, USPAP requires the valuer or appraiser to advise the client of the most profitable concept that is feasible for a given site. That concept of feasibility is judged against three criteria:

- *Technical feasibility*
- *Legal feasibility*
- *Market feasibility*

The new RICS Global Valuation standard moves the methodology closer to this concept.

The method

The *highest and best use* represents an explicit, logical, and robust approach to refining the potential uses for a site. In theory, any given site has a range of possible uses. The extent of this range is limited by various considerations, such as site area, accessibility, slope, stability, and surrounding uses. It may be quite possible, therefore, to build housing, shops, leisure facilities, factories, or warehouses. In fact, any mix of these uses could be considered on the site. However, it should be recognised that:

- Some of the proposed uses will produce a higher rate of return than others.
- Not all represent feasible solutions in this location.
- Not all types of development will be granted planning permission.

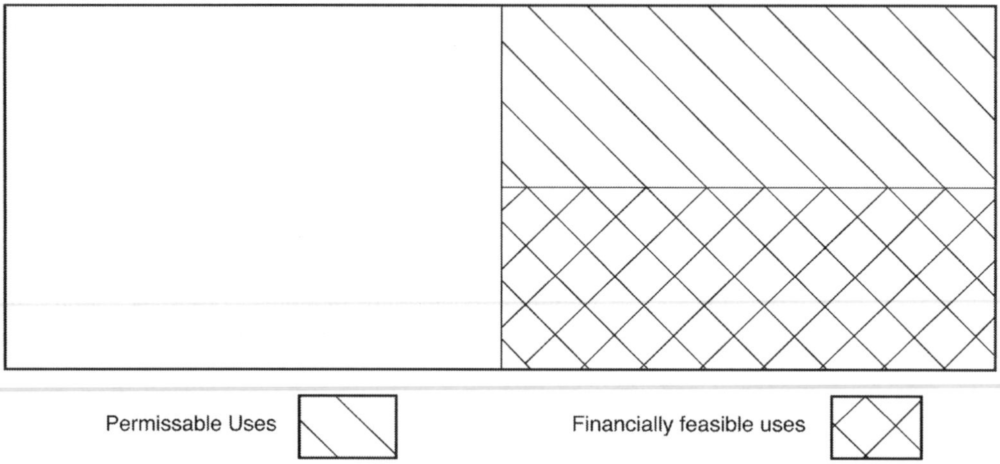

Figure 10.2 The highest and best use process.

Appraisers need, therefore, to refine their list of possible solutions to take these parameters into account. Figure 10.2 demonstrates the process.

In the diagram, the outer bold rectangle illustrates all the possible uses of the site, with no restrictions.

The process requires a filtering process to refine the choices in a systematic way to find the highest and best choices from the list of all possible uses in terms of being able to get regulatory approvals, including planning, and what use (or, more commonly, mix of uses) will provide the best return for the appropriate level of acceptable risk.

Starting from the total possible uses (the largest box), which can be produced by brainstorming, the uses can be refined first by identifying those possible uses that are permissible (the two right-hand boxes). This process involves consideration of the development plan and perhaps discussion with the planning authority, as well as recognition of any other legal constraints on the site, such as covenants, rights of way, etc. Under the US system, this represents the legal feasibility evaluation stage. Next, we narrow our choice further by considering financial feasibility. At this point, we would need to examine the macro-economic climate and local market and submarket indicators, taking into account interest rates, the property cycle, current and projected supply and demand, and local market conditions. The calculation would also involve consideration of development costs, including site acquisition, clearance, construction, and/or refurbishment set against the development value of the completed scheme. In simple terms, a consideration of the ratio of value to building costs on a per-square-metre basis gives an indication of the *highest and best uses* in financial terms. The higher the value for the same costs, the more viable a use is. However, of course, this ignores the fundamental demand-and-supply backdrop to such calculations. Retail uses may, for example, produce the highest amount of revenue compared to their building costs. However, developing the whole area of a mixed-use scheme with retail, even if permissible, is likely to flood the market and therefore not be financially feasible despite the most positive cost-to-value ratio.

Only those possible uses which fall within the bottom right-hand box would therefore be worthy of detailed consideration. In this way, appraisers and development consultants can demonstrate to their client that they have taken a rational approach and considered all the angles. It is, of course, what many practitioners/developers do unconsciously and implicitly, but we would advocate an explicit approach using the aforementioned methodology.

The highest and best use as defined in IVS 2025

The latest version of IVS defines *highest and best use*, in A90, as

> the use, from a participant perspective, that would produce the highest *value* for an *asset.*

The standard goes on in A90.6 to reinforce our description of the process given:

> The determination of the *highest and best use* involves consideration of the following:
>
> (a) to establish whether a use is physically possible, regard will be had to what would be considered reasonable by participants
> (b) to reflect the requirement to be legally permissible, any legal restrictions on the use of the *asset*, eg, town planning/zoning designations, need to be taken into account as well as the likelihood that these restrictions will change
> (c) the requirement that the use be financially feasible takes into account whether an alternative use that is physically possible and legally permissible will generate sufficient return to a typical participant, after taking into account the *costs* of conversion to that use, over and above the return on the existing use.

We explore the practical application of the highest and best use later in this chapter.

RICS Professional Standard: Valuation of development property (Global), effective 1 February 2020

The RICS has consolidated previous guidance notes into the first professional standard for the valuation of development property. The principles set out in the standard apply to all valuations of development property and are designed to be read in conjunction with *RICS Valuation Global Standards 2017* (*Red Book Global Standards*), which incorporates the IVS, in particular IVS 410.

Definitions

The RICS embeds the principle of highest and best use immediately in the introduction with reference to the definition of *development property* in IVS 410 as

> *interests where redevelopment is required to achieve the highest and best use, or where improvements are either being contemplated or are in progress at the valuation date and include:*
>
> *a the construction of buildings*
> *b previously undeveloped land which is being provided with infrastructure*
> *c the redevelopment of previously developed land*
> *d the improvement or alteration of existing buildings or structures*

e land allocated for development in a statutory plan, and
f land allocated for a higher value use or higher density in a statutory plan.

Basis of valuation

The standard makes it clear that market value, or market value subject to assumptions/special assumptions, will be the most frequent basis of valuation. *Market value* is stated as the value of the development property assuming optimum development, taking into account current and prospective economic and market circumstances and planning conditions. This may include alternative development solutions for the site. It notes that a valuer may require specialist assistance from other professionals to form a judgement on the optimum development, that is, the *highest and best use*.

Development-based valuations often require the consideration of changes in circumstances of the asset, and this should be reflected in the market value examples given in the standard:

- The prospect of development where there is no current permission for that development
- The prospect of marriage value arising from merger with another property or asset, or interests within the same property or asset, at a future date

The valuation approach

Valuations are normally undertaken in two ways:

- The market comparison approach
- The residual method (see IVS 410 paragraph 40.1)

The standard clearly states that

> in the case of the valuation of development property, valuations are normally undertaken in two ways: the market comparison approach and the residual method. Best practice avoids reliance on a single approach or method of assessing the value of development property.

In many cases, a market approach looking at comparable land transactions can be used as a reality check against a residual land calculation. However, where a complex development appraisal involving new permissions, uses, and products is being undertaken. existing use values may not be consistent with the result.

The professional standard sets out in section 3 the main requirements for establishing the facts for development valuation. It also states that a number of documents should be consulted, including:

- The *Red Book Global Standards* and the latest IVS
- *RICS property measurement*, RICS
- *Environmental risks and global real estate*, RICS
- *IPMS*
- *International Construction Measurement Standards (ICMS): Global Consistency in Presenting Measurement Costs*

It states that

> the extent of factual information necessary for a valuation is determined by a range of factors, including the stage at which the valuation is being prepared, the purpose and the individual characteristics of the property being valued and any assumptions or special assumptions made.

Appendix A sets out a useful checklist of facts and information to be considered in preparing a development valuation through both physical inspection and desk-based research to support the valuation.

It starts by reminding valuers that sites for potential development may contain many hazards, and reminds them to refer to guides to safe working practice, such as the current edition of *Surveying safely: health and safety principles for property professionals*.

Factual issues to be considered include thus:

- The presence of archaeological features. These may be evident, or there may be a high probability of their presence due to the site location (for instance, close to city centres).
- Evidence of waste management obligations and whether these obligations have been fulfilled.
- Water or mineral extraction rights that may be available.
- Geotechnical conditions, including potential for contamination or other environmental risks (see RICS *Environmental risks and global real estate*).
- Limitations, encumbrances, or conditions imposed on the relevant interest by private contract
- Rights of access to public highways or other public areas
- The availability of and requirements to provide or improve necessary services, for example, water, drainage, or power
- The need for any off-site infrastructure improvements and the rights required to undertake this work

It also provides a thorough examination of existing planning matters and the local planning regime to be considered from the national to the local level, including special controls, such as tree preservation orders; listing; and environmental issues, such as sites of special scientific interest (SSSIs). These include:

- The existence of any particular development plan or elements of zoning of land for different uses.
- The existence of any current permissions to undertake development. This may be in outline or in full and may include conditions or reserved matters.
- Where the permission is time-limited, it should be established whether it is still valid and, if close to expiry, whether a similar permission would be granted again.
- The existence of regulations that specify the extent to which development of the site might be permissible without the need for a planning application or consent.
- The permitted use of existing buildings (if to be retained) or the possibility of identifying an established use.
- Legally binding agreements that have been, or are to be, documented in order to secure the grant of planning permission.
- Any special controls that may apply to the site or buildings (for example, these may relate to heritage and conservation of sites or buildings).

- Requirements to protect or enhance environmentally sensitive features, such as watercourses or wildlife, and to comply with the relevant environmental protection legislation.
- Any requirements for view corridors, sight lines, or buffer zones.

Section 4 discusses the assessment of development potential and recognises that where the existing use or permissions are not optimum (the highest and best use), the valuer needs to form a view as to the form and extent of development that could be approved. The standard requires valuers to consult emerging consultative planning and relevant development policies depending on the institutional structures of the country – national, regional, local, federal, state, etc., which should be recognised.

It also sets out a list of matters that should be considered in detail, which include:

- Permissible land uses within the particular planning regime.
- Potential land uses within the particular planning regime.
- Density of development, establishing the bulk, scale, and massing, particularly in urbanised areas subject to different property types.
- Topography and site development factors, including availability of services and infrastructure, ground conditions, and development restraints.
- Building-related issues, such as the period of time estimated to complete the new buildings, achieving optimum occupational efficiency ratios, car parking standards and/or restrictions, regulations concerning energy-efficiency, and the extent to which the development control system is being used to help deliver climate change obligations.
- Development consent issues, such as requirements as to the provision of developer contributions, or planning obligations attached to the permission to develop.
- Adjacent land: although a valuation is required of the actual subject property, there may be a possibility of increasing the development potential by acquisition of, or merger with, adjacent land. Conversely, it may be necessary to acquire adjacent land, or rights over adjacent land, including oversailing rights, before the proposed development can take place.
- Accessibility and developability of the subject property.
- Environmental issues that may have a material bearing on the success of the project.

Sufficient enquiries should be made to establish whether the presence of on-site or neighbouring environmental features influences the development process, the density, or even the viability of the project. For more information, see the current edition of the RICS *Environmental risks and global real estate*.

The standard recognises that some development projects may be large and complex and contain multiple phases, and the timing and strategy for development may provide four key options:

1. Develop
2. Develop in phases
3. Sell or dispose
4. Defer or wait

Readers are encouraged to read the *Professional Standard* as it provides practical guidance and lists of inputs to discounted cash flow–based appraisals, which will help you understand the examples provided in this chapter.

RICS *FVIP: conduct and reporting*, professional standard 2023

In the UK, a system was introduced to examine the viability of development projects from both the developer and the local authority perspectives to support greater transparency in the justification and negotiation of planning gain requirements. The 2023 professional standard builds upon a 2012 guidance note and is designed to support appropriate and professional processes around the assessment of development value and the assessment of the viability of planning gain requirements, particularly social housing, as part of the planning process. It provides a framework and professional guidance on how development appraisals should be undertaken, and how the assessment of the provision of planning gain can be incorporated in a viability assessment. It is designed to make the development appraisal process more transparent, so that developers cannot evade appropriate provision of social housing and other planning gain requirements, and to provide an objective approach to assessing viability and a mechanism for assessments of viability to be scrutinised.

The provisions of FVIP are beyond the scope of this book, but what is interesting is that they provide the same rigour and guidance for development-based valuations to create consistency, fairness, and appropriate conduct within a public sector context.

IVS 2025

IVS 2025 includes a significant section on development property (IVS 410: Development Property). It is outside the scope of this chapter to present these provisions in detail, and the main provisions' relation to property value and the implications for the latest RICS *Red Book* and the valuation standard are discussed elsewhere. We have referred to the definition of *development valuations* earlier as it is incorporated into the RICS valuation standard. We recommend readers consult IVS 401: Development to see the full scope of this standard.

IVS 410 on development property does, however, emphasise some key issues discussed in this chapter, and those that resonate with the key messages of this chapter are set out in what follows.

The latest standard continues to emphasise that:

> *The residual value or land value of a development property can be very sensitive to changes in assumptions or projections concerning the income or revenue to be derived from the completed project or any of the development costs that will be incurred. This remains the case regardless of the method or methods used or however diligently the various inputs are researched in relation to the valuation date.*
>
> (20.04)

> *This sensitivity also applies to the impact of significant changes in either the costs of the project or the value on completion of the current value. If the valuation is required for an intended use where significant changes in value over the duration of a construction project may be of concern to the user (eg, where the valuation is for loan security or to establish a project's viability), the valuer must highlight the potentially disproportionate effect of possible changes in either the construction costs or end value on the profitability of the project and the value of the partially completed property. A sensitivity analysis may be useful for this intended use provided it is accompanied by a suitable explanation.*
>
> (20.05)

The standard confirms that there are three main *valuation* approaches and one main *valuation* method in relation to the *valuation* of development value:

1. The market approach (see section 70)
2. The income approach (see section 80)
3. The cost approach (see section 90)
4. The residual method, which is a hybrid of the market approach, the *income approach*, and the cost approach (sections 100) (60.01)

Applying the *highest and best use* in practice

The *highest and best use* methodology is not just a theoretical concept or a professional aspiration; in fact, it is widely used in international practice. We particularly like the way in which Jones Lang LaSalle (JLL) applies the concept; the authors have worked with this methodology in a number of student projects, mainly in the Middle East.

The JLL approach has five distinct stages that are worked through to build up the highest and best use analysis and build client reports.

Step 1: A general market analysis which examines thus:

- The macro-economic context
- The context of the sector(s) being examined – for example, tourism for hotel and hospitality
- The investment context and sentiment
- The legal framework, context, and any significant issues/changes
- Demographics (age, income, population growth, immigration)
- Macro demand drivers, such as airport capacity, infrastructure, and visa policies (for example, the growth of digital nomad visas)

Step 2: A site analysis which examines thus:

- The location
- Proximity to demand generators, for example, tourist attractions for the hotel sector
- Current and proposed land use and surrounding land uses
- Specific site access and egress for pedestrians, vehicles, and utilities, such as fire, refuse, etc.
- General transport and infrastructure provisions

Step 3: A submarket analysis for each submarket being considered in the highest and best use For example, in the hospitality market, it would examine:

- Factors influencing the hotel and hospitality submarkets
- Existing hotel stock (current competition)
- Future hotel supply (future competition)
- Demand – segregated into sub-sectors, for example, 3-star, 4-star, etc.
- Price structures, rack, and average room rates; seasonal and event factors
- Occupancy rates

Step 4: Development advisory, which includes the following:

- The project vision and objectives
- The site concept

- The development brief
- Specifications of each component of the project
- Critical success factors for each component/use

Step 5: A financial analysis, which examines thus:

- The appraisal assumptions
- The revenues and costs
- The financing of the project, debt (types and structures of loans), equity (opportunity costs), and contemporary options, such as unitisation and crowdfunding
- The analysis outputs, including IRR and NPV
- Sensitivity analysis and examination of alternative scenarios

The residual method

In the valuation of sites and/or property with development potential, an adaptation of the *income approach*, known as the residual method, is the most widely used technique. The conventional approach to a residual valuation is based upon the very simple concept introduced earlier: that the value of the completed project less all the costs incurred in the project is equal to the residual value. The value of the completed project is usually known as the gross development value. This is the capital value of the finished development and is usually found by capitalisation of the projected income from the scheme. To confirm:

Gross development value **(GDV)** ***less*** all costs plus the developer's profit = **The site value**

The costs of development include all the costs of construction, including professional fees (architects' fees, building cost, consultant/quantity surveyors' fees, etc.), the cost of borrowing the capital to undertake the scheme, and the developer's profit. The surplus that remains after deducting all these costs from the projected value is the residual value of the bare site or site and buildings to be redeveloped.

Exploring the main inputs to a residual appraisal

It has already been suggested that one of the difficulties in applying the residual method revolves around the number and variability of the inputs. These are introduced in what follows.

Components to establish the completed development value of a project

This is the market value of the completed development, which would be arrived at by the comparison or investment method as appropriate, depending upon the type of property. In the case of commercial developments, it is assumed that on completion, the developer will sell the development to an investor. This will incur agents' fees and advertising and legal fees, which may amount to approximately 3% of the value of the completed development. In addition, it will be easier to sell if the building is already occupied, and so letting agents' fees, which might account for some 10% of the first year's rent, also need to be taken into account.

Rent and yield

In the case of most commercial/industrial buildings, the completed development value will be calculated using the investment method of valuation. This will require an estimation of the rent of the development and the likely investment yield. Both will be based on current market evidence. The rent is normally based on the lettable area of the completed development. This will usually be the NIA, now known globally as IPMS3. It should be noted that in a residual appraisal, it is the current rent that is used, notwithstanding the fact that rental levels may have changed by the time the development is completed. It is regarded by some as extremely dangerous to base an appraisal on inflated future rents, as there can be no guarantee that these will be achieved. Others would argue that a full explicit cash flow approach requires the modelling of rents based upon projections, both up and down, over the anticipated period of construction and letting. The possibility of rental increase/decrease over the period of the development is something that can be reflected in the risk or profit allowance.

Incidental value–orientated costs

It is assumed that the developer will dispose of the completed development and will thus incur incidental disposal costs, which will include agent and legal fees amounting to around 2.5%. The developer will also incur costs in acquiring the site, including any property/land taxes due, such as the UK's SDLT, currently at 1–4%; legal fees, typically between 0.25 and 0.5%; and agents' fees, typically of 1–2%.

Components to establish the total costs of a project

The main costs to be deducted from the completed development value will include pre- and post-development costs.

Pre-development costs

The pre-development costs are those occurring before the commencement of construction and typically include:

- Demolition
- Environmental protection during demolition and construction
- Provision of infrastructure, such as roads, electricity, and gas supplies
- Planning gain, including cash contributions or construction works, for example, a road junction
- Compensation to tenants to remove them if they have security of tenure
- Site investigation
- Ground investigation
- Land surveys
- Planning fees
- Planning consultant fees
- Legal costs for appeals
- Heritage building consents and negotiations
- Archaeological surveys or digs

- Building regulations, scale fee based on final building cost
- Funding fees

Construction costs

Construction costs in initial appraisals or residual valuations are typically assessed on a price-per-square-metre basis. Sources of cost information include Spon's *Architects' and Builders' Price Book*, which is published annually, and the RICS Building Cost Information Service. In practice, advanced appraisals will be based upon detailed building costs identified by a quantity surveyor. Appraisals should recognise that, for most significant developments, building costs are paid in instalments over the construction period, and this should be mirrored in any cash flow–based appraisal.

The RICS professional standard refers to the two most common shapes for the distribution of costs within the cash flow as:

- *Straight line.* This assumes that the preliminary costs are incurred near to/at the beginning of the development period, and the principal development costs are incurred in equal tranches at regular and equal intervals throughout the development period.
- *S-curve.* The weighting of the construction costs may be incurred irregularly within a project, and different property types may require a different pattern of delivery of construction costs. Rather than being distributed equally over the development period, generally the costs are quite small at the beginning of a construction project, relatively accelerate in the middle, and reduce towards the end of the construction period. The purpose of an S-curve is to reflect more accurately the incidence of the costs in a particular project and may require expert advice from other construction professionals involved with the development.

Fees

The professional fees for architects, quantity surveyors, engineers, etc. need to be added to the construction costs. It is typical to add 10–15% to the construction costs, but this will be dependent upon the nature and scale of the development. In more complex schemes, it may be necessary to make additional allowance for project manager's fees and for structural and engineering consultants and specialists, such as mechanical engineers, if the building is air-conditioned or, for example, heated by sustainable technology, such as ground source heat pumps.

Finance

The developer will often need to take out short-term loans to fund the development. Even if the developer is able to fund the development from capital, this will incur an opportunity cost. Funding needs to cover the length of the development period, which must reflect the pre-contract period at the outset and any letting or void period once the development is completed. It is only on successful completion and sale that the developer will be able to recoup costs.

The basic residual usually adopts a *rule of thumb* approach to the calculation of finance charges, by taking either half the development period or half the building costs. This is usually adopted to reflect the average cost. Clearly, the developer will not borrow the whole of the cost of the development at the beginning of the development period, as payments to contractors will be staged as the project progresses.

The costs of short-term finance vary dramatically with economic conditions and the strength of the borrower in terms of risk. They have historically been between 1 and 6% above the prevailing base rate. The actual rate will depend upon the status and track record of the developer, as well as the type, quality, and location of the development.

The developer will also need to allow for the cost of interest payments on the site acquisition price. Where a site is being purchased at the outset, the cost of this must be funded over the whole period of development, from purchase through to disposal. In valuations, finance is typically assumed to be 100% debt-based; in reality, it is more likely to be a combination of equity and debt.

Marketing/letting, promotion

It is necessary to allow for both the letting and sale of the completed development. This will include both traditional and contemporary methods of promotion, especially where off-plan sales are involved, where an understanding of the completed project can be established using virtual and augmented reality.

- Television, radio, and editorial advertising
- Particulars and brochures, both paper and virtual formats
- Site boards and billboards
- Mail shots
- Websites and virtual reality
- Augmented reality
- Drone footage
- Video, YouTube, and cinematic video
- Launch ceremonies
- Physical and virtual show suites
- Public relations
- Testimonials
- Letting incentives

Miscellaneous costs

Other costs that need to be reflected may include the following:

- Party wall agreements
- Right to light agreements
- Void periods – maintenance and insurance
- Rent-free periods
- Site owner–based property taxes

Some costs are difficult to identify or estimate accurately, and these are sometimes reflected in a contingency of around 3–5%. It is sometimes argued that these elements are included in the profit/risk element. Thus, in the case of a development with major cost uncertainty, the developer will seek to make a higher allowance for profit.

Developer's profit or risk

If no allowance was made for profit, the developer would be undertaking the project for nothing. This allowance is normally related either to the cost of the development or to the completed development value; typically in the UK, this might be 15–25% cost, or 10–20% value. The amount of profit will largely be determined by the risk inherent in the project; very risky, marginal schemes will only proceed on the basis of a developer's profit at the high end of this range. In large schemes and in other countries, these rates may be less.

If the market is highly competitive, developers may be forced to trim profit margins; however, if the risks are high – for example, if the time taken to let or sell the development is extended in a weak market – then the profit element would need to be increased.

Example 10.1

A plot of land has planning permission for the erection of 7,000 m^2 (gross) of office space on five floors. The development will be completed two years from now, and rents are expected to average £120/m^2 (pa). Building costs are expected to average £400/m^2, excluding fees.

Prepare a valuation to advise a prospective developer of the maximum bid to be made for the site. (Comparable evidence of prices obtained for similar sites is not available.)

Gross development value (GDV)

7,000 m^2 (gross) $\times$ 85%* = (say) 6,000 m^2 (net)
6,000 m^2 at £120/m^2 pa = £720,000 pa
$\times$ capitalisation factor at 7% 14.29
£10,285,714
*Reduction for non-usable space of 15%.
Gross development value (GDV) **£10,285,714**
Less development costs:

(a) Building costs:

7,000 m^2 at £400/m^2 £2,800,000

(b) Architects and building cost:

Consultant/quantity surveyors'
fees at 12.5% of building costs £350,000

Total building costs: £3,150,000
(c) Finance costs:

15% pa for 2 years on 50% of total
building costs
(£1,575,000 $\times$ (1.15)^2) − £1,575,000 = £507,937

(d) Legal fees, estate agents' fees,
and advertising upon disposal:
at 2% GDV = £205,714

(e) Promotion, say: £50,000
(f) Developer's profit: at 15% GDV £1,542,857
Total costs: **£5,456,508**

Residual: (GDV − costs) = **£4,829,206**
Let site value = x
Legal and valuation fees on site purchase at 3% = $0.03x$
Total accumulated debt after 2 years at 15% pa = $1.03x\ (1.15)^2$
$1.03x\ (1.15)^2$ = £4,829,207
$1.362x$ = £4,829,207
Site value = £3,545,673

As stated earlier, the use of the residual method varies according to the stage of the development process. The preceding calculation is much generalised in its assumptions, and the technique is often criticised as a method of determining land value. Example 10.1 can be used to illustrate the dangers of the generalised assumptions built into the simple residual valuation. For example, the finance for development is calculated on only 50% of the total construction costs to reflect the phased drawing-down of borrowing over the development period. This assumption will hold true if the payments to contractors follow an equal distribution pattern similar to that shown in Figure 10.3 for the traditional office building. However, if the building involves a steel portal frame, that brings the development costs forward; the expenditure pattern may then change to that shown in Figure 10.3 for an industrial building. This destroys the 50% assumption, and an expenditure of this pattern should assume funding on around 75% of the total building costs.

The need for accurate, cash flow–based projections of building costs is therefore essential for accurate residual valuations. The *RICS VIP 12* (now the *Valuation Standard*) recognises this issue and states that an *S-curve* may be applied; the S-curve is the shape of the preceding figure when the amounts are made cumulative. Each development will have its own 'signature' S-curve depending upon its use, construction, and procurement options, and these can be evaluated by a quantity surveyor or cost consultant.

Sensitivity analysis

In addition to being subject to generalised assumptions, the final value is very sensitive to small variations in key variables, particularly the yield, rental values, and building costs. This has always been a key criticism of development appraisal; we see it more as just a reality. To illustrate this problem, note the effects on the residual value calculation in Example 10.1 with the changes to the key variables as shown in Figure 10.4.

Critics would argue that creating a plethora of alternative values is unhelpful and undermines some of the principles of valuation. However, sensitivity analysis is more than just creating alternative outcomes; it is a technique for helping analyse the key risks in a project and

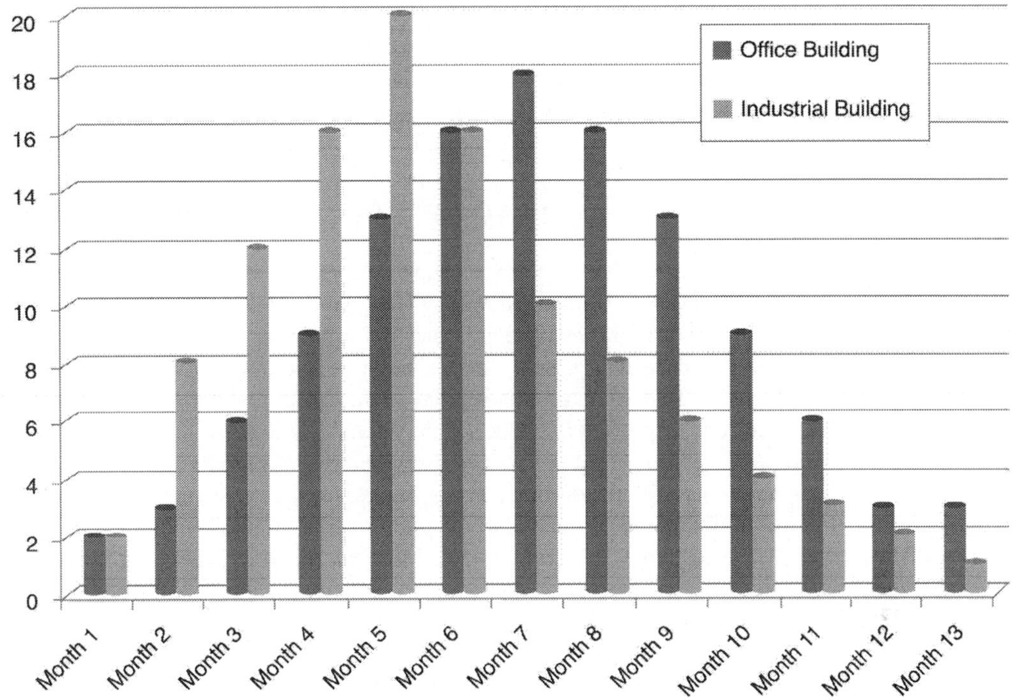

Figure 10.3 Distribution of building costs for construction of an office building using a traditional approach and an industrial building using a steel portal frame.

Original site value	£3,545,673
Site value when yield only increased by 0.5%	£3,120,825
Site value when rents only increased by 10%	£4,171,945
Site value when building costs only increased by 10%	£3,276,680

Figure 10.4 Illustration of the sensitivity of the residual site value calculated in Example 10.1 to changes in key variables.

formulating explicit, known parameters within which a developer or investor can make decisions. Furthermore the *RICS Professional Standard* specifically directs the consideration of sensitivity analysis in section 7: '[R]isk analysis should be used to evaluate how changes to individual inputs, such as construction cost or sales values, might affect the valuation of development property and to help model various different scenarios'.

Sensitivity analysis is also discussed and regarded as useful within **IVS 20.05**.

The use of 'star diagrams' to analyse relative sensitivity is an increasingly common practice, and one advocated by the authors. Following financial modelling principles, it maps onto a single line chart the variation in returns against changes in a number of variables. The chart is therefore a quick and easy way to read what variables a project is most sensitive to through the steepness of the lines, which indicates the change in profit against the change in variables. This methodology requires all changes in variables to be measured in percentage terms using the same percentage step for each variable. Many computer packages, such as EstateMaster, Argus, and ProDev, allow you to do this and create a table of values or returns against changes in variables.

Using the information in Example 10.1 again, we can create a table, in Figure 10.5, of site values using changes in the main underlying variables.

These can then be turned into a star diagram, as shown in Figure 10.6, to demonstrate the relative sensitivity of the site value to changes in the underlying variables. The chart shows that rents and yields are the most sensitive variables, as the line is steeper, indicating that for the same change in the variable, there is a greater change in the residual value than for building costs. Complex developments will have very different sensitivities, depending on timescales, cash flows, and the nature of the property products being delivered. Sensitivity analysis helps focus the investor and developer on the key risks and provides a more robust way of comparing different schemes and project configurations in terms of the risk profile created by the sensitivity to change in the underlying variables and appraisal inputs.

A further problem facing valuers has been the difficulty of handling an increasing number of variables, such as VAT, fees, etc., in the valuation exercise when carried out manually. Currently, many development surveyors use spreadsheets or computer software packages which do not remove the basic criticism, but their flexibility and speed of operation enable valuers to check each appraisal for sensitivity and avoid generalisations in the assumptions to achieve greater confidence in their opinions of value. (An example of this approach is provided in Appendix B.)

Example 10.1 has been reappraised using a package originally known as EstateMaster, which is one of several programmes now available to development valuers. EstateMaster has been taken over by Argus (an Altus Group company) and is now sold as Argus EstateMaster. For the sake of simplicity and continuity, our examples are produced using the original EstateMaster version, and the latest Argus version includes further sophistications, including an enterprise-level approach.

We set out in what follows the output from the Argus EstateMaster system using the appraisal data from Example 10.1.

Note that the EstateMaster system, as with many others, uses a more sophisticated approach than the simple appraisal set out in Example 10.1. Also, the methodology is slightly different,

	Building costs	Site value	Office rents	Site value	Yield	Site value
–20%	320	4082290	96	2291759	5.6%	5112039
–15%	340	3948021	102	2605123	6.0%	4589765
–10%	360	3813753	108	2918488	6.3%	4241582
–5%	380	3679485	114	3231852	6.7%	3825841
Original figures	**400**	**3545673**	**120**	**3545673**	**7.0%**	**3545673**
+5%	420	3410948	126	3858581	7.4%	3206444
+10%	440	3276680	132	4171945	7.7%	2975463
+15%	460	3142412	138	4485310	8.1%	2694103
+20%	480	3008143	144	4798674	8.4%	2500669

Figure 10.5 A table of sensitivity of the site value in Example 10.1 to changes in the key variables of the appraisal.

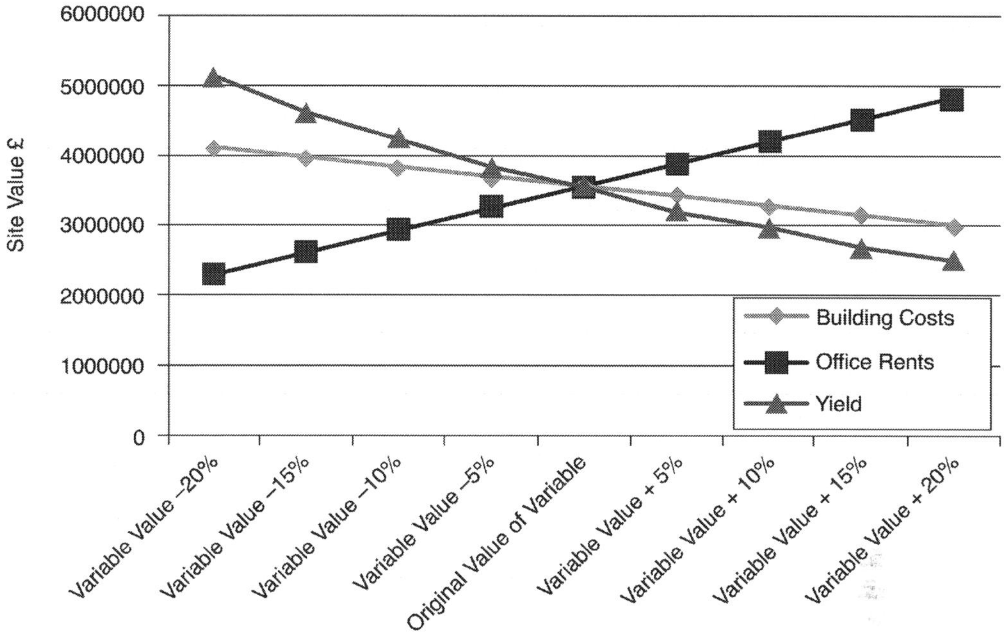

Figure 10.6 A star diagram showing the sensitivities in key variables for Example 10.1.

with the agent's letting fees deducted from the rental to give a net rental income. As this is the figure capitalised, it gives a different value than the GDV calculated in the manual version. Another key difference is the way interest is rolled up into one amount for both the development costs and the land purchase. It is a good exercise to compare and contrast the manual calculation and the output from EstateMaster based on the same inputs. Whilst the numbers are similar, they are not the same because of the differences indicated earlier.

We have included the majority of the report pages for completeness (although one page was excluded as it had no entries). They illustrate both the comprehensive and flexible nature of the software to manage complex multi-phase schemes, complex cash flows, multiple financing methods, and both sales (and pre-sales) and lettings within a scheme. A summary report of a more complicated development project produced by the system is set out in case study 1, together with notes on the assumptions and methodologies used. The traditional residual approach can be significantly improved with the application of computers through sensitivity analysis and more complex treatment of the underlying assumptions. However, although packages such as those produced by EstateMaster allow for complex phasing of developments by the linking of parts and areas of schemes together, they still rely heavily on assumptions made about the distribution of payments and receipts.

The need for adoption of the cash flow approach

Cash flow modelling is now a standard feature of most development appraisal systems, and its consistent use is encouraged to counter the difficulties and criticisms described earlier. The use of cash flows is specifically referenced in **IVS 80.01**, where it states establishing the residual value of a development property may involve the use of a cash flow model in some markets.

Main Inputs for 2 Year Construct

PRELIMINARY

Cash Flow Title	2 Year Construction	Description of Option/Stage	Nick Nunnington
Date of First Period:	Sep-2010		
Cash Flow Rest Period:	Monthly		
Enter Project Size (a)			
Enter Project Size (b)	6,000.00	GFA	
Enter Site Area	7,000.00		0 :1

Type	Miscellaneous
Status	Under Review

1000 **LAND PURCHASE & ACQUISITION COSTS**

Land Purchase Price	3,533,726

Code	Stage		% of Land Purchase Price % paid	Amount	AND/OR Lump Amount	Month Start	Month Span	Cash Flow Period	Total Current Costs	Total Escalated Cost
1002	-	Deposit in Trust Account ¹	0.00%	-	-	0	-	-	-	-
1003	-	Payment 1	0.00%	-	-	0	-	-	-	-
1004	-	Payment 2	0.00%	-	-	0	-	-	-	-
1005	-	Payment 3	0.00%	-	-	0	-	-	-	-
1006	-	Payment 4	0.00%	-	-	0	-	-	-	-
1007	-	Settlement (Balance)	100.00%		3,533,726	0	1	Sep-10 - Sep-10	3,533,726	3,533,726
1008	-	Stamp Duty ²	NIL		-	0	1	Sep-10 - Sep-10	-	-
		Interest on Deposit in Trust Account	0.00%	Interest from deposit shared between parties						
		Profit Share to Land Owner	0.00%	Paid progressively as project makes a profit						
								TOTAL	3,533,726	3,533,726

(The label "Stamp Duty" is printed beside the Total Current Costs row for code 1008.)

Code	Stage	Other Acquisition Costs	% of Land Purchase Price % paid	Amount	AND/OR Lump Amount	Month Start	Month Span	Cash Flow Period	Remarks	Total Current Costs	Total Escalated Cost
1011	-	Stamp Duty	1.00%	35,337	-	0	1	Sep-10 - Sep-10		35,337	35,337
1012	-	Agents Fees	1.00%	35,337	-	0	1	Sep-10 - Sep-10		35,337	35,337
1013	-	Legal Fees	1.00%	35,337	-	0	1	Sep-10 - Sep-10		35,337	35,337
1014	-		0.00%	-	-	0	-	-		-	-
1015	-		0.00%	-	-	0	-	-		-	-
									Manual Input (refer to Cash Flow)	-	-
									TOTAL	106,012	106,012

¹ (No VAT credit available for Stamp Duty)

² Pro-rata with Land Payments ('L')

COST ESCALATION Monthly Compounded Escalation - based on Cashflow Period Years commencing

Escalation Rates (Monthly Compounded Escalation) based on Cashflow Period Years commencing

Code		Sep-10	Sep-11	Sep-12	Sep-13	Sep-14	Sep-15	Sep-16	Sep-17	Sep-18	Sep-19
	Professional Fees	0.00%	0.00%	0.00%	0.00%	0.00%	0.00%	0.00%	0.00%	0.00%	0.00%
	Construction Costs (Uncategorised)	0.00%	0.00%	0.00%	0.00%	0.00%	0.00%	0.00%	0.00%	0.00%	0.00%
SUB	Subdivision Costs	0.00%	0.00%	0.00%	0.00%	0.00%	0.00%	0.00%	0.00%	0.00%	0.00%
STG	Stage Costs	0.00%	0.00%	0.00%	0.00%	0.00%	0.00%	0.00%	0.00%	0.00%	0.00%
BUI	Built Form	0.00%	0.00%	0.00%	0.00%	0.00%	0.00%	0.00%	0.00%	0.00%	0.00%
OT1	Other	0.00%	0.00%	0.00%	0.00%	0.00%	0.00%	0.00%	0.00%	0.00%	0.00%
OT2	Other	0.00%	0.00%	0.00%	0.00%	0.00%	0.00%	0.00%	0.00%	0.00%	0.00%
	Statutory Fees	0.00%	0.00%	0.00%	0.00%	0.00%	0.00%	0.00%	0.00%	0.00%	0.00%
	Miscellaneous Costs 1	0.00%	0.00%	0.00%	0.00%	0.00%	0.00%	0.00%	0.00%	0.00%	0.00%
	Miscellaneous Costs 2	0.00%	0.00%	0.00%	0.00%	0.00%	0.00%	0.00%	0.00%	0.00%	0.00%
	Miscellaneous Costs 3	0.00%	0.00%	0.00%	0.00%	0.00%	0.00%	0.00%	0.00%	0.00%	0.00%
	Land Holding Costs	0.00%	0.00%	0.00%	0.00%	0.00%	0.00%	0.00%	0.00%	0.00%	0.00%
	Selling and Leasing Costs	0.00%	0.00%	0.00%	0.00%	0.00%	0.00%	0.00%	0.00%	0.00%	0.00%
	Finance Costs	0.00%	0.00%	0.00%	0.00%	0.00%	0.00%	0.00%	0.00%	0.00%	0.00%

2000	PROJECT CONTINGENCY	-	And / Or	0.00%	of Construction, Professional (exc Development Management), Statutory Fees, Miscellaneous Costs 1, Miscellaneous Costs 2, Miscellaneous Costs 3	TOTAL	-

Figure 10.7a–k Indicative output from EstateMaster development feasibility software.

Main Inputs for 2 Year Construct

3000	PROFESSIONAL FEES

Code	Stage	Description	% of Construct [1]	AND / OR No. Units	Base Rate / Unit	Escalate (E,R,N)	S-Curve	Month Start[2]	Month Span	Cash Flow Period	Remarks	Total Current Costs	Total Escalated Cost
3001	-	Architect	9.00%	-	-	-	-	C	-	Sep-10 - Aug-12		252,000	252,000
3002	-	Quantity Surveyor	3.50%	-	-	-	-	C	-	Sep-10 - Aug-12		98,000	98,000
3003	-	.	0.00%	-	-	-	-	0	-	-		-	-
3004	-	.	0.00%	-	-	-	-	0	-	-		-	-
3005	-	.	0.00%	-	-	-	-	0	-	-		-	-
3006	-	.	0.00%	-	-	-	-	0	-	-		-	-
3007	-	.	0.00%	-	-	-	-	0	-	-		-	-
3008	-	.	0.00%	-	-	-	-	0	-	-		-	-
3009	-	.	0.00%	-	-	-	-	0	-	-		-	-
3010	-	.	0.00%	-	-	-	-	0	-	-		-	-
3011	-	.	0.00%	-	-	-	-	0	-	-		-	-
3012	-	.	0.00%	-	-	-	-	0	-	-		-	-
3013	-	.	0.00%	-	-	-	-	0	-	-		-	-
3014	-	.	0.00%	-	-	-	-	0	-	-		-	-
3015	-	.	0.00%	-	-	-	-	0	-	-		-	-
3099	-	Development Management	0.00%	% of Project Costs (exc Land, Finance & Tax)			-	0	-	-		-	-
											Manual Input (refer to Cash Flow)	-	-
											TOTAL	350,000	350,000

[1] % Based on Net Costs

[2] Pro-rata with Construction ('C')

[3] Dev Mgmt Fee: Pro-rata with Construction ('C'), Settlements ('S'), Project Costs inc Land ('P1') or exc Land ('P2')

4000	CONSTRUCTION COSTS

Code	Stage	Description	Cost Type	Units	Base Rate / Units	Escalate (E,R,N)1	S-Curve	Month Start	Month Span	Cash Flow Period	Remarks	Total Current Costs	Total Escalated Cost
4001	-	Office	-	7,000	400	-	S	0	24	Sep-10 - Aug-12		2,800,000	2,800,000
4002	-	.	-	-	-	-	-	0	-	-		-	-
4003	-	.	-	-	-	-	-	0	-	-		-	-
4004	-	.	-	-	-	-	-	0	-	-		-	-
4005	-	.	-	-	-	-	-	0	-	-		-	-
4006	-	.	-	-	-	-	-	0	-	-		-	-
4007	-	.	-	-	-	-	-	0	-	-		-	-
4008	-	.	-	-	-	-	-	0	-	-		-	-
4009	-	.	-	-	-	-	-	0	-	-		-	-
4010	-	.	-	-	-	-	-	0	-	-		-	-
4011	-	.	-	-	-	-	-	0	-	-		-	-
4012	-	.	-	-	-	-	-	0	-	-		-	-
4013	-	.	-	-	-	-	-	0	-	-		-	-
4014	-	.	-	-	-	-	-	0	-	-		-	-
4015	-	.	-	-	-	-	-	0	-	-		-	-
4016	-	.	-	-	-	-	-	0	-	-		-	-
4017	-	.	-	-	-	-	-	0	-	-		-	-
4018	-	.	-	-	-	-	-	0	-	-		-	-
4019	-	.	-	-	-	-	-	0	-	-		-	-
4020	-	.	-	-	-	-	-	0	-	-		-	-
											Manual Input (refer to Cash Flow)	-	-
4099		Construction Contingency		-	And / Or	0.00% of Construction Costs					Construction Contingency	-	-
											TOTAL	2,800,000	2,800,000

[1] Escalation ('N' = no escalation, 'E' = escalation to start period, 'R' = escalation to start period and through span)

5000	Statutory Fees

Code	Stage	Description	Units	Base Rate / Units	Escalate (E,R,N)	S-Curve	Month Start	Month Span	Cash Flow Period	Remarks	Total Current Costs	Total Escalated Cost
5001	-	.	-	-	-	-	0	-	-		-	-
5002	-	.	-	-	-	-	0	-	-		-	-
5003	-	.	-	-	-	-	0	-	-		-	-
5004	-	.	-	-	-	-	0	-	-		-	-
5005	-	.	-	-	-	-	0	-	-		-	-
5006	-	.	-	-	-	-	0	-	-		-	-
5007	-	.	-	-	-	-	0	-	-		-	-
5008	-	.	-	-	-	-	0	-	-		-	-
5009	-	.	-	-	-	-	0	-	-		-	-
5010	-	.	-	-	-	-	0	-	-		-	-
5011	-	.	-	-	-	-	0	-	-		-	-
5012	-	.	-	-	-	-	0	-	-		-	-
5013	-	.	-	-	-	-	0	-	-		-	-
5014	-	.	-	-	-	-	0	-	-		-	-
5015	-	.	-	-	-	-	0	-	-		-	-
										Manual Input (refer to Cash Flow)	-	-
										TOTAL	-	-

Figure 10.7a–k (Continued)

Main Inputs for 2 Year Construct

6000 Miscellaneous Costs 1

Code	Stage	Description	%of Construction[1]	AND / OR No. Units	Base Rate / Unit	Escalate (E,R,N)	S-Curve	Month Start[2]	Month Span	Cash Flow Period	Remarks	Total Current Costs	Total Escalated Cost
6001	-		0.00%	-	-	-	-	0	-	-		-	-
6002	-		0.00%	-	-	-	-	0	-	-		-	-
6003	-		0.00%	-	-	-	-	0	-	-		-	-
6004	-		0.00%	-	-	-	-	0	-	-		-	-
6005	-		0.00%	-	-	-	-	0	-	-		-	-
6006	-		0.00%	-	-	-	-	0	-	-		-	-
6007	-		0.00%	-	-	-	-	0	-	-		-	-
6008	-		0.00%	-	-	-	-	0	-	-		-	-
6009	-		0.00%	-	-	-	-	0	-	-		-	-
6010	-		0.00%	-	-	-	-	0	-	-		-	-
											Manual Input (refer to Cash Flow)	-	-
											TOTAL	-	-

[1] Based on net costs [2] Pro-rata with Construction ('C') or Settlements ('S')

6000 Miscellaneous Costs 2

Code	Stage	Description	%of Construction[1]	AND / OR No. Units	Base Rate / Unit	Escalate (E,R,N)	S-Curve	Month Start[2]	Month Span	Cash Flow Period	Remarks	Total Current Costs	Total Escalated Cost
6001	-		0.00%	-	-	-	-	0	-	-		-	-
6002	-		0.00%	-	-	-	-	0	-	-		-	-
6003	-		0.00%	-	-	-	-	0	-	-		-	-
6004	-		0.00%	-	-	-	-	0	-	-		-	-
6005	-		0.00%	-	-	-	-	0	-	-		-	-
6006	-		0.00%	-	-	-	-	0	-	-		-	-
6007	-		0.00%	-	-	-	-	0	-	-		-	-
6008	-		0.00%	-	-	-	-	0	-	-		-	-
6009	-		0.00%	-	-	-	-	0	-	-		-	-
6010	-		0.00%	-	-	-	-	0	-	-		-	-
											Manual Input (refer to Cash Flow)	-	-
											TOTAL	-	-

[1] Based on net costs. [2] Pro-rata with Construction ('C') or Settlements ('S')

6000 Miscellaneous Costs 3

Code	Stage	Description	%of Construction[1]	AND / OR No. Units	Base Rate / Unit	Escalate (E,R,N)	S-Curve	Month Start[2]	Month Span	Cash Flow Period	Remarks	Total Current Costs	Total Escalated Cost
6001	-		0.00%	-	-	-	-	0	-	-		-	-
6002	-		0.00%	-	-	-	-	0	-	-		-	-
6003	-		0.00%	-	-	-	-	0	-	-		-	-
6004	-		0.00%	-	-	-	-	0	-	-		-	-
6005	-		0.00%	-	-	-	-	0	-	-		-	-
6006	-		0.00%	-	-	-	-	0	-	-		-	-
6007	-		0.00%	-	-	-	-	0	-	-		-	-
6008	-		0.00%	-	-	-	-	0	-	-		-	-
6009	-		0.00%	-	-	-	-	0	-	-		-	-
6010	-		0.00%	-	-	-	-	0	-	-		-	-
											Manual Input (refer to Cash Flow)	-	-
											TOTAL	-	-

[1] Based on net costs. [2] Pro-rata with Construction ('C') or Settlements ('S')

7000 LAND HOLDING COSTS

Code	Stage	Description	No. Units	Base Rate /unit/term	Term[1]	Escalate (E,R,N)	Month Start	Month Span[2]	Cash Flow Period	Remarks	Total Annual Costs	Total Escalated Cost
7001	-		-	-	M	-	0	-	-		-	-
7002	-		-	-	M	-	0	-	-		-	-
7003	-		-	-	M	-	0	-	-		-	-
7004	-		-	-	M	-	0	-	-		-	-
7005	-		-	-	M	-	0	-	-		-	-
7006	-		-	-	M	-	0	-	-		-	-
7007	-		-	-	M	-	0	-	-		-	-
7008	-		-	-	M	-	0	-	-		-	-
7009	-		-	-	M	-	0	-	-		-	-
7010	-		-	-	M	-	0	-	-		-	-
7011	-		-	-	M	-	0	-	-		-	-
7012	-		-	-	M	-	0	-	-		-	-
										Manual Input (refer to Cash Flow)	-	-
										TOTAL	-	-

[1] Y=Yearly, BA=BiAnnually, Q=Quarterly, BM=BiMonthly, M=Monthly [2] Diminish proportionally with Leasing ('DR') or Settlements ('DS')

Figure 10.7a–k (Continued)

Main Inputs for 2 Year Construct

REVENUE ESCALATION Monthly Compounded Escalation - based on Cashflow Period Years commencing

Code	Category	Sep-10	Sep-11	Sep-12	Sep-13	Sep-14	Sep-15	Sep-16	Sep-17	Sep-18	Sep-19
RS1	Residential - 1 Bedroom Units	0.00%	0.00%	0.00%	0.00%	0.00%	0.00%	0.00%	0.00%	0.00%	0.00%
RS2	Residential - 2 Bedroom Units	0.00%	0.00%	0.00%	0.00%	0.00%	0.00%	0.00%	0.00%	0.00%	0.00%
RS3	Residential - 3 Bedroom Units	0.00%	0.00%	0.00%	0.00%	0.00%	0.00%	0.00%	0.00%	0.00%	0.00%
RDD	Detached Dwellings Lots	0.00%	0.00%	0.00%	0.00%	0.00%	0.00%	0.00%	0.00%	0.00%	0.00%
RTH	Townhouse Lots	0.00%	0.00%	0.00%	0.00%	0.00%	0.00%	0.00%	0.00%	0.00%	0.00%
COM	Commerical Office	0.00%	0.00%	0.00%	0.00%	0.00%	0.00%	0.00%	0.00%	0.00%	0.00%
RET	Retail Shops	0.00%	0.00%	0.00%	0.00%	0.00%	0.00%	0.00%	0.00%	0.00%	0.00%
IND	Industrial Units	0.00%	0.00%	0.00%	0.00%	0.00%	0.00%	0.00%	0.00%	0.00%	0.00%
STW	Storage & Warehousing	0.00%	0.00%	0.00%	0.00%	0.00%	0.00%	0.00%	0.00%	0.00%	0.00%
OTH	Other	0.00%	0.00%	0.00%	0.00%	0.00%	0.00%	0.00%	0.00%	0.00%	0.00%

Rental escalation occurs up to lease start date. For rent review escalation during lease period refer to the Tenants sheet.

8000 **SELLING & LEASING COSTS**

		Sales Commission	Sales Comm[1]	% of Comm Pre-sales[2]	Deposits (% of Price)[3]	Remarks	Total Current Costs	Total Escalated Cost
8001	RS1	Residential - 1 Bedroom Units	0.00%	0.00%	0.00%		-	-
8002	RS2	Residential - 2 Bedroom Units	0.00%	0.00%	0.00%		-	-
8003	RS3	Residential - 3 Bedroom Units	0.00%	0.00%	0.00%		-	-
8004	RDD	Detached Dwellings Lots	0.00%	0.00%	0.00%		-	-
8005	RTH	Townhouse Lots	0.00%	0.00%	0.00%		-	-
8006	COM	Commerical Office	1.00%	0.00%	0.00%		102,857	102,857
8007	RET	Retail Shops	0.00%	0.00%	0.00%		-	-
8008	IND	Industrial Units	0.00%	0.00%	0.00%		-	-
8009	STW	Storage & Warehousing	0.00%	0.00%	0.00%		-	-
8010	OTH	Other	0.00%	0.00%	0.00%		-	-
						TOTAL	102,857	102,857

[1] % of Gross Purchase Price
[2] Percentage of Sales Commission paid at exchange date for pre-sales
[3] Percentage of price deposited on exchange (for pre-sales)

Pre-sale Comm are reported as a	Project Cost
Interest Rate on Deposits Invested in Trust Account	0.00%
% of Interest retained by Developer upon Settlement	0.00%

Code	Stage	Other Selling Costs	% of Gross Sales	AND / OR No. Units	Base Rate / Unit	Escalate (E,R,N)	Month Start	Month Span	Cash Flow Period	Remarks	Total Current Costs	Total Escalated Cost
8101	-		0.00%	-	-	-	0	-	-		-	-
8102	-		0.00%	-	-	-	0	-	-		-	-
8103	-		0.00%	-	-	-	0	-	-		-	-
8104	-		0.00%	-	-	-	0	-	-		-	-
8105	-		0.00%	-	-	-	0	-	-		-	-
8106	-		0.00%	-	-	-	0	-	-		-	-
8107	-		0.00%	-	-	-	0	-	-		-	-
8108	-		0.00%	-	-	-	0	-	-		-	-
8109	-		0.00%	-	-	-	0	-	-		-	-
8110	-		0.00%	-	-	-	0	-	-		-	-
8111	-		0.00%	-	-	-	0	-	-		-	-
										Manual Input (refer to Cash Flow)	-	-
										TOTAL	-	-

Pro-rata with Settlements ('S') or Exchanges ('E')

Code	Stage	Other Leasing Costs	% of Gross Rent	AND / OR No. Units	Base Rate / Unit	Escalate (E,R,N)	Month Start	Month Span	Cash Flow Period	Remarks	Total Current Costs	Total Escalated Cost
8201	-	Marketing	0.00%	1	50,000	-	R	-	Sep-12 - Sep-12		50,000	50,000
8202	-	0	0.00%	-	-	-	24	1	Sep-12 - Sep-12		-	-
8203	-		0.00%	-	-	-	0	-	-		-	-
8204	-		0.00%	-	-	-	0	-	-		-	-
8205	-		0.00%	-	-	-	0	-	-		-	-
8206	-		0.00%	-	-	-	0	-	-		-	-
8207	-		0.00%	-	-	-	0	-	-		-	-
8208	-		0.00%	-	-	-	0	-	-		-	-
8209	-		0.00%	-	-	-	0	-	-		-	-
8210	-		0.00%	-	-	-	0	-	-		-	-
8211	-		0.00%	-	-	-	0	-	-		-	-
										Manual Input (refer to Cash Flow)	-	-
										TOTAL	50,000	50,000

Pro-rata with Rental Income ('R')

Figure 10.7a–k (Continued)

Main Inputs for 2 Year Construct

9000	**SALES**

Code	Stage	Description	No. Units	Total Area SqM	Current Sale Price	Sales Calc Method	Pre-Sale Exchange Month Start	Pre-Sale Exchange Month Span	Settlements Month Start	Settlements Month Span	Settlements Cash Flow Period	Sales Rate Units / SqM per Month	Land Use Code	Total Current Sales Revenue	Total Escalated Sales Revenue
9001	-		-	-	-	Per Unit	0	-	0	-	-	-	-	-	-
9002	-		-	-	-	Per Unit	0	-	0	-	-	-	-	-	-
9003	-		-	-	-	Per Unit	0	-	0	-	-	-	-	-	-
9004	-		-	-	-	Per Unit	0	-	0	-	-	-	-	-	-
9005	-		-	-	-	Per Unit	0	-	0	-	-	-	-	-	-
9006	-		-	-	-	Per Unit	0	-	0	-	-	-	-	-	-
9007	-		-	-	-	Per Unit	0	-	0	-	-	-	-	-	-
9008	-		-	-	-	Per Unit	0	-	0	-	-	-	-	-	-
9009	-		-	-	-	Per Unit	0	-	0	-	-	-	-	-	-
9010	-		-	-	-	Per Unit	0	-	0	-	-	-	-	-	-
9011	-		-	-	-	Per Unit	0	-	0	-	-	-	-	-	-
9012	-		-	-	-	Per Unit	0	-	0	-	-	-	-	-	-
9013	-		-	-	-	Per Unit	0	-	0	-	-	-	-	-	-
9014	-		-	-	-	Per Unit	0	-	0	-	-	-	-	-	-
9015	-		-	-	-	Per Unit	0	-	0	-	-	-	-	-	-
9016	-		-	-	-	Per Unit	0	-	0	-	-	-	-	-	-
9017	-		-	-	-	Per Unit	0	-	0	-	-	-	-	-	-
9018	-		-	-	-	Per Unit	0	-	0	-	-	-	-	-	-
9019	-		-	-	-	Per Unit	0	-	0	-	-	-	-	-	-
9020	-		-	-	-	Per Unit	0	-	0	-	-	-	-	-	-
													Capitalised Sales (refer to Tenants)	10,285,714	10,285,714
													Manual Input (refer to Cash Flow)	-	-
													TOTAL	10,285,714	10,285,714

9100	**OTHER INCOME**

Code	Stage	Description	Land Use Code	Units	Base Rate / Units	Month Start	Month Span	Cash Flow Period	Remarks	Total Current Income	Total Escalated Income
9101	-		-	-	-	0	-	-		-	-
9102	-		-	-	-	0	-	-		-	-
9103	-		-	-	-	0	-	-		-	-
9104	-		-	-	-	0	-	-		-	-
9105	-		-	-	-	0	-	-		-	-
9106	-		-	-	-	0	-	-		-	-
9107	-		-	-	-	0	-	-		-	-
9108	-		-	-	-	0	-	-		-	-
9109	-		-	-	-	0	-	-		-	-
9110	-		-	-	-	0	-	-		-	-
									Manual Input (refer to Cash Flow)	-	-
									TOTAL	-	-

10000	**FINANCING** (Advanced Mode)

General Notes: All Line Fees are paid during period of debt, in arrears.
All Profit Share is Paid progressively as project makes a profit.

Equity Notes: Equity is paying outstanding debt
Equity is repaid at project end.

Equity

Developer's Equity Contribution	Fixed Amount	Percentage	
Injected in total upfront.	-	0.00%	Fixed Amount

Code			
10001	Interest Charged on Equity	0.00%	per annum Nominal - Capitalised (Compounded)
10002	Interest received on Surplus Cash	0.00%	per annum received in arrears.
	% of Available Funds to Repay Equity Before Debt	0.00%	

Opening Balances		Equity Totals
	Developer's Injections	-
-	Interest Charged	-
-	Interest Received	-

Figure 10.7a–k (Continued)

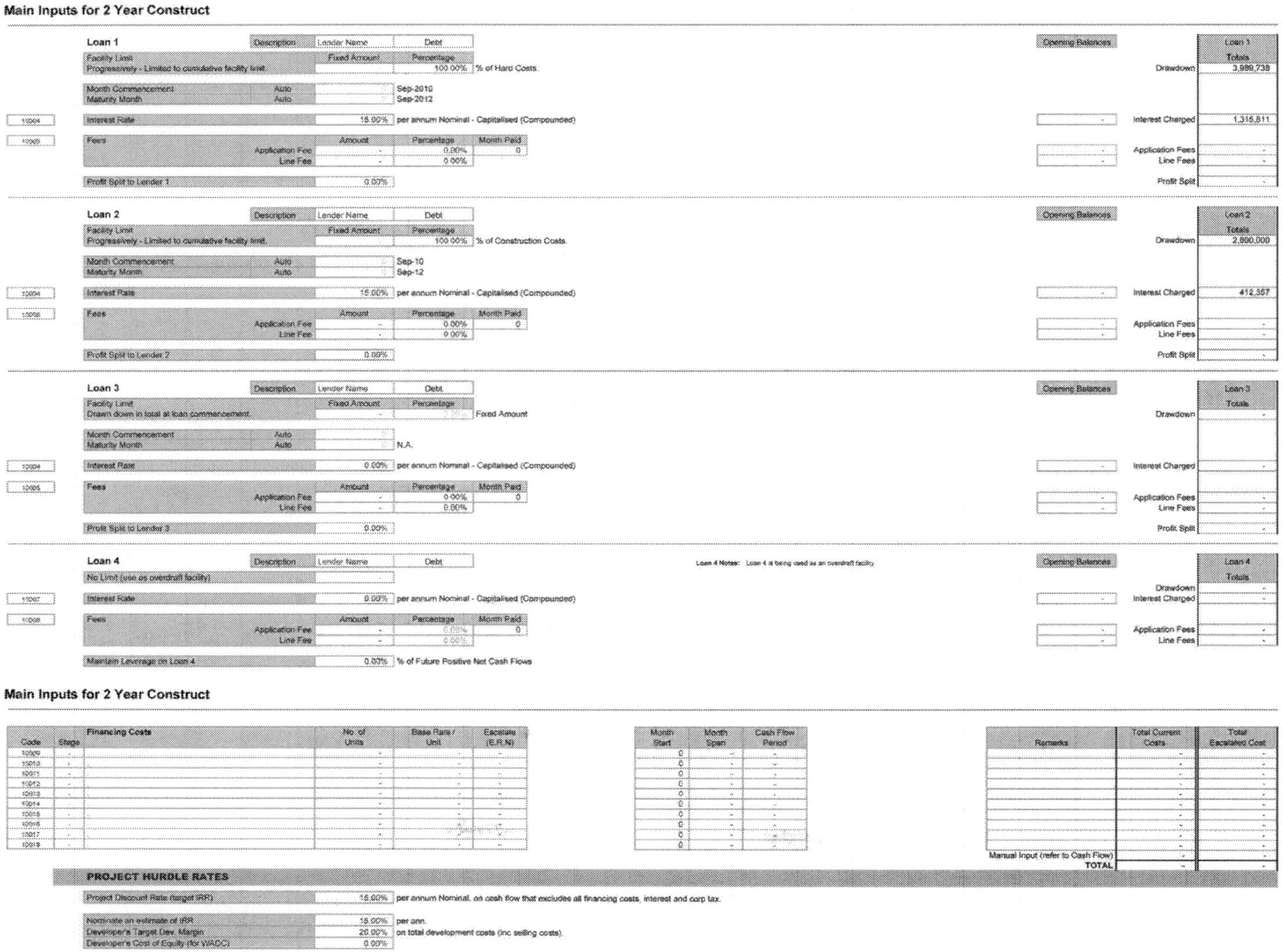

Main Inputs for 2 Year Construct

Code	Loan 1					Opening Balances		Loan 1 Totals
	Description	Lender Name	Debt					
	Facility Limit – Progressively - Limited to cumulative facility limit.	Fixed Amount:	Percentage: 100.00%	% of Hard Costs.			Drawdown	3,989,738
	Month Commencement – Auto		Sep-2010					
	Maturity Month – Auto		Sep-2012					
10004	Interest Rate	15.00%	per annum Nominal - Capitalised (Compounded)			-	Interest Charged	1,315,811
10005	Fees	Amount	Percentage	Month Paid				
	Application Fee	-	0.00%	0		-	Application Fees	-
	Line Fee	-	0.00%			-	Line Fees	-
	Profit Split to Lender 1	0.00%					Profit Split	-

Code	Loan 2					Opening Balances		Loan 2 Totals
	Description	Lender Name	Debt					
	Facility Limit – Progressively - Limited to cumulative facility limit.	Fixed Amount:	Percentage: 100.00%	% of Construction Costs.			Drawdown	2,800,000
	Month Commencement – Auto		Sep-10					
	Maturity Month – Auto		Sep-12					
10004	Interest Rate	15.00%	per annum Nominal - Capitalised (Compounded)			-	Interest Charged	412,357
10006	Fees	Amount	Percentage	Month Paid				
	Application Fee	-	0.00%	0		-	Application Fees	-
	Line Fee	-	0.00%			-	Line Fees	-
	Profit Split to Lender 2	0.00%					Profit Split	-

Code	Loan 3					Opening Balances		Loan 3 Totals
	Description	Lender Name	Debt					
	Facility Limit – Drawn down in total at loan commencement.	Fixed Amount: -	Percentage: 0.00%	Fixed Amount			Drawdown	-
	Month Commencement – Auto	0						
	Maturity Month – Auto	0	N.A.					
10004	Interest Rate	0.00%	per annum Nominal - Capitalised (Compounded)			-	Interest Charged	-
10005	Fees	Amount	Percentage	Month Paid				
	Application Fee	-	0.00%	0		-	Application Fees	-
	Line Fee	-	0.00%			-	Line Fees	-
	Profit Split to Lender 3	0.00%					Profit Split	-

Code	Loan 4					Opening Balances		Loan 4 Totals
	Description	Lender Name	Debt	Loan 4 Notes: Loan 4 is being used as an overdraft facility.				
	No Limit (use as overdraft facility)						Drawdown	-
10007	Interest Rate	0.00%	per annum Nominal - Capitalised (Compounded)			-	Interest Charged	-
10008	Fees	Amount	Percentage	Month Paid				
	Application Fee	-	0.00%	0		-	Application Fees	-
	Line Fee	-	0.00%			-	Line Fees	-
	Maintain Leverage on Loan 4	0.00%	% of Future Positive Net Cash Flows					

Main Inputs for 2 Year Construct

Code	Stage	Financing Costs	No. of Units	Base Rate / Unit	Escalate (E.R.N)	Month Start	Month Span	Cash Flow Period	Remarks	Total Current Costs	Total Escalated Cost
10009	-		-	-	-	0	-	-		-	-
10010	-	.	-	-	-	0	-	-		-	-
10011	-	.	-	-	-	0	-	-		-	-
10012	-	.	-	-	-	0	-	-		-	-
10013	-	.	-	-	-	0	-	-		-	-
10014	-	.	-	-	-	0	-	-		-	-
10015	-	.	-	-	-	0	-	-		-	-
10016	-	.	-	-	-	0	-	-		-	-
10017	-	.	-	-	-	0	-	-		-	-
10018	-	.	-	-	-	0	-	-		-	-
									Manual Input (refer to Cash Flow)	-	-
									TOTAL	-	-

PROJECT HURDLE RATES

Project Discount Rate (target IRR)	15.00%	per annum Nominal, on cash flow that excludes all financing costs, interest and corp tax.
Nominate an estimate of IRR	15.00%	per ann.
Developer's Target Dev. Margin	20.00%	on total development costs (inc selling costs).
Developer's Cost of Equity (for WACC)	0.00%	

Figure 10.7a–k (Continued)

EstateMaster Development Feasibility

SUMMARY OF PROJECT RETURNS

2 Year Construct
2 Year Construction
Nick Nunnington

Estate Master Licensed to: Estate Master

Time Span:	Sep-10 to Sep-12
Type:	Miscellaneous
Status:	Under Review
Site Area:	7,000
#N/A	:1
Project Size:	
	6,000 GFA 1 per 1.16 of Site Area

				Total GBP	GBP Per 0	GBP Per of Site Area	GBP Per Total Net Costs
Revenues							
	Quantity	SqM	GBP/Quantity				
Gross Sales Revenue	-	6,000.00	-	10,285,714		1,469	120.8%
Commerical Office	-	6,000.00	-	10,285,714			
Less Selling Costs				(102,857)		15	-1.2%
Less Purchasers Costs				-		-	0.0%
NET SALES REVENUE				10,182,857		1,455	119.5%
	Average Yield	SqM	GBP/SqM/annum				
Gross Rental Income	7.00%	6,000.00	120.00	60,000		9	0.7%
Commerical Office	7.00%	6,000.00	120.00	60,000			
Less Outgoings & Vacancies				-		-	0.0%
Less Letting Fees				(144,000)		21	-1.7%
Less Incentives (Rent Free and Fitout Costs)				-		-	0.0%
Less Other Leasing Costs				(50,000)		7	-0.6%
NET RENTAL INCOME				(134,000)		19	-1.6%
Interest Received				-		-	0.0%
Other Income				-		-	0.0%
TOTAL REVENUE				**10,048,857**		**1,436**	**118.0%**
Costs							
Land Purchase Cost				3,533,726		505	41.5%
Land Acquisition Costs				106,012		15	1.2%
Construction Costs				2,800,000		400	32.9%
Professional Fees				350,000		50	4.1%
Statutory Fees				-		-	0.0%
Miscellaneous Costs 1				-		-	0.0%
Miscellaneous Costs 2				-		-	0.0%
Miscellaneous Costs 3				-		-	0.0%
Project Contingency (Reserve)				-		-	0.0%
Land Holding Costs				-		-	0.0%
Pre-Sale Commissions				-		-	0.0%
Finance Charges (inc. Fees)				-		-	0.0%
Interest Expense				1,728,168		247	20.3%
TOTAL COSTS				**8,517,906**		**1,217**	**100.0%**

Performance Indicators			Per 0	Per of Site Area
[1] **Net Development Profit**		**1,530,951**		219
[3] **Development Margin (Profit/Risk Margin)**	Based on total costs (inc selling costs)	**17.76%**		
[4] Residual Land Value	Based on Target Margin of 20%	3,417,782		488
[5] **Net Present Value**	Based on Discount Rate of 15% p.a. Nominal	**1,136,268**		
[6] Benefit Cost Ratio		1.1797		
[7] **Project Internal Rate of Return (IRR)**	Per annum Nominal	**25.70%**		
[8] Residual Land Value	Based on NPV	4,636,899		662
Equity IRR	Per annum Nominal	N.A.		
Equity Contribution		-		
Peak Debt Exposure		8,412,747		
Equity to Debt Ratio		N.A.		
[9] Weighted Average Cost of Capital (WACC)		15.00%		
[10] Breakeven Date for Cumulative Cash Flow	Month 24	Sep-2012		
[11] Yield on Cost		8.35%		
[12] Rent Cover		2 Yrs, 2 Mths		
[13] Profit Erosion		1 Yrs, 3 Mths		

Footnotes:

1. Development Profit: is total revenue less total cost including interest paid and received
2. Note: No redistribution of Developer's Gross Profit
3. Development Margin: is profit divided by total costs (inc selling costs)
4. Residual Land Value: is the maximum purchase price for the land whilst achieving the target development margin.
5. Net Present Value: is the project's cash flow stream discounted to present value. It excludes all financing costs, interest and corp tax.
6. Benefit Cost Ratio: is the ratio of discounted incomes to discounted costs and excludes all financing costs, interest and corp tax.
7. Internal Rate of Return: is the discount rate where the NPV above equals Zero.
8. Residual Land Value (based on NPV): is the purchase price for the land to achieve a zero NPV.
9. The Weighted Average Cost of Capital (WACC) is the rate that a company is expected to pay to finance its assets.
10. Breakeven date for Cumulative Cash Flow: is the last date when total debt and equity is repaid (ie when profit is realised).
11. Yield on Cost is Current Net Annual Rent divided by Total Costs, including all Selling Costs.
12. The total net development profit divided by the current net annual rental expressed as a number of years/months.
13. The period of time post practical completion that it can remain unsold (but leased out) until finance and land holding costs erodes the profit for the development to zero.

Figure 10.7a–k (Continued)

EstateMaster Development Feasibility — **SUMMARY OF PROJECT RETURNS**

2 Year Construct
2 Year Construction
Nick Nunnington

Estate Master Licensed to: Estate Master

Returns on Funds Invested	Developer's Equity	Loan 1 Debt Lender Name	Loan 2 Debt Lender Name	Total Equity	Total Debt
1 Funds Invested (Cash Outlay)	-	3,989,738	2,800,000	-	6,789,738
% of Total Funds Invested	0.00%	58.76%	41.24%	0.00%	100.00%
2 Peak Exposure	-	5,240,048	3,172,698	-	6,412,747
Date of Peak Exposure	N.A.	Aug-12	Aug-12	N.A.	Aug-12
Month of Peak Exposure		Month 23	Month 23	Month 0	Month 23
Weighted Average Interest Rate	N.A.	15.00%	15.00%	N.A.	15.00%
Interest Charged	-	1,315,811	412,357	-	1,728,168
Line Fees Charged	-	-	-	-	-
Application Fees Charged	-	-	-	-	-
Profit Share Received	-	-	-	-	-
3 Total Profit to Funders	1,530,951	1,315,811	412,357	1,530,951	1,728,168
4 Margin on Funds Invested	N.A.	32.98%	14.73%	N.A.	25.45%
5 Payback Date	N.A.	Sep-12	Sep-12	N.A.	Sep-12
Month of Payback	N.A.	Month 24	Month 24	N.A.	Month 24
6 IRR on Funds Invested	N.A.	15.00%	15.00%	N.A.	15.00%
7 Equity to Debt Ratio		N.A.	N.A.		N.A.
8 Loan to Value Ratio	0.00%	50.94%	30.85%	0.00%	81.79%
9 Loan Ratio	0.00% of Land Purchase Price	132.98% of Hard Costs.	114.73% of Construction Costs.	0.00% of Land Purchase Price	241.05% of Land Purchase Price.

Footnotes:

1. The total amount of funding injected into the project cash flow.
2. The maximum cash flow exposure of that equity/debt facility including capitalised interest.
3. The total repayments less funds invested, including profit share paid or received.
4. Margin is net profit divided by total funds invested (cash outlay).
5. Payback date for the equity/debt facility is the last date when total equity/debt is repaid.
6. IRR on Funds Invested is the IRR of the equity cash flow including the return of equity and realisation of project profits.
7. Equity to Debt Ratio is the amount of equity contributed into the project as a percentage of debt funding.
8. Loan to Value ratio is the Peak Equity/Debt Exposure divided by Total Sales Revenue.
9. Loan Ratio is the total funds invested by the lender (cash outlay) divided by the nominated ratio calculation method. It includes capitalised interest and fees.

Tenancy Schedule for 2 Year Construct

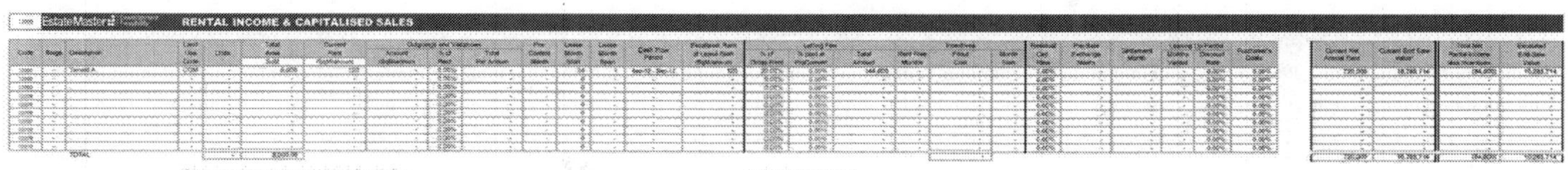

Figure 10.7a–k (Continued)

Assumptions Report for 2 Year Construct as at Aug-2016
2 Year Construction - Nick Nunnington

General	
Taxation Format	Nil Tax
Currency Format	British Pound (£ - GBP)
Stamp Duty Location	N/A
Cash Flow Rest Period	Monthly
Cash Flow Duration	Sep-2010 to May-2014 (45 Months)
Financial Year End Month	June
Cost and Sales Escalation % Rates	Monthly Compounded Escalation - Based on Cashflow Period Years
Development Management Percentage Basis	% of Project Costs (exc Land, Finance & Tax)
Miscellaneous Costs 1 Percentage Basis	% of Construction Costs (exc Tax)
Miscellaneous Costs 2 Percentage Basis	% of Construction Costs (exc Tax)
Miscellaneous Costs 3 Percentage Basis	% of Construction Costs (exc Tax)
Sales Revenue Collection Profile	Disabled
Stage Cash Flow Global Cost Allocation Method	As per Global Costs Cash Flow
Stage Cash Flow IRR/NPV Calculation	Calculate from Start of Project

Taxation	
Nominated Tax Type	N/A
Project Costs Entered	N/A
Rent and Leasing Costs Entered	N/A
Sales Revenue Entered	N/A
Other Income Entered	N/A
Tax Liability Calculation Method	N/A
Developer's Tax Liabilities	N/A
Developer's Tax Input Credits	N/A
Developer's Tax Input Credits on Land Purchase	N/A

Joint Venture	
Development Type	Single Entity
Tax Liability Apportionment	N/A
Land Owner's Tax Liabilities	N/A
Land Owner's Tax Input Credits	N/A

Hurdle Rates	
Project Discount Rate (target IRR)	15.00%
Land Owner's Discount Rate (Target IRR)	N/A
Annual to Rest Period Conversion Method	Converted to Monthly % Rate using Nominal method.
IRR and NPV Calculation	Calculated on cash flow that excludes all financing costs, interest and corp tax.
Developer's Target Dev. Margin	20.00%
Development Margin Calculation	Calculated on total development costs (inc selling costs).
Key Performance Indicators	Calculated on Net Development Profit (After Profit Share)

Financial Reporting			
Financial Reporting and Corporate Tax	Disabled		
Revenue Recognition Type	% Completed		
Revenue Recognition % Sold Method	Based on % of Revenue Sold		
Land Included in WIP Calculation	Yes		
Works in Progress, Expensed or Operating Cost			
Land and Acquistion	WIP	Project Contingency	WIP
Professional Fees	WIP	Land Holding Costs	WIP
Construction Costs	WIP	Selling Costs	WIP
Statutory Fees	WIP	Leasing Costs	WIP
Miscellaneous Costs 1	WIP	Finance Costs (exc Interest & Fees)	WIP
Miscellaneous Costs 2	WIP	Application and Line Fees	WIP
Miscellaneous Costs 3	WIP	Interest Charges	WIP
Tax Benefit	Roll forward loss until profit is realised		
Tax Payment	Paid Monthly		
Project Equity Treatment	Shareholders Equity (Project Capital)		

General Financing			
Annual Interest Rates	Converted to Monthly % Rate using Nominal conversion method.		
Total Debt to Loan Ratio Calculation (Summary Report)	Based on % of Land Purchase Price.		
Running Debt to Loan Ratio Calculation (Cash Flow Report)	Based on Cumulative Loan Drawdowns		
Static Debt to Loan Ratio Calculation (Summary Report)	Includes Capitalised Interest and Fees		
Profit Share Payments	Paid progressively as project makes a profit.		
Hard Costs			
Land Payments	Included	Statutory Fees	Included
Land Acquisitions Costs	Included	Land Holding Costs	Included
Project Contingency	Included	Miscellaneous Costs 1	Included
Professional Fees	Included	Miscellaneous Costs 2	Included
Construction Costs	Excluded	Miscellaneous Costs 3	Included
Tax on Hard Costs	N/A		

Equity		
Loan Facility Limit	Based on Fixed Amount of 0	
Equity Drawdown Method	Injected in total upfront.	
Interest Charged on Equity	0.00%	Capitalised (Compounded)
Interest Received on Suplus Cash	0.00%	
Equity Ratio Calculation (Summary Report)	Based on % of Land Purchase Price.	
Equity Repayment Method	0% of Available Funds to Repay Equity Before Debt then balance repaid at project end.	
Equity Pays Outstanding Debts at Project End	Yes	

Loan 1 (Debt)		
Loan Facility Limit	Based on 100% of Hard Costs.	
Loan Drawdown Method	Progressively - Limited to cumulative facility limit.	
Interest Charged on Loan	15.00%	Capitalised (Compounded)
Loan Ratio Calculation (Summary Report)	Based on % of Hard Costs.	
Loan Commencement	Automatic	
Loan Maturity	Automatic	
Refinancier at Manual Loan Maturity	Not Applicable as loan is automatically repaid as funds are available.	

Loan 2 (Debt)		
Loan Facility Limit	Based on 100% of Construction Costs.	
Loan Drawdown Method	Progressively - Limited to cumulative facility limit	
Interest Charged on Loan	15.00%	Capitalised (Compounded)
Loan Ratio Calculation (Summary Report)	Based on % of Construction Costs.	
Loan Commencement	Automatic	
Loan Maturity	Automatic	
Refinancier at Manual Loan Maturity	Not Applicable as loan is automatically repaid as funds are available.	

Loan 3 (Debt)		
Loan Facility Limit	Based on Fixed Amount of 0	
Loan Drawdown Method	Drawn down in total at loan commencement.	
Interest Charged on Loan	0.00%	Capitalised (Compounded)
Loan Ratio Calculation (Summary Report)	Based on % of Land Purchase Price.	
Loan Commencement	Automatic	
Loan Maturity	Automatic	
Refinancier at Manual Loan Maturity	Not Applicable as loan is automatically repaid as funds are available.	

Loan 4 (Debt)		
Loan Facility Limit	No Limit (use as overdraft facility)	
Interest Charged on Loan	0.00%	Capitalised (Compounded)
Loan Ratio Calculation (Summary Report)	Based on % of Land Purchase Price.	

Figure 10.7a–k (Continued)

These are especially useful when dealing with large phased developments, such as business parks or major residential schemes. The cash flow approach also allows the valuer to take into account the financial status of the client and to provide a more accurate and realistic financial appraisal. For example, joint ventures or schemes involving equity finance may require differential interest rates and complex flows of income, which cannot be accommodated in a traditional appraisal.

As the valuation profession is being encouraged to be more conscious of the needs of its clients, the cash flow approach should be more widely adopted to ensure that the financial situation of the client is accurately modelled within the appraisal.

The cash flow approach explained

Testing the traditional residual results against cash flow models suggests that the greater variability occurs with phased developments, such as low-rise residential schemes, where cash inflows coincide with cash outflows. The variation in result is less pronounced with the relatively simple one-off development, especially those with a project term of 12 months or less. In the absence of alternative development schemes, it is almost certain that the cost of borrowing money will be at least as high as the rate of interest that the developer could earn on excess funds. It is therefore reasonable to conclude that the developer should pay off debts as soon as income is produced by the development.

The developer will pay considerable attention to the likely cash flow and the likely maximum borrowing requirement over the building period. A cash flow table will provide full information concerning these two points and will allow for the accurate estimation of finance charges. In very complex multi-use schemes, the development surveyor and building cost consultants/quantity surveyors may be able to use cash flow analysis to reschedule the detailed timing of development to improve the overall return on capital.

Systems such as Argus allow the user to construct a monthly cash flow from the original residual valuation, which can then be edited to reflect more accurate data provided by building cost consultants/quantity surveyors, or to explore the effects of adjusting the cash flow. The programme provides the user with a full DCF analysis of the project and can also be converted into a spreadsheet files.

The following example illustrates the cash flow approach using a simple manual calculation.

Example 10.2

A site has the benefit of planning permission for the erection of a block of 20 apartments, totalling 1,100 m^2 of space.

The value of each completed apartment is estimated to be £100,000, using comparable transactions, but apartment prices are expected to continue to rise at 5% pa.

Building costs are estimated at £800/m^2 and are expected to rise at 0.5% per month. It is expected that five apartments will be sold in each of months 9–12 and that building costs will be evenly distributed over a 12-month building period.

Architects' and building cost consultants'/quantity surveyors' fees will be payable in two instalments in months 6 and 12 at 10% of the building costs.

Agents' and solicitors' fees will be charged on sale at 1% of the sale price of each apartment, and the developer will require a profit of 15% of the sale price.

Advise the developer as to the maximum bid for the site.

Step 1: *Inflate the apartment prices in line with forecasted growth.* Apartment prices are increasing at 5% pa or 0.4% per month.

In month 9, five flats will sell for £103,660 each = £518,300
In month 10, five flats will sell for £104,070 each = £520,350
In month 11, five flats will sell for £104,500 each = £522,500
In month 12, five flats will sell for £104,910 each = £524,550

Step 2: *Inflate the building costs (if applicable) by appropriate amounts based on projected increases.* Building costs are estimated at £800/m^2, spread evenly over 12 months, rising at 0.5% per month. Total building costs for the entire block of £880,000 divided by 12 gives a cost of £73,330 in month 1, which will increase by 0.5% in month 2, etc. This results in a building cost flow of:

month 1: £73,330
month 2: £73,700
month 3: £74,060
month 4: £74,440
month 5: £74,810

In *month 6*, fees are payable on the costs to date at a rate of 10% (10% of £445,490 = £44,549), *so* £75,180 + £44,549 = £119,729.

month 6: £119,729
month 7: £75,560
month 8: £75,940

In *month 9*, the first of the sales of five flats is achieved, adding fees at 1% and profit at 15% of the sale price to the costs (1% of £518,300 = <u>£5,183</u> and 15% of £518,300 = <u>£77,745</u>).
£76,320 + £5,183 + £77,745 = £159,248

month 9: £159,248

In *month 10*, £76,700 + £5,204 + £78,053 (calculated as the preceding but on month 10 sale prices).

month 10: £159,957

In *month 11*, £77,080 + £5,225 + £78,375 (as earlier).

month 11: £160,680

In *month 12,* £77,470 + £5,246 + £78,683 + £45,900 (arch/QS fees at 10% of costs months 7–12).

month 12: £161,399

From this information, a cash flow table can be constructed incorporating the effects of interest charges at 1% per month.

Month	*Sales (£)*	*Less*	*Building Costs (£)*	*Net Capital (£)*	*Cumulative Cash Flow*	*Interest (£)*
1		–	73,330	−73,330	−73,333	733
2		–	73,700	−73,700	−147,760[1]	1,480
3		–	74,060	−74,060	−223,301	2,233
4		–	74,440	−74,440	−299,974	3,000
5		–	74,810	−74,810	−377,784	3,778
6		–	119,729	−119,729	−501,291	5,013
7		–	75,560	−75,560	−581,863	5,819
8		–	75,940	−75,940	−663,622	6,636
9	518,300		159,248	+359,052	−311,206	3,112
10	520,350		159,957	+360,393	+46,075[2]	+461
11	522,500		160,680	+361,820	+408,355	+4,084
12	524,450		207,299	+317,251	+729,690[3]	

Residual	=	£729,690
Let site value	=	x
Let fees on site purchase	=	3% = 0.03x

then

$1.03x$ + interest for 12 months at 1% per month	=	£729,690
$1.03x(1.01)^{12}$	=	£729,690
$1.03x(1.1268)$	=	£729,690
$1.1606x$	=	£729,690

Therefore:

X (***the site value***)	=	**£628,718**

Note:

1. Already outstanding from month 1 is a debt of £73,330 and an interest charge of £733. Added to these is a new loan of £73,700, the total being £147,763, on which the interest charge for month 2 will be levied.
2. This figure is positive, as the total debt of £311,206 plus £3,112 interest is more than repaid by the receipt in month 10 of £360,393. Hence, a surplus of £46,075 remains. This can be invested to earn interest at 1% per month.
3. This final surplus remains after paying all costs, except the cost of the site itself, and fees and finance on its purchase. Calculated as before, the developer's maximum site bid is £628,718.

Example 10.3

In Example 10.2, the cash flow has been undertaken on a terminal value basis.

In this example, the same development is valued using a present value approach.

Month	*Sales (£)*	*Less*	*Building costs (£)*	*Net capital (£)*	*PV £1 at 1%*	*PV outstanding*
1		–	73,330	−73,330	0.99001	−72,604
2		–	73,700	−73,700	0.98030	−72,248
3		–	74,060	−74,060	0.97060	−71,882
4		–	74,440	−74,440	0.96098	−71,535
5		–	74,810	−74,810	0.95147	−71,179
6		–	119,729	119,729	0.94205	−112,270
7		–	75,560	−75,560	0.93272	−70,476
8		–	75,940	−75,940	0.92348	−70,129
9	518,300		159,248	+359,052	0.91434	+328,296
10	520,350		159,957	+360,393	0.90529	+326,259
11	522,500		160,680	+361,820	0.89632	+324,308
12	524,450		207,299	+317,251	0.88745	+281,544

Adding the PV outstanding gives the NPV **648,134**

NPV = **£648,134**

Let site value = x

Let fees on site purchase = 3% = 0.03x

then

1.03x = £648,134

x = £629,256 **(variation from Example 10.2 due to rounding)**

the site value = ***£629,256***

An important prerequisite for a cash flow approach is an accurate scheduling of construction activities.

This uses network analysis, or critical path analysis, whereby critical sequential events become the critical path, and other non-sequential events can be scheduled to run in parallel; in turn, some of the parallel costs may also be sequential.

The end result is the production of a project network with a start and finish date, with all the intermediate critical dates. The actual costs can then be estimated and transferred to a cash flow. The two together will then become part of the project manager's management tools for monitoring the development programme. Such an approach, if applied to Example 10.1, would highlight the need to reduce the construction period in order to reduce the high finance costs. The need to reduce finance costs in periods of high interest rates has played its part in developing innovative 'fast track' construction techniques.

Cash flows and residuals can be calculated on a current cost basis or on a future cost basis, building in variations in rents, as well as variations in labour and materials. None of the computerised sophistications can, however, overcome the problem that the acceptability of residual and cash flow development appraisal methods rests not with their rationale, which is irrefutable, but with the quality of the evidence used by the appraisal team to estimate costs and benefits.

Viability studies

Earlier in this chapter, it was stated that the residual method of valuation is based upon a simple equation:

GDV	*less*	all costs plus the developer's profit	=	**The site value**

The aim of the residual valuation is to find the unknown in this equation: the residual value. Often, however, a prospective developer will be aware of the likely site cost and, consequently, of the cost of fees and finance, either because the vendor has stated an asking price or because negotiations have revealed the minimum figure which the vendor will accept. This is particularly likely in the case of an existing building that is to be improved or replaced. In such a case, the single unknown in the equation can now become the developer's profit. The aim of a viability study is to assess the profit likely to be made from a development scheme, given the cost or asking price of the subject site.

Most appraisal software systems allow the user to select from a drop-down menu whether the system is to calculate residual land value or developer's profit. The input screens and processing are then adjusted according to the user's selection.

The cash flow approach is advocated by most software manufacturers, if not valuers, because of its greater accuracy and ability to model the payment of contractors, architects, and finance charges much more closely to the actual payment profile than the smoothed assumptions often contained in a traditional appraisal. In addition, a detailed cash flow approach allows the modelling of complex financial arrangements that may often underpin major property developments around the world: For example, equity sharing and joint ventures where the interest rates may differ for each project partner, depending upon their risk profile.

The cash flow table is particularly adaptable as a viability study, and the authors recommend that valuers using appraisal software always utilise a cash flow approach and only use the initial appraisal as a way of inputting data and obtaining a summary picture of the development analysis.

Example 10.4

Using the facts of Example 10.2, advise the developer of the potential profit if the total cost of the site, including fees, is £500,000.

Month	*Sales (£)*	*Less*	*Building costs (£)*	*Net capital (£)*	*Cumulative cash flow*	*Interest (£)*
1	–		573,330	−573,330	−573,333	5,733
2	–		73,700	−73,700	−652,763	6,528
3	–		74,060	−74,060	−733,351	7,334

(Continued)

(Continued)

Month	*Sales (£)*	*Less*	*Building costs (£)*	*Net capital (£)*	*Cumulative cash flow*	*Interest (£)*
4	–		74,440	−74,440	−815,124	8,151
5	–		74,810	−74,810	−898,086	8,981
6	–		119,729	−119,729	−1,026,796	10,013
7	–		75,560	−75,560	−1,112,623	11,126
8	–		75,940	−75,940	−1,199,690	11,997
9	518,300		81,503	+436,797	−774,890	7,749
10	520,350		81,904	+438,446	−344,193	3,442
11	522,500		82,305	+440,195	+92,561	+926
12	524,450		128,616	+395,934	**+489,420**	
Totals	2,085,600	1,515,897				

£489,420 represents the developer's profit.

This can be expressed as $\frac{£489,420}{£2,085,600}$ = **23.47% of GDV**

or as $\frac{£489,420}{£1,515,897}$ = **32.29% of total costs**

The residual method has become a straightforward DCF exercise. Nevertheless, there will still be areas of variability, and the criticism that the result is sensitive to changes in inputs remains. The next stage is therefore to incorporate probability measures in the analysis, such as Monte Carlo risk simulations, as advocated in the *RICS Valuation Standard*. Rents, costs, and interest charges may be weighted according to the valuer's expectation of possible changes, such an approach being eminently suited to computer analysis.

Discount rates in DCF analysis for development projects

When analysing a real estate development project using the DCF method, one of the most critical decisions is the selection of appropriate discount rates. Many practitioners make the mistake of using a single discount rate for both construction costs and income from sales, which can lead to inaccurate valuations. This section explains why different discount rates should be used for construction costs and sales revenues, how to determine appropriate rates, and how this distinction improves the accuracy of investment decision-making.

Understanding the differences between construction costs and sales income

In real estate development, cash flows can be divided into two major categories:

- *Construction costs.* Capital expenditures incurred for land acquisition, construction, materials, and labour.
- *Sales income.* Revenues generated from selling the completed units or leasing the property.

Key differences in the risk profiles of costs v income

Aspect	*Construction costs*	*Sales income*
Nature	Cash outflows for land, materials, labour	Cash inflows from unit sales or leasing
Timing	Short-term (1–3 years)	Mid- to long-term (varies by project)
Certainty	Higher certainty (contractor estimates, financing terms)	Higher uncertainty (market demand, absorption rates)
Risk profile	Lower risk, based on planned expenditures	Higher risk; depends on market conditions, unforeseen changes
Discount rate	Cost of capital: WACC or cost of debt	Investment return required by real estate investors

Since these two categories exhibit different risk profiles, it is now widely argued that they ***should not be discounted at the same rate***.

Selecting the right discount rate for each cash flow

Discounting construction costs

Construction costs are incurred over a relatively short period and usually financed through a mix of debt and equity. The appropriate discount rate should reflect the developer's actual financing costs.

Example 10.5 Discounting construction costs

A developer plans to build a residential complex with an estimated total construction cost of **€10 million**, spread over **two years**. The developer secures financing at an annual interest rate of **6%**, while the project's WACC is estimated at **8%**.

- If fully debt-financed: discount at **6%**.
- If financed with a 50–50 mix of debt and equity: use **WACC of 8%**.

Discounted cost calculation

Year 1: construction cost = €5 million
Year 2: construction cost = €5 million
Total discounted construction cost at 8% = **€8.92 million**.

Discounting sales income

Income from sales is highly uncertain and depends on market conditions at the time of project completion. The discount rate should reflect the required return for real estate investors, which is typically higher than the financing cost.

Example 10.6 Discounting sales income

The developer expects to sell **all units for €15 million** equally in three years (**€5 million per year)**. Given the market uncertainty, an investor requires a **12% return** on real estate investments.

Total discounted income at 12% = **€12.01 million.**

Why using a single discount rate is incorrect

Applying the same discount rate to both construction costs and income from sales can lead to inappropriate valuations:

- *Understating costs.* If a high discount rate is used for costs, it may overvalue the project due to the discounting of costs as it reduces them.
- *Understating risk in revenues.* If a low discount rate is used for sales income, it overestimates future revenues.

Example 10.7 Impact of incorrect discounting

Scenario	*Discount rate for costs*	*Discount rate for sales*	*NPV calculation*
Correct approach	**8%**	**12%**	**€3,092,833**
Same discount rate used (8%)	8%	8%	€3,969,161
Same discount rate used (12%)	12%	12%	€3,558,901

This demonstrates how using the **same discount rate** can **inflate** the perceived project value.

The correct approach in a DCF model

Step 1: *Discount construction costs separately.* Use the cost of debt or WACC to discount costs (typically 5–8%, depending on financing structure).

Step 2: *Discount income from sales separately.* Use a higher discount rate reflecting market risk (typically 10–15%, based on investor expectations).

Step 3: *Calculate the NPV.*

$$NPV = \sum \frac{Sales\ Income}{(1+r_{sales})^\wedge t} - \sum \frac{Construction\ Costs}{(1+r_{costs})^\wedge t}$$

Where:

r_{sales} = discount rate for sales income

r_{costs} = discount rate for construction costs

When can the same discount rate be used?

In rare cases, a single discount rate **may** be used:

- If the project is **entirely self-financed** with a fixed required return
- If **pre-sales contracts** are secured, reducing sales risk
- If the market has **no volatility** (highly unlikely)

In conclusion, using separate discount rates for construction costs and sales income is essential for accurate DCF analysis.

Construction costs should be discounted using the **WACC,** whilst **sales income** should be discounted at a **higher rate reflecting investment risk**.

Applying **a single discount rate misrepresents project value**, leading to poor investment decisions. By correctly differentiating discount rates, developers and investors can make more informed and financially sound decisions.

A practitioner's perspective

The biggest questions for practitioners are:

- Should we build residential units, an office scheme, or a hotel?
- What price should we bid for the site?
- What might our returns be if we developed this site?

All these questions can be answered using development appraisals. In practice, depending upon the size and scale of the project, they can become highly complex. However, it is always important to remember that, at their core, all appraisals are based upon the same simple formula. But beware: Don't mistake them for being easy. When it comes to development appraisals, the devil is in the detail, and small changes in inputs can often have significant impact upon the end results. 'Garbage in, garbage out', as they say.

Not all appraisals are created equally, and for this reason, situations where experienced and knowledgeable professionals come to different development appraisal outcomes are far from rare. As a result, sensitivity analysis is a crucial component in the development appraisal *toolkit* (in addition to the experience gathered from simply 'knowing your market').

Summary

- Residual valuations or development appraisals are used to assess the value of land and buildings with latent development potential.
- The method is based upon established economic principles, and the method itself is logical.
- However, the accuracy of the residual method relies on the valuer's ability to estimate all the variables within the calculation. The costs and benefits must be carefully estimated with respect to the development market, using hard evidence where possible to support the rents.
- Capitalisation rates and building costs adopted.
- Development appraisal is an ideal application for information technology, allowing the valuer to assess the sensitivity of the valuation to changes in the underlying variables and to apply simulation techniques to model the probability of outcomes.

- The final opinion of development value should, wherever possible, be checked against comparable market evidence. However, as each development opportunity is unique in terms of its highest and best use, market comparison will only provide the valuer with a basic 'yardstick' with which to cross-reference the appraisal.

Using Excel to build an appraisal

Here we provide a step-by-step approach to building a simple development appraisal in Excel.

Value

We will now explore a straightforward development appraisal using a cash flow approach by building an Excel spreadsheet to undertake the appraisal.

Appraisal data

- Construction of 2,000 m^2 (net lettable area) offices
- Estimated net rental value per square metre on completion: £1,000 paid in advance
- Estimated time from completion of construction: 12 months to full occupation.
- Estimated take-up rate: 500 m^2/quarter
- Construction complete: 12 months' time
- Construction commences immediately

Using this data, we can begin to construct a development appraisal spreadsheet.

Step 1: Enter quarterly time periods across the spreadsheet:

0	1	2	3	4	5	6	7	8

Step 2: Enter the rental income, which commences in quarter 4 (remember that in Excel, payments are treated as occurring at the end of the period when using NPV and IRR functions).

Step 3: Assuming that an investor has made an agreement to purchase the development when it is fully let at an agreed all-risks yield of 8%, enter the sale proceeds. (Calculated using what you have learnt so far.)

We have now completed the (positive) value inputs into the development, and if using Excel, your spreadsheet should be similar to the following screenshot:

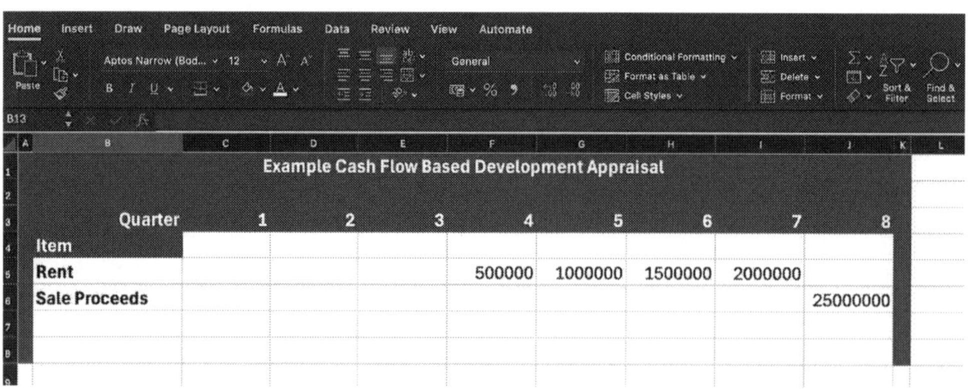

Example Cash Flow Based Development Appraisal

Quarter	1	2	3	4	5	6	7	8
Item								
Rent				500000	1000000	1500000	2000000	
Sale Proceeds								25000000

Note: The sale proceeds have been calculated by applying a capitalisation factor (or all-risks yield) to the total income when the development is fully let, that is:

$2{,}000\ \text{m}^2$ at $£1{,}000/\text{m}^2 = £2{,}000{,}000$

$$\text{Capitalisation Factor}\ \left(\frac{1}{i}\right) = \left(\frac{1}{0.08}\right) = 12$$

Step 4: Multiply the income with the capitalisation factor to find GDV, which equals £25,000,000.

We have now considered the value side of the residual equation by considering the income flow during the development period. Now we need to consider the costs of development.

Estimating the costs of development

The costs of development have been discussed in detail in the preceding text, and examples of costs during the pre-construction, principal, or main construction period, and post-construction period are given under both the *RICS Valuation Standard* and the section titled 'The residual method' in this chapter.

Step 5: We must now enter all the costs of development into our Excel spreadsheet. These are:

a. **Pre-development costs** (all occurring in the first month)

1. A planning fee of £10,000
2. A building permit fee of £5,000
3. A site investigation of £20,000

b. ***Construction costs*** (which are spread over the construction period)

1. Construction costs:

   i. £2,000,000 end of quarter 0
   ii. £6,000,000 end of quarter 1
   iii. £6,000,000 end of quarter 2
   iv. £2,000,000 end of quarter 3

2. Contingency:

   a. 5% of construction costs

3. Professional fees:

   10% of construction costs (excluding contingency)

Finance

Finance is available at an interest rate (cost of capital) of 2.5% per quarter.

Marketing

Agent's letting fees: 7.5% of initial rent.

Promotion

£50,000 quarter 0
£20,000 quarter 1
£10,000 quarter 2
£10,000 quarter 3
£10,000 quarter 4
£20,000 quarter 5

Site costs (all payable in quarter 0)

1 The site costs of £4,500,000.
2 Agents' and solicitors' fees amount to £100,000.
3 Property tax of £50,000.

When you have entered the data into the spreadsheet, it should look something like the screenshot set out in the following:

Example Cash Flow Based Development Appraisal

	Quarter								
	0	1	2	3	4	5	6	7	8
INCOME									
Rent					500000	1000000	1500000	2000000	
Sale Proceeds									25000000
COSTS									
Planning Fees	-10000								
Building Approval Fees	-5000								
Site Investigation	-20000								
Construction Costs	-2000000	-6000000	-6000000	-2000000					
Contingency	-100000	-300000	-300000	-100000					
Professional Fees	-200000	-600000	-600000	-200000					
Letting Fees					-37500	-37500	-37500	-37500	
Promotion Activities	-50000	-20000	-10000	-10000	-10000	-20000			
Land Costs	-4500000								
Fees on Land acquisition	-100000								
Taxes on Land Acquistion	-50000								

Step 6: Now we have to calculate the cash flow.

Use the Excel Sum function to add the figures in each column.
In our spreadsheet, in cell C23, type

=SUM(C4:C21). (or use Autosum)

This can then be copied across to all the rows for quarter 0 to quarter 8 using the copy and paste buttons/keys or the auto-drag option.

You now have the undiscounted cash flow.

Step 7: Calculating the net present value (PV).

Next, enter the formula to calculate the PV of the cash flow in cell C24. Use the formula:

$$\frac{1}{(1+i)^n}$$

Remember you will need appropriate brackets and that we are using quarterly periods and a quarterly interest rate. (Also if you are working in Europe and are using a European-configured computer, you will be using 0,025.)

The contents of our cell C24 are:

= 1/((1 + 0.025)^C3)*C23

This formula can then be copied across all the quarters (cells D23 to K2).

You can then use the sum function to add all the quarterly discounted cash flow figures to give you the NPV.

= SUM(C24:K24)

Alternatively, you can use the Excel NPV function [= NPV (discount rate, range of cash flow)] by typing in:

= NPV (0.025, C23:K23)

If you use the preceding, you will note that it does not give the same figure as calculating it using the PV formula and summing the discounted cash flows. This is because the function assumes the investment begins one period before the date of the first value and that the initial payment is made at the end of the first time period. In fact, in our appraisal, as in most property cash flows, the site is purchased at the start of the first period. Therefore, the NPV function must be modified in our case, in cell C26, to:

= NPV (0.025,D23:K23) + C23

This function should now give you the **£2,165,410** expected result.

You are advised to print out, read, and file in this section the Excel help sheet on the NPV function.

The NPV represents the developer's profit and can be converted to a return on capital costs or value.

Step 8: Calculating the internal rate of return.

The IRR can now be calculated using the Excel IRR function: IRR(range of cash flow, trial rate).

In our spreadsheet, we have entered into cell C27 the formula:

= IRR (C23:K23;0.05)

Remember, a trial rate is needed (we have used 0.05) to initialise the process of iteration.

You are advised to print out, read, and file in this section the Excel help sheet on the IRR function.

When you have calculated the NPV and IRR, your spreadsheet should look something like the following.

Example Cash Flow Based Development Appraisal

	Quarter 0	1	2	3	4	5	6	7	8
INCOME									
Rent					500000	1000000	1500000	2000000	
Sale Proceeds									25000000
COSTS									
							FINANCE RATE =		2,50%
Planning Fees	-10000								
Building Approval Fees	-5000								
Site Investigation	-20000								
Construction Costs	-2000000	-6000000	-6000000	-2000000					
Contingency	-100000	-300000	-300000	-100000					
Professional Fees	-200000	-600000	-600000	-200000					
Letting Fees					-37500	-37500	-37500	-37500	
Promotion Activities	-50000	-20000	-10000	-10000	-10000	-20000			
Land Costs	-4500000								
Fees on Land acquisition	-100000								
Taxes on Land Acquistion	-50000								
CASH FLOW	-7035000	-6920000	-6910000	-2310000	452500	942500	1462500	1962500	25000000
P.V @ 2.5%	-7035000	-6751220	-6577037	-2145065	409942,7	833032,7	1261109	1650983	20518664
NPV =	€2 165 410,18								
IRR =	3,96%				Effective Annual Rent			=	16,81%

Step 9: Converting the result to an effective annual rate.

Remember, this is a quarterly IRR, and to convert to an effective annual rate, you must use the formula:

$$(1+i)^{nth} - 1 \times 100$$

Where n is the number of periods pa.

Therefore, in our example:

$$(1+0.03960)^{4} - 1 \times 100\,(4 = \text{quarterly periods})$$

In our spreadsheet we have entered into cell I26 the following formula:

$$(((1+(C26))\wedge 4)-1)$$

This cell has been set to a percentage format; if your cell K27 is not in a percentage format, your formula must be adjusted by adding *100 to the end of the formula.

Chapter 11

The cost method

Introduction

The cost method is widely regarded as a method of last resort in property valuation. It is typically applied under specific circumstances:

- Where – and this can be frequent – there is an absence of directly comparable evidence, attributable to variations in land area relative to the building, the building's area, the grade of materials employed, the building's age, its condition, and other factors
- When the property is not income-generating, ruling out the investment and the profits method
- When the property is not intended for development, eliminating the use of the residual method
- When construction works are not yet completed

Discrepancies may also arise from a lack of available information or transparency in the comparable data, for example, in terms of internal area, the inclusion or otherwise of covered verandas, uncovered verandas, basements, and attics. Therefore, there may be no accurate way of adjusting comparable transaction figures before applying them to the subject property.

It is also frequently applied where the permissible building density is not completely used, either for the property in question or for certain comparable evidence, which impedes a meaningful comparison.

The cost method is used for owner-occupied, public, or specialised properties rarely sold or rented on the open market. Examples of properties that often fall into this category include cemeteries, prisons, public libraries, public hospitals, town halls, art galleries, and social centres. Although the cost method is primarily applied to these types of properties, it can also be used as a supplementary check against other valuation methods for different property types.

When is the cost method used?

The cost method is primarily used for two key applications: rating assessments and accounting purposes. When used to estimate the rateable value of specialised properties for tax purposes, it is known as the contractor's basis in UK rating valuations (Isaac 2002; Scarrett and Osborn 2014). For accounting purposes, the method is referred to as the DRC. While the contractor's basis and DRC are fundamentally similar, they differ in terms of rationale, purpose, and application. A variation of the cost method is commonly used to assess reinstatement costs for insurance purposes. However, it is important to note that although valuers are often tasked with performing this assessment, it is not considered a formal valuation method.

DOI: 10.1201/9781003490357-13

Depreciated Replacement Cost (DRC)

The DRC method is based on the economic theory of substitution – that a potential buyer would pay no more for the subject property than the cost of acquiring an equivalent new one – and it is defined in the *RICS Red Book Global Standards* (2021) as 'the current cost of replacing an asset with its modern equivalent asset less deductions for physical deterioration and all relevant forms of obsolescence and optimisation' (RICS 2021, p. 9). The method consists of the following steps:

- **Step 1:** Estimate the gross replacement cost.
- **Step 2:** Adjust replacement cost to account for depreciation and/or obsolescence.
- **Step 3:** Assess the value of the land.
- **Step 4:** Estimate depreciated replacement cost = replacement cost − depreciation and/or obsolescence + land value.

Step 1: Estimate the gross replacement cost

The replacement cost for a modern equivalent building includes all the usual costs associated with constructing a building but excludes developer's profits. Examples of costs that may be expected to be incurred in replacing the building include:

- Site improvements
- Fitting out
- Materials
- Labour
- Professional fees (such as for architects and engineers)
- Contingency
- Planning permissions
- Finance charges

In assessing the replacement cost, all costs that a potential buyer would incur on the valuation date should be considered, unless otherwise directed, such as in the case of the instant build concept commonly applied in the UK public sector. When such concept is used, there is no need to include provisions for finance costs or a contingency allowance.

Construction cost estimates can be sourced from contractors, quantity surveyors, and builders or through professional cost estimators, such as the BCIS in the UK, RSMeans Data by Gordian in the USA, and Custo Unitário Básico (CUB) in Brazil, to mention a few examples.

Step 2: Adjust replacement cost to account for depreciation and/or obsolescence

The replacement cost must be adjusted to account for the difference in utility between the modern equivalent and the actual subject property. Therefore, allowances for depreciation and/or obsolescence must be factored into the valuation. The key principle is that the hypothetical buyer can choose to acquire either the modern equivalent or the actual asset. If the modern equivalent better meets the buyer's needs, the price paid for the actual asset should reflect all the disadvantages it has in comparison.

There are three main types of depreciation and obsolescence (Blackledge 2009):

Physical deterioration. This refers to the wear and tear that occurs over time due to use and lack of proper maintenance. For example, a prison that has suffered from rusted structures, cracked walls, and outdated plumbing due to years of neglect would experience physical deterioration.

Functional obsolescence. This occurs when the design or specifications of an asset no longer meet the needs or expectations for which it was originally intended. For instance, a library with a design that is no longer suitable for modern needs, such as a lack of spaces for digital learning or social space for interaction.

Economic obsolescence. This is caused by external factors, such as changes in market demand, economic conditions, or technological advancements, which reduce the asset's value. For example, a social centre located in an area where the local population has decreased or shifted, reducing the demand for such a facility.

There are various methods to assess depreciation, including straight-line, reducing balance, and S-curve approaches. The straight-line method is the most used due to its simplicity. It entails taking account of the building's economic life and remaining useful economic life. RICS (2024) defines *economic life* as 'how long a succession of owners including the current entity could derive economic benefit from using the asset for its designed purpose, having regard to its constituent parts and ignoring the impact of any potential replacement of parts, refurbishment or reconstruction'. Straight-line depreciation can be calculated as:

$$\textit{Straight line depreciation} = \frac{\textit{Age of the building}}{\textit{Total estimated life}}$$

Example 11.1 Straight-line depreciation

Suppose a building has an estimated total lifespan of 80 years, and it currently has 75 years of remaining life. To calculate the percentage of the replacement cost that should be deducted for depreciation, we have thus:

- Age of the building: 80 years − 75 years remaining = 5 years
- Total estimated life: 80 years

$$\textit{Straight line depreciation} = \frac{5}{80} = 6.25\%$$

Since the building is 5 years old, 6.25% of its replacement cost is accounted for as depreciation.

The annual rate of depreciation using the straight-line method is:

$$\textit{Annual depreciation rate} = \frac{100\%}{80\ \textit{years}} = 1.25\%\ \textit{per year}$$

Thus, as we calculated before, over five years, the accumulated depreciation is:

$$1.25\% \times 5 = 6.25\%$$

If the replacement cost of the building is estimated at $1,000,000, the depreciated cost is:

$$\textit{Depreciated cost} = 1{,}000{,}000 \times (1 - 0.0625) = 937{,}500$$

A key weakness of the straight-line method is its assumption of a constant rate of depreciation, which is rarely accurate. In contrast, the reducing balance method applies a fixed percentage to the asset's net book value each year, leading to higher depreciation charges in the early years and lower charges as the asset ages.

Example 11.2 Reducing balance depreciation

Now, let us estimate the depreciated cost of the building using the reducing balance method with a fixed annual depreciation rate of 1.25%.

Year	*Replacement cost*	*Depreciation*	*Depreciated value*
0	1,000,000	−1,000,0000 × 0.0125 = −12,500	987,500
1	987,500	−987,500 × 0.0125 = −12,343.75	975,156.25
2	975,156.25	−975,156.25 × 0.0125 = −12,189.45	962,966.80
3	962,966.80	−962,966.80 × 0.0125 = −12,037.08	950,929.72
4	950,929.72	−950,929.72 × 0.0125 = −11,886.62	939,043.09
5	939,043.09	−939,043.09 × 0.0125 = −11,738.04	927,305.06

After five years, using the reducing balance method, the estimated depreciated value of the building is $927,305.06.

Depreciation often follows an S-curve pattern, where it is low in the early years, accelerates during the middle years, and then slows down again towards the end of the asset's lifespan. This S-curve model offers a more realistic depiction of how an asset loses value over time compared to straight-line or reducing balance methods. It is particularly useful when sufficient historical data is available to inform the rate of change at different stages. However, without empirical evidence to support these inputs, the shape of the S-curve may rely heavily on subjective assumptions, potentially making it no more accurate than the other methods discussed.

Step 3: Assess the value of the land

The value of land itself does not depreciate, and it is typically assessed using the comparison method. Careful consideration must be given to which land is deemed appropriate for comparison. Whilst the goal of the DRC method is to produce a valuation of the subject property in its actual location, if the actual site is one that a prudent buyer would no longer consider appropriate because it would be commercially wasteful or an inappropriate use of resources, the modern equivalent site can be located elsewhere, assuming it has the appropriate characteristics to deliver the required service potential.

For example, consider a hospital located in a city centre where land is both scarce and expensive. The current site may be considered unsuitable for modern healthcare due to limited room for expansion, insufficient parking, and heavy traffic congestion. In such a case, a prudent buyer would likely avoid this location for a new hospital, as it could be viewed as commercially inefficient and impractical for its intended purpose. Instead, a more appropriate modern equivalent site could be located in a suburban area, offering more space for future growth, lower traffic, and improved accessibility for both patients and staff.

Similarly, imagine a prison located near the city centre, where land is again scarce and costly. The site may be constrained by limited space and surrounding residential or commercial developments that pose security concerns. In this scenario, a prudent buyer would likely avoid this location for a new prison, as it would be considered commercially wasteful and inefficient for its purpose. Instead, a more suitable modern equivalent site could be located in more remote areas, offering more space and the ability to incorporate proper security features.

The fundamental principle in choosing what land to consider is that 'the hypothetical buyer for a modern equivalent asset would purchase the least expensive site that would realistically be suitable and appropriate for its proposed operations and the envisaged modern equivalent facility' (RICS 2024, p. 13).

Step 4: Estimate DRC

The final step of the DRC method is to reconcile the inputs and calculate the DRC as the following:

$$DRC = Replacement\ cost - depreciation\ and\ /\ or\ obsolescence + land\ value$$

Example 11.3 DRC valuation of a library

The property is a 30,000 ft^2 public library that opened 20 years ago, situated on a 1.5 ac site. The replacement cost is \$200/ft^2, and the building has a remaining useful life of 30 years. Comparable sites are currently trading at \$500,000/ac.

Replacement cost = 30,000 × 200	\$6,000,000
Less depreciation at 20/50 = 40%	\$2,400,000
Depreciated replacement cost of the building	\$3,600,000
Plus land value = 1.5 × 500,000	\$750,000
Depreciated replacement cost	\$4,350,000

Example 11.4 DRC valuation of a church

A church is situated in the city centre of a large metropolitan area. The estimated replacement cost of the building is £3,000,000, and the site covers 0.5 ha. Comparable land in the suburbs is valued at £200,000/ha. An allowance for obsolescence of 65% is deemed appropriate. The fees for constructing the replacement building are estimated at 10% of the building cost, contingency is estimated at 3% of the building cost, and the rebuilding is expected to take two years, with a finance rate of 7% to cover construction costs.

Replacement cost	
Building cost	£3,000,000
Construction fees = 0.1 × 3,000,000	£300,000
Contingency = 0.03 × 3,000,000	£90,000
Finance charges = $\left(\left(1+0.07\right)^2-1\right)\times\frac{3,390,000}{2}$	£245,605.50
Gross replacement cost	£3,635,605.50
Less obsolescence at 65%	£2,363,143.6
DRC of the building	£1,272,461.93
Plus land value = 0.5 × 200,000	£100,000
DRC	**£1,372,461.93**

Observation

Finance charges are applied to half the sum of building costs plus construction fees and contingency using the formula $Finance\ charges=\left(\left(1+r\right)^n-1\right)\times\frac{Total\ construction\ costs}{2}$.

The contractor's basis

As previously mentioned, the primary use of the contractor's basis is to determine annual values for rating assessments. The contractor's basis operates on the premise that a hypothetical tenant has the option to construct a similar property for their own use, instead of renting the subject property. Therefore, the rental bid for the subject property would correspond to the annual equivalent of the capital cost of constructing this comparable property, including the land value.

The steps involved in the contractor's basis can be set out as follows:

Step 1: Estimate the gross replacement cost.
Step 2: Adjust replacement cost to account for depreciation and/or obsolescence.
Step 3: Assess the value of the land.
Step 4: Calculate the effective capital value (ECV).
Step 5: Assess the annual equivalent.

As shown, the first three steps of the contractor's basis align with the DRC method. The effective capital value calculated in step 4 is likely to be equivalent to the depreciated replacement cost determined in step 4 of the DRC method. It represents the gross replacement cost minus deductions for depreciation and/or obsolescence, plus the land value. Thus, the key difference lies in the final step, which involves determining the annual equivalent.

To assess the annual equivalent, once the ECV is established, it is multiplied by a decapitalisation rate. Prior to 1990, selecting the appropriate rate was largely based on the valuer's judgement (RICS 2023). However, this often led to disputes and legal challenges, and to address these issues, statutory decapitalisation rates have been set by the respective UK jurisdictions since 1990. For example, in England, the government set two distinct rates for the 2017–2023 rating list: 2.6% for educational, healthcare, and defence properties, and 4% for all other properties valued using the contractor's basis (Sandford 2023).

$$\textit{Annual equivalent} = \textit{Effective capital value} \times \textit{Decapitalisation rate}$$

Example 11.5 Contractor's basis

A public art gallery with a total floor area of 10,000 ft² is being assessed for property taxation purposes. The gallery, built 30 years ago, has a remaining economic life of 20 years. The estimated replacement cost for the building is $360/ft². Based on comparable evidence, the land value is estimated at $830,000. A decapitalisation rate of 4% will be used for this valuation.

Replacement cost = 10,000 × 360	$3,600,000
Less depreciation at 30/50 = 60%	$2,160,000
Depreciated replacement cost of the building	$1,440,000
Plus land value	$830,000
Effective capital value (ECV)	**$2,270,000**
Annual equivalent at 4%	**$90,800**

Reinstatement Cost Assessment (RCA)

A variation of the cost method is applied in RCAs for insurance purposes, though they are not valuations and therefore not covered by the requirements of the *RICS Valuation – Global Standards* (2023). RCAs aim to estimate the cost to rebuild or replace a property in the event of total loss or significant damage, where the entire building would require demolition and reconstruction. It is important to note that the land value is excluded from this assessment, as it is assumed the land would remain unaffected even if the building was destroyed.

In the UK, most buildings are insured based on a 'day one reinstatement' figure, also known as the declared value. According to RICS (2018, p. 5), the *declared value* is defined as the

> equivalent of a fixed price, lump sum, competitive tender submitted by a suitable and competent contractor for works to commence on site on the first day of the period of insurance, including allowances for demolition, associated costs, and all relevant professional and statutory fees.

Regardless of the building's age or condition, if insurers have agreed to indemnity by reinstatement cover, the damage will be repaired to a state similar to its original condition without any enhancement or betterment. The 'day one reinstatement' figure typically aligns with the gross replacement cost in the DRC but includes provisions for demolition, debris removal, and associated costs.

Although depreciation is not factored into the 'day one reinstatement' basis, in the 'obsolete buildings' basis, the valuer would need to account for depreciation. This and additional bases for reinstatement cost assessments are detailed in the RICS professional standard *Reinstatement Cost Assessment of Buildings* (2018).

Example 11.6 RCA of an office building

The property is an office building with a gross internal area of 5,000 m², situated in a regional city. The estimated rebuilding cost is £2,500/m², with demolition costs calculated at 15% of the rebuilding cost. Professional fees are estimated at 15% of the combined total of rebuilding and demolition costs. Additional expenses, including building control and planning, amount to £50,000.

Rebuilding cost = 5,000 × 2,500	£12,500,000
Demolition cost at 15% rebuilding cost	£1,875,000
Professional fees at 15% rebuilding cost + demolition cost	£2,156,250
Other fees	£50,000
Total RCA	**£16,581,250**

Further detailed and applied discussion of the DRC method

The cost approach (VPS 5 and elaborated in IVS 105) is characterised as 'the method of last resort', as previously noted. However, this method is commonly employed in many countries, particularly in the assessment of individual housing projects or specialised structures.

It also has been observed that, in several cases, the DRC method might be preferred compared to other methods by a court during legal disputes due to its simplicity. Furthermore, some UK Inland Revenue and other tax offices may require a separation between land and buildings to be made for tax purposes. For the reasons noted at the introduction of the chapter, and because, in certain cases, stakeholders require reporting separate figures for the land and the buildings, the cost approach is often used to value the buildings on the land. This must consider the specifics of the property in question: its age, materials, state of repair, developers' profit (where applicable), etc.

It should be made clear in the report that the figures reported with this method are indicative and may not represent the market value, and that they may also not reflect the current cost of construction of the building, due to adjustments made to reflect the circumstances of the existing structure.

For the land element, comparable evidence is more often accessible and must be modified to account for the particular location and other attributes of the subject property, using the same methodologies as set out in Chapter 7. In numerous instances, the current land use of the property in question does not represent the 'highest and best use'. In such instances, the land value cannot be ascertained from comparable transactions, necessitating valuers to make additional adjustments, such as subtracting the value of the land in relation to undeveloped plots.

It is important that the basis of valuation be made clear, both in the terms of engagement and in the report, for the following reasons:

1. The resultant aggregate figure (land and buildings) may not accord with the market value of the property as a single asset. For example, the existing use of the buildings might not be the highest and best use of the property.
2. An apportionment of the market value of the entire property between its constituent parts may not necessarily accord with the individual figures arrived at for those parts in isolation.

Example 11.7 Valuation of houses with the cost method and unused density

A residential property with a two-storey house on a 500 m² plot sits within a residential zone, with a building density **of 100%**. The house occupies 200 m² (ground floor, 100 m²; first floor, 100 m²), which represents 40% of the allowable density.

If the market value of the land is €500/m² and the construction cost is €1,200/m² for residential buildings and the building under study is 20 years old, with a depreciation rate of 1% per year (physical depreciation), then calculate the value of the property.

Step 1: *Calculate the unused density.* Unused density is 60% of the allowable density (100% − 40% = 60%), equaling 300 m² of unused density.

Step 2: *Calculate the land value (assuming that the unused can be utilised).*

$$\textit{Land value} = \textit{land area} \times \textit{land market value per } m^2$$
$$\textit{Land value} = 500\,m^2 \times €500/m^2 = €250{,}000$$

Step 3: *Calculate the replacement cost of the house.*

$$\textit{Replacement cost} = \textit{built area} \times \textit{construction cost per } m^2$$
$$\textit{Replacement cost} = 200\,m^2 \times €1{,}200/m^2 = €240{,}000$$

Step 4: *Account for depreciation.*

$$\textit{Depreciated cost of house} = \textit{replacement cost} - (\textit{depreciation rate} \times \textit{replacement cost})$$
$$\textit{Depreciated cost of house} = €240{,}000 - (20\% \times €240{,}000) = €240{,}000 - €48{,}000 = €192{,}000$$

Step 5: *Calculate the total value of the land plus the DRC.*

$$\textbf{\textit{House value}} = \textit{land value} + \textit{depreciated cost of house} = €250{,}000 + €192{,}000 = €442{,}000$$

BUT does this figure of €442,000 **reflect the market value?**

Suggestions for the valid application of the DRC

- The €442,000 figure may be accurate, provided that the house was constructed correctly and that extra storeys or nearby structures may be built, fully using the density.
- If the house is located at the centre of the land and the foundation construction does not accommodate additional floors, then the remaining density of 300 m² cannot be exploited, and its value cannot reach €500/m². In this scenario, the 200 m² may be appraised at €500/m², whereas the unused 300 m² can solely serve as supplementary

space or yard for the residence, necessitating a substantially lower valuation from the appraiser (say, €100/m² – 20% of the land value as ready to be built).

Therefore, the valuation is then expected to be something like:

$$House\ Value = Land\ Value(properly\ used) + Depreciated\ Cost\ of\ House + Land(yard)$$
$$= 200\,m^2 \times €500/m^2 + €192{,}000 + 300\,m^2 \times €100/m^2 = €322{,}000$$

In this case, the €322,000 might be closer to the market value, as the unused density can only be used as an additional amenity, in this case a larger yard.

The differences between cost and value in property valuation

Cost is the total expenditure required to construct, acquire, or replace a property. It includes materials, labour, fees, and other inputs.

Formula: $\boldsymbol{Cost} = direct\ costs + indirect\ costs + profit\ margin$

Value is the monetary worth of a property as determined by market forces (demand and supply).

Formula: $\boldsymbol{Value} = comparable\ sales\ adjusted\ price$

or

$\boldsymbol{Value} = rental\ income \times market\ capitalisation\ rate$

Understanding the differences between cost and value is critical in property valuation for valuers and stakeholders. Cost represents the expenditure required to create or replace a property, including materials, labour, fees, and other associated costs. Value, on the other hand, reflects the worth of the property in the market, determined by what buyers are willing to pay under prevailing conditions.

The distinction lies in that cost focuses on the supply side (production or replacement), while value is driven by the demand side (market perception).

Set out in what follows is a detailed comparison of the two concepts, to highlight their differences and applications.

Table 11.1 The key differences between cost and value in property valuation

Aspect	*Cost*	*Value*
Basis	Reflects the expenditure required to produce or replace a property	Reflects the property's worth in the open market
Perspective	Supply-side metric (construction/development cost)	Demand-side metric (market perception of worth)

(*Continued*)

(Continued)

Aspect	*Cost*	*Value*
Depreciation impact	Accounts for all depreciation (physical, functional, external)	May reflect demand regardless of physical depreciation
Market influence	Does not directly depend on market trends or conditions	Heavily influenced by market dynamics, demand, and location
Timing	Based on current or historical costs	Based on current market conditions
Usage	Used for insurance purposes, cost-based valuations, and feasibility analysis	Used for investment decisions, sales, and market analysis

Example 11.8 Application of the DRC in a house with 'unnecessary features'

Location and market context

The property is located in an area where typical houses have the following characteristics:

- Are **200 m²**
- Have **four bedrooms**
- Are around **10 years old**
- Fully utilise the **permitted building density** of the plot
- Average price for such a house is **€310,000**, including **land and building**

Property under study

- **Size:** 300 m² (50% larger than typical houses).
- **Age:** 10 years old (similar to comparables).
- **Land:** The property fully uses the allowed **building density** (same as comparables).
- **Design inefficiencies:** The additional 100 m² is primarily used for **large corridors**, **auxiliary spaces**, and a **second sitting room**, which are unnecessary for the average buyer in the area.

Application of the steps to formulate the DRC

Step 1: *Estimate the replacement cost (building).*

Replacement cost of a typical **200 m² house** (building only) = €1,250/m² × 200 m² = **€250,000**
Replacement cost of the **300 m² house** (building only) = €1,250/m² × 300 m² = **€375,000**

Step 2: *Adjust for inefficient design.* The additional 100 m² includes **wasted space** (large corridors, auxiliary rooms, and a second sitting room). We assume that buyers would pay only **50% of the cost** for these unnecessary features:

Wasted space adjustment = 100 m² × €1,250 × 50% = **€62,500**

Step 3: *Include the land value.* The land value for a typical house (200 m² built area) is **€60,000** (based on the comparable method). Since both the comparables and the house under study fully utilise the building density, the land value for the **300 m² house** should be **€90,000**.

Step 4: *Apply depreciation (building only).* Assume **10% depreciation** for the 10-year-old building:

$$\text{depreciation} = (€375{,}000 - €62{,}500) \times 10\% = \mathbf{€31{,}250}$$

Step 5: *Calculate the final DRC value using the following:*

Replacement cost (building only)	€375,000
Adjustment for wasted space	−€62,500
Depreciation (building only)	−€31,250
Land value	+€90,000

Final DRC value = €375,000 − €62,500 − €31,250 + €90,000 = **€371,250**

Conclusion

- The inefficient design reduces the building's contribution to the property's value.
- Despite the house being larger than typical properties, the **wasted space** and **functional obsolescence** lead to a lower effective value.
- The **land value remains consistent**, as the property utilises the allowed density like other comparables.
- In this case, the **final DRC value (€371,250)** reflects the adjustments for inefficient design and depreciation, highlighting how unnecessary features and wasted space can detract from overall value.

Questions

1. A public library built ten years ago has an estimated total lifespan of 70 years. The current replacement cost for a modern equivalent is estimated to be $2,500,000.

   a. Calculate the annual rate of depreciation using the straight-line method.
   b. Determine the total depreciation over the ten years of the building's age.
   c. Estimate the depreciated replacement cost of the building.

2. A courthouse has a replacement cost of $5,000,000 and an expected useful life of 50 years. A reducing balance depreciation rate of 2% is applied annually. Calculate the depreciated value of the building after five years using the reducing balance method.
3. A community centre sits on a 3 ac site in a suburban area. The replacement cost for the building is estimated to be $3,000,000, and the economic life of the building is assessed at 40 years. It has been in use for 25 years. The value of comparable land in the area is $150,000/ac.

   a. Calculate the total depreciation using straight-line depreciation.
   b. Estimate the depreciated replacement cost of the building.
   c. Calculate the final DRC.

4. A local fire station, built 20 years ago, has a total replacement cost of $6,000,000 and a remaining economic life of 30 years. Based on comparable evidence, the land value is estimated at $1,200,000. The decapitalisation rate for rating purposes is 3%.

   a. Calculate the accumulated depreciation using straight-line depreciation.
   b. Determine the effective capital value (ECV).
   c. Calculate the annual equivalent rent using the decapitalisation rate.

5. Assume that construction costs have increased by 30% over the past year.
   Discuss how this affects the market values of the following types of transactions:

   - New properties under construction
   - Newly completed properties
   - Resales

6. Why would a DRC valuation **not** normally be acceptable for secured lending purposes? Under what special circumstances can this be accepted?
7. A valuer estimated the value of a house at €827,500 with the use of the cost method. The house is located in a residential area with 140% density and was recently erected. The replacement cost is given from the quantity surveyor, and the land value is estimated based on comparable transactions from sales of vacant land plots in the vicinity. Do you agree with this approach? What might be wrong?

Land value	550 *sq.m.* × 800€/sq. = €440,000
Replacement cost	250 *sq.m.* × 1,550€/sq. = €387,500
Market value	€387,500

Hint: Is allowable density appropriately used?

References

Blackledge, M. (2009). *Introducing property valuation*. Routledge, London.
Isaac, D. (2002). *Property valuation principles*. Palgrave, Basingstoke.
RICS (2018) *Reinstatement cost assessment of buildings*. 3rd ed. Royal Institution of Chartered Surveyors (RICS), London.
RICS (2021). *RICS valuation – global standards*. Royal Institution of Chartered Surveyors (RICS), London.
RICS (2023). *The contractor's basis of valuation for rating purposes*. 2nd ed. Royal Institution of Chartered Surveyors (RICS), London.
RICS (2024). *Depreciated replacement cost method of valuation for financial reporting*. Royal Institution of Chartered Surveyors (RICS), London.
Sandford, M. (2023). *Business rates: The 2023 revaluation*. House of Commons, London.
Scarrett, D., Osborn, S. (2014). *Property valuation: The five methods*. 3rd ed. Routledge, London.

Part III

Advanced topics and applications of valuation methods

Chapter 12

Automated valuation models

Mass appraisals: history, definitions, and methodologies

Mass appraisals were originally developed for property taxation purposes. Property tax is an essential source of fiscal revenue in many developed economies. *Property tax* is a type of asset tax which is categorised under the direct tax category within the comprehensive tax system. However, the establishment of a property tax system is complicated and not only involves relevant policies and laws but also requires a valuation mechanism and methods.

Key milestones in AVM development include thus:

- **1980s** – early computerised valuation models used for mass taxation assessments
- **1990s** – introduction of CAMA systems
- **2000s** – financial institutions beginning to adopt AVMs for mortgage lending
- **2010s–present** – AI-driven AVMs utilising neural networks, big data analytics, and GIS integration

Generally speaking, for a property taxation system to be established, many tax base assessments of real estate assets need to be carried out in a relatively short period of time. This assessment should conform to the laws of property valuation. In practice, it is essential to adopt a mass appraisal model which fits a specific country's real estate market structure and that is also adaptable to changes in this structure over time. In many cases, it is observed that such systems (CAMAs) that are built for taxation purposes focus on consistency and uniformity rather than accuracy. Despite AVMs and CAMAs existing in harmony with traditional valuation methods throughout the years, recently there has been an increased discussion surrounding them that can be attributed to several key factors.

Technological advancements. Rapid advancements in technology, including big data analytics, artificial intelligence, and machine learning, have significantly enhanced the capabilities of AVMs. These improvements have made AVMs more accurate and reliable, prompting wider adoption and greater scrutiny.

Real estate market dynamics. Fluctuations in real estate markets worldwide have heightened the need for precise, real-time property valuations. AVMs offer a quick and cost-effective way to obtain these valuations, which is particularly valuable in fast-moving markets.

Regulatory interest. As AVMs become more internal tools to financial decision-making, especially in lending and investment, regulators are paying closer attention. This is to ensure that these models do not inadvertently introduce systemic risks or biases, prompting discussions about standards, oversight, and compliance.

DOI: 10.1201/9781003490357-15

Accessibility of data. The availability of vast amounts of property data has increased due to digitisation. This makes it easier to develop and refine AVMs, leading to more companies entering the market with new solutions, thus spurring discussion and competition.

Market penetration and dependence. As more sectors, including real estate and banking, rely on AVMs for a variety of functions, from pricing to portfolio management, their impact on the market has grown, leading to debates about their role, benefits, and potential limitations.

Public and professional scrutiny. The increased use of AVMs has raised questions amongst professionals about the future role of human appraisers and the potential of AVMs to replace traditional methods. This has led to a critical examination of the technology from both a practical and an ethical standpoint.

The question today is no longer whether AVMs, potentially enhanced with AI, will become a standard tool in real estate market analysis but, rather, when and how this transition will occur to improve efficiency. However, there is still some resistance and reluctance to adopt AVMs on a broader scale and in the continued reliance on 'traditional' numerical methods within the real estate valuation industry.

Despite their utility, AVMs have sparked considerable controversy and remain at the centre of much debate. Discussions revolve around their application, quality benchmarks, and a variety of both specific and overarching topics linked to property valuation and market analysis. These conversations range from detailed examinations of key issues to broader overviews, underscoring the complex and multifaceted aspects of the ongoing discourse. In an era when society seems to have access to unlimited information and many feel confident in their understanding of modern valuation and market analysis solutions, nearly everyone has an opinion on the matter.

Some valuation clients, in extreme cases, perceive valuations as subjective outcomes manipulated by valuers, leading to scepticism about their findings. Even valuers themselves often lack a clear perspective. While many would welcome tools to support their work, they also express significant concerns that these innovations could reduce their income or even render their roles obsolete, as automation and machines might replace human expertise.

Last but not the least, some worry about their ability to keep up with the rapid changes reshaping the industry.

Definitions of mass appraisals

There are many definitions of *mass appraisal*. Most of the definitions vary and depend on the purpose of the system being introduced. After an extensive literature review, the most popular and accurate definitions found are the following:

- Eckert et al. (1990) conclude that 'mass appraisal is the systematic appraisal of groups of properties as of a given date, using standardised procedures and statistical testing'.
- The **IAAO (2017)** defines *mass appraisal* or *mass valuation* as the 'process of valuing a group of properties as of a given date and using common data, standardised methods, and statistical testing'.
- The relatively newly formed association the European AVM Alliance defines *mass valuation* as 'the practice of valuing large numbers of properties as of a given effective date by the systematic and uniform application of valuation methods and techniques that allow for statistical review and analysis of the results'.

- Labropoulos (2013) adds that 'the purpose of mass appraisal is to provide an equitable and efficient appraisal of all the properties that fall within the specific administrative boundaries for ad valorem tax purposes'.

In the process of mass appraisal, a valuation model capable of replicating the force of supply and demand over a wider area is created, and appraisal judgements relate to groups of properties rather than to single properties. The quality, accuracy, and consistency control of these assessments are verified by statistical methods. Since mass appraisals are also used by the owners of a significant number of properties, they need to demonstrate the equitability of the taxation system.

Many institutions, organisations, and scholars have contributed significant research regarding mass appraisal theory and the mass appraisal criteria foundations. Not only have they provided criteria regarding the process and methods of mass appraisal in the practical application process, but also they have provided their own detailed suggestions that include analytical guidance on complaint treatment following the assessment.

The most popular forms of mass appraisals (or mass valuations) are CAMAs and AVMs.

AVMs are mathematical-based computer software programmes that market analysts use to produce an estimate of the market value of a property. The distinguishing feature of an AVM is that it is a market appraisal produced through mathematical modelling. They are based on a variety of datasets that are collected separately, such as market analysis of location, market conditions, and real estate characteristics. The credibility of an AVM is dependent on the data used and the abilities of the modeler who created the AVM.

The IVSC (2022) defines a *valuation model* as follows: 'A quantitative implementation of a method in whole or in part that converts inputs into outputs used in the development of a value'. In addition, IVS (effective 31 January 2025) defines an *AVM* as a 'type of model that provides an automated calculation for a specified asset at a specified date, using an algorithm or other calculation techniques without the valuer applying professional judgement over the model, including assessing, and selecting inputs or reviewing outputs'.

Key characteristics of AVMs according to IVSC are:

- *Mathematical modelling.* AVMs utilise various calculation techniques, often incorporating complex statistical and mathematical methods, to estimate property values.
- *Automated process.* The process is automated – meaning, it requires minimal human intervention once the model is set up and data is inputted.
- *Data-driven.* Quantitative and qualitative information available to the valuer. Data, assumptions, and adjustments determined to be relevant and assessed or selected by the valuer to be used in the valuation, based upon professional judgement.
- *Role of professional judgement.* The use of accumulated knowledge and experience, as well as critical reasoning, to make an informed decision.

IVS 105 (*Valuation Models*) further states that '[n]o model without the valuer applying professional judgement, for example an automated valuation model (AVM), can produce an IVS-compliant valuation'. Further to these requirements, the following may apply to compliance with the IVS.

Human input is required. A purely automated AVM without valuer input cannot produce an IVS-compliant valuation. Valuers must ensure that all requirements of IVS are met, which include understanding the model, verifying its assumptions, and applying professional judgement throughout the valuation process.

Documentation and transparency. Valuers must keep appropriate records supporting the selection or creation of the AVM, understand and verify the outputs, and ensure the model's assumptions are consistent with the basis and scope of the valuation.

> However, it should be noted that though an AVM cannot in itself provide an IVS-compliant valuation, this does not mean that an AVM cannot be used as a tool to assist the valuer in providing an IVS-compliant valuation. Furthermore, standard setting is a continual process, and standards are developed to meet market needs, and with the current rapid development of artificial intelligence, this situation may change in the future.

Mass appraisal involves a broad spectrum of issues pertaining to large-scale property value determination. AVMs allow for effective handling of the intricate factors that affect property values, resulting in greater accuracy and reliability, especially in the face of rapidly evolving real estate markets.

Mass appraisal systems

According to McCluskey et al. (1997), in mass appraisal modelling,

> the aim is to construct a representative mathematical model that replicates the market within which the real estate is traded and to be clearly defined in accordance with the principles of microeconomics and at the same time being adaptable to groups of different types of properties.

Eckert et al. (1990) adds that 'mass appraisal builds on the same principles as single-property appraisal'.

Mass appraisal techniques, however, emphasise valuation models, standardised practices, and statistical quality control. For the execution of mass appraisal, given the scale of valuations, statistical methods are used for quality control, to measure accuracy variations in the assessed value from actual sale prices. For most of the appraisal models, if the average deviation from sales prices falls within a predetermined range, the model and quality are considered to be good. On the other hand, in single-property valuation, this procedure is simpler, as the quality can be measured by direct comparison with specific comparable sales.

Renshow (1958) says that it may not be possible to identify and consider all the factors that affect the selection and purchase of a property. However, it is possible to select some of them and examine their correlation with the values of these properties. Although this selection may seem arbitrary, it nevertheless seems to be sufficient to specify the value of the under-study properties with the required statistical accuracy. His view is actually the main idea behind all modern mass appraisal systems.

Eckert et al. (1990) discusses that mass appraisals evolved out of the need for uniformity and consistency in ad valorem appraisals. The first attempts of mass appraisals took place in 1896 in the city of St Paul, Minnesota, by William A. Somers. The necessity for the creation

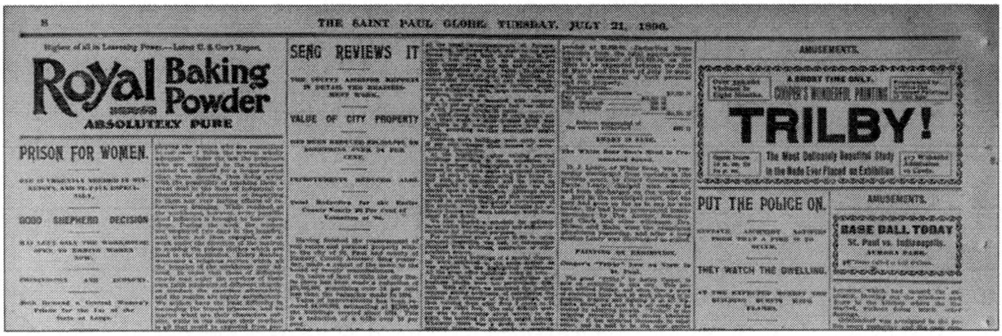

THE SAINT PAUL GLOBE: TUESDAY, JULY 21, 1896.

Royal Baking Powder

ABSOLUTELY PURE

PRISON FOR WOMEN.

SENG REVIEWS IT

VALUE OF CITY PROPERTY

AMUSEMENTS.

A SHORT TIME ONLY.

TRILBY!

PUT THE POLICE ON.

THEY WATCH THE DWELLING.

AMUSEMENTS.

BASE BALL TODAY

Figure 12.1 Publication of the first mass appraisal attempt.

Source: Labropoulos (2013).

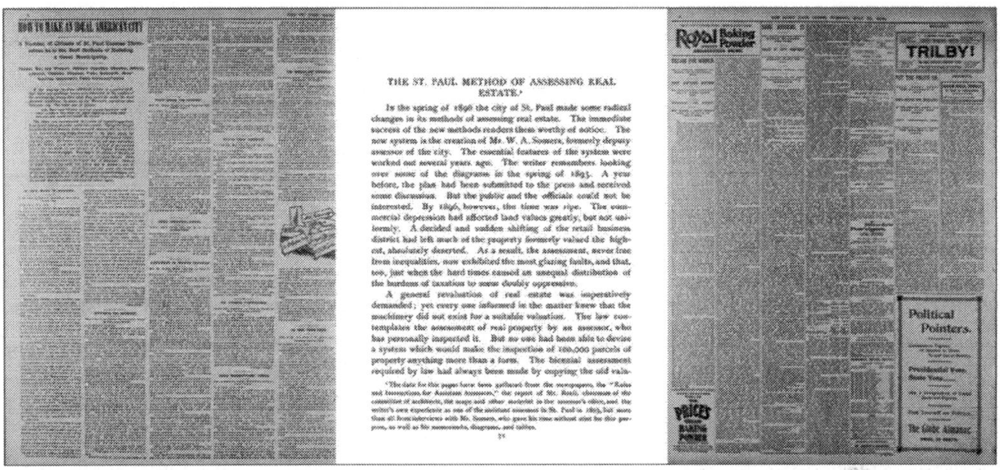

THE ST. PAUL METHOD OF ASSESSING REAL ESTATE.[1]

In the spring of 1896 the city of St. Paul made some radical changes in its methods of assessing real estate. The immediate success of the new methods renders them worthy of notice. The new system is the creation of Mr. W. A. Somers, formerly deputy assessor of the city. The essential features of the system were worked out several years ago. The writer remembers looking over some of the diagrams in the spring of 1893. A year before, the plan had been submitted to the press and received some discussion. But the public and the officials could not be interested. By 1896, however, the time was ripe. The commercial depression had affected land values greatly, but not uniformly. A decided and sudden shifting of the retail business district had left much of the property formerly valued the highest, absolutely deserted. As a result, the assessment, never free from inequalities, now exhibited the most glaring faults, and that, too, just when the hard times caused an unequal distribution of the burdens of taxation to seem doubly oppressive.

A general revaluation of real estate was imperatively demanded; yet every one informed in the matter knew that the machinery did not exist for a suitable valuation. The law contemplates the assessment of real property by an assessor, who has personally inspected it. But no one had been able to devise a system which would make the inspection of 100,000 parcels of property anything more than a form. The biennial assessment required by law had always been made by copying the old valu-

TRILBY!

Political Pointers.

Figure 12.2 Publication and explanation of the first mass appraisal attempt.

Source: Labropoulos (2013).

of such a system arose, as described in the relevant article at that time (Clow 1896), when an economic recession had affected land values greatly but not uniformly. Meanwhile, as reported in *The Saint Paul Globe* on 21 July 1896, issued by William A. Somers, the changes in existing land and building uses, in combination with many overpriced commercial properties that remained vacant, led to the imperative need for a mass revaluation of the area's approximately 100,000 properties (Labropoulos 2013).

Components, functions, and principles of mass appraisal systems

Eckert et al. (1990) divide a mass appraisal system into four key interdependent subsystems:

- The data management
- The sales analysis

- The valuation
- The administrative subsystem

Each of the four components may be further divided into smaller sub-categories and are briefly outlined in what follows.

The *data management system* has components for collection, entry, editing, organisation, conversion, and storage of property and ownership data. This data management system is the heart of the mass appraisal system and should carefully be planned and designed but also be subjected to quality controls to assure the accuracy of the values. The author would add that the integration with a GIS system is more than necessary as it ensures all the aforementioned. The quality of the results of an AVM strongly depends on the data used. Dimopoulos et al. (2018) emphasise the importance of high-quality, accurate, and comprehensive data to ensure reliable assessments. Key data sources for AVMs include:

- *Transaction data.* This includes historical sales prices of properties, which form the basis for comparative valuations.
- *Property characteristics.* Physical attributes, such as plot size, building condition, number of bedrooms, and fixtures and fittings are important inputs.
- *Geographical data.* Using GIS, AVMs can consider location-based factors, such as proximity to transportation hubs, schools, and shopping centres.
- *Economic indicators.* Macroeconomic factors, such as interest rates, inflation, and unemployment rates, are also considered as they can affect real estate values.

The inclusion of geospatial data has become particularly important in the development of more sophisticated AVMs. Many researchers have highlighted the integration of GIS in AVMs, pointing out that spatial data greatly enhances the model's ability to capture neighbourhood-level variations that affect property values.

The *sales analysis system* should have mechanisms for sales data collection, sales screening and processing, ratio studies, and sales reporting. The main component of this system is the ratio studies component, which has the main objective of providing the best available measures of appraisal performance and is a valuable tool for monitoring appraisal results, identifying reappraisal priorities, adjusting valuations to the market, and assisting management in planning and scheduling.

The *valuation system* consists of mass appraisal applications for the comparison of sales, costs, and income approaches to valuation. Sales comparison applications include multiple regression analysis, adaptive estimation procedure (AEP) or feedback, and automated comparable sales analysis. The cost approach requires maintenance of computerised cost schedules and equations, derivation of depreciation schedules from market data, and reconciliation of cost-generated values with the market. Mass appraisal applications of the *income approach* include the development and use of income multipliers and overall rates, such as market yields, discount rates for different property types (hotels, businesses, etc). Values produced by these three approaches should be reviewed and reconciled to select a final value for assessment purposes.

The *administrative system* is composed of a variety of functions and activities subject to varying degrees of automation. One of the first functions to be automated was the preparation of assessment notices and tax bills – it is oriented for property taxation purposes. Many jurisdictions have also computerised other administrative functions. Finally, the administration system manages the appeals process. It should be added that the support of the appropriate legal framework is a prerequisite, especially for tax purposes. On the other hand, if the purpose is for portfolio management or loan management, the regulations from the central bank of a country or the institutional internal procedures are required.

Within the same body of work that is considered as the cornerstone of mass appraisals, Eckert et al. (1990) distinguish three basic functions in a typical MAS: *reappraisal*, *data maintenance*, and *general principles*, all of which are discussed in this chapter.

Reappraisal. A mass appraisal system that must be constantly kept up-to-date and reflect the market conditions as much as possible. For this reason, a periodic reappraisal of the values of the properties under study is required, with a proposed revaluation frequency of once per annum or a different periodicity. However, today, the application of complex AI algorithms may allow for instant updates. Even one new transaction may change the values of the under-appraisal properties instantly. The aforementioned is one of the main advantages that AI presents in comparison with traditional approaches that are based in MRA.

Principles of mass appraisal

Although mass appraisal builds on the same basic principles as single-property appraisal, mass appraisal techniques emphasise more in equations, tables, and schedules, collectively called ***models***.

Constructing such models can be viewed as a two-step process:

Step 1: *Specifying the general model.* This provides a framework to simulate supply and demand forces operating in a real estate market and is adaptable to many uses. The model builder specifies the variables, or property characteristics, to be used in the model and their relationships. A simple general model can be expanded to reflect the complexities of the market by expansion of each variable. Model specification is the first step in the development of any mass appraisal model. Mass appraisers must understand the models they use and be sure they reflect the way property is valued in the local market. When the model structure is imposed, as it often is in a commercial mass appraisal software, the user may not be able to select the property characteristics and their relationships in the model. In contrast, other software, particularly general-purpose statistical software, afford the user complete freedom in determining model structures. The choice of such model is not standardised but is completely different for a different dataset and shall be adjusted as per the provisions of the regulatory jurisdiction.

Step 2: *Model calibration.* This is the process of adjusting mass appraisal formulas, tables, and schedules to the current market. Although the structure of a mass appraisal model may be valid for many years, the model is usually calibrated or updated every year. To update for short periods, trend factors may suffice. Over longer periods, complete market analyses are required. At this point, it is stated again the importance of an in-house index that time adjusts the sales transactions.

Mass appraisal models are divided into three categories: the cost, the income, and the comparative approach, similar to the three valuation approaches explained in earlier chapters.

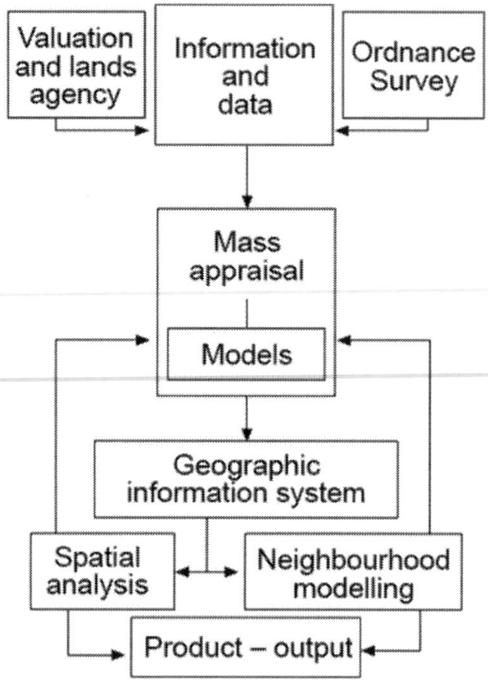

Figure 12.3 Graphical representation of a mass appraisal model.

Source: McCluskey et al. (1997).

Elements of the cost approach model

1. *Cost tables.* Include base rates, per square foot adjustments, and lump sum adjustments that are used to calculate the replacement cost as new.
2. *Depreciation schedules.* Developed for each type of property and then tested to ensure that they reflect the local market, considering the condition and age of the property.
3. *Time and location modifiers.* Used to adjust cost data for local variations and changes in costs over time.
4. *Market adjustment factors.* Used to adjust the values obtained from the cost approach to reflect market conditions. These adjustments should be applied by type of property and area based on sales ratio studies or other types of market analysis.

   NOTE: Accurate cost schedules, condition ratings, and depreciation schedules will minimise the need for the market adjustment factor.

Elements of the income approach model

1. *Market and existing/passing rents.* Can be developed from income data obtained from the owners or, in some cases, local third-party sources.
2. *Vacancy and expense ratios.* Used to adjust potential gross income to typical net income. These ratios must reflect the local market and differentiate according to the types of commercial properties and the age groups.
3. *Income rates and multipliers.* Used to convert income to market value.

Elements of the comparables sales model

1. *Land valuation tables.* Contain land values per unit, along with standard adjustments for topography, depth, water supply, frontage, and other locational features.

(Continued)

(Continued)

Elements of the comparables sales model

2. *Sales comparison adjustments.* Derived from local sales analysis and can be based on statistical techniques, such as multiple regression analysis and feedback.
3. *Multiple regression analysis equations.* Separate equations should be developed for each market area in a jurisdiction, or a general model should be adjusted for locational variations.
4. *Feedback equations.* Another automated version of the sales comparison approach, while the 'prices' assigned to each property characteristic are based on sales analysis.
5. *Base home table.* They repackage a statistical equation, derived through multiple regression analysis equations.

It should, however, be noted that, in many cases, the models are hybrid, where more than one approach is utilised.

Mass appraisal model building and calibration

According to Eckert et al. (1990), a *model* serves as a representation of how a system or process operates. Building effective models necessitates a solid foundation in theory, data analysis, and research methodologies. The most successful valuation models are characterised by their accuracy, logical consistency, and clarity. The development of these models involves two distinct steps: model specification, which relies on economic and appraisal theory and market analysis, and model calibration.

In practice, specification and calibration occur through an iterative process (IAAO 2018), which typically includes:

1. Model specification
2. Model calibration
3. Testing the model
4. Adjusting model specifications
5. Recalibrating the model
6. Retesting the model
7. Repeating the process until the model satisfies quality assurance standards

Eckert et al. (1990) further note the availability of diverse data types, variables, and structures for mass appraisal models.

Types of data

- Qualitative data (e.g. neighbourhoods, cooling systems, roof types)
- Quantitative data (e.g. building age)

Data transformations

Variables derived from property data may remain untransformed or undergo mathematical transformations. These can include converting qualitative data into:

- Binary variables
- Scalar variables

Quantitative data transformations may involve:

- Reciprocal transformations
- Exponential transformations
- Logarithmic transformations
- Multiplicative transformations
- Quotient transformations

Types of models

A model typically consists of one dependent variable (e.g. property value) and one or more independent variables (predictors). Common model structures include:

- *Additive models.* These use a linear form, where property value is expressed as the sum of coefficients multiplied by variables.

Example:

$$\mathbf{MV} = B_0 + B_1 \times X_1 + B_2 \times X_2 + \ldots$$

Where:

$\boldsymbol{MV}$ = the dependent variable
$\boldsymbol{B_0}$ = a constant
$\boldsymbol{X_1}$ = represents the independent variables in the model
$\boldsymbol{B_1}$ = corresponding rates or coefficients

- *Multiplicative models.* Variables are raised to powers and multiplied, capturing curvilinear relationships and proportional adjustments.

Example:

$$\mathbf{MV} = B_0 \times X_1^{B1} \times X_2^{B2} \times \ldots$$

Where:

MV = the dependent variable
B_0 = a constant
X_1^{B1} = where X represents the independent variables in the model
X_1^{B1} = where $B1$ represents the corresponding rates or coefficients

- Multiplicative models consist of a constant B_0 and percentage adjustments. They have several advantages, including the ability to capture curvilinear relationships more effectively and the ability to proportionately adjust the value of the property being appraised. Multiplicative models are usually calibrated using linear regression packages. This requires some of the variables to be mathematically transformed for calibration. In this case, the market value would require a transformation (IAAO 2018).
- *Hybrid (non-linear) models.* The combination of additive and multiplicative models. The following example of a hybrid model is specified to the same as a cost model and demonstrates

the flexibility of the hybrid model. A hybrid (non-linear) model specification that separates value into building, land, and 'other' components (e.g. outbuildings) follows.

Example:

$$MV = \pi GQ \times [(\pi BQ \times \Sigma BA) + (\pi LQ \times \Sigma LA) + \Sigma OA]$$

Where:

πGQ = the product of general qualitative components
πBQ = the product of building qualitative components
πLQ = the product of land qualitative components
ΣBA = the sum of building additive components
ΣLA = the sum of land additive components
ΣOA = the sum of other additions additive components

Data analysis for model structures

Successful mass appraisal models align with market behaviour and begin with market analysis. Tools for data analysis include:

- *Data profiles*. Statistical summaries (e.g., median, range) and frequency distributions for qualitative data.
- *Two-way analyses*. Tools like cross-tabulations and scatter plots, to examine relationships between variables.
- *Three-way analyses*. Techniques, such as contingency tables, to explore relationships among three variables.

Whilst traditional methods like multivariate regression analysis (MRA) are essential, the complexity of real estate markets has driven the adoption of advanced data science and machine learning techniques for improved accuracy and performance.

Model specification

Model specification is used to determine which valuation models yield the best possible results by reviewing the data used. Model specification is based on data analysis and appraisal theory.

Sample model specifications are provided in the following.

Cost approach specification

> The cost approach is an indirect method of arriving at market value, based on specification of replacement cost new, less depreciation, plus market derived land value. The cost approach is calibrated by review of construction costs and sales. This approach generally produces more acceptable results for newer properties, specialised properties and properties with insufficient sales. Model specification for the cost approach requires the estimation of separate land and building values.
>
> (IAAO 2018)

Two cost approach formulas for model specification are:

$$MV = \pi GQ \times [(1 - BQD) \times RCN + LV]$$

Where:

MV = the market value estimate
πGQ = represents the general qualitative variables, such as location, economic adjustments, and time of sale
BQD = a building qualitative variable representing depreciation
RCN = the replacement/reproduction cost new
LV = the land value

and

$$MV = \boldsymbol{\pi GQ \times [(\pi BQ \times \Sigma BA) + (\pi LQ \times \Sigma LA) + \Sigma OA]}$$

Where:

MV = the market value estimate
πGQ = the product of general qualitative variables
πBQ = the product of building qualitative variables
ΣBA = the sum of building additive variables
πLQ = the product of land qualitative variables
ΣLA = the sum of land additive variables
ΣOA = the sum of other additive variables

If a third party provides the cost tables, it is the responsibility of the market analyst to calibrate the cost tables to the local market to provide a valid indicator of value by the cost approach.

> It is also the responsibility of the analyst to fully understand the assumptions made by the third party in constructing the cost tables, the original source of the construction costs used, and the way the unit-in-place costs were aggregated by the third party to arrive at the published square foot rates.
>
> (IAAO 2018)

The specification of the cost approach includes the following steps (Eckert et al. 1990):

Step 1: *Stratification improvements into homogenous groups.* There is usually a different market for each group. For example, for residential properties, properties are usually stratified by occupancy, storey height, and construction grade or wall type, while for non-residential properties, properties are stratified based on structure type and number of floors.

Step 2: *Specification of the typical characteristics, or base specifications, of each model.* Using the unit-in-place or quantity survey method, reproduction or replacement cost as new is then estimated. In mass appraisal, these costs are converted to comparative units of cost. For residential structures, costs can usually be divided into horizontal, vertical, building addition, constant building component, qualitative, and addition costs, while the non-residential cost models contain the six basic cost components mentioned, but with some important differences in the format of the horizontal and vertical components.

Sales comparison models specification

According to the IAAO (2018), the sales comparison approach can be implemented as a comparable sales model or as a direct market model. The comparable sales model routine includes two steps. The first step involves the selection of the sales most comparable to the subject properties and a series of data filters. The second step adjusts the sale prices of the comparable sale properties to the subject properties based on differences in data characteristics. The adjusted sale prices are then used to determine an estimation of the value for the subject properties. Both steps in the comparable sales routine may be supported by statistics or a statistically based direct market model.

The model specification for the comparable sales method can be summarised as follows:

$$\mathbf{MV = SP_C + ADJC}$$

Where:

MV = represents the market value estimate
SP_C = represents the selling price of comparable sale properties
ADJ_C = represents adjustments to the comparable sales

The direct market model may use one of the three model structures mentioned earlier – the additive, the multiplicative, or the hybrid model.

According to Eckert et al. (1990), the specification of the sales comparison approach includes the following approaches:

Approach 1: *Stratification and location analysis.* There are three basic ways to analyse location in mass appraisal model building: multiple models based on geographic stratification, multiple models based on cluster analysis, and a single model with location adjustments.
Approach 2: *Geographic stratification.* Boundaries are drawn along rivers and other natural barriers, major streets, and subdivision lines to reflect major differences in location. Separate models are then specified for each such area.
Approach 3: *Stratification using cluster analysis.* This combines properties into relatively homogenous strata-based categories, such as location, and physical characteristics, such as age, size, style, and construction quality.
Approach 4: *A single model with location adjustments.* In this method, a single model with locational variables is used for the entire jurisdiction. A more sophisticated technique is used to identify value influence centres (VICs), a technique that adjusts location smoothly, virtually eliminating boundary problems.
Approach 5: *Selection of the appropriate model considering the property type.* Model specification varies by property type and can be additive, multiplicative, or hybrid. The properties can be divided into three categories: residential properties, income-generating properties, and land. Each category can be further subdivided (e.g. the residential properties can be subdivided into single-family residential, condominiums/townhouses, etc.), using different types of models.

Income approach specification

The *income approach* is an indirect method for determining market value. The appraiser evaluates income for quantity, quality, direction, duration, and the expense incurred to earn this income and then converts it by means of a capitalisation rate into an estimate of market value (IAAO 2018).

In an income model, the dependent variable may be income, income per unit (net rentable area), expenses, and expenses per unit or capitalisation rate.

The steps in this approach are (Eckert et al. 1990):

Step 1: *Modelling gross income.*
Step 2: *Modelling net income.*
Step 3: *Modelling a gross income multiplier (GIM) – or gross rental multipliers (GRM) (IAAO 2018) – which expresses the relationship between gross income and rent.*
Step 4: *Modelling an overall rate model (OAR), which expresses the relationship between NOI and property value.*

All these can be estimated in one of two ways: stratification and typical units of comparison, often using spreadsheet software, or statistical models using MRA or alternative techniques.

Model calibration

Eckert (1990) suggests that the calibration of each model should be as follows:

Cost models calibration. The cost approach is applied in four steps: estimate RCN, apply depreciation, estimate and add land value, and apply general qualitative adjustments. The calibration of the model includes the calibration of RCN and the development of formula-driven cost models, cost-trend factors, depreciation schedules, and market adjustment factors.

Sales comparison calibration. The two more common techniques available for calibrating sales comparison models are MRA and feedback. MRA is older and more common. Feedback was first used in property appraisal in the late 1970s. MRA models can be additive, multiplicative, or hybrid. Their objective is to model the relationship between property characteristics and value so that the latter can be estimated from the former. The results are evaluated using eight statistics: the coefficient of determination (R^2), the standard error of the estimate (*SEE*), the coefficient of variation (*COV*), the average percent error, the coefficient of correlation, the t-statistic, the F-statistic, and the beta coefficients.

Income models calibration. In mass appraisals, income and expense analysis is conducted mainly by microcomputer spreadsheet software. Entering available data, the appraiser establishes typical income figures and expense ratios by property type and estimates unreported figures. In mass appraisal, direct capitalisation models are used in the form of GIMs and OARs. These models are developed in two ways, stratification and MRA. In general, income models should include data collected prior to property taxes being subtracted as an expense, with an effective property tax rate added to OARs and GIMs after calibration.

However, these suggestions currently seem very generic and outdated. Model calibration is the second task in building models, in which the values for the coefficients in a model are estimated. According to the IAAO (2018), many mass appraisal systems in use today rely on statistical models as the method of calibration. Some example techniques may be based on regression, geographic-weighted regression, or neural networks. All regression programmes are based on statistics, while neural networks draw on the analogies of adaptive neurons learning.

Therefore, the following calibrations are suggested.

Model calibration: *statistical methods*

MRA is a statistically based analysis that evaluates the linear and/or curvilinear relationship between a dependent (response) variable and several independent variables (predictors). Models produced using MRA come with a rich set of diagnostic statistics that provide evaluation tools for the market analyst to compare results between and among specified models. These include 'goodness of fit' statistics, such as R^2, adjusted R^2, SEE, COV, and 'measures of variable significance', such as R^2, t, F, etc. (IAAO 2018). According to Kauko (2008), the importance of mass appraisal may also be seen from the possibility of linking the relationship between the property value, the property characteristics, and urban social and economic problems. The literature available on mass appraisal modelling tools (including AVMs) is rich and ever-evolving. While the standard MRA-based hedonic price models may not be the most suitable tools for capturing all the necessary information involved in the formation of value, MRA currently remains the most important theoretical framework regarding mass appraisals.

Tabales et al. (2013) add that

> their aim is to estimate the price of a complex good, such as a dwelling, as a function of its characteristics. These characteristics cannot be separated and should be treated as a whole. They include attributes such as the size of the residence, or its internal structure, and variables related to its environment, such as the building properties, its surroundings, and so on.

Mooya (2017) adds that the 'basic principle of hedonic models is that the value of a property is a based on the sum of its constituent characteristics'. This suggests that a property being valued can be broken down into its fundamental components, and the market values of those components. For example, the market value of a specific residential property is regarded as a summation of the values of its size, location, age, and so on, all of which hold a particular value in the market. Typically, hedonic models use regression techniques to estimate the contribution of each characteristic of the property to its overall value. Using the same methodology, hedonic AVMs use regression analysis for the estimation of property value. The term *regression analysis* is used to describe the various mathematical methods (including econometrics) that aggregate observations into a form in which a dependent variable is a mathematical function of independent variables ($y = f(x1, x2, \ldots, xn)$), often in a way that allows a statistical inference regarding the parameters of the function outside the specific sample (Mooya 2017).

The popularity of hedonic price modelling for mass appraisal purposes stems from several factors. First, it rests on MRA. Thériault et al. (2003) suggest that this conceptually sound and very powerful analytical tool combines probability theory with calculus, thereby allowing the separation of crossed influences that affect property values. Second, it perfectly fits the very definition of *market value*, expressed as 'the most probable price' that should be paid in a competitive and transparent market setting. A market value, being a probability distribution, calls for a statistical treatment that conveys objectivity through direct market reading, which is easily reproducible and offers adequate testing of result reliability; MRA provides for such qualities. Third, the hedonic approach is not confined to producing value estimates, as it adds very useful insights into the causal dimensions of property value determination. Thus, it is viewed as a decision-making tool that brings about market intelligence, even more so when used in combination with GIS, a 'natural', and increasingly imperative, complement for adequately handling spatial issues.

Model calibration: AI and artificial neural networks (ANNs)

Another way of calibrating real estate valuation models is with ANNs, also known as machine learning models. Machine learning plays a central role in the advancement of AVMs. These models are designed to 'learn' from historical data and improve their predictive accuracy over time. The history of the ANN technique, as presented in Abidoye and Chan's (2016) work titled *Research Trend of the Application of Artificial Neural Network in Property Valuation*, can be categorised into four stages:

- Beginning of neural networks, the 1940s
- The first golden age, the 1950s to 1960s
- The quiet years, the 1970s
- The renewed enthusiasm period of the 1980s to now

The final stage is the present-day neural network research area, where the application of the technique exploded and received more attention from scholars. The study of McCulloch and Pitts (1943) was the first to employ the ANN technique to demonstrate the 'threshold logic' in the field of mathematics.

In the real estate sector, the seminal study of Borst (1991) was the first to apply the ANN technique to property appraisal. The author found that ANN is reliable and accurate for property value estimation but recommended that more research efforts be invested in the application of ANN. Since it has been established that the ANN technique handles the shortcomings of most of other appraisal techniques (Do and Grudnitski 1992), researchers in different real estate markets around the world have investigated its application in their domains and have mostly reported a positive result (Limsombunchai 2004).

> Neural networks can calibrate models that consist of both linear and nonlinear terms simultaneously. The user inputs each variable with assigned weights (coefficients). The software exposes the data using an algorithm in a hidden layer where the weights are adjusted (calibrated) in a manner that reduces the squared error. This is an iterative process much like those found with hybrid (nonlinear) regression. The final output results in a single estimate of value with the exact algorithm remaining hidden from the market analyst.
>
> IOAA (2018)

An ANN is like a non-linear regression or a multivariate regression model with non-observable linking variables. Once the topology and the parameters of the network are specified, it can be presented as an ordinary statistical or econometric model. Neural networks are used with different purposes, such as the estimation of models, classification, forecasting, and so on. Here it is used as a modelling tool and as a practical alternative to the well-known econometric hedonic models. Also, it aims to forecast the value of properties, so it can be used to measure the downturn that occurred over the last few years within the real estate market (Tabales et al. 2013).

Tabales et al. (2013) add that the marginal prices obtained with ANN are more realistic than the classical hedonic prices. The price to be paid, using this more complex methodology, is linked both to the mathematical difficulties in obtaining the marginal prices, even with simple network topologies, and to the need of a fairly large sample in order to obtain stable parameter estimates.

In general, the ANN approach presents clear advantages over the classical hedonic methodology, despite its complexity and the large samples required in order to obtain reasonable forecasting capabilities.

Classical hedonic models have been used for several years in real estate valuation to estimate prices of complex properties. The availability of neural network software packages has allowed for the use of this alternative methodology, with significantly improved results. The use of ANN is more flexible than the use of classical econometric models, when there is suffice sample data available. Some common problems in hedonic models, such as non-linearities in the extreme range of prices of the real estate market, are resolved by using ANN and even adapt to properties that could be labelled as outliers, although some authors criticise this 'black box' approach. Neural networks allow for a more appropriate fit with the usual measures, such as the determination coefficient, or with error-related statistics (root mean square error [RMSE], mean absolute error [MAE], and so on). The 'dark side' of this methodology is the sample of data required. This can be realistically overcome only with the collaboration of realtors with a broad presence in the market, with large internal databases that contain real data regarding properties, where a random sample can be selected, and the uncertainty about the exact identification of each dwelling is eliminated, allowing the inclusion of particular data belonging to each building and its location (Tabales et al. 2013).

Model adjustments

According to IAAO (2018):

> in the process of specification and calibration of AVMs, adjustments should be considered for location and market trends. Additionally, the data must be reviewed to ensure representativeness, and variables should be selected using advanced statistical analysis. These requirements can be met through the following processes.

Time series analysis. Time series analyses are techniques that can be used to measure the cyclical movements, random variations, seasonal variations, and cyclical trends observed over time. In property valuation, these analyses can be used to develop a multiplier or index factor to update existing appraised values or to adjust sale prices for individual properties to a valuation date. Since values can change at different rates in different markets, separate factors should be tested for each property type and market area. Methods used to develop time trend factors for property valuation include:

- Value per unit analysis
- Re-sales analysis
- Sales/assessment or sales/AVM value ratio trend analysis
- Median/mean value per period in the form of a moving average/median sale price
- Inclusion of time sale variables in MRA/AVM models

Once a time trend is established, it can be used to adjust values to any point within the sales period. Trend factors can be extrapolated for a brief period beyond the sales period, but this should be done with caution and grows increasingly unreliable as the period is lengthened.

Independent variable selection for models. Modelers should include statistically significant and reasonable variables. Variables (e.g. property characteristics) that are highly correlated with other variables or have a statistically insignificant effect should be excluded or used with caution in the model. Developing a comprehensive criterion for variable selection requires in-depth knowledge of appraisal of various property types, advanced statistical analysis, and mathematics.

Location adjustments. Variables to express the influence of location are critical in any model. The effect of factors external to the property should be determined. There are several methods of accounting for location. Two common methods for developing location adjustments are the creation or use of existing geo-economic areas and use of geocoordinates. *Geo-economic areas* are the traditional and most usual form of location analysis. In AVMs, geo-economic areas may be based upon streets and natural boundaries and government-designated areas.

Geographic coordinate techniques. Relate common elements such as price to each property's unique location. Software uses a variety of smoothing techniques to compute a unique location adjustment for each property. This variable is then included, along with other variables, in a multiple regression or other model to capture location influences.

Geographically weighted regression. Another method of accounting for location works by allowing model coefficients to vary by location to pick up spatially heterogeneous effects.

Model frameworks that incorporate these methods for accounting for location include the following:

- Spatial regression models (i.e. spatial lag models, spatial error models)
- Inclusion of locational variable in the model specification (i.e. regression with locational binaries, random forests with a locational variable as a feature)
- Multi-stage locational adjustment models (i.e. regression model paired with locational value response surface, market-adjusted cost by geo-economic area)
- Spatial interpolation models (e.g. Kriging)
- Locally weighted regression model (i.e. geographically weighted regression)
- Segmented models (i.e. separate regression models by market area)
- Cluster analysis

Where can these mass appraisal models be applied?

According to the *RICS Information Paper* (2013), which refers to the use of mass appraisals as AVMs, they can be used in both the public and private sectors (Amato 2017), as described in what follows.

- *By banks (lenders), for granting a new loan or subsequent revaluation for adjustment decisions regarding an existing loan.* Lenders can obtain an AVM that can sometimes be used prior to processing a case, to see if the proposed figures being discussed are likely to be adequate, without incurring the cost of a full valuation by a valuer. Alternatively, an AVM can be used midway through a mortgage term to check how a property's value may be fluctuating.
- *In arrears assessment and planning.* Banks may require checking the value of a mortgaged property provided as security for a loan to establish if the property value still covers the amount of the mortgage, and what scope there is in arrears planning. An AVM can provide a cost-effective and quick indication.
- *In an audit of valuations.* Banks, and those who audit valuations, sometimes require a second valuation to audit the original valuation for a property. An AVM can provide this second valuation for a range of properties, as well as in individual cases.
- *For the identification of fraudulent activity.* An AVM can be applied to a range of property valuations to determine whether there is any activity that does not follow normal market trends.

- *For estimating potential compensation payments that need to be paid to owners of properties that have been affected by public construction work, such as the creation of a new road network that affects the living conditions of a neighbourhood for a certain period of time.* Public authorities can use AVMs as a quick and cost-effective way to estimating the likely cost of compensation as part of a total 'project' cost. An appropriate model can provide a pre-project value estimate of affected properties and keep this up to date as required.
- *For lending (capital to risk-weighted assets).* AVMs can be applied to a portfolio of properties that are mortgaged to obtain an indication of how well they are performing. This can also assist the bank and their accountants in determining capital adequacy ratios.
- *To provide a valuation estimate for large-scale assets.* This could include large portfolios of properties or 'packages' of properties that were obtained by a bank through bad debt and are now being sold off. In these cases, an AVM can be used to provide cost-effective sample valuations for portfolio purposes or for the valuation of a whole portfolio.
- *To provide a valuation for individual capital tax purposes.* An AVM can be used in these situations for individual properties, or for a portfolio of properties, to establish a quick estimate of likely tax implications. This can aid with tax planning.
- *To analyse the cost and benefits that are associated with potential public expenditure.* When estimating the cost or benefit of public works, an AVM can provide a total cost estimate of the residential value of properties likely to be affected, in a quick and cost-effective manner.
- *For tax purposes.* Mass appraisals can be used for local taxation purposes. A model called CAMA is a sophisticated AVM that can provide a cost-effective valuation estimate for thousands or millions of properties.

Benefits of the use of AVM, and its limitations

The main benefits of AVM use are presented in what follows:

1. *Speed and efficiency.* One of the main advantages of AVMs is their speed. Traditional real estate appraisals can take days or weeks, especially if a physical inspection is required. In contrast, AVMs can produce valuations within minutes. This speed is particularly advantageous for large institutions, such as banks and government agencies, that need to process thousands of valuations quickly.

   Several other studies found that the speed of AVMs improves not only operational efficiency but also decision-making in markets where real estate values fluctuate rapidly. By providing real-time valuations, AVMs allow stakeholders to make informed decisions faster than ever before.

   In Dimopoulos's 2020 study, AVMs were used to assess property tax for over 10,000 properties in Nicosia. The model's ability to process such large amounts of data simultaneously was a key factor in the successful implementation of the mass appraisal system.
2. *Cost-effectiveness.* By automating much of the valuation process, AVMs significantly reduce the costs associated with traditional valuations. McCluskey and Borst (2011) argue that AVMs eliminate the need for physical inspections in many cases, reducing labour costs. In addition, AVMs allow large-scale property assessments to be conducted with fewer resources, making them ideal for applications such as property taxation, where municipalities may not have the budget to conduct individual assessments for each property.

   In the context of mass appraisals, the cost savings are significant. Dimopoulos (2020) found that AVMs in Cyprus reduce the administrative costs of property tax assessments by almost 40% while improving the accuracy and consistency of assessments.

3. *Consistency and objectivity.* One of the biggest challenges of traditional valuation methods is the subjectivity involved. Different appraisers may interpret market data or property characteristics differently, resulting in different valuations. AVMs address this problem by applying standardised models to large datasets, ensuring consistent and objective results. Pagourtzi et al. (2003) emphasise the importance of consistency in property valuation, especially in mass valuations, where uniformity is crucial. AVMs reduce the risk of human error and bias and provide more reliable valuations for large datasets.

   In Dimopoulos's (2020) mass appraisal study, the implementation of AVMs led to a significant reduction in variance between property valuations. This consistency is particularly important in large-scale applications, such as property tax assessments, where uniform valuation of thousands of properties is essential to ensure fairness and compliance with legal standards. By eliminating the variability associated with human judgement, AVMs create a more objective and transparent valuation process.

On the other hand, there are significant limitations that have been observed, and the main are listed in what follows.

Data quality and completeness

The effectiveness of an AVM is highly dependent on the quality and completeness of the data used. AVMs require access to accurate and up-to-date transaction data, property characteristics, and market trends. However, in regions where data is sparse, outdated, or inaccurate, AVMs can hardly provide reliable results.

Dimopoulos (2020) found that in regions with limited real estate transaction data, such as rural areas or emerging markets, AVMs can provide less accurate valuations compared to traditional methods. He also emphasises that data gaps, inconsistencies, or errors can undermine the accuracy of AVM results, leading to inaccurate valuations and, in extreme cases, significant financial losses for stakeholders. To address this limitation, Dimopoulos et al. (2018) recommend the integration of GIS and remote sensing techniques to supplement missing or incomplete data in certain markets. The combination of GIS and spatial econometrics can help AVMs fill data gaps and improve accuracy in regions where transaction data may be lacking.

Applicability to unique or complex properties

While AVMs perform well in the valuation of standardised residential or commercial properties, they often struggle with unique, high-value, or highly complex properties. Geltner et al. (2014) found that properties such as historic buildings, luxury estates, or industrial facilities do not necessarily fit the data patterns that AVMs rely on. As a result, AVMs may produce valuations that do not fully capture the nuances of such properties, resulting in under- or overvaluation. Dimopoulos (2020) also noted that in cases involving properties with rare or specialised features, human expertise is still required to ensure accurate valuations. For example, the valuation of a listed building with unique architectural features may require a combination of AVM results and expert judgement to capture the full value of the property.

Market volatility and economic shifts

AVMs often rely on historical data when preparing valuations, which can limit their ability to react to sudden changes in the real estate market. In a volatile economic environment, where real estate values can change rapidly due to factors such as recessions, natural disasters, or geopolitical events, AVMs may struggle to keep up with real-time changes.

Dimopoulos (2020) highlighted that AVMs are less effective in times of market volatility as they are unable to quickly adapt to changes in buyer behaviour, interest rates, or other external economic factors. He also points out that while AVMs can consider current sales data, they may not be as good at forecasting future market trends, especially in unstable or rapidly changing economic conditions.

Ethical challenges and guidelines for ethical use of AVMs

- Transparency and fairness
- Accountability
- The role of humans

There are issues with opaque valuation methods and the potential for algorithmic discrimination. It is crucial that AVMs are transparent in their methodologies to ensure fairness and avoid biases. Determining responsibility for misevaluations can be challenging. Clear guidelines are needed to assign accountability in cases of errors. While AVMs automate much of the valuation process, human oversight is essential to validate and interpret the results, ensuring they are reasonable and accurate.

Developing comprehensive guidelines for the ethical use of AVMs is essential.

Key points should include:

- *Data access and quality.* Ensuring access to high-quality data is critical for accurate valuations.
- *Transparency.* AVMs should clearly disclose their methods, the data used, and the models employed.
- *Predictor impact.* Understanding and communicating the influence of various predictors used in AVMs.
- *Fairness and non-discrimination.* Ensuring that AVMs do not perpetuate biases or unfair practice.
- *Accountability.* Establishing clear lines of accountability for the outputs of AVMs.

Best practices for implementing AI and ML in AVMs

- Ensure data quality
- Transparency with customers
- Continuous training and updates
- Data security

It is of the utmost importance to ensure that data inputs are updated and validated on a regular basis to maintain the accuracy and reliability of AVMs. It is evident that transparent communication regarding the capabilities and limitations of AI and machine learning (ML) technologies to customers is instrumental in fostering trust and establishing realistic expectations. It is imperative that AI and ML models undergo continuous training and updates to adapt to evolving market conditions and customer behaviours. It is of the utmost importance to implement robust data protection measures to safeguard sensitive customer information, thereby ensuring compliance and maintaining trust. By adhering to these best practices and addressing the ethical challenges inherent in the process, AVMs can be effectively integrated into property valuation processes, thereby enhancing the accuracy and fairness of the resulting valuations while maintaining trust and accountability.

The future of property valuations and AVMs

The future of AVMs lies in their ability to integrate with advanced AI, blockchain, and Internet of Things (IoT) technologies.

Key developments include:

- *AI-enhanced valuations.* Improved machine learning algorithms capable of real-time market adaptation, incorporating deep learning models for nuanced property assessments.
- *Blockchain for data integrity.* Secure, tamper-proof property records enhancing AVM reliability, reducing fraud, and ensuring data transparency.
- *Hybrid models.* Combining AVMs with professional valuer input to ensure accuracy and compliance, leveraging AI-driven recommendations for final human assessment.
- *IoT integration.* Real-time property condition monitoring using smart sensors to provide up-to-date property data, improving valuation accuracy.
- *AR and VR.* Enhanced property inspections and simulations for remote valuation assessments.

AI-enhanced valuations

AI-enhanced valuations leverage deep learning algorithms, real-time market data, and predictive analytics to improve accuracy, efficiency, and transparency in property assessments. These systems utilise automated feature recognition to analyse property images and descriptions, integrate real-time data from IoT-enabled smart sensors and blockchain property records, and employ predictive market trends to forecast future property values based on macroeconomic indicators and local real estate patterns. Explainable AI (XAI) enhances trust and regulatory compliance by making valuation decisions more transparent, while hyper-personalised valuations assess factors such as expected rental yield, appreciation potential, and investment risk based on specific buyer profiles. Additionally, AI-powered fraud detection algorithms identify anomalies in valuation patterns, flagging potential fraudulent property listings or price manipulations. As AI technology continues to evolve, the real estate industry will experience deeper integration of AI-enhanced valuations, optimising valuation accuracy and democratising access to reliable property insights for both institutional investors and individual buyers.

Blockchain for data integrity

Blockchain technology is emerging as a game-changer in real estate valuation by enhancing the accuracy, security, and transparency of property data. Blockchain functions as a decentralised digital ledger that records transactions in an immutable manner, ensuring that property-related information remains secure, verifiable, and resistant to manipulation. When integrated with AVMs, blockchain helps eliminate fraudulent data, reduces errors, and enhances confidence in valuation results.

One of the primary advantages of blockchain in valuation is the creation of a secure, tamper-proof registry of property records. Traditional property valuation relies on databases that are often fragmented, outdated, or susceptible to inaccuracies due to human error or deliberate manipulation. Blockchain addresses these issues by maintaining an incorruptible record of property transactions, including ownership history, mortgage data, zoning regulations, and previous valuations. This level of transparency significantly reduces disputes and enhances trust among buyers, sellers, lenders, and regulatory authorities.

Another key application of blockchain in AVMs is with smart contracts, which automate processes such as property transactions, valuation updates, and compliance verifications. These contracts are self-executing and triggered when predefined conditions are met, ensuring that all stakeholders in a valuation process adhere to agreed-upon terms. By integrating smart contracts into AVMs, real estate professionals can streamline workflows, minimise administrative overhead, and improve the accuracy and efficiency of valuations.

Blockchain also plays a vital role in data security and privacy. Since property valuation involves sensitive financial and personal information, securing data against breaches and unauthorised access is crucial. Blockchain's cryptographic protection mechanisms ensure that property-related data remains confidential while still being accessible to authorised parties. This approach mitigates risks associated with data corruption, cyberattacks, and identity fraud, which have become increasingly prevalent in the digital age. Moreover, blockchain technology enhances interoperability between different real estate market participants. Currently, data silos across multiple institutions, such as banks, appraisal firms, government agencies, and real estate platforms, often lead to inefficiencies and inconsistencies in valuation data. Blockchain resolves this issue by providing a unified, decentralised data repository accessible to authorised stakeholders. This seamless data exchange improves the reliability of AVMs and ensures that property valuations reflect real-time market conditions.

The integration of blockchain with AI-driven AVMs also enhances predictive analytics. With access to immutable historical property data, AI models can generate more precise forecasts of property values based on transaction trends, demographic shifts, and macroeconomic indicators. The availability of verified, high-quality data enables machine learning algorithms to refine valuation models continuously, leading to more reliable and accurate property assessments. Furthermore, blockchain enhances regulatory compliance by ensuring that valuation practices adhere to legal and ethical standards. Real estate markets worldwide are subject to stringent regulations, requiring precise documentation and compliance with valuation methodologies. Blockchain's transparent audit trails provide regulators with a clear and immutable record of property valuations, reducing the potential for disputes and legal challenges.

Despite its numerous benefits, blockchain adoption in real estate valuation still faces challenges. High implementation costs, technological complexities, and regulatory uncertainty pose barriers to widespread adoption. However, as technology advances and regulatory frameworks adapt, the integration of blockchain with AVMs is expected to become more prevalent, leading to a more secure, efficient, and transparent valuation ecosystem.

As blockchain technology continues to evolve, its impact on property valuation will become increasingly significant. By ensuring data integrity, enhancing security, and streamlining valuation processes, blockchain is poised to revolutionise the way AVMs operate, ultimately fostering greater trust and efficiency in the real estate market.

Hybrid valuation models

Hybrid valuation models represent a balanced integration of AVMs and traditional human-led appraisals, ensuring a comprehensive approach that leverages both technology and professional expertise. These models harness the speed, efficiency, and data-driven insights of AVMs while incorporating the nuanced judgement, contextual understanding, and critical analysis of human valuers. Hybrid models are particularly valuable in complex property transactions, where automated systems alone may struggle to capture unique property attributes, market sentiment, and local regulatory considerations.

One of the key advantages of hybrid models is their ability to mitigate AVM limitations, such as inaccuracies arising from outdated data, overreliance on algorithmic assumptions, and the inability to assess qualitative property features. By incorporating human oversight, hybrid models allow valuers to interpret AVM-generated outputs, validate property conditions through physical inspections, and adjust valuations to reflect market trends and neighbourhood dynamics more accurately. This collaborative approach enhances reliability and ensures that valuation results align with industry standards and regulatory requirements.

Hybrid models also facilitate greater adaptability in volatile real estate markets. While AVMs efficiently process large datasets to provide instant valuations, human valuers bring interpretive skills that allow them to adjust valuations in response to unexpected market fluctuations, economic downturns, or localised property developments. This adaptability makes hybrid models particularly valuable for lending institutions, property investors, and tax assessors, who require precise and contextually relevant valuation insights.

Additionally, hybrid models support enhanced transparency and accountability. By integrating professional judgement with machine-generated results, they provide a clear audit trail that justifies valuation decisions. This transparency fosters confidence among stakeholders, including buyers, sellers, financial institutions, and regulatory bodies, ensuring that property valuations reflect a well-rounded assessment rather than purely algorithmic determinations. Despite their advantages, implementing hybrid models presents certain challenges. The integration of human expertise with AI-driven AVMs requires efficient workflow management, training for valuation professionals to interpret AVM outputs effectively, and the development of standardised protocols to ensure consistency. Moreover, hybrid models must be continually refined to accommodate advancements in AI, machine learning, and data analytics, ensuring that they remain accurate and effective as real estate markets evolve.

Looking forward, hybrid valuation models are expected to play an increasingly significant role in the real estate industry. As AVMs continue to improve with the integration of blockchain, IoT, and AI-driven enhancements, the synergy between automated systems and human expertise will provide the most reliable, accurate, and compliant property valuations. By striking a balance between technological advancements and professional judgement, hybrid models represent the future of valuation, offering a solution that is both innovative and grounded in industry best practices.

IoT integration

The integration of IoT into property valuation is revolutionising the way data is collected and analysed, providing real-time insights that enhance the accuracy of AVMs. IoT technology enables the continuous monitoring of property conditions through smart sensors embedded in buildings, capturing key data such as temperature, humidity, structural integrity, energy-efficiency, and security system status. This wealth of real-time information significantly improves the reliability of property assessments by offering a dynamic understanding of a property's condition beyond static historical data.

IoT devices facilitate more precise property valuations by detecting maintenance issues, environmental risks, and energy performance, factors that influence market value. For instance, smart sensors can identify moisture accumulation that may indicate mould risks or structural weakening, providing valuers with an early warning system that traditional assessment methods may overlook. Moreover, IoT-enabled energy monitoring systems can assess a property's sustainability and efficiency, factors increasingly relevant in modern valuation practices. These insights allow for more informed pricing models that better reflect the true condition of a property, making AVMs more robust and adaptable to changing real estate dynamics.

The integration of IoT with AVMs also enhances transparency and accountability within the valuation process. By utilising blockchain technology in conjunction with IoT data, property records can be stored securely and immutably, ensuring that property condition reports remain verifiable and tamper-proof. This level of transparency is particularly beneficial in real estate transactions, mortgage lending, and regulatory compliance, where accurate and up-to-date property data is crucial. Additionally, IoT-driven valuation models can help insurers assess risk more effectively by analysing historical sensor data related to fire hazards, flood risks, or mechanical failures, leading to more accurate insurance premium calculations.

Another key benefit of IoT integration in AVMs is its ability to support remote property assessments. In situations where in-person inspections are costly or impractical, IoT sensors provide continuous, real-time property monitoring, allowing valuation professionals to access up-to-date information remotely. This capability is particularly useful in large-scale property portfolio management, where monitoring multiple assets across various locations is essential for investors and financial institutions.

While IoT integration in property valuation presents numerous advantages, challenges remain in its widespread adoption. Issues such as data standardisation, privacy concerns, and technological costs must be addressed to ensure seamless implementation. Nevertheless, as IoT technology advances and adoption increases, its integration with AVMs will become an indispensable tool in the valuation industry, improving accuracy, efficiency, and trust in property assessments.

AR and VR

AR and VR technologies are revolutionising property valuation by providing immersive and interactive experiences that enhance the accuracy and efficiency of real estate assessments. AR overlays digital information onto the real-world environment, allowing valuers and potential buyers to visualise property enhancements, modifications, or planned developments in real time. VR, on the other hand, enables remote property inspections through fully immersive 3D virtual tours, eliminating the need for physical site visits and allowing stakeholders to explore properties from anywhere in the world.

These technologies improve the valuation process by offering detailed visualisations of property interiors, layouts, and potential renovations, reducing ambiguities in property assessments. For instance, VR allows valuers to conduct comprehensive walkthroughs of properties, assess spatial arrangements, and identify potential defects or structural issues without being physically present. AR applications, when integrated with AVMs, can enhance decision-making by providing real-time data overlays on property features, zoning regulations, and market trends.

Furthermore, AR and VR facilitate enhance client engagement and transparency in property transactions. Buyers, investors, and lenders can interact with digital property representations, gaining a better understanding of valuation determinants, projected returns, and asset conditions. This capability is particularly valuable for large-scale commercial properties, real estate developments, and international transactions where on-site inspections may be logistically challenging.

Whilst AR and VR integration in AVMs presents numerous advantages, adoption challenges remain, including technological costs, data accuracy concerns, and the need for standardisation in virtual property representation. However, as these technologies continue to advance and become more accessible, their role in property valuation will become increasingly essential, bridging the gap between automated assessments and human expertise while enhancing accuracy, efficiency, and user experience.

Example 12.1

A real estate company wants to develop an AVM to estimate the market value of residential properties in a specific urban area. The company has collected data on 500 recently sold properties in the area, including some key property characteristics (features). Multiple linear regression will be used to create a model that predicts property value based on these features.

Property	*Sale_price (EUR)*	*Living_area (sq. m.)*	*Bedrooms*	*Bathrooms*	*Age (years)*	*Location_ score*
1	220,000	120	3	2	10	8
2	260,000	150	3	3	5	9
3	200,000	110	2	1	20	7
4	290,000	170	4	3	5	8
5	310,000	180	4	3	2	10
6	190,000	100	2	1	25	6
7	350,000	200	4	4	5	10
8	270,000	140	3	2	8	8
9	240,000	130	3	2	15	7
10	230,000	120	2	2	1	9
...						
...						
500	252,000	129	4	3	3	7

Although the process that will be described here aligns with typical software tools (like Python, R, or Excel), the focus is on the methodology rather than specific programming code.

In what follows, we set out a proposed AVM system with MLR in ten steps:

Step 1: Gather and organise your data

1.1 **Compile a dataset** of 500 recently sold properties. Each record should have:

- **Sale** price (the dependent variable)
- A set of features (independent variables) that might include:
  - Living area (square metres)
  - Bedrooms
  - Bathrooms
  - Age (years since construction)

- Location score (a numeric score or indicator reflecting neighbourhood quality)
- Additional variables as relevant (e.g. garage size, lot size, property condition, proximity to amenities)

1.2 **Standardise data formats** so that each column has a consistent numeric or categorical representation.

Step 2: Clean and pre-process the data

2.1 **Check for missing values** and decide on a strategy:

- *Imputation.* Replacing missing values with the mean or median of that variable.
- *Removal.* Eliminating records with too many missing fields, if they are few.

2.2 **Identify outliers:**

- Extreme sale prices or features might excessively skew the model.
- Decide whether to remove them or transform the data (e.g. using a logarithmic transformation on **Sale price**).

2.3 **Encode categorical variables** (if any):

- If you have features like **Property type** (apartment/house) or **Neighbourhood**, transform them into numeric representations (e.g. dummy variables) so that they can be used in a linear regression.

Step 3: Exploratory data analysis (EDA)

3.1 Summary statistics:

- Calculate mean, median, and standard deviation for each feature and for sale price.

3.2 Correlation checks:

- Produce a correlation matrix to see how strongly each feature correlates with **Sale_price** (and with each other).
- Identify potential multicollinearity (where two or more independent variables are highly correlated).

3.3 Visualisation:

- Plot scatter charts of **Sale_price** vs **Living_area**, **Sale_price** vs **Age**, etc. to gauge the shape of relationships.

Step 4: Split the dataset into training and test sets

4.1 Train set (e.g. 80% of the data, 400 properties) to fit (train) the model.
4.2 Test set (e.g. 20% of the data, 100 properties) to evaluate the model's predictive accuracy on unseen data.

Splitting ensures that you don't evaluate your model's accuracy on the same data used to train it, which can give overly optimistic results.

Step 5: Apply a multiple linear regression model

$$Y = \beta_0 + \beta_1 (Living_{Area}) + \beta_2 (Bedrooms) + \beta_3 (Bathrooms) + \beta_4 (Age) + \beta_5 (Location\_Score) + \ldots + \in$$

Where:

β_0 = intercept (baseline sale price if all other features are zero)
$\beta_1 \ldots \beta_k$ = regression coefficients for the respective features
ϵ = error term (difference between predicted and actual prices)

5.1 Determine model coefficients. Regression algorithm (ordinary least squares [OLS]) finds the set of coefficients (β\betaβ values) that minimise the sum of squared errors between actual and predicted sale prices.

Interpretation

- β_1 (coefficient for **Living_area**). The average change in sale price for each additional square metre, holding other variables constant.
- Negative coefficient values (e.g. for **Age**) imply a decrease in **Sale_price** as that feature's value increases.
- Positive coefficients imply an increase in **Sale_price** as that feature's value increases.

Step 6: Evaluate model performance

6.1 Predict on the test set (the 100 properties not used in training).
6.2 Compare predictions vs actual sale price using performance metrics:

- *MAE.* Average absolute difference between predicted and actual prices.
- *RMSE.* Penalises large errors more heavily.
- *R^2 (coefficient of determination).* Measures the proportion of variance in sale price explained by the model.

6.3 Look for bias:

- Plot predicted vs actual sale prices to see if your model tends to systematically under- or overestimate certain property types or price ranges.

Step 7: Refine the model

If the initial model performance is unsatisfactory, you can improve it by:

7.1 Feature engineering:

- Incorporate new variables, like distance to city centre, recent renovations, views, or accessibility if relevant data is available.

7.2 Non-linear transformations:

- If relationships are non-linear, you can apply polynomial or logarithmic transformations to some features or to **Sale_price** itself.

7.3 Interaction terms:

- Interactions between variables (e.g. the effect of living area might differ in high-end neighbourhoods vs lower-end ones).

7.4 Regularisation:

- Techniques such as Lasso (L1) or Ridge (L2) can help reduce overfitting if you have many correlated predictors.

Step 8: Example of final model (hypothetical coefficients)

After training your MLR model, you might end up with results like:

$$Y = \beta_0 + \beta_1 (Living_{Area}) + \beta_2 (Bedrooms) + \beta_3 (Bathrooms) + \beta_4 (Age) + \beta_5 (Location\_Score) + \ldots + \in = 80{,}000 + 800 \times Living_{Area} + 5{,}000 \times Bedrooms + 7{,}500 \times Bathrooms - 1{,}500 \times Age + 12{,}000 \times Location\_Score = \ldots$$

(Numbers are hypothetical and for illustration only.)

Step 9: Use the model to predict new property values

9.1 Collect feature data for a new property listing (**Living_area**, **Bedrooms**, **Bathrooms**, **Age**, **Location_score**, etc).
9.2 Plug these values into the final equation to get the predicted sale price.
9.3 Report the predicted price and, ideally, the confidence interval or prediction interval to indicate uncertainty.

Step 10: Monitor and update the AVM

An AVM is not a static tool; it requires continuous monitoring and updating to ensure accuracy, reliability, and relevance in an evolving property market. This step ensures that the AVM remains statistically sound, market-responsive, and operationally efficient.

Example of bad practice of AVM (Zillow) and the threats of misplaced applications

Through its 'Zillow Offers' programme, Zillow, a well-known real estate website, purchased homes using its in-house AVM, the *Zestimate*. The business primarily used its algorithm to set purchase prices in the hopes of making money by reselling houses at a slight premium.

What went wrong

Launched in 2018, Zillow's ambitious Zillow Offers initiative was based on the Zestimate, the company's automated valuation model (AVM), which finds, buys, and resells homes at a profit. But by 2021, the scheme had failed, resulting in losses of over $500 million, the termination of 25% of Zillow's employees, and the complete closure of Zillow Offers. The risks of relying too much on AVMs in real estate deals, especially under erratic market conditions, are brought to light by Zillow Offers' collapse.

Zillow made a major error by relying too heavily on its AVM to calculate property prices. The business thought that, by using its algorithm, Zillow could make automated purchasing decisions with little assistance from humans and estimate property prices with high accuracy. Real estate assessments are complicated, though, and AVMs frequently find it difficult to take into consideration particular property attributes, regional market quirks, and outside economic variables.

The primary purpose of Zillow's AVM was to provide online shoppers with home value estimates, not to make crucial investment decisions. In the hopes that it would reliably forecast price increases, Zillow nevertheless utilised it to purchase thousands of homes. Unfortunately, Zillow overpaid for homes that it later found difficult to sell for a profit since the Zestimate often overstated home values.

Key takeaways

Zillow's downfall serves as a warning for any company or investor looking to fully automate real estate transactions without understanding the limitations of AVMs.

Therefore, it is important to note:

- *AVMs are not foolproof.* They should be used as a guide, not an absolute decision-maker.
- *Market conditions change rapidly.* AVMs often fail to adjust in real time.
- *Local expertise is essential.* No algorithm can fully replace on-the-ground knowledge.
- *Real estate transactions require human oversight.* AVMs should complement, not replace, professional valuation and risk analysis (as is mentioned in several IVSC publications).

A synopsis of the popular AVM models and associated practices

To further support the discussion of the development and implementation of AVMs and CAMA systems, the following tables provide a structured analysis of key technological advancements, methodologies, and their respective advantages and challenges. These tables summarise critical aspects of AI-driven valuation models, including ANNs, machine learning techniques, and spatial analysis tools, as well as their impact on valuation accuracy, efficiency, and reliability. By presenting a comparative overview, this section aims to highlight both the strengths and limitations of various AVM methodologies, offering valuable insights into their applicability within different valuation contexts.

In Table 12.1 we set out the advantages and obstacles of applying AI-based technology in AVMs. Table 12.2 sets out the advantages and obstacles of ANN technology in AVM and CAMA stages.

Table 12.1 The advantages and obstacles of applying AI-based technology in AVMs

Advantages/ obstacles	*Technology used*	*Challenges in automated valuation and market analyses*	*AVM stages**	*CAMA steps***
	ANN	**Value assessment**	**8**	**V**
Advantages	• *High accuracy.* ANNs can model complex, non-linear relationships in data, leading to precise property valuations. • *Ability to manage different types of data and relationships.* Making them suitable for diverse property datasets and valuation purposes. • *Ability to process missing data.* ANN can handle missing input data without imputation. • *Learning from large datasets.* Well-suited for large-scale data, which is common in mass property valuation. • *Incorporating uncertainty.* Can effectively integrate uncertainty in predictive models, improving robustness. • *Reduction in computation time.* Once trained, ANNs can make predictions quickly, which is beneficial for mass valuative tasks. • *Adaptation to market volatility.* They can be trained on both historical and continuously updated data, allowing them to adapt to changing market conditions and better predict future property values. • *Identifying hidden patterns.* Uncovering 'hidden' patterns and relationships that may be unnoticeable or difficult to spot with traditional property valuation models.			
Obstacles	• *Computationally intensive training.* Training ANNs requires significant computational resources and time, especially with large datasets. • *Black box nature.* Lack of transparency in how ANNs derive their predictions, making it difficult to interpret and explain the results. • *Lack of understanding, and therefore trust, on the part of beneficiaries.* • *Need for large datasets.* ANNs often require large amounts of data for effective training. • *Overfitting risk.* Prone to overfitting, especially if not properly regularised or if the model is too complex relative to the data. • *Hyperparameter tuning.* Require careful tuning of numerous hyperparameters (e.g., number of layers, learning rate), which can be challenging and time-consuming. • *Lack of training data for specific, unique cases.* This can lead to problems in accurately predicting values. • *Dependency on data quality.* Highly sensitive to the quality of the training data; noisy or biased data can significantly impact performance.			

Table 12.2 The advantages and obstacles of ANN technology in AVM and CAMA stages

Technology integration	*ANN provides opportunities to gain additional benefits or eliminate certain drawbacks inherent in the classical NN approach upon integration with a number of technologies*
1. Random forest (RF)	
Advantages	• *Effectiveness.* For data with many features (square footage, number of rooms, location, etc.), which is typical in AVM tasks. • *Resistance to overfitting.* Random selections of features and observations during tree construction reduce the risk of model overfitting. • *Feature importance assessment.* Allows assessing which property features are most relevant to valuation, making it easier to interpret the model. • *RF does not require scaling or transformation of independent variables.* This simplifies data preparation. • *Ability to process missing data.* RF can handle missing input data without imputation. • *Parallel processing.* The construction of individual trees in the forest can be performed in parallel, speeding up the learning process. • *Scalability.* Efficiently processes large datasets, which are common in real estate applications.
Obstacles	• *Difficulty of interpretation.* Although RF allows the importance of individual features to be assessed, understanding the interactions between hundreds of decision trees in a forest can be difficult and less transparent than with single models. • *Computational requirements.* Building and storing many decision trees in memory requires significant computational resources. • *Processing time.* Many trees in the RF can slow down the valuation process (parallel computing is the solution). • *Need for large training datasets.* RF often requires large amounts of data for effective training.
2. Extreme gradient boosting (XGBoost)	
Advantages	• *High accuracy of prediction.* XGBoost combines predictions of many weak models of decision trees, which often leads to better accuracy of real estate valuation than in the case of individual models. • *Effectiveness for data with many features (square footage, number of rooms, location, etc.)* • *Feature importance assessment.* Allows assessing which property features are most relevant to valuation, making it easier to interpret the model. • *Parallel processing.* The construction of individual trees in the forest can be performed in parallel, speeding up the learning process. • *Scalability.* Efficiently processes large datasets, which are common in real estate applications. • *Model adjustment.* It offers adjustment mechanisms, such as limiting the number of parameters. • *Flexibility.* It can be used with different cost functions and others – adapting the model to the specific needs of real estate valuation. • *Handling missing data.* Can internally manage missing data. • *Regularisation techniques.* To prevent overfitting.

(*Continued*)

Table 12.2 (Continued)

Technology integration	*ANN provides opportunities to gain additional benefits or eliminate certain drawbacks inherent in the classical NN approach upon integration with a number of technologies*
Obstacles	• Resource-intensive. Can be computationally intensive, requiring significant memory and processing power, especially for very large datasets. • *Sensitivity to noisy data.* Although robust, XGBoost can still be affected by noisy or irrelevant features if not properly managed. • *Training time.* While efficient, the training time can still be substantial for very large datasets or complex models. • *The complexity of configuring the model and tuning numerous hyperparameters.* The need to understand and tune many model hyperparameters, such as tree depth, number of estimators, regularisation coefficients, etc. • *Debugging complexity.* Difficulties in identifying and repairing problems in the model.
3. Deep learning neural network (DLNN)	
Advantages	• *High predictive power.* DLNNs can model complex non-linear relationships between features and the target variable, leading to high prediction accuracy. • *Transfer learning.* Deep learning can easily adapt to new data. • *Ease of updating.* Capability to analyse unstructured data and support multimodal inputs. • *The ability to learn complex patterns and non-linear relationships.* In real estate data, this can lead to higher prediction accuracy than linear models or decision trees. • *Automatically learns feature representations from raw input data (e.g. image, text).* Eliminating the need to pre-process it. • *Scalability.* Efficiently processes large datasets, which are common in real estate applications.
Obstacles	• *Increased complexity.* More complex than traditional ANNs, requiring specialised knowledge and computational resources for training and deployment. • *Large data requirements.* Typically require large amounts of labelled data for effective training, which may not always be available in property valuation contexts. • *Computational requirements.* Resource-intensive, requiring significant computational power, often involving the graphics processing unit (GPU), tensor processing unit (TPU), application-specific integrated circuit (ASIC), cloud or edge computing for effective training and inferences. • *Interpretability challenges.* Viewed as 'black box' models, making it difficult to interpret and explain their decisions, which can be problematic in property valuation, where transparency is important. • *Risk of overfitting.* Susceptible to overfitting, especially when trained on small datasets or when the model architecture is overly complex relative to the data.

(Continued)

Table 12.2 (Continued)

Technology integration	*ANN provides opportunities to gain additional benefits or eliminate certain drawbacks inherent in the classical NN approach upon integration with a number of technologies*
4. Spatial neural network (SNN)	
Advantages	• *Localised predictions.* SNNs can provide localised predictions by considering the spatial context of each property, enabling more accurate valuations tailored to specific regions or neighbourhoods (precise micro-localisation analysis). • *Robustness to spatial autocorrelation.* SNNs are robust to spatial autocorrelation – meaning they can capture and account for spatial patterns and trends in property values. • *Ability to learn and combine complex spatial patterns.* Includes relationships in property data, such as location, neighbourhood, infrastructure availability, etc. • *Automatic extraction of relevant spatial features.* From raw data, such images or maps. • *Potentially higher accuracy of property value.* Prediction due to dedicated consideration of spatial context. • *Can incorporate concepts.* From graph neural networks (GNNs) and convolutional neural networks (CNNs).
Obstacles	• *Complexity of spatial data.* Handling spatial data requires specialised techniques and pre-processing steps, adding complexity to the model development process. • *Data availability and quality.* Spatial datasets may be limited or of varying quality, which can impact the effectiveness of SNNs and require careful data pre-processing and validation. • *Interpretability challenges.* Like other neural network models, SNNs may lack interpretability, making it difficult to explain how spatial features contribute to property valuations. • *Computationally intensive.* Processing spatial data and training SNNs can be computationally intensive, requiring adequate computational resources and time for model training and optimisation. • *Risk of overfitting.* On training data if the model is not properly regularised. • *Inclusion of spatial data additionally introduces further elements of uncertainty.* Quality, effectiveness of results due to the difficulty of integrating spatial data with other numerical data – assessing their accuracy in relation to these other data.
5. Support vector machines (SVM)	
Advantages	• *Effective in high-dimensional spaces.* Perform well in high-dimensional feature spaces, making them suitable for property valuation tasks with many input variables. • *Memory efficiency.* Only use a subset of training data points (support vectors) to define the decision boundary, resulting in memory efficiency, particularly with large datasets. • *Robustness to overfitting.* Contain effective regularisation techniques (e.g. kernel tricks, soft margin) that help prevent overfitting, ensuring good generalisation performance. • *Explicit control over trade-offs.* Allow explicit control over the trade-off between margin maximisation and classification error, providing flexibility in model tuning.

(Continued)

Table 12.2 (Continued)

Technology integration	*ANN provides opportunities to gain additional benefits or eliminate certain drawbacks inherent in the classical NN approach upon integration with a number of technologies*
	• *Global optimisation.* Solve a convex optimisation problem, guaranteeing convergence to the global optimum, which ensures reliable and consistent result. • *Interpretability.* Provide relatively simple decision functions, making them more interpretable compared to the complex neural network model. • *Stability of results.* Important with frequent updates of property valuations. • *Transparency.* Simple, linearly defined decision boundaries. • *Efficiency.* In detecting market anomalies and outliers.
Obstacles	• *Limited scalability.* May become computationally intensive and memory-demanding with very large datasets, especially when using non-linear kernels. SVMs are effective when working with small- and medium-sized datasets, poor performance with very large datasets. • *Difficulty handling noisy data.* Sensitive to noisy data and outliers, which can affect the positioning of the decision boundary and degrade performance. • *Complexity in kernel selection.* Choosing an appropriate kernel function and its parameters can be challenging and may require expert knowledge or extensive experimentation. • *Inefficient with unbalanced datasets.* May produce biased results when dealing with highly imbalanced datasets, where one class dominates the other. • *Less effective with non-linear relationships.* Whilst they can model non-linear relationships using kernel tricks, they may not perform as well as neural networks in capturing complex non-linear patterns. • *Limited probabilistic outputs.* SVMs originally provide binary classification; obtaining probabilistic outputs requires additional calibration, which can be less straightforward compared to neural networks. • *Need for careful normalisation of data.* • *Sensitivity to data scaling.* Input data should be scaled appropriately to avoid dominance of certain features. • *Non-resilience to sequential data, such as time-dependence, cannot be considered.*
6. General regression neural networks (GRNN)	
Advantages	• *Fast training and prediction.* They have a simple structure and require minimal tuning, leading to faster training and prediction times compared to more complex neural networks. Rapid learning in a single pass-through training data. • *Non-parametric regression.* Perform non-parametric regression – meaning, they can capture complex relationships between input and output variables without assuming a specific functional form, which is advantageous in property valuation tasks with non-linear relationships. • *Adaptability to new data.* Can quickly adapt to new data by updating the kernel bandwidth, enabling continuous learning and adaptation to change property market conditions. • *Variance reduction.* May reduce the variance of results, leading to more stable valuations.

(Continued)

Table 12.2 (Continued)

Technology integration	*ANN provides opportunities to gain additional benefits or eliminate certain drawbacks inherent in the classical NN approach upon integration with a number of technologies*
Obstacles	• *Limited interpretability.* Like other neural network models, GRNNs may lack interpretability, making it difficult to understand how input features contribute to property valuations. • *Memory inefficiency.* As they store training data in memory and perform fast inference using a radial basis function (RBF) kernel, making them memory-efficient, especially with large datasets. • *Dependency on kernel bandwidth.* Sensitive to the choice of kernel bandwidth, which requires careful tuning and may impact prediction accuracy. • *Scalability.* Efficient with small- to medium-sized datasets; they may become computationally intensive and memory-demanding with very large datasets. • *Requires data normalisation.* Like other neural network methods, require appropriate normalisation of the input data. • *Difficulty with high-dimensional data.* May struggle to handle high-dimensional data efficiently, as the number of required training data points increases exponentially with the number of dimensions. • *Risk of overfitting.* Can be prone to overfitting, especially if the training dataset is small or noisy, necessitating appropriate regularisation techniques and validation strategies. This risk is especially true with small datasets. • *Less flexible than DLNN.* • *High calculation costs.* For large datasets.

In all the following tables, we refer to the various stages and steps in the AVM and CAMA approaches, as defined in what follows:

*The main stages of AVM procedures are:

1. Information definition
2. Data collection
3. Database creation
4. Data verification and selection
5. Feature significance indication
6. Homogeneity market definition
7. Comparable properties set definition
8. Initial value determination
9. Model calibration
10. Model testing
11. Quality control
12. Indication of representative sets of assessed properties
13. Value determination of assessed properties

**According to the IPTI Workshop 2024, the main steps of the CAMA system are:

I. Definition of appraisal assignment
II. Data collection
III. Data review and validation
IV. Exploratory data review and analysis
V. Model development and calibration
VI. Model testing and quality control
VII. Model application and value review

Behavioural perspectives on the utility of the previous solution

Humans have a limited capacity to analyse vast datasets, whereas ANNs can quickly and accurately process this data, identifying patterns and trends that may elude human analysts. Human minds often struggle with the analysis of complex, non-linear relationships. ANNs are designed to effectively handle non-linearity, leading to more precise property valuations. Moreover, humans learn slowly and on a smaller scale. ANNs can rapidly learn from large sets of historical data, continuously improving their forecasts and adapting to changing market conditions. Additionally, humans are susceptible to cognitive biases and fatigue. ANNs operate impartially and without fatigue, reducing the risk of errors, resulting in more reliable valuations. These capabilities underscore the importance of integrating ANNs into mass property valuation processes, providing accuracy, speed, and adaptability, which are difficult to achieve with traditional methods and human analysts alone.

Table 12.3 The advantages and obstacles of VR and ET and CVT in both AVM and CAMA stages

Advantages/ obstacles	*Technology used*	*Challenges in automated valuation and market analyses*	*AVM stages**	*CAMA steps***
	VR, eye tracking (ET), and computer vision technology (CVT)	**Collection of the information**	**1** **2**	**II**
Advantages	• *Inclusion of behavioural factors.* Human perception analyses. • *Property features fusion measurement.* In an environment close to reality. • *The multisensory human perception.* • *Objective lab analyses.* • *Pattern extraction.* Based on complex dataset. • *Time and cost savings.* VR eliminates the need for physical visits to properties; CV, in conjunction with convolutional neural networks (CNNs) and graph neural networks (GNNs), provides automatic analysis of images; and video reduces the need for manual review of material.			
Obstacles	• *High cost of technology and equipment purchase.* • *Health state–based limitations.* For VR use. • *Limiting access to infrastructure.* For example, VR, eye tracking, and their standardisation. • *Divergent interpretations.* Of the determinants of people's emotional responses, for example, based on pupil diameter. • *Need for expertise knowledge.* Implementing and operating these technologies require expertise in machine learning and image processing and psychology in addition to knowledge of property valuation and IT.			

Behavioural perspective on the utility of the preceding solution

The integration of VR with emotions tracking (e.g. pupil tracking) and computer vision (CV) technologies significantly enhances the collection and analysis of information in the real estate market, minimising human limitations. VR with pupil tracking enables immersive property inspections, providing a realistic spatial experience. By tracking pupil movements, this technology reveals user preferences and areas of interest that may not be visible in traditional surveys. CV technologies automatically assess visual aspects of properties, precisely measuring spaces and evaluating their conditions. CV can also detect potential issues that may escape human attention, leading to more accurate and reliable assessments. This technology reduces human errors and biases, ensuring that the gathered data is comprehensive, precise, and useful. As a result, the property valuation process becomes more efficient and reliable, better adapting to dynamic market conditions.

Table 12.4 The advantages and obstacles of genetic algorithms in AVM and CAMA stages

Advantages/ obstacles	*Technology used*	*Challenges in automated valuation and market analyses*	*AVM stages**	*CAMA steps***
	Genetic algorithm	**Collection of the information**	**1** **2**	**II**
Advantages	• *Ability to find the 'best overall solution': global extreme.* Global multi-criteria optimisation. • *Big data processing.* • *Diverse size and form.* Of coding the data. • *Synergistic features.* Analysis results. • *No need to define an objective function.* • *Numerical solutions can be performed in a parallel manner.* • *Resistance to data noise.* • *Adaptability.* Can adapt to changing market conditions. • *A simplified approach to non-linear problems that are difficult to model with traditional methods.* • *Can be applied to various stages of AVM.* From defining information to data collection and model calibration.			
Obstacles	• *Finding quasi-optimal solutions.* Often, finding solutions that are only good enough, that is, quasi-optimal. • *Risk of overtraining.* • *Problems with convergence.* Of analysis results for very complex optimisation problems. • *Results may differ significantly from those obtained by classical AVM methods.* Which may require additional analysis and verification. • *Time-consuming.* Especially with large datasets and complex problems. May require considerable computing power. • *Complex implementation.* Requiring specialist knowledge. • *Sensitivity to initiating parameters.* Results can be strongly dependent on the choice of parameters, such as population size, mutation rate, etc. • *Need for calibration.* Often require calibration and adjustment to specific problems, which can be time-consuming and non-automatic.			

Behavioural perspective on the utility of the preceding solution

Genetic algorithms stand out due to their construction method based on principles of natural selection, which introduce a new quality to real estate market analysis. Traditional human analytical approaches often exhibit a susceptibility to errors and limitations stemming from personal preferences or biases. In the case of genetic algorithms, the process of solution selection and refinement follows the principles of evolution, thereby eliminating subjective factors and potentially leading to the discovery of optimal solutions in a manner challenging to achieve for the human mind. By leveraging this methodology, genetic algorithms can detect non-linear dependencies and account for various factors influencing property value in a more objective and comprehensive manner. Additionally, their ability to iteratively refine solutions allows for adaptation to changing market conditions and the detection of patterns that may be overlooked in traditional analyses. Thus, genetic algorithms introduce a novel perspective to real estate market analysis by mitigating human errors and biases and facilitating a more objective approach to valuation and analysis.

Table 12.5 The advantages and obstacles of RST and SOM in AVM and CAMA stages

Advantages/ obstacles	*Technology used*	*Challenges in automated valuation and market analyses*	*AVM stages**	*CAMA steps***
	Rough set theory (RST) and self-organising maps (SOM)	**Data selection/ preference indication**	**2**	**II**
Advantages	• *Synergistic features.* Provide enhanced analysis possibility. • *RST effectively identifies and reduces irrelevant features.* Improving the quality of input data. • *A combination of RST and SOM.* Enables the detection of complex, non-linear patterns in real estate data. • *Highly interpretable.* RST identifies features, and SOM spatially visualises the results of the analysis (identification of patterns and anomalies). • *Increased transparency of analysis.*			
Obstacles	• *Require larger datasets.* For comprehensive analysis. • *Results may be unpredictable.* Dependent on initiating assumptions. • *Need for careful tuning of parameters.* • *Scalability.* • *High operating memory requirements.* • *Risk of overtraining.* Especially in relation to SOM. • *RST is sensitive to noise in the data.* Requires pre-processing.			

Behavioural perspective on the utility of the preceding solution

This approach effectively handles the imprecision and subjectivity inherent in data selection and preference indication. Rough set theory helps manage vague and uncertain data without requiring additional information, while self-organising maps facilitate the visualisation and clustering of complex data, mirroring human cognitive processes. Together, they provide a robust framework that accommodates human fallibility, offering a more intuitive and flexible method for analysing and interpreting data.

Table 12.6 The advantages and obstacles of ERT and N in AVM and CAMA stages

Advantages/ obstacles	*Technology used*	*Challenges in automated valuation and market analyses*	*AVM stages**	*CAMA steps***
	Emotion recognition technology (ERT) and neural network (NN)	**Attributes significance indication**	**2** **5**	**IV**
Advantages	• *Objectivity.* Of the obtained results (low probability of manipulation). • *Higher flexibility.* In response to changes in trends and market preferences. • *Speed of data acquisition.* With indirect (remote) contact with the respondent. • *Precision.* In understanding customer preferences. • *Adaptability.* Rapid adaptation of the prioritisation of features in relation to the changing preferences of real estate market participants. • *Faster identification of relevant features.* In real estate market analysis.			
Obstacles	• *Cost-effectiveness and technical accessibility.* • *Big data processing.* Calculation requirements. • *Legal and ethical constraints.* Safeguarding requirements for sensitive data. • *Interpretation difficulties.* In the results and their accuracy.			

Behavioural perspective on the utility of the preceding solution

Humans are susceptible to subjective judgements and cognitive biases that can affect data interpretation. Emotion recognition technology eliminates these subjective elements by providing an objective analysis of the emotional responses of potential buyers. This allows for the identification of hidden preferences, used to gauge market sentiment and consumer preferences, thereby increasing the accuracy of property valuations through the analysis of emotional reactions to various property attributes. Furthermore, this technology enables faster adaptation to changing market trends and reduces the time and costs associated with market research. As a result, emotion recognition technology not only enhances the precision and objectivity of assessments but also provides deeper and more valuable insights.

Table 12.7 The advantages and obstacles of RST and RG in AVM and CAMA stages

Advantages/ obstacles	*Technology used*	*Challenges in automated valuation and market analyses*	*AVM stages**	*CAMA steps***
	RST and robust geo-estimation (RG)	**Homogeneity market indicating**	**6** **7**	**III** **IV**
Advantages	• *No need to define precise boundary.* Markets a priori. • *No need to define weights.* Significance of features a priori. • *The area is interpreted as an adjacency and mutual phenomenon.* Described by the features developed in the buffer mode. • *Indicated homogeneous areas.* Understood as the indiscernibility areas that consider specificity of the properties and markets related to their fundamental differentiation. • *High flexibility and scalability.* Of the algorithm related to the boundary conditions of the model (k-coefficient and entropy weight). • *Automated selection of comparable properties.* • *Reducing human subjectivity in market area indication.*			
Obstacles	• *Adjustable comparability.* Coefficient and minimum significance weight based on analytics. • *Probability of obtaining large groups of non-characters with a small number of objects.* • *Manual supervision strongly required for optimal results obtained in an iterative procedure.* • *Objectivity.* Of the obtained results, depends on the amount of information contained in the database. • *High computational requirements.* • *Implementation complexity.*			

Behavioural perspective on the utility of the preceding solution

Humans often struggle with interpreting large, intricate datasets that may contain ambiguities and uncertainties. Rough set technology enables the analysis of such data more precisely, eliminating subjective errors and cognitive biases. This allows for a more accurate determination of market homogeneity by identifying patterns and relationships that may be challenging for human analysts to detect. Robust geo-estimation further supports this process by providing tools for accurately assessing spatial parameters of properties. This enables more reliable and precise estimations that consider both locational and spatial market characteristics. This technology minimises errors stemming from human interpretation and allows for better adaptation to dynamic changes in the real estate market. As a result, the application of rough set technology and robust geo-estimation not only enhances the precision and objectivity of assessments but also provides deeper and more valuable insights.

Table 12.8 The advantages and obstacles of RST and Kohonen NN in AVM and CAMA stages

Advantages/ obstacles	*Technology used*	*Challenges in automated valuation and market analyses*	*AVM stages**	*CAMA steps***
	RST and Kohonen NN (SOM)	**Representative indication**	**7 8 10**	**IV V**
Advantages	• *Particularly useful.* For limited transaction markets. • *Tolerant.* Of attribute variability and missing data. • *Supports any functional correlation.* Between attributes and price. • *Does not require preliminary analysis.* • *No limits.* Related to the number of transactions and their attributes. • *Identification of hidden correlations.* In high-dimensional datasets. • *Precise selection of comparable properties.* RST identifies key features; SOM clusters properties based on these features. • *Automatic data analysis and segmentation.* • *Effective model calibration.* RST eliminates unobserved variables. SOM allows continuous learning from new data. • *High transparency.* Of the mathematical apparatus of the model. • *Scalability.* RST effectively processes large sets of information.			
Obstacles	• *Multiple possible alternative choices.* For the representative decisional rule, posing the challenge of having to select one. • *Conditional attributes of properties can be repeatedly duplicated.* By multiple decisional rules. • *High calculation requirements.* For both RST and SOM. • *Risk of overtraining.* Especially the SOM component. • *Difficult to adapt to different markets.* RST for different markets requires separate models. • *Sensitive.* To parameter tuning.			
	RST and fuzzy logic (FL)	**Insufficient number of comparable properties**	**7 8**	**V**
Advantages	• *Effective in the absence of comprehensive information on the property market.* • *Effective in the inaccurate and 'fuzzy' character of real estate data.* • *Effective in complex methods of data description.* Differences in the scale of attribute description. • *Absence of the restrictive assumption.* Of linearity in data relation. • *Absence of homogenous functional dependencies.* Between real estate attributes.			
Obstacles	• *High design requirements.* Of both RST and FL • *Implementation complexity.* RST as well as FL requires specialist knowledge. • *Risk of overtraining.* FL models can be susceptible to overtraining.			

Behavioural perspective on the utility of the preceding solution

This approach accommodates the inherent imprecision and subjectivity in real estate valuations. It allows for the modelling of complex decision-making processes where crisp boundaries between valuation criteria are unsuitable from a rational and realistic perspective. By handling degrees of truth rather than binary true/false values, fuzzy logic provides a more nuanced and flexible approach to property valuation, reflecting the real-world complexities and subjective judgements that often influence market values.

Table 12.9 The advantages and obstacles of CS and MLT in AVM and CAMA stages

Advantages/ obstacles	*Technology used*	*Challenges in automated valuation and market analyses*	*AVM stages**	*CAMA steps***
	Cognitive systems (CS) and machine learning technologies (MLT)	**Preserve the original data structure/avoid decreasing the information capacity**	**1** **2** **3**	***II***
Advantages	• *Automation.* The system automatically identifies similar properties, significantly speeding up the valuation process and reducing manual effort. • *Scalability.* Most ML algorithms are scalable, allowing the system to handle increasing amounts of data without a significant rise in computational power demands. • *Relationship detection.* Adept at identifying complex relationships that are difficult to define a priori. • *Handling incomplete data.* Can manage incomplete or inconsistently recorded data, ensuring robust analysis even with imperfect datasets. • *Multi-threaded search.* ML performs multi-threaded and multi-epoch searches to find optimal solutions efficiently. • *Accuracy control.* A well-trained ML model ensures precise valuations, leading to more accurate property comparisons. • *Automatic parameter adjustment.* Adjusts model parameters automatically based on data and the specific analysis purpose. • Model analysis. Evaluates multiple models and their parameters to find the optimal solution. • *Continuous retraining.* Models are continuously updated with new data, allowing adaptation to new requirements, emerging data, and changing trends.			
Obstacles	• *Complex development.* Building, training, and fine-tuning machine learning models require specialised knowledge and expertise. • *Data requirements.* Constructing a flexible and precise model necessitates a large, semantically diverse dataset, which can be difficult to obtain. • *Overfitting risk.* Models may become overly fitted to training data, reducing their ability to generalise and predict outcomes accurately, making them sensitive to noise and nuances. • *Interpretability.* Complex models can be difficult to interpret, complicating the explanation of decisions made by the ML. • *Resource-intensive.* Training and maintaining effective models demand significant computational resources and time.			

Source: Renigier-Biłozor, M., Ache, P., Dimopoulos, T., & Janowski, A. (2025).

Behavioural perspective on the utility of the preceding solution

The utility of this technology lies in its ability to preserve the original data structure and avoid diminishing information capacity, particularly due to human limitations or inadequacies. Cognitive systems and machine learning (ML) technologies excel in processing vast amounts of data with minimal human intervention, ensuring that the integrity and richness of the dataset are maintained. By leveraging these technologies, we can mitigate the risk of data loss or distortion caused by human error or insufficient analysis. Additionally, cognitive systems and ML can uncover patterns and insights within the data that may elude human perception, enhancing the overall quality and depth of analysis.

Questions

1. What is the basic concept of mass appraisal, and how does it differ from traditional single-property valuation?
2. Why have mass appraisals and AVMs become increasingly important in recent years?
3. Which types of stakeholders tend to benefit the most from large-scale automated property valuations?
4. Contrast CAMA and AVM. What core characteristics do they share, and how do they differ?
5. Why is data quality so critical for the success of an AVM?
6. What kinds of data are typically fed into an AVM, and which data-related challenges might undermine a model's reliability?
7. How do professional valuation standards (for example, IVS 105, RICS guidance, TEGoVA) address automated valuation methods?
8. In what ways are valuer input and professional judgement still required for AVM outputs to meet standard-compliant valuations?
9. What are the main advantages and limitations of mass appraisals as compared to single-property, traditional appraisal approaches?
10. What ethical challenges or concerns could arise from the increasing use of AVMs for large-scale property valuations?
11. How might regulators or professional bodies mitigate these concerns?
12. The chapter highlights a movement toward AI-driven AVMs. In your view, what emerging technologies or data sources might further transform automated valuation in the next five to ten years?
13. Imagine you are tasked with implementing an AVM for a municipal tax authority. What practical steps would you recommend to ensure both accuracy and public acceptance?
14. AVMs can generate quick results at lower costs. However, what arguments might a professional valuer present if they believe human expertise cannot be replaced?
15. How would you weigh speed and cost saving versus the potential for systematic error or bias?

References/Further reading

Abidoye, R. B., & Chan, C. A. P. (2016). Research Trend of the Application of Artificial Neural Network in Property Valuation. *33rd CIB W78 Conference 2016, November, 1–9.*

Amato, M. (2017). *Advances in Automated Valuation Modeling* (Vol. 86). https://doi.org/10.1007/978-3-319-49746-4

Borst, R. A. (1991). Artificial neural networks: The next modelling/calibration technology for the assessment community. *Property Tax Journal*, 10(1), 69–94.

Clow, Frederick R. (1896). The St. Paul method of assessing real estate. *Journal of Political Economy*, 5(1), 71–80.

Dimopoulos, T. (2020). Critical Investigation of Novel Computational Techniques for Automated Valuations of Real Estate Properties in Cyprus. *PhD Thesis*. https://hdl.handle.net/20.500.14279/18838

Dimopoulos, T, Tyralis, H., Bakas, N. P., & Hadjimitsis, D. (2018). Accuracy measurement of Random Forests and Linear Regression for mass appraisal models that estimate the prices of residential apartments in Nicosia, Cyprus. *Advances in Geosciences*, 45, 377–382. https://doi.org/10.5194/adgeo-45-377-2018

Do, A. Q., & Grudnitski, G. (1992). A neural network approach to residential property appraisal. *The Real Estate Appraiser*, 58(3), 38–45.

Eckert, J. K., Gloudemans, R. J., & Almy, R. R. (Eds.). (1990). *Property Appraisal and Assessment Administration*. International Association of Assessing Officers.

European AVM Alliance. (n.d.). *Glossary of Terms & Definitions*.

European Commission. (2021). *European Commission's Proposal for the AI Act*. https://eur-lex.europa.eu/legal-content/EN/TXT/?uri=CELEX%3A52021PC0206

Europäische Kommission, High level Expert Group on Artificial Intelligence. (2018). *A Definition of AI: Main Capabilities and Scientific Disciplines*.

européenne. Commission européenne, U., & EUROSTAT. (2013). *Taxation Trends in the European Union: Data for the EU Member States, Iceland and Norway*. Office for Official Publications of the European Communities.

Frey, C. B., & Osborne, M. A. (2017). The future of employment: How susceptible are jobs to computerisation? *Technological Forecasting and Social Change*, 114, 254–280. https://doi.org/10.1016/j.techfore.2016.08.019

International Association of Assessing Officers. (2014). Guidance on international mass appraisal and related tax policy. *Journal of Property Tax Assessment & Administration*, 11(1), 5–33. https://search.proquest.com/docview/1524709595?accountid=14775%0Ahttp://openurl.ac.uk/athens:uls/?genre=article&issn=13571419&volume=11&issue=1&title=Journal+of+Property+Tax+Assessment+%26+Administration&spage=5&date=2014-01-01&atitle=Guidance+on+Internat

International Association of Assessing Officers. (2017). *Standard on Mass Appraisal of Real Property International Association of Assessing Officers*. www.iaao.org

International Association of Assessing Officers. (2018). *Standard on Automated Valuation Models*. www.iaao.org/media/standards/AVM_STANDARD_2018.pdf

International Federation of Surveyors (FIG), Commission 9 Valuation and the Management of Real Estates (expected publication at the end of 2024), *FIG- Viewpoint on Transparency in Real Estate Marktets*.

International Valuation Standards Council (IVSC). (2022). *Perspectives Paper: Automated Valuation Models and Residential Valuations*.

Kauko, T., & Amato, M. (2008). *Mass Appraisal Methods: An International Perspective for Property Valuers* (T. Kauko & D. Maurizio, Eds.), 1st ed. Wiley-Blackwell.

Labropoulos, T. (2013). *Αυτοματοποίηση μαζικών εκτιμήσεων αξιών ακινήτων για τον Ελληνικό χώρο, Διδακτορική Διατριβή*. National Technical University of Athens.

Limsombunchai, V. (2004). House Price Prediction: Hedonic Price Model vs. Artificial Neural Network. *New Zealand Agricultural and Resource Economics Society Conference, 25–26*.

McCluskey, W. J., & Borst, R. A. (2011). Detecting and validating residential housing submarkets: A geostatistical approach for use in mass appraisal. *International Journal of Housing Markets and Analysis*, 4(3), 290–318. https://doi.org/10.1108/17538271111153040

McCluskey, W. J., Cornia, G. C., Walters, L. C., Kitchen, H., & Slack, E. (2013). *A Primer on Property Tax Administration and Policy Edited by* (W. J. Mccluskey, G. C. Cornia, L. C. Walters, H. Kitchen, & E. Slack, Eds.)). Blackwell Publishing Ltd.

McCluskey, W. J., Deddis, W., Mannis, A., McBurney, D., & Borst, R. (1997). Interactive application of computer assisted mass appraisal and geographic information systems. *Journal of Property Valuation & Investment*, 15(5), 448–465.

McCulloch, W. S., & Pitts, W. (1943). A logical calculus of the ideas immanent in nervous activity. In R. S. Wurman (Ed.), *History of Information*. Retrieved June 12, 2025, from https://www.historyofinformation.com/detail.php?entryid=782

Mooya, M. M. (2017). Automated valuation models and economic theory. In *Advances in Automated Valuation Modeling* (pp. 33–57). Springer International Publishing. https://doi.org/10.1007/978-3-319-49746-4_3

Pagourtzi, E., Assimakopoulos, V., Hatzichristos, T., & French, N. (2003). Real estate appraisal: A review of valuation methods. *Journal of Property Investment & Finance*, 21(4), 383–401. https://doi.org/10.1108/14635780310483656

Renigier-Biłozor, M., Ache, P., Dimopoulos, T., & Janowski, A. (2025). The future of property valuation: AVMs and CAMAs in the era of AI. *Economics and Environment* (Article submitted and under review).

Renigier-Biłozor, M., Chmielewska, A., Walacik, M., Janowski, A., & Lepkova, N. (2021). Genetic algorithm application for real estate market analysis in the uncertainty conditions. *Journal of Housing and the Built Environment*, 36, 1629–1670. https://doi.org/10.1007/s10901-020-09815-

Renigier-Biłozor, M., Janowski, A., & d'Amato, M. (2019). Automated valuation model based on Fuzzy and Rough Set Theory for real estate market with insufficient source data. *Land Use Policy*, 78, 104–115. https://doi.org/10.1016/j.landusepol.2018.06.012

Renshow, E. F. (1958). Scientific appraisal. *National Tax Journal*, 11(4), 314–322.

RICS. (2013). *Automated Valuation Models*, 1st ed. RICS Information Paper.

Tabales, J. N., Carmona, F. R., & y Ocerin, J. M. (2013). Implicit prices in urban real estate valuation. *Revista de La Construcción*, 12(2), 116–126. https://doi.org/10.4067/s0718-915x2013000200009

Thériault, M., Des Rosiers, F., Villeneuve, P. and Kestens, Y. (2003), Modelling interactions of location with specific value of housing attributes, *Property Management*, 21(1), 25–62. https://doi.org/10.1108/02637470310464472

Chapter 13

Contemporary impacts on markets and values

Introduction

Modern-day valuations sit in a world of uncertainty and rapid change. In this chapter, we discuss some of the factors that may require greater consideration by valuers, including the movement to more and more sustainable development, occupiers' need for more productive environments, and the introduction of new property products, such as flexible workspaces.

The rationale for keeping and extending this chapter is to illustrate that the valuation process must be dynamic and respond to change in the many frameworks and market forces within which a property valuation sits. A valuer should be aware of such change(s) in the business environment and how they drive the changing requirements of property occupiers and investors, and therefore the associated risk and returns. If a property is no longer providing what customers want – that is, it has functional obsolescence – it is unlikely to find tenants, and the cash flow will be compromised. In a rapidly changing world of work, especially following the disruptive impacts of post-COVID-19 and hybrid working, understanding these rapid shifts has shown to be critical.

In many ways, a valuer is working implicitly within a classic PESTEL business environment framework.

So in this chapter, we examine different aspects of this framework, for example:

- The importance of sustainability and its influence on value is both a **political** and **environmental** driver.
- The change of work patterns and its influence on the design of office buildings for greater productivity is an **economic** (business) change factor.
- The use of artificial intelligence and mass valuation is a **technological** driver (considered in depth in Chapter 12).
- The development of new property products such as flexible workspaces and co-living is a **social** driver.
- And finally, we explore **political** and **legal** factors by examining the local considerations a valuer must engage with using a diverse selection of three very different countries kindly provided by eminent local practitioners.

Whilst a continuing commitment to global valuation standards and the introduction of global measurement standards are supporting greater consistency, it is important to recognise that local factors, practice, and procedures will be significant elements of valuations in different countries.

DOI: 10.1201/9781003490357-16

The strap line 'Think Global, Act Local' is used in many contexts by companies such as JLL, and we want to illustrate this need to balance the consistency brought about by global standards for valuation practice and measurement with the awareness of how local factors influence the valuation process and procedures. We do this by providing a brief review of local factors for four very different countries: China, the UAE, Cyprus, and Greece.

ESG in property valuation

The property industry and the professions that support it have a huge role to play in reducing carbon emissions and maximising energy efficiencies as the world struggles to tackle climate change. According to the UK Environmental Audit Committee (House of Commons, 2022), the built environment accounts for 25% of the UK's total carbon footprint, and therefore, the property sector has a huge role in achieving the national emission reduction targets. This is likely to be similar to impacts in other countries.

According to IVSC (2024), *ESG* refers to the criteria used to assess the impact of sustainability and ethical practices on a company, asset, or liability, considering financial performance and operations. ESG encompasses three key pillars – environmental, social, and governance – which collectively influence market performance and broader societal outcomes.

Sustainability in a property context has rapidly matured in level of awareness, and it demands attention from valuers in most, if not all, valuations. Legislation and practice have driven the embedding of sustainability to such a degree that to ignore it is extremely dangerous, if not negligent.

In September 2009, the RICS published a valuation information paper (13) called *Sustainability and Commercial Property Valuation* to help develop a standardised approach to include sustainability in the valuation process.

Section 2.2 stated:

> In forming a judgment as to the extent to which sustainability factors will impact on valuation, the attitudes of those likely to be in the market for the subject property, both as occupier and owner, are relevant. Often it may be difficult for such factors to be quantified; nevertheless it may fall within the remit of the valuer to provide some qualitative comments.

Section 2.3 stated:

> Valuers will need to develop knowledge of such possible implications to advise building owners and occupiers, particularly as governments introduce legislation aimed at mitigating the impacts of climate change and reducing carbon emissions. Indeed, it is the responsibility of the valuer to be aware of these trends and be knowledgeable about how they feed through to rental and capital values.

This made it clear that valuers must be aware of changing legislation on sustainability issues if they are to advise clients competently and consider the potential impact upon valuation. The bottom line is that clients need to know whether the building they are buying today will be hit hard by changes to sustainability legislation and market sentiment in the future. In terms of the income approach, this will be in respect of how future cash flows may be impacted by such changes.

By 2013, when the second edition of the guidance note had been produced, things had moved on to recognise the growing evidence of the impact of sustainability.

Section 1.4 states:

> Recent empirical research has shown that many national property markets increasingly distinguish and differentiate between buildings that exhibit different sustainability-related building features and associated performance. This is recognised not only by responsible investors and their professional advisers, but also by the wider investment community. There is much greater awareness of the risks to value of building obsolescence in relation to climate change, energy shortages and price volatility, occupier demand, their occupation costs and corporate social responsibility objectives. It is also the case that many owner-occupiers are developing awareness of how sustainability impacts on their decision-making processes throughout their supply chains, including their real estate purchasing decisions.

However, the advice to valuers remains cautious and conservative, differentiating between valuations for financial reporting and investment valuations.

As Section 2.2 states:

> Where a market value or fair value, for use in financial statements prepared under IFRS, is being prepared, only those factors that can be evidenced from analysis of the market should be included. However, where a valuer is providing an investment value (commonly known as 'worth'), factors not yet reflected in market value, but that may influence an investor's decision-making, should be considered and a view taken as to whether they are relevant over the proposed holding period. Similarly, where an estimate of fair value is made that differs from the market value, it may be relevant to incorporate consideration of sustainability factors that are not yet evidenced through market transactions.

Two examples underpin the reality of the importance of being aware of the impact of sustainability regardless of the type of valuation:

- MEES in the UK regulate the energy performance of privately rented properties. Since April 2018, new leases and renewals require an EPC rating of at least E, with the rule extending to all existing commercial leases from April 2023. The UK government plans to tighten standards further, aiming for a minimum EPC rating of C by 2027 and B by 2030. This creates an unprecedented level of risk to the cash flow of substandard properties and clearly impacts drastically on value.
- In many countries, public occupiers are required to only take leases of buildings with specified minimum environmental credentials – in the UK, this relates to BREEAM, and in the United States, LEED.

Some UK examples include thus:

- The UK's GPA, set up to create a transformed, shared, sustainable, and value-for-money government estate that supports civil servants working productively in every nation and region of the UK, has now formally established several red lines:

Project type	*BREEAM*	*NABERS UK*	*EPC*
New build	>Excellent	5* >	A
Major refurbishment	>Very good	4* >	B
Minor refurbishment	Out of scope	Out of scope	Out of scope

- The Welsh Assembly requires BREEAM Very Good for all non-residential buildings with a floorspace of 1,000 m^2 or more. In addition, they must achieve the minimum standards for Excellent under the code of Ene1, "Reduction of CO_2 emissions credit" issue. (Applies to land sold by the Assembly.)
- The Northern Ireland Executive requires all new or refurbished buildings occupied by their departments to meet at least BREEAM Very Good.
- All health authorities in the UK require occupiers to only lease properties with BREEAM Excellent for new buildings and Very Good for refurbishment, subject to certain capital cost thresholds.

So from a valuation perspective, a building not meeting the required standard(s) could preclude many (if not most) public sector occupiers from taking the lease of a commercial building. This restriction on letability clearly increases risk and must therefore impact negatively upon value.

Many planning authorities worldwide also require minimum sustainability standards as part of their development policy and/or processes.

In addition to this public sector regulation–based 'push' towards sustainability, many corporate occupiers and their advisors are becoming increasingly aware of the reduced total occupancy costs, well-being, and productivity benefits of sustainable buildings. This also, as we will explain here, has tangible and significant influences on the market demand for buildings and the justification for paying more for a sustainable building in return for the increasingly quantifiable benefits of those buildings. This represents a 'pull' towards sustainable buildings, which will also impact upon value.

In the *Business Case for Green Building* (World Green Building Council 2013), several studies are reported which evidence productivity benefits for employees from features associated with property designed to achieve high sustainability rankings. These include:

- Individual temperature control (up to +3%)
- Improved ventilation (up to +11%)
- Lighting design (up to +23%)
- Access to natural daylight (up to +18%)

Within the same report, potential annual savings and productivity gains are estimated, in billions of dollars, from reduced respiratory disease, reduced allergies and asthma, and reduced sick building syndrome symptoms, as well as from direct improvements in worker performance that are unrelated to health.

The valuation of property with 'sustainability' aspects is not a new phenomenon, and much has been written in other textbooks and academic articles. Incorporating and integrating sustainability is not a deviation from traditional valuation methods, particularly the *income approach*, or contrary to global standards, it is an increased emphasis on and awareness of the proven benefits that sustainability brings that have changed. Associated with this change are the increased risks associated with a building not meeting the expectations and aspirations of occupiers.

Austin (2012) reviews the work of Lorenz and Lützkendorf (2011), who identify four main arguments for the importance of the integration of sustainability aspects into the valuation process, which are summarised in what follows.

1. The integration of sustainability aspects into the valuation process is increasingly supported by market behaviour, transactions, and foreseeable developments, such as rising energy costs, evolving user requirements, environmental legislation, and corporate social responsibility reporting. Empirical studies highlight the financial benefits associated with green certifications. For example, Chegut et al. (2010) recorded a 20% price premium and 22–27% rent premium for UK office buildings with a BREEAM Excellent rating. Similarly, Porumb et al. (2020) found a 19% price premium for certified office buildings across Finland, France, and Germany, while Fuerst and McAllister (2011) reported rental and sales price premiums for LEED and Energy Star certifications in the United States. The financial appeal of green buildings, reflected in both higher capital and rental values and lower operational costs, has drawn the attention of institutional investors. According to Nareit's 2024 REIT Industry Sustainability Report, 76% of REITs by market capitalisation held LEED certification in 2023, further underscoring the growing importance of sustainability in real estate investment decisions.
2. The professional ethics of the real estate and valuation profession demand real estate professionals to have an increased responsibility to society to promote sustainable development.
3. Poor property valuation practice and a reluctance to change delay the acceptance of sustainability and can promote a misallocation of resources. For this reason, Muldavin (2010) argues that, in the past, the failure of property investors and their professional advisors to appropriately incorporate revenue and risk considerations into sustainable investment decisions has resulted in an under-investment in sustainable real estate.
4. The lack of awareness of stakeholders regarding the relationships between the sustainability performance of buildings on the one hand and the resultant property risks and financial performance on the other lead to a situation where individual property assets are mispriced. One example of this we will explore in more detail in this chapter is the lack of transparent information on total occupancy costs within the marketing of real estate.

Lorenz and Lützkendorf (2011, p. 654) purposefully state:

> Sustainability issues play a role in any valuation assignment, including the valuation of conventional buildings, because such buildings are already associated with higher risks (eg faster obsolescence and shorter remaining economic lifespan), which must be considered and priced today; this is particularly true whenever an income-producing property is valued by discounting and/or capitalizing the property's net income. The nature of underlying logic of this valuation method simply requires that future risks (which are already foreseeable today) be considered when selecting appropriate discount or capitalisation rates.

This quote confirms that sustainability should be an inherent, integrated, and consistent component of the *income approach*.

Austin (2012) also provides an excellent review of sustainable practice and a detailed checklist of the ways in which sustainability should be considered and incorporated into the *income approach*. The following summary is based on the information provided in his substantive paper.

The growth of future rental income, vacancy, and collection loss

Growth of the inputs to the *income approach* may be influenced by the following aspects linked to sustainable developments (primarily in the office sector):

- Increasing energy costs, which, in turn, focus the occupier's mind on the total occupancy cost of their buildings. Reduced occupation costs of sustainable buildings may increase the amount available for rent. Eccles et al. (2017) provide an evidence-based case study of how a major development in Poland has proven this argument.
- Other positive influences on productivity and convenience that make sustainable buildings attractive to occupiers, such as the convenience of being able to walk to work, use and store bicycles (and shower before starting work), and/or public transportation access.
- Combining both the previous points, buildings closer to public transport may reduce occupiers' travel costs and time spent commuting, generally associated with being unproductive.
- Negative influences, such as a building being inconsistent with an organisation's corporate social responsibility (CSR) statements and position or compromising a company's ability to attract and recruit appropriate employees; younger employees have been shown to be highly aware of sustainability and more demanding of sustainability features in the buildings they work in.
- Tax incentives, grants, and loans associated with sustainable development.
- The image of the building and its consistency with an occupier's brand and CSR position.
- Building design and selection of appropriate materials (e.g. thermal, acoustic, and visual comfort, such as glazing and shading), as well as other design issues, have been proven to promote a healthy building, worker productivity, and reduced worker sickness and absenteeism, and can be related to the rental rate (e.g. Boyd 2005).
- Developments where landlords commit to green leases will attract customers, such as Marks & Spencer in the UK, who commit to a target of occupation of buildings on green lease terms; without such a commitment, some tenants may be prevented from taking a building, increasing market-based risks.
- Buildings designed with sustainable standards may be less prone to environmental 'disasters', such as flooding, thereby reducing future risks.
- Capital expenditure required by occupiers where a building does not meet their requirements/expectations.

Operating expenses may be impacted by:

- Enhanced structural aspects of sustainable buildings, which may reduce future maintenance.
- Building materials that have a longer life and reduced physical obsolescence.
- Energy-efficiency, which reduces total occupancy costs.
- More emphasis on planned maintenance inherent in sustainable design.
- Green leases – they improve the allocation and efficiency of operating expenses/tenant reimbursement.
- Potential for better space flexibility.
- Mitigation of environmental costs, taxes, and future liabilities.

Terminal capitalisation rate may be impacted by:

- A market's perceived future risk for buildings that do not meet corporate occupiers' expectation and aspirations.
- Market expectations of future supply/demand for sustainable buildings.
- Expectations of income and value changes over a typical holding period.
- Risk reduction due to a building's resilience to flooding and other environmental conditions.
- Market expectations regarding economic life and obsolescence.

It is not the remit or intention of this book to fully explore the link between value and sustainability. What we do want to do is raise awareness of the compelling arguments that suggest sustainability should always be considered as one of the market factors in most valuations, especially those of office buildings.

There should be no doubt that sustainability is now a fundamental concern of commercial real estate, affecting both long-term value generation and short-term profitability, especially in the context of mature markets. As we have briefly illustrated here, the combined demands of more socially aware occupiers, employees, investors, and regulators are such that tangible impacts upon many aspects of the income approach are evident, including rental growth, capitalisation rates, total occupancy costs, and voids.

We conclude this section with a quote from Jon Lovell of Deloitte: '[T]he tide is most definitely turning, and those that choose to remain agnostic and defensive in their approaches to sustainability will more than likely see their investment returns underperform the wider market' (Deloitte 2014, p. 12). Further updating this, research from Knight Frank in conjunction with Active Capital provides evidence of this 'green premium' for highly rated sustainable buildings. They found an 8–18% sales price premium for green-rated buildings compared to equivalent buildings without a BREEAM or NABERS rating across the markets studied, depending on the level of green rating.

Sustainability is becoming embedded in and integral to corporate real estate decision-making, investment selection, and corporate social responsibility programmes and to secure lower occupancy costs and promote productivity and well-being. The valuation profession seems to be in a process of catch-up with all this, still seeking more evidence of linkages to value. But many aspects of the *income approach*, especially the assessment of risks associated with not having a sustainable asset, are proven, and therefore the valuation must inherently reflect these factors. The challenge is to make them more explicit and be more in tune with the importance occupiers and investors are placing on sustainability.

RICS (2024b) highlights ESG factors that may be relevant during inspection, investigation, reporting, and documentation, making them important considerations in valuation,

To comply with best practice in reporting, RICS (2024a) recommends that valuers:

- Provide a clear description of the sustainability and ESG-related property characteristics and attributes that have been collected, which may include items not directly reflected in the final advice as to value.
- Assess the extent to which the subject property currently meets sustainability and ESG criteria expected within the context of its market standing.
- Provide a statement of their opinion on the relationship between sustainability/ESG factors and the resultant valuation, including a comment on the current benefits/risks that are associated with these characteristics, or the lack of benefits/risks.

Table 13.1 ESG factors relevant to property valuation

Environmental	*Social*	*Governance*
• Energy rating • Energy consumption • Type of energy used • On-site energy generation • Renewable energy systems • Labels and certificates (e.g. BREEAM, LEED, Well Building Standard [WELL]) • Greenhouse gas emissions • Physical climate risk factor (e.g. flood, heat, drought, sea level) • Water usage • Construction materials	• Connectivity and infrastructure • Building accessibility (e.g. ramps) and mobility features (e.g. bicycle parking) • Indoor air quality • Community impact • Adaptability • Wellness	• Safety considerations • Risks related to ownership, occupation, and transaction funding sources • Diversity, equity, and inclusion • Leases incorporating sustainability/ESG provisions

- Form an informed assessment of the potential impact on valuation, considering how a well-informed purchaser would factor these aspects into their offer price decision.
- Provide a statement of the valuer's opinion on the potential impact of these benefits and/or risks to relative property values over time.

In February 2024, the RICS published the ESG data list for real estate valuations, which describes itself as a 'practical reference document on legislative, market-driven and future ESG requirements for valuers and financial clients in the EU A practical reference document on legislative, market-driven and future ESG requirements for valuers and financial clients in the EU'. The document sets out a comprehensive list and explanations of the data to be captured and the units of measurement to be used for 12 ESG indicators. In Table 13.2, we illustrate some of these indicators.

Finally, it is worth noting that many countries such as the UK are working towards *net zero*, which can be defined as a target of completely negating the amount of greenhouse gases produced by human activity, to be achieved by reducing emissions and implementing methods of absorbing carbon dioxide from the atmosphere. Clearly, buildings have a significant role to play in these targets. In a 2023 Knight Frank ESG Property Investor Survey, 41% of investors surveyed were looking to achieve net zero by 2030, and the rebalancing of portfolios with buildings that can support that aim means that buildings that do not achieve BREEAM Excellent or Outstanding or EU Energy rating of B and above are unlikely to meet the criteria for investment.

To bring things fully up to date, in the UK, the pilot version of the UK Net Zero Carbon Buildings Standard was launched in September 2024. It is the first time that all the relevant components have been brought together into a single standard. It contains the technical details on how a building should meet the Standard, including what limits and targets it needs to meet, the technical evidence needed to demonstrate this, and how it should be reported. In the future, projects will be able to use the tool to verify that a project conforms to the Standard.

Flexible workspaces and their impact on property valuation

Office evolution

The office building became a defining architectural feature of the 20th century, symbolising the post-industrial age. In the 1920s, influenced by Frederick Taylor's principles of efficiency, office

Table 13.2 ESG indicators, data requirements, and metrics

	ESG indicator	*Examples of data to be captured and analysed*	*Unit of measurement/indicative performance measure*
01	**Energy rating**	• Energy performance certificate (EPC) • Other energy ratings in the market • Any improvements to the building made since the energy rating occurred	• Energy label (A–G) • kWh/m^2 • Expiry date of EPC
02	**Energy consumption**	• Primary and final energy consumption • Energy intensity	• kWh/m^2/year • kWh/m^2
03	**Renewable energy production (on-site)**	• Method of energy generation • Quantity and specification of renewable energy systems (e.g. solar panels, heat pumps, biomass, wind turbines) • Heating source • Usage	• kWh/m^2/year • % of primary/final energy demand met by renewable energy produced on-site • % used on-site versus % delivered back to the grid
04	**Labels and certificates**	• Green building certification schemes • National-level certificates • BREEAM, LEED, WELL, Fitwel, BOMA360, SHORE, DGNB, HQE	• Label/certificate – yes or no; if yes, specify the level • Date of issue and expiry
05	**Greenhouse gas emissions**	• CO_2 emissions, both excluding and including refrigerant gases, based on real energy consumption	• kgCO_2e/m^2/year
06	**Emissions pathway analysis**	• Carbon risk real estate monitor (CRREM) pathway analysis • Other pathway analysis (examples include ParisProof in the Netherlands, DGNB SYSTEM for Buildings in Use (or DGNB AWARD 'Climate Positive') in Germany and beyond, BREEAM in the United States, and UK's Net Zero Carbon Buildings Standard	• Whether current property performance is on the pathway and in line with future targets • Decarbonisation capex and updated stranded date
07	**Physical climate risk**	• Climate risk (such as flood, heat, drought, sea level, precipitation) analysis issued by a recognised source (e.g. MSCI, Moodies, R4R) • Mitigation measures already in place?	• Has the client done a study indicating whether the property is subject to climate risk by 2050 (yes/no)? • If no/not entirely, are mitigation measures considered in the capex budget?
08	**Location characteristics**	• Local infrastructure (including utilities for its occupants) • Connectivity (e.g. highways, distance from public transport, mass transit services/routes, frequency)	• Amenities in and around the building walkability score (how easy it is to reach certain essential places on foot) • Types of buildings in the proximity • Public transport proximity and frequency

(Continued)

Table 13.2 (Continued)

	ESG indicator	*Examples of data to be captured and analysed*	*Unit of measurement/indicative performance measure*
09	**Mobility**	• Number of EV charging points • Bicycle parking spaces for residents/occupiers	• Charging points per FTE/ total number of parking spaces • Number of bicycle parking spaces per FTE
10	**Building access**	• Access for people with disabilities	• Confirmation of compliance
11	**Landlord–tenant relationship**	• Tenant activity • Rental contract types • Green leases in place	• Description of current tenants, contracts, and/or green leases in place
12	**Material use**	• Materials used for construction or renovation	• % of material by total weight/volume/value • % of material certified for its sustainable qualities

design embraced expansive open-plan layouts with orthogonally arranged desks, all oriented toward the supervisor, while managers occupied separate cellular offices (Van Meel 2000; Duffy 2003; Fuzi 2015).

Following World War II, rectangular high-rise office buildings with glazed facades became a hallmark of commercial architecture for the next five decades. The introduction of air-conditioning in the 1930s and fluorescent lighting in the 1940s eliminated reliance on natural ventilation and daylight, allowing for deeper floor plates. This shift was seen as both cost-effective and more adaptable for subdivision and leasing (Van Meel 2000).

In the 1960s, office design evolved to foster workplace interaction. The German Quickborner team introduced *Bürolandschaft*, or 'office landscape', an open-plan concept with randomly arranged desks to encourage communication and collaboration (Van Meel 2000; Walsh 2015; Rumpfhuber 2017). While this approach gained traction across Europe, it declined in the 1970s due to economic constraints, leading to a resurgence of cellular offices and 1920s-inspired layouts (Van Meel 2000).

The 1980s brought rapid advancements in information technology, necessitating significant office infrastructure upgrades such as data cabling, enhanced cooling systems, deep floor plans, raised flooring, suspended ceilings, substantial riser capacity, and expansive trading rooms. In the 1990s, alternative practices like desk-sharing gained popularity (Van Meel 2000). The decade also saw a shift towards user-centric office design, with the emergence of the 'office hotel' concept, which introduced flexible work formats and improved space management (Hickey 2015).

At the turn of the 21st century, globalisation and advancements in information and communication technology facilitated greater workplace flexibility, particularly for self-employed and knowledge workers (Harris 2015). New office solutions, such as flexible workspaces, were developed to meet the evolving needs of occupants by enhancing productivity, cost efficiency, flexibility, and collaboration, while contemporary office design increasingly integrates hospitality-style amenities to improve the user experience (Harris 2015; IPF 2020; Sankari 2019).

The development industry has, until the last decade, been slow to respond to the dramatic changes in office design and the evolving needs of occupiers (Akkerman & Nunnington, 2025). For a long time, developers have been focusing on attracting investors before, and buyers after, a project was delivered. This meant efficient construction and easy maintenance had to be on the front page. It also created many disconnects between the developer and the end user. Most developers had little interest in the occupier's needs beyond the initial leases before the exit towards institutional landlords, burdening them with the pressure of maintaining occupancy and dealing with any design-related maintenance or service issues.

The trend to shorter and shorter leases has been ongoing since the 1990s but accelerated in recent times through agile business models, technology, and of course, most recently, by the pandemic. Figure 13.1 demonstrates the trend towards not only shorter leases but also new models and greater choices for occupiers.

The COVID-19 pandemic brought forward significant shifts in these trends, accelerating the adoption of flexible work models, increasing interest in co-working spaces, and changing perceptions about the desirability of long-term leases versus the flexibility of shorter-term arrangements. Consequently, occupiers are now gravitating toward (1) reducing overall office space and (2) demanding more flexibility in lease terms, favouring shorter durations.

Both these elements directly challenge the traditional financing model of investors and have led to a re-evaluation of what constitutes a valuable real estate investment from the investor's perspective – more concretely, a contented occupier whose real estate delivers personal and/or organisational satisfaction and/or performance, is the prime component that now gives any real estate asset their true value.

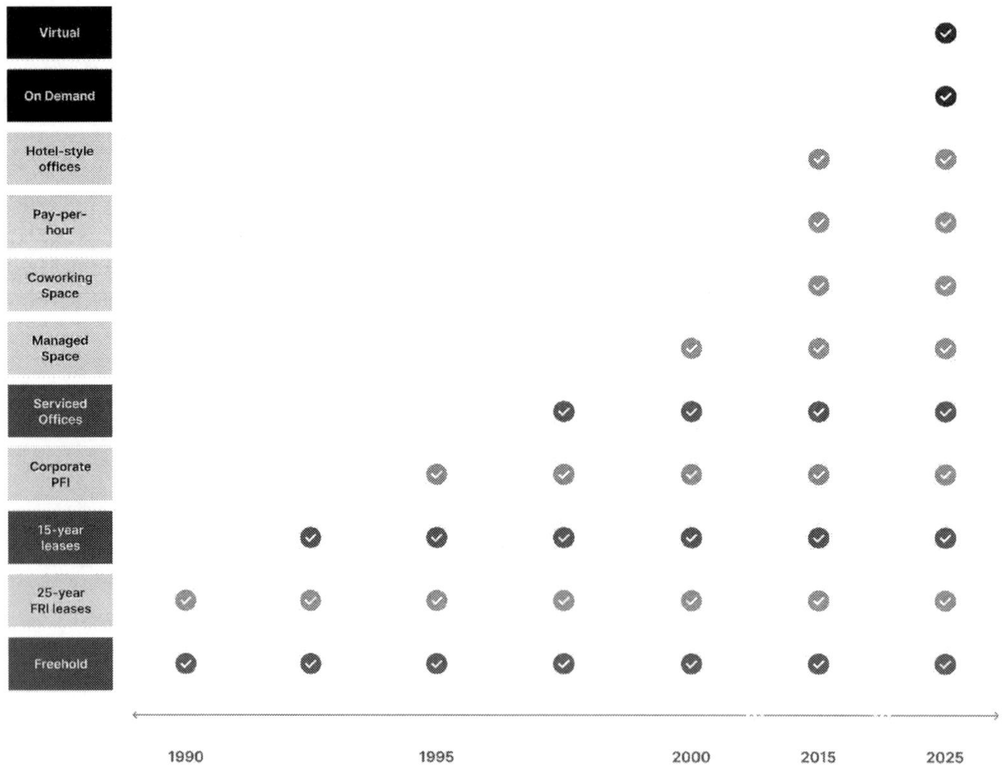

Figure 13.1 The evolution of flexible occupation.

With occupiers at the helm, it was high time for developers to up their user/workplace/tenant (increasingly customer) experience game and make the occupiers want to take up space and remain in it. Without these enhancements, developers end up with bare blocks of concrete and steel, sitting in their portfolio shredding value each year instead of producing income.

Commercial structures are not merely made of concrete and glass: They signify value for property owners and their lenders – value, that is, if they are utilised as intended. Should they become redundant, a significant portion of the economy stands at risk, given Statista's estimate of the global commercial real estate market being estimated to reach a staggering US$120 trillion in 2025. Easier to manage is Nareit's 2021 estimate that the office market hovers around $3.2 trillion in value in the United States. Even if half of it falters, the ripple effects could be monumental. There are many occupiers large and small reporting that they were moving to around 50% of their portfolio, which is not surprising, given a global average occupancy rate, which remains below 50% in all global regions except APAC (62%), Basking 2024.

In a post-pandemic real estate paradigm, the focus has shifted from designing office spaces primarily for investors to prioritising the needs and preferences of occupiers. This change recognises that the true value and ROI of a property hinge on its appeal and utility to those who use it daily.

Consequently, the new measure of success in real estate is creating spaces that are not just financially viable but also desirable and functional for tenants and end-users. Over the past 30 years, there has been a transition, nicely summed up by Steven Skinner, the UK CEO of HB Reavis, from what he calls, 'lazy wholesaling of office space, to the retailing of office space' (Bisnow 10 April 2019). So as well as a shift in occupation terms, we have also seen a dramatic rise in the service level expectations of occupiers as demonstrated in Figure 13.2.

When long leases, upward-only rent reviews, onerous repairs, and other lease clauses are stripped away, and agile businesses demand flexibility, developers must understand their end users so much more, as the stable cash flows of the past are no longer guaranteed. Armed with

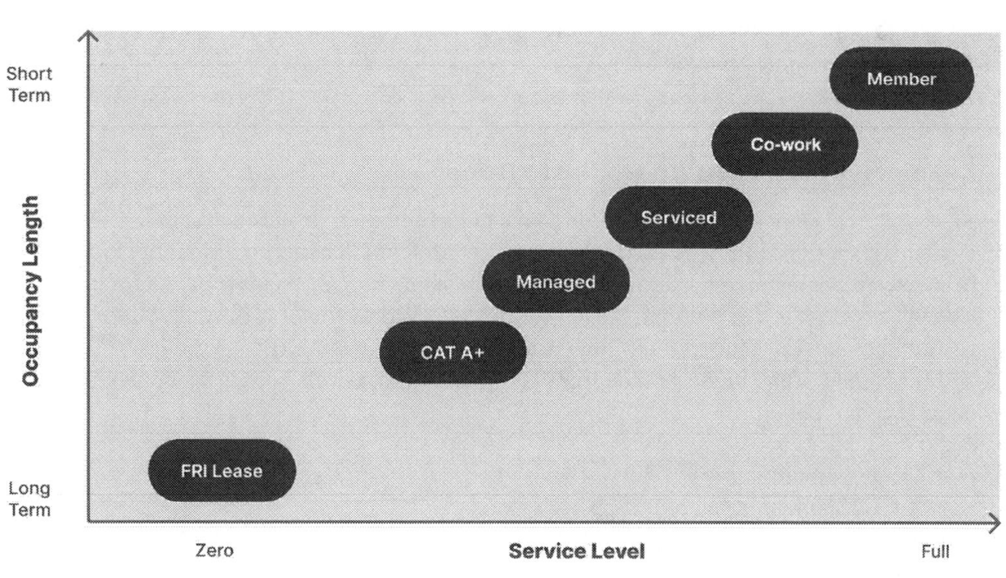

Figure 13.2 The evolution of customer service expectations.

this understanding, they can provide targeted, added-value, flexible products that drive customer satisfaction, tenant retention, and additional revenues.

A good example of these shifts in practice is the well-publicised decision of HSBC in the UK to vacate their landmark 45-floor, 1.1 million ft^2 (102,000 m^2) tower in Canary Wharf, occupied under a traditional 20+ years UK-style lease to a 556,000 ft^2 (52,000 m^2) refurbished office (former headquarters of British Telecom) near St Paul's in the City of London. HSBC is reported as saying the shift would support digital innovation, help it meet net zero carbon commitments, and integrate a high level of wellness and other amenities to support employee satisfaction and retention.

Meanwhile, their old building has been the subject of much speculation about radical plans to refocus it, in keeping with the discussion in this chapter, emphasising flexibility, integration of co-working, high levels of amenities, and achieving the highest credentials in sustainability. This is all part of a larger plan to redevelop Canary Wharf into a vibrant mixed-use neighbourhood, offering workspace, retail, homes, leisure, and amenities all in one location, 'a true 15-minute city', according to Shobi Khan, the chief executive of Canary Wharf Group.

As subtext to this story, it is worth noting that areas such as Canary Wharf in London and the financial districts of Manhattan have suffered from the continuing reduced occupation of offices. The knock-on impact to external amenity and food and beverage providers, particularly operators such as Pret-a-Manger, challenges the vitality of the areas, and according to research by Urbanite Advisors and their Vibrancy Index, occupiers are moving to areas with more choice and variety of amenities and the associated 'buzz'.

All these issues are important shifts in the business environment which have an impact on the attractiveness of office developments and the ability to retain tenants (customers) and, therefore, resilient cash flows. Valuers should be aware of the evolving demands of occupiers and question if the assets they are valuing remain *fit for purpose* in a world of increasing demands for flexibility, amenitisation, and customer service.

Set out in what follows are two examples of this new paradigm which illustrate what occupiers are seeking and set a kind of benchmark for end user satisfaction. The question must be asked: What adjustments need to be made for buildings that do not meet these expectations within the Grade A office market?

Example 1 ICD Brookfield Place, Dubai

The asset managers of ICD Brookfield Place talk about their project as 'curating the best customer experience in Dubai'. Their success in attracting tenants and retaining them has been exceptional, and this is largely due to an understanding of the new paradigm that we have discussed. The Sir Norman Foster–designed 53-storey tower is the tallest and largest LEED Platinum building in the Middle East and North Africa (MENA) region.

It has some of the highest credentials for most of the aspects sought by contemporary occupiers, including:

- **LEED** Platinum
- **WELL** Health and Safety Rating
- **Wiredscore** Platinum
- **Smartscore** Platinum

Several floors of high-quality, flexible offices and bookable spaces known as The Suites and Sky Suites are available on a range of flexible terms to tenants. A wide range of amenities are provided within the project, including a Waitrose supermarket, three gyms catering to different demographics, seven restaurants, a private members club, a florist, dry cleaning, and more.

What is interesting is that ICD Brookfield, in their own marketing material, refers to many corporate real estate and occupier metrics and evidence, indicating the closing of the gap and the shared understanding necessary in the new paradigm. For example, in their main marketing brochure, they quote: 'Improvements in workplace environment could yield 20% increase in employee productivity' (Gensler). Their attention to air quality, views, temperature, and other ambient controls addresses many aspects known to influence productivity.

'Poor health costs US employees $530 billion and 1.4 billion workdays of absence and impaired performance per annum' (The Integrated Benefits Institute). Their UV filtration system and other wellness benefits are designed to address this workplace issue.

Despite the exceptional quality and premium rent, a range of lease lengths and flexibility are available to meet occupier demand. However, the asset managers closely monitor and proactively manage the WALE, to moderate risk. It is perhaps no surprise that the building maintains exceptional occupancy rates, and there is a waiting list.

Example 2 HB Reavis

In addition to a similar approach as ICD Brookfield Place, in focusing on customer engagement and retention through high levels of sustainability and amenitisation in its projects, HB Reavis embeds, across its portfolio, a range of unique features, amenities, and occupation options in keeping with the new occupier paradigm, including:

Symbiosy, a tech sensor–driven real-time information platform that delivers information about occupation, air quality, and temperature and assesses space utilisation, indoor environmental quality, and in-company collaborative networks. It enables excellent workspace management, boosts cooperation, and ensures user comfort, helping companies connect and thrive. The system can facilitate personalisation of the indoor environment and comfort with automatic adjustments to personal preferences.

Qubes provides fully serviced office spaces on short-term leases that flex to a company's changing needs. They can complement leased accommodation in the same building or be stand-alone. They are fully customisable and provide a range of services, including reception, security, cleaning, catering, and meeting rooms, all in a simple short-term contract arrangement.

HubHub is what we can call a true co-working space, dedicated as much to knowledge exchange as it is to flexibility. HB Reavis has been a pioneer of the new investor-occupier paradigm and, through innovation, developed a new type of relationship with its customers. An example of its pioneering approach was to create the PropTech-specific co-working space called HubHub to develop new ideas and potential projects, from which Symbiosy was born.

At the heart of this shift in office occupation is the need for businesses to be ever agile and flexible. With businesses finding it impossible to plan for more than a few years, signing a five- or even a three-year lease is inconsistent with the fluidity and volatility of the business environment. To respond to this need, flexible workspace, especially co-working, has exploded globally, with one study quoting 7,000 co-working centres in 2015 compared to a projected 30,000+ in 2025 (Statista/Gitnux).

Flexible workspace

Flexible workspace is an umbrella term for various space-as-a-service solutions, including co-working spaces, serviced offices, managed offices, and hybrid spaces. These spaces typically offer shorter contractual terms or an all-inclusive membership model, where users pay a daily, weekly, or monthly fee to access workstations – such as a hot desk in a shared office or a private office – technology, and several amenities. In many ways, flexible workspace providers operate similarly to hotel operators, managing physical spaces on a short-term basis and offering a diverse range of amenities and services to a broad clientele (Wheaton and Krasikov 2019; IPF 2020).

The rise of flexible workspaces reflects a shift in the business landscape, driven by growing uncertainty and the need for greater adaptability. As companies face shorter planning horizons and prioritise flexibility, traditional commercial leases – with their long-term commitments, fixed terms, and stringent covenant requirements – have become less appealing. In contrast, flexible workspaces prioritise accessibility and scalability, catering to a wide range of occupiers without the burden of rigid contracts. This model has reshaped occupier expectations, offering businesses the agility to scale up or down as needed, making it an attractive option for freelancers, start-ups, and even large corporations.

Market commentators such as Knight Frank (2020) highlight that flexible workspace operators focusing on corporate clients tended to fare better due to the extended membership commitments from these entities and their tendency to take more space than individual occupiers or small and medium enterprises. Large corporations use flexible workspaces for many reasons, including thus:

- *Swing space.* Temporary space to accommodate staff during specific projects, office renovations, or new office construction (British Council for Offices 2019; Echeverri, Jylhä and Koppels 2021).
- *Satellite office.* Part of a hub-and-spoke (Sargent et al. 2018) or 'core and flex' strategy (Echeverri et al. 2021), where the flexible workspace complements a core headquarters location (British Council for Offices 2019).
- *Overflow space.* Additional workspace when demand exceeds a company's main office capacity (Weinbrenn 2016; Echeverri et al. 2021).
- *Market testing.* Providing a low-commitment entry point to test new markets (Arora 2017; Echeverri et al. 2021).
- *An alternative to traditional leasing or ownership.* Replacing conventional office space with flexible workspace (Josef and Back 2019 in Heinzel et al. 2021).

Examples of flexible space applications

Flexible space, especially co-working, can be deployed in many different scenarios, including the following:

Table 13.3 Flexible space scenarios and examples of occupiers

Scenario	*Companies using the approach*
In-house flexible space for employees	NatWest bank in the UK developed a flexible space in 250 Bishopsgate for employees, converting outdated meeting rooms to provide a variety of work settings. Aviva and others are adopting a similar approach.
In-house flexible space open to the public	This risky approach can support branding, innovation, and customer relations. In 2015, Barclays repurposed space and created Eagle Labs, a start-up incubator.
In-house alliance with an operator	Companies can partner with an operator to create and manage flexible space in underutilised space. Verizon, the global telecoms company, has partnerships with Alley in New York, Work. Life in London, and JusCo in Singapore.
Individual memberships in a co-working space	In New York, Microsoft has, in the past, given 300 sales employees access to all WeWork locations in the city.
Private office in a flexible workspace	Some companies take private offices to allow quick expansion or contract-specific projects. Alibaba has used this approach at CoWrks, a provider in Bangaluru, to establish a presence in the Indian start-up capital. Goldman Sachs used WeWork in Birmingham to dip a toe in the UK market.
Showroom/customer care centre	In this approach, corporations act as a sponsor/member of a flexible space to showcase their products. AT&T created an engagement centre to showcase their technology in Indiana, and Samsung has experimented with care centres in three US locations, offering services to both WeWork members and Samsung customers.

Operational models

The growth of the flexible workspace sector has prompted a significant realignment of traditional roles and relationships within the property sector. The conventional role of landlords as rent collectors and service charge overseers is increasingly threatened by the ascendancy of hospitality-oriented flexible workspace providers; consequently, landlords are exploring diverse roles, whether as lessors, partners, or even owners (IPF 2020).

According to the RICS publication *Insights into the valuation of flexible workspace* (Golding 2019), flexible workspace providers typically operate under the following models:

Freehold ownership. The flexible workspace provider owns the building outright and directly manages the workspace. This model gives the operator full control over the property and its operations.

Leasehold ownership. The provider leases space from a freeholder under conventional office letting terms. This is the most common model, as noted by IPF (2020), and allows operators to focus on managing the workspace without the burden of property ownership.

Property company/operating company (propco/opco) split. The freehold ownership is held by a separate property company (PropCo), while the operating company (OpCo) delivers the flexible workspace services. Golding (2019) emphasises that valuers must understand the company and ownership structure, as well as the terms of the OpCo's letting or management agreement.

Management agreement. Landlords and operators enter into profit-sharing agreements rather than traditional leases. This model is gaining popularity, as highlighted by IPF (2020) and Savills (2024), as it aligns the interests of both parties and allows them to share in the success of the flexible workspace. Savills (2024) reports a significant increase in the adoption of management agreements in the UK, rising from just 9% of flexible office operator transactions before the COVID-19 pandemic to 41% in 2024. This trend underscores the growing preference for collaborative and risk-sharing arrangements in the sector.

Valuation implications

Income considerations

A building leased to a flexible workspace provider under a conventional lease arrangement can be valued using traditional investment methods or a DCF approach, where the rent paid by the provider to the landlord is capitalised at an appropriate capitalisation rate, like any other office tenant, while accounting for risk differences, to be discussed later. In contrast, valuing freehold ownership, propco/opco, or management agreement models requires a more detailed assessment of inputs, including the following:

- *Occupancy rate.* The percentage of space occupied over time. Stoyanchev (2024) notes that successful co-working spaces typically maintain occupancy rates between 70% and 90%. As an illustration, International Workplace Group (IWG), formerly Regus, reported an 85.3% workstation occupancy rate in the Americas in 2012, which declined to 69% in 2021 (Bohne 2023).
- *Membership fees.* Revenue from different membership tiers, such as hot desks, dedicated desks, and part-time usage plans.
- *Private office revenue.* Income from enclosed office spaces contracted within the flexible workspace.
- *Meeting room revenue.* Fees charged for short-term or hourly meeting room rentals.
- *Ancillary income.* Additional revenue sources, such as renting out event spaces, locker rentals, hosting events.
- *Food and beverage sales.* Revenue from in-house cafés, vending machines, or catering services.
- *Fixed costs*. Recurring operational expenses, including staff salaries, core technology infrastructure, utilities, and maintenance.
- *Variable costs.* Costs tied to occupancy levels, such as energy usage, consumables, and event expenses.
- *Capital expenditure.* Fit-out costs, furniture, and ongoing refurbishment expenses
- *Percentage paid to landlord (in management agreement).* The proportion of revenue allocated to the property owner under a revenue-sharing structure.

To illustrate the range of membership fees flexible workspace providers charge, the following table presents illustrative pricing derived from several UK operators to illustrate the wide range of options available.

Immediately, it can be seen that a traditional area-based valuation is not applicable. A co-working space has a very complex cash flow with a high degree of volatility, a range of tiered pricing, and a high degree of uncertainty, as members can switch between pricing models

Table 13.4 Illustrative pricing structures for co-working options

Type of membership	*Occupation (all options include coffee and Wi-Fi)*	*Cost per month per person (UK£)*
Hotdesking	8 hours/month	25
Hotdesking	16 hours/month	50
Hotdesking	40 hours/month	100
Hotdesking	Unlimited access	200
Fixed desk	Unlimited access	250
Resident	Unlimited access plus locker, meeting rooms, etc.	325
Private office	Unlimited access	on request

with little or no notice. Whilst some operators take space on a traditional lease and then adapt the space and rent it out on this more complex basis, increasingly, co-working is being created by direct investment, so a direct approach to the underlying complex cash flow will be required.

Risk premium considerations

As seen in Chapter 3, the *risk premium* is the component of the capitalisation rate that encompasses all the risks associated with the individual property, including the tenant covenant. In the case of flexible workspace providers, their presence as tenants introduces both opportunities and risks for landlords. Benefits include attracting a diverse occupier base, from freelancers to multinational corporations, enabling landlords to reach a segment of the market that conventional leases may not capture, and thus enhancing the tenant mix, a crucial factor in mitigating risk, particularly during periods of economic uncertainty (Green 2014; Avison Young 2019). Additionally, the high-quality fit-out and extensive amenities typically associated with flexible workspaces can slow the functional, economic, and physical depreciation of a building (Weijs-Perrée et al. 2019; Chegut and Langen 2019). Having a flexible workspace provider as a tenant in a multi-let office building might also be seen as an additional amenity for existing tenants by providing overflow space (Avison Young 2019; Kern 2019). A key advantage of leasing to flexible workspace providers lies in the length of their lease agreements and the size of the space they occupy. These tenants typically sign longer leases and take up large floor plates (Chegut and Langen 2019; Silva et al. 2022), which, in theory, contributes to income stability by reducing the frequency of lease renegotiations and limiting periods of vacancy.

Leasing space to flexible workspace providers presents notable risks for landlords, primarily concerning the financial sustainability of the operator. Some researchers argue that the rent arbitrage business model underpinning flexible workspace providers is inherently fragile as, traditionally, the model relies on operators securing long-term leases from landlords while offering short-term, flexible memberships to occupants, exposing operators to fluctuating occupancy rates (Green 2014; Seo et al. 2017; Wheaton and Krasikov 2019). An illustration of this mismatch is the discrepancy between lease commitments and revenue streams. In 2019, WeWork's lease obligations in the United States averaged 15 years, whereas its memberships had an average duration of only 15 months (The We Company 2019). From a landlord's perspective, this structural vulnerability, combined with substantial capital expenditures required for fit-outs and the low-margin nature of the business, heightens the risk of tenant default (Spinuzzi 2012;

Morgan Stanley 2018; Jamal 2018; Golding 2019; Wheaton and Krasikov 2019). In addition, flexible workspace providers frequently structure their leases through SPV, a legal entity created to accomplish a well-defined and often temporary objective that protects the parent company from financial risks (Golding 2019). If the tenant is an SPV with no other assets, then the landlord can only pursue that SPV, as there will likely be no other assurance, such as a deposit or corporate guarantee.

The risks associated with flexible workspace (FW) providers are reflected in the underwriting of office investment. For instance, investment research company MorningStar reports the risk factors associated with buildings that have a significant FW component. In a similar way to underwriting properties with significant single-tenant exposure, FW, specifically co-working, requires adjustments such as higher tenant improvement packages and lower occupancy assumptions (Johnston and D'Souza 2019). As reported by JLL (2021), the creditworthiness and the finance of FW providers have been the most visible concerns in the wake of the pandemic, and as the industry evolves, landlords are integrating more clauses in case of non-performance and requiring more financial clarity from FW operations.

Research on the impact of FW providers as tenants on capitalisation rates remains limited, yet the proportion of space they occupy within a building appears to influence capitalisation rates to some extent. CBRE's Global Intentions Survey 2019 indicates that most investors believe that the higher the proportion of flexible workspace in a building, the worse for long-term capital value. However, for a fraction up to 20% of the building, investors see the presence of flexible workspaces as a value enhancement (CBRE 2019a). The results from studies in the United States corroborate such investors' sentiment toward flexible workspace operators. Greenwood et al. (2018) analysed the sales of 17 properties in the United States with co-working tenants and found the cap rates to be lower or within the range of other comparable sales when 40% or less of the building was leased as co-working; however, as this proportion increases, so does the difference between the cap rates. Similarly, CBRE (2019b) analysed 31 sales transactions of office buildings with at least 10% flexible occupancy across 13 US markets and found the same pattern – as the proportion of flexible space within a building grows, capitalisation rates initially decrease, then level off, and ultimately rise. Wheaton and Krasikov's (2019) findings resemble the previously mentioned studies; they found little impact on prices when the co-working share is small; however, as the proportion increases, prices fall, and cap rates rise. The results from the studies indicate that investors might pay a premium for buildings that house FW operators up to a certain inflexion point; beyond this, investors may demand a discount.

The net impact on cap rates is contingent upon a variety of factors, including the location of the office building, the overall market conditions, and the financial health of the flexible workspace providers. As this trend continues to evolve, its influence on cap rates underscores the need for a nuanced understanding of the dynamic interplay between flexibility, risk, and real estate valuation.

So what are the practicalities of valuing a co-working space?

Calculating the value of a co-working rental model is often based on a profits method–style approach, assessing the multiple income streams of the co-working business. It requires a comprehensive understanding of financial performance, including revenue streams, operating expenses, and profitability. Measuring the worth of a co-working rental business should also involve considering its potential for generating steady and sustainable income in the future. This may lead to a DCF analysis using the data gathered on past performance, appropriate comparable data, and projections based on the factors set out in what follows.

What valuers should undertake as part of their assessment

- An evaluation of the occupancy rates and membership retention rates
- An analysis of the local market competition and demand for co-working spaces
- Consideration of the potential for upselling additional services or amenities
- The scalability of the business model and its ability to adapt to changing market trends
- The financial projections and previous financial performance of the business

The valuation of a co-working business has at its core an analysis of the profitability and cash flow generated by the business. This includes assessing the average revenue per customer, occupancy rates, and the cost of acquiring and maintaining office space and amenities. To do this effectively, the following elements should be incorporated:

- *Revenue streams.* Analyse the different sources of revenue, such as monthly memberships, day passes, and additional services offered.
- *Expenses.* Determine the fixed and variable costs associated with running the co-working space, including rent, utilities, salaries, and marketing expenses.
- *Occupancy rates.* Evaluate the average occupancy rates to gauge the demand for the co-working space and the potential for future growth.
- *The market.* Understand the competitive landscape and the demand for co-working spaces in the specific geographical area.

Essentially, co-working, whilst having more complex and volatile income streams, is approached in the same way as many other income-producing property assets, and the valuation methods employed will be as set out in what follows (or a combination of):

Income approach. Estimates the value based on the projected future earnings and cash flow of the business. This could involve DCF or be based on a multiplier of EBITDA.

Comparable sales approach. Comparing the co-working rental business to similar businesses that have recently been sold; adopting appropriate adjustments for income, expenses, and capitalisation rates.

Asset-based approach. Focusing on the value of the physical assets owned by the business, such as office space and amenities.

The need to be aware of local factors in a world of global standards

In this section, we aim to illustrate, with the help of leading practitioners in three countries, the need for valuers to balance a commitment to global standards, including IPMS (2023) and IVS, with an understanding of how the local context influences value and the valuation process. It is dangerous to assume that, because we live in a global environment, the world is the same. Each of the three countries selected has unique aspects which may impact upon the *income approach*. To help emphasise the differences, the points are listed in a consistent framework for each country. We have taken the opportunity here to illustrate some of the practical, unusual, and anecdotal aspects of valuation that make it interesting.

China

This section is an update by Catherine Kain, PhD candidate, Programme Director, Master of Property, at the University of Adelaide, of the original entry kindly provided by Sunny Han, CFA, MRICS, MCIREA, Director CBRE/Valuation and Advisory Services, Shanghai, China.

Measurement

The adoption of the International Property Measurement Standards (IPMS) in China's property valuation processes remains limited. Consequently, a variety of terminologies continue to be used to describe floor areas, including *gross floor area* (GFA), *net floor area* (NFA), *gross lettable area* (GLA), *net lettable area* (NLA), and *usable area*.

The registered areas on building title certificates vary across different regions, despite the existence of nationwide measurement standards in China. For instance, the traditional Chinese unit **'mu' (亩)**, equal to **666.67 m^2** (or **0.0667 ha**), is commonly used for measuring land area, especially in agricultural and real estate contexts.

There are also regional differences in how unit rates are quoted for residential and office properties. For example, unit rates for residential units are quoted based on GLA in Chongqing and GFA in Shanghai and Beijing. Similarly, unit rentals for office premises are quoted based on GFA in Shanghai and GLA in Guangzhou. These inconsistencies highlight the need for valuers to clearly connect the quoted unit rates with the appropriate area measurements to ensure consistency.

In 2019, the Ministry of Housing and Urban-Rural Development issued a notice stating that residential spaces should be traded based on dwelling floor space, or usable area, rather than GFA. This shift aimed to provide a more accurate and scientific measurement of liveable area, reducing confusion associated with the conventional use of GFA.

Despite these efforts, the widespread adoption of IPMS in China has not yet been realised. This ongoing lack of standardisation underscores the importance for valuation professionals to meticulously align quoted unit rates with the correct area measurements, thereby promoting consistency and transparency in property valuations.

Culture and history

The real estate valuation industry in China continues to require specific licenses for both individual valuers and valuation organisations. According to the Asset Appraisal Law of the People's Republic of China enacted in 2016, all appraisers must be certified by the China Appraisal Society to practice legally. Additionally, valuation firms are mandated to employ a minimum number of registered appraisers to operate.

Specifically, an appraisal firm structured as a corporation must have at least eight registered appraisers, while a partnership requires a minimum of two. These stringent licensing requirements aim to uphold the quality and credibility of valuation services in China's real estate sector.

The acceptance of RICS valuation standards in China has seen notable progress. While, traditionally, local banks, government authorities, state-owned enterprises (SOEs), and the stock exchange primarily accepted valuation reports from locally licensed valuation firms, there has been a shift towards integrating international standards. This change aligns with China's broader economic reforms and efforts to enhance transparency and efficiency in the real estate sector. The IVSC has been actively engaging with Chinese authorities and valuation professionals to promote the adoption of IVS, which are closely aligned with RICS standards. This collaboration aims to bridge the gap between local practices and global valuation standards, facilitating greater consistency and reliability in property valuations across China.

Moreover, the IVSC's efforts to promote IVS have contributed to a more consistent and reliable valuation framework in China. This alignment with global standards enhances the credibility of Chinese valuations in international markets.

These developments indicate a positive trend towards the broader acceptance and implementation of RICS and the International Valuation Standards within China's financial and regulatory institutions.

Land use pattern and planning requirements

Inconsistencies between actual land use and planning requirements persist in China's urban development, often leading to environmental, social, and economic challenges. One notable example to illustrate these discrepancies is Dongtan Eco-City, Shanghai.

Dongtan Eco-City was conceived in 2005 as a model for sustainable urban development. Dongtan Eco-City aimed to be the world's first truly sustainable new urban area. However, by 2010, the project had stalled due to various challenges, including its proposed location in a highly valued wetlands area and loss of political support. The project has been described as a failure because it was not built, highlighting the gap between ambitious planning and practical implementation.

Market maturity

China's real estate market has undergone significant transformations, yet certain practices from previous years persist, particularly in tier 2 and tier 3 cities.

In tier 2 and tier 3 cities, the practice of offering strata retail units with guaranteed returns for extended periods, such as five or ten years, continues. These arrangements often promise higher rental yields compared to standard market transactions. Valuation professionals must carefully assess these guaranteed structures to determine the genuine market value of such retail properties.

Despite advancements, China's real estate market, especially in lower-tier cities, continues to face challenges related to overdevelopment and market saturation. A study indicates that third-tier cities, accounting for about 60% of China's GDP, have experienced significant overbuilding, leading to a surplus of unoccupied properties and declining housing prices. This overdevelopment poses financial risks for developers and investors, necessitating careful regulation and strategic planning to ensure sustainable growth.

The complexities arising from guaranteed return structures, government interventions, and market saturation require valuation professionals to exercise heightened diligence. Accurately distinguishing between nominal and actual market values is essential to provide reliable assessments that reflect the true economic conditions of the real estate sector.

While China's real estate market has matured in various aspects, ongoing challenges necessitate meticulous analysis by valuation professionals to navigate the intricacies of guaranteed returns, governmental influences, and market dynamics.

Transparency

China's real estate market has made strides toward greater transparency, yet challenges persist in accessing accurate transaction information, particularly in tier 2 and tier 3 cities.

Valuers continue to face difficulties in obtaining precise rental data for shopping centres in these cities. Valuers must navigate complexities and significant transparency challenges in various sectors by employing diligent research and leveraging available data sources to provide accurate property assessments.

In the residential sector, underreporting of transaction prices for tax minimisation remains an issue. Buyers and sellers often register property sales at values lower than the actual transaction price to reduce tax liabilities. This practice obscures true market prices, posing significant challenges for valuers aiming to determine accurate property values.

Accessing reliable data on ADR and occupancy rates for hotels continues to be problematic. Such operational information is typically considered confidential, hindering the ability of valuers to obtain accurate and up-to-date data for comparable properties. This lack of transparency complicates the valuation process for hospitality assets.

Despite these challenges, there have been efforts to improve transparency in China's real estate market. Globally, transparency improved in 2024, with Asia rising as a standout as markets made progress toward sustainability goals and AI adoption. However, specific issues in China's commercial property sector, such as opaque practices and limited data accessibility, persist. The lack of transparency in credit markets, particularly concerning the property sector, continues to hinder investors' confidence and the development of a robust bond market.

Legal tenancy, security of tenancy (income), and land law

China's real estate landscape continues to be shaped by its unique land ownership and leasing structures, as well as challenges related to tenant stability.

All land is state-owned, and land use rights are allocated for specific durations (up to 70 years for residential; commercial/office for 40–50 years) based on the property's intended use.

These terms are standardised nationally and are not subject to negotiation with local authorities. The absence of a clear regulatory framework for renewing land use rights upon expiration has been a long-standing concern among investors and property owners. Uncertainty regarding renewal procedures can significantly impact real estate pricing and investment decisions.

The security of tenancy in China's commercial real estate sector remains relatively low. Legal and financial penalties for early termination of leases by tenants are minimal, leading to potential volatility in rental income for property owners. For instance, there have been cases where a shopping centre experienced a high occupancy rate of 95% in the first year, only to see half of its tenants vacate in the second year. Such fluctuations pose challenges for valuers, who must carefully analyse tenancy schedules and account for potential income instability in their assessments.

Given these dynamics, valuers must exercise caution and conduct thorough analyses when assessing property values in China. Understanding the specific terms of land use rights, anticipating potential challenges related to lease renewals, and accounting for the possibility of tenant turnover are crucial for accurate property valuations.

Other local issues that impact upon valuation

Deal structure

In the context of en bloc investment transactions in Shanghai, the deal structure – whether equity-based or asset-based – significantly influences the transaction price and valuation process. For instance, an office building in downtown Shanghai might be marketed at RMB 52,000/m^2 for an equity-based deal, whereas the price could rise to RMB 57,000/m^2 for an asset-based deal. This disparity arises due to differences in tax implications, liability considerations, and transaction complexities associated with each structure.

Valuers must thoroughly understand the background, structure, and pricing nuances of en bloc transactions to ensure accurate, like-for-like comparisons. Given the evolving nature of the real estate market, staying informed about current market conditions and regulatory changes is crucial for precise valuations.

Tax impact

China's property tax policies continue to apply two primary taxation methods on real estate holdings, particularly affecting rental income and property ownership:

1. *Tax on rental income.* 12% of the annual rental income.
2. *Tax on property value.* An annual tax levied at 1.2% of the original value of buildings. Local governments commonly offer a reduction of 10% to 30% on this tax.

The application of these differing tax policies can significantly impact the valuation of similar buildings. Valuers must exercise caution and thoroughly assess the applicable tax regime when analysing market transactions to ensure accurate property valuations.

In an effort to stimulate the struggling property sector, China introduced new tax incentives effective 1 December 2024. These measures include expanding eligibility for a reduced 1% deed tax to apartments up to 140 m^2, up from the previous limit of 90 m^2. Additionally, the minimum pre-collection rate for the land value-added tax has been reduced by 0.5 percentage points. Such changes aim to reduce transaction costs and encourage property investments, potentially influencing property valuations.

By diligently staying informed, analysing tax implications, and ensuring accurate comparisons, valuers can provide more accurate and reliable property assessments in China's dynamic real estate market.

Cyprus and Greece

In the aftermath of earlier financial crises, Cyprus and Greece both embarked on paths of recovery and reform. In 2025, new developments, especially large-scale projects in coastal and prime urban areas, cater increasingly to foreign investors. Yet issues such as fragmented measurement practices, bureaucratic delays in planning and permitting, and unstable or complex tax regimes continue to hamper market transparency and maturity. Both countries rely heavily on overseas demand – facilitated by Golden Visa programmes – while local buyers often face stricter lending criteria and significant property-related taxes.

Measurement practices

Cyprus and Greece lack a mandatory measuring code despite IPMS awareness. The lack of consistency in Cyprus causes professionals to use their own rules – *gross external* vs *gross internal* area – making direct comparisons difficult. Greece has faced similar challenges, exacerbated by partial land registry modernisation and complex, centuries-old inheritance traditions that often cloud titles and areas. Valuers must calculate floor areas using rigorous internal procedures, explain them in reports, and reconcile official or developer-supplied measurements with on-site observations.

Cultural and historical backdrop

Cyprus, with its British-influenced legal framework, has a mostly freehold market and faces ongoing valuation complications for Turkish-occupied territory or Turkish Cypriot–owned properties in non-occupied areas. Greece, meanwhile, only formally recognised and regulated professional valuers in 2013; RICS or REV (TEGoVa) qualifications have come to represent the professional gold standard. Both countries retain strongly decentralised land planning systems, so approval and development processes can vary widely among regions or municipalities.

Professional framework and valuation standards

In Cyprus, the Scientific and Technical Chamber (ETEK) sets membership rules but emphasises academic credentials rather than post-qualification experience. Banks, however, generally require RICS-qualified valuers for mortgage and loan security purposes, as RICS membership substantiates professional training. Greece lacks a single ETEK-like authority as the Technical Chamber does not manage or want to become the regulatory authority, and licensing can involve professional associations. Nonetheless, legislation recognising RICS and REV qualifications has placed them at the heart of high-level valuation work. Most formal valuations adhere to IVS or EVS, though local practices do not always align perfectly with these frameworks.

Land use, planning, and development

Cyprus has separate local town plans for each major city (Nicosia, Limassol, Paphos, Larnaca, Paralimni), revised roughly every three to five years. High-rise developments, especially in Limassol, are typically approved through case-by-case allowances that can override standard zoning density. This unpredictability forces valuers to run multiple scenarios when appraising land or unfinished projects. Greece, by contrast, regulates development through municipal- or prefecture-level planning, leading to significant variations in density limits and approval times. Although Athens and some popular islands have experienced a wave of regeneration and tourism-driven building, rural municipalities can lag in updating their planning regulations, slowing investment and complicating valuations.

Comparative taxation

Taxation in both countries directly affects property values and investor sentiment. Cyprus imposes transfer fees at a sliding scale of 3–8%, plus a 20% capital gains tax on most property sales, though there are exemptions and allowances for primary residences. Immovable Property Tax (IPT) was abolished in 2017 and replaced by generally lower municipal charges, easing the annual burden on owners.

Greece takes a different approach: Property transfers ordinarily incur a rate of around 3%, supplemented by minor municipal surcharges. However, when purchasing new builds issued a construction license after 2006, buyers may face VAT at 24% instead of the transfer tax – though the government periodically suspends or reduces this measure to stimulate the market. Annual property taxes, including the Unified Property Ownership Tax (ENFIA), can become costly. ENFIA is calculated on 'objective values' set by the state, which have gradually been raised in popular neighbourhoods, significantly increasing the recurring tax obligations for higher-value or multiple-property owners.

Transparency

Market transparency remains a critical concern in both countries. In Cyprus, official transactions are recorded at the Department of Lands and Surveys, but persistent under-declaration of sale prices can distort the data. Details such as property age, condition, or inclusion of VAT are not consistently itemised, and the declared purchase amount often diverges from the real sum exchanged. Greece, despite ongoing efforts to modernise its land registry, still suffers from fragmented or incomplete public records, especially in rural and island regions, where historical inheritance patterns complicate title clarity. Many valuations rely on private data, local agent input, or short listings of comparable properties offered for sale. In both jurisdictions, municipalities or national authorities periodically revise the government-assigned property values that serve as a tax basis, yet accurate, up-to-date information is not always easily accessible. Professional bodies like RICS have advocated for more uniform, transparent reporting; however, wide-ranging adoption remains a work in progress. The result is that valuers, investors, and financiers often must cross-reference multiple sources to build a reliable market snapshot, leaving room for uncertainty and inconsistent appraisal outcomes.

Property indices

Cyprus is served by a few well-known indices that attempt to track market performance from slightly different perspectives. The RICS, with KPMG Cyprus Property Index, uses standardised hypothetical properties to provide quarterly snapshots of price trends, while the Central Bank of Cyprus tracks valuations used for mortgage purposes, offering insight into bank-influenced metrics. A transaction-based index produced by the Cyprus Statistics Authority relies on data from the Department of Lands and Surveys, though under-declaration remains a concern. An additional reference, the Neapolis University Property Index, has emerged from academic research to capture broader market signals by combining transaction records, listing data, and localised economic factors. Each index portrays the real estate sector with a different sampling or methodological focus, leading to occasional discrepancies in reported trends but collectively providing valuable indicators of overall direction. Greece relies primarily on the central bank's Residential Property Price Index, which draws on valuation data for mortgage lending, although certain private or region-specific research efforts also generate supplementary datasets. The relative scarcity of fully transparent, transaction-based data has so far limited the availability of more localised or niche indices in the Greek market.

Market conditions in 2025

Both Cyprus and Greece have continued to attract foreign investors, even after the termination of the Citizenship by Investment programme in Cyprus and an increased minimum threshold for Greece's Golden Visa in selected locations. Major urban centres in Cyprus have seen further construction of high-rise complexes, mixed-use waterfront districts, and resort-style facilities designed to enhance tourism and residential demand. In parallel, smaller coastal cities in Cyprus have pursued large-scale marina expansions, local hotel clusters, and family-oriented holiday communities, providing opportunities for mid-tier international buyers. Greece has undergone multiple urban renewal efforts, notably in Athens, where significant portions of older industrial or underused land near the coast have been redeveloped into upscale residential blocks, large green spaces, and waterfront leisure zones. Tourism-oriented locations on the mainland and on various islands have likewise expanded or upgraded existing resorts, integrating luxurious amenities and timeshare villa concepts, but in some areas, greater restrictions have emerged to address overtourism and protect local heritage. Meanwhile, the second-largest city in Greece has improved its waterfront zones, consolidating a cultural and commercial district that draws interest from local and regional entrepreneurs. Although rising interest rates since 2023–2024 have moderated some of the speculative enthusiasm, these major undertakings indicate that the property markets in both countries have embraced strategic, high-profile developments intended to attract international capital and rejuvenate key urban and coastal locales. Signs of market evolution remain uneven, with distressed sales persisting due to non-performing loans, yet the emphasis on large integrated complexes and resort-style projects illustrates a shift toward higher-value, long-term investments in prime and peripheral regions.

Valuers' challenges

Valuers must navigate data gaps, measurement inconsistencies, and the presence of distressed sales from NPLs. In both countries, banks and asset management firms manage portfolios of repossessed properties and frequently auction them at below typical market values, complicating comparable data analysis. The under-declaration of transaction prices is also widespread, as buyers and sellers sometimes reduce official amounts to minimise transfer fees or capital gains taxes. Municipalities in both nations update their objective or official values sporadically, creating sudden shifts in recurring property tax liabilities (notably ENFIA in Greece) and, in turn, altering how properties are perceived in the market. Where planning is fluid, such as in Cyprus's high-rise approvals or Greece's local zoning variations, valuers often resort to multiple-scenario valuations that account for potential extra densities and lengthy permit timelines.

Looking ahead

Future prospects depend on several interrelated factors. In Cyprus, foreign demand is likely to remain centred on coastal developments and prime city cores. Although the termination of the Citizenship by Investment programme reduced the volume of high-end international buyers, a stable flow of visitors, retirees, and regional investors continues to support the market. Greece's broad-based recovery remains dependent on tourism, urban regeneration projects (including the long-awaited Hellinikon development in Athens), and a strong appetite for Golden Visa–eligible properties. At the same time, enforcement of capital gains tax rules, repeated adjustments to ENFIA, and ongoing bureaucracy could affect investor enthusiasm.

Overall, these Mediterranean markets have made progress since 2017. They offer appealing opportunities, especially for overseas buyers, but valuers and their clients must remain vigilant. Key demands include reliable measurement, consistent tax policy, expedited planning approvals, and robust, publicly accessible transaction data. As of 2025, investors and valuation professionals alike find themselves treading a path toward greater transparency, though it remains a work in progress, shaped by local tradition, new legislative initiatives, and the broader global economic climate.

United Arab Emirates

This section is an update of material kindly provided by David Cockerton, BSc, FRICS, FInstCPD, FCMI, MCIArb, Principal, Hyperion Consulting Group Limited, and Lecturer, Heriot Watt University, Edinburgh and Dubai.

Culture and history

The UAE comprises seven separate emirates, each governed individually with increasing but still limited consolidation at federal level. Collectively, the UAE has been one of the world's most rapidly growing economies over the past 20 years, with real estate playing an integral part in this growth. Although still a commodity-based economy, with shipments of oil and natural gas accounting for around 40% of total exports and GDP, increasing investment in the tourism, financial, transportation, and construction sectors has enabled the UAE to both grow and diversify its economy.

As of January 2024, Dubai's GDP was estimated at United Arab Emirates dirham (AED) 429 billion (US$ 116.779 billion). Over the past few decades, the city has transformed into a global business and financial hub. *The International Herald Tribune* once described Dubai's economic model as 'centrally planned free-market capitalism'. Oil production, which once contributed 50% of Dubai's GDP, now accounts for less than 1%.

Looking at the UAE as a whole, using data from the Federal Statistics Centre, demonstrates the dramatic shift from oil to other drivers of GDP. In 1975, oil accounted for 57% of GDP, whilst in 2020, it only accounted for 17%, with unprecedented growth in other sectors over the same period, especially manufacturing (0.53 to 127.62 billion AED) and construction (8.96 to 122.672 billion AED).

Over this period, the UAE, and particularly Dubai, has seen unprecedented real estate development, initially through infrastructure development; then in the retail, hospitality, and residential asset classes; and, more recently, industrial, commercial office, and theme park development. The other emirates have adopted a similar approach to Dubai, but the latter remains the most significant emirate from a real estate perspective.

As of 2024, Dubai's population is estimated at 3,051,016, reflecting a 1.44% annual growth rate. Notably, only 15% of the population comprises native Emiratis, while 85% are expatriates. This constant growth in GDP, population, and wealth has been a huge stimulus to property development. Much of the real estate development has been government-driven either directly or through government-owned development companies. This has included much of the speculative development undertaken both prior to and following the global financial crisis in 2008. This is partly due to the use of real estate to stimulate further economic growth, but also a reflection of the tight control government has over real estate as an asset.

Market maturity

Historically, most of the real estate development in the UAE, particularly residential, was undertaken both speculatively and through the use of off-plan sales. The latter was used as the principal mechanism for funding many projects with well-publicised consequences when the market contracted in the wake of the global financial crisis.

The off-plan sales approach was a contributory factor in the development of quantity rather than quality. As the market has matured and become more competitive, there has been a greater focus on both occupier and investor needs in response to their 'flight to quality' now that the market provides a choice. Until recently, only locals could purchase property outright and expats visas were cancelled at retirement age, typically 60. However, with the introduction of free zones and other areas that allow international ownership and retirement visas, the demand for many new forms of residential development has fuelled further growth and increasing maturity, especially in Dubai and Abu Dhabi.

The UAE residential real estate market was valued at around US$36.32 billion in 2024 and is projected to reach US$52.32 billion by 2030. Along with this, the market is estimated to grow at a compound rate of around 5.1% in 2025–2030.

Another consequence of the global financial crisis was the development of a greater understanding of the various stages in the real estate life cycle. Master developers often took on the role of subdevelopers to see the completion of master-planned developments and, in turn, also became owner-investors and asset and property managers, there being little or no opportunity for sales. In parallel, banks found themselves with significant real estate product in their portfolios.

The UAE has been at the forefront of expecting high standards of sustainability in commercial development, and there are many landmark buildings with the highest credentials, such as LEED Platinum. Whereas in the past, developments were very fragmented and often suffered from the lack of integration of infrastructure, this has now evolved, and master-planning and placemaking are both common and expected. Most office towers in Dubai and Abu Dhabi now have extensive amenities, including co-working as an essential component.

In 2024, the government launched a 'Golden Visa' programme, granting long-term residency to investors, entrepreneurs, and skilled workers. This move has made it more attractive for foreign nationals to buy property in Dubai and other cities, especially with the promise of residency rights linked to property ownership. Predictions for 2025 show that this programme will continue to attract high-net-worth individuals and will likely help bring more stability to the market.

At the other end of the market, the recent introduction of digital nomad visas, allowing both temporary residency and easy company formation, has fuelled a lot of interest in contemporary real estate products, such as co-working and co-living spaces, especially in free zones such as Expo City.

Transparency

Historically, the UAE real estate market has been classified as opaque, with very limited market data available. This is partly due to the nature of the principal players being government-based but also, in part, cultural.

Although there is no single market-wide property performance index, following the post-crisis introduction of various regulatory frameworks and registration requirements, transaction data as well as various market tracking indices are now starting to become available, albeit almost exclusively focused on the residential market.

Increasingly wider access to market information is expected not only to ease and assist market activity but also to develop market confidence, which, in time, should lead to increased investor interest in the market, particularly from global investors who, to date, have not invested heavily in the region.

According to JLL (2024), the UAE has made significant advances in global real estate market transparency, with Dubai and Abu Dhabi securing a place among the top five improvers in the transparency score globally, according to JLL's 2024 Global Real Estate Transparency Index (GRETI). Dubai was in third place, and Abu Dhabi fifth, for improvements made from 2022 to 2024, and both have been recognised among the top global improvers for scaling digital capabilities, implementing strong AML and beneficial ownership (BO) regulations, and increasing the availability of market data for established and start-up market providers.

Legal tenancy, security of tenancy (income), and land law

Given the importance of real estate in the UAE economy, the challenge for government has been to introduce the legislation and regulations necessary to provide a robust and comprehensive framework from which to protect and legislate for all parties' rights and obligations.

Dubai has been at the forefront of both legislative and regulatory development, including:

- The creation of the Real Estate Regulatory Authority (RERA), responsible for developing and administering the real estate regulatory framework
- The introduction of escrow law to provide protection for purchasers' money invested in off-plan sales in developments under construction, supplemented by an Interim Real Estate Register requiring off-plan sale contracts to be registered on a government register
- The establishment of a framework for registration of lenders' pre-mortgage interests on the Interim Real Estate Register and mortgages on the main Real Estate Register, together with mechanisms for enforcement
- The introduction of Strata Law, empowering property owners to jointly manage their communities and buildings and, at the same time, providing transparency and fairness on issues such as service charge levels, contracts with service providers, and the use of common areas and facilities
- The introduction of a requirement for both individuals and entities to obtain a licence prior to undertaking real estate activities

Other emirates have introduced similar legislation and regulatory frameworks, each largely with the objective of providing a platform for increased confidence in the real estate market.

Two other legal areas of interest are:

- *Security of tenure.* No such rights exist for tenants throughout the UAE.
- *Rent cap.* With variations in place for each emirate, rent increases are generally controlled by legislation, being either subject to a cap or relative to an index.

Land use pattern, planning, and sustainability

Unlike other countries in the region, most emirates now permit 'foreigners' to own real estate, but only in certain areas ('investment zones'). Outside these areas, ownership is restricted to Gulf Cooperation Council (GCC) nationals/entities, although ownership on the main island in Abu Dhabi remains restricted to UAE nationals/entities.

Each emirate has a well-developed land use and planning framework, the most sophisticated being in place for Abu Dhabi following the publication of the Abu Dhabi Plan 2030 administered by the Urban Planning Council.

Sustainability requirements are now included in the planning framework, Estidama for Abu Dhabi and Emirates Green Building Council for Dubai. In Abu Dhabi, all developments must meet minimum local Estidama requirements.

Professional practice

Whereas several global real estate professional bodies have provided input and advice to UAE governments, the RICS has probably developed the closest ongoing working relationships throughout the region.

In addition to providing input into real estate legislation and regulation, the RICS has, for the past ten years, also been at the forefront in encouraging each emirate to embrace global best practice and standards, an example being IPMS now fully adopted by both the Dubai and Abu Dhabi governments and gradually being rolled out into each market.

Although significant work is still required in this area, the RICS has made significant progress across all areas of professional practice and standards to establish a platform for enhanced investor confidence in the region.

Real estate practitioners require a government license to operate, and Abu Dhabi has introduced its own code of conduct and professional ethics for real estate practitioners. Whilst mirroring many of the RICS professional regulations, it also focuses on community values.

Other local issues that impact upon valuation

Other areas of interest in the UAE real estate market include:

- *Indirect property investment vehicles.* Although REIT legislation has been in place in Dubai since 2010, take-up has been generally slow with Emirates REIT being the only operational REIT in the UAE. However, with REIT legislation now in place in Abu Dhabi and its international financial centre, Abu Dhabi Global Market (ADGM), also established to rival its Dubai equivalent, Dubai International Finance Centre (DIFC), the UAE has recently seen a proliferation of new REITs being established. This trend is expected to continue as investors seek the liquidity, diversity, and low price point benefits supported by a highly regulated operational framework which provides investor confidence.
- *Taxation.* The UAE introduced a 5% rate of VAT on 1 January 2018. The UAE's VAT law differentiates between residential and commercial properties:
    - Commercial properties are taxed at the standard 5% VAT rate.
    - Residential properties and bare land are typically VAT-exempt.

References/Further Reading

Akkerman, A., & Nunnington, N. (2025) *White Paper: Navigating the Investor-occupier Paradigm Shift in a Rebalancing of Powers in the Real Estate Value Chain.* www.workplaced.com/insights/whitepapers

Apostolidis, S., Dimopoulos, T., & Katafygiotou, M. (2023) A proposal for a residential housing price index in cyprus through analysis of transaction-based data and comparison with existing indices. *Journal of Real Estate Literature*, 31(2), 73–124. https://doi.org/10.1080/09277544.2023.2236175

Arora, S. (2017) Changing dynamics of corporate real estate: The rise of coworking spaces. *Corporate Real Estate Journal*, 7(2), 127–136.
Austin, G.W. (2012) Sustainability and income producing property valuation: North American status and recommended procedures. *JOSRE*, 4(1), 78–122.
Avison Young (2019) *Avoid a Void: How Landlords are Responding to the Serviced Offices Boom*. www.avisonyoung.co.uk/documents/38901/59345308/Avoid+a+void.pdf/403143d4-1083-422d-8963-6310f4a2225b (Accessed: 7 April 2021).
Bohne, R. (2023) *Workstation Occupancy Rate of International Workplace Group (IWG) Worldwide from 2012 to 2021, by Region*. www.statista.com/statistics/554048/occupancy-rate-of-regus-by-region/
Boyd, T.P. (2005) 'Assessing the triple bottom line impact on commercial buildings', A. Sidwell (ed.), *QUT Research Week, Conference Proceedings*, Brisbane, Australia.
Britannica (2025) 'Mou' Chinese unit of land measurement. *briticannica.com*. www.britannica.com/science/mou (Accessed: 10 March 2025).
British Council for Offices (2019) *The Rise of Flexible Workspace in the Corporate Sector*. www.bco.org.uk/Research/Publications/The_Rise_of_Flexible_Workspace_in_the_Corporate_Sector.aspx (Accessed: 17 May 2020).
CBRE (2019a) *Investor Intentions Survey 2019*. www.cbre.com/research-and-reports/Global-Investor-Intentions-Survey-2019 (Accessed: 22 October 2019).
CBRE (2019b) *The Property Value Implications of Flexible Space*. www.cbre.us/research-and-reports/US-Property-Value-Implications-of-Flexible-Space-January-2019 (Accessed: 22 October 2019).
Chegut, A., Eichholtz, P., Kok, N., & Quigley, K.M. (2010) *The value of green buildings new evidence from the United Kingdom*. Working paper. www.propertyfinance.it/sitoeres/contents/papers/id150.pdf
Chegut, A., & Langen, M. (2019) The financial impacts of coworking: rental prices and market dynamics in the commercial office market. *SSRN Electronic Journal* [Preprint]. https://doi.org/10.2139/ssrn.3481142
Deloitte (2014) *Breakthrough for Sustainability in Commercial Real Estate*. https://www2.deloitte.com/content/dam/Deloitte/us/Documents/financial-services/us-fsi-breakthrough-for-sustainability-in-real-estate-051414.pdf (Accessed: 5 December 2016).
Dimopoulos, T., Labropoulos, T., & Hadjimitsis, D.G. (2014, 12 August) 'Comparative analysis of property taxation policies within Greece and Cyprus evaluating the use of GIS, CAMA, and remote sensing techniques', *Proceedings of SPIE – Second International Conference on Remote Sensing and Geoinformation of the Environment (RSCy2014)*, p. 9229. https://doi.org/10.1117/12.2070457
Duffy, F. (2003) 'Office buildings and organisational change', in *Buildings and Society: Essays on the Social Development of the Built Environment*, pp. 140–156. www.taylorfrancis.com/chapters/edit/10.4324/9780203480755-18/office-buildings-organisational-change-francis-duffy
Eccles, T., Haynes, B.P., & Nunnington, N. (2017) *Corporate Real Estate Asset Management: Strategy and Implementation*, 2nd edition, Taylor & Francis, Abingdon.
Echeverri, N., Jylhä, T., & Koppels, P. (2021) Searching for flexibility in corporate real estate portfolio: Six co-working strategies for user corporations. *Buildings*, 11(3), 115. https://doi.org/10.3390/buildings11030115
Fuerst, F., & McAllister, P. (2011) The impact of energy performance certificates on the rental and capital values of commercial property assets. *Energy Policy*, 39(10), 6608–6614.
Fuzi, A. (2015) Co-working spaces for promoting entrepreneurship in sparse regions: The case of South Wales. *Regional Studies, Regional Science*, 2(1), 462–469. https://doi.org/10.1080/21681376.2015.1072053
Golding, C. (2019) *Valuation of Flexible Workspace*. London. www.rics.orgISBN9781783213771 (Accessed: 22 October 2019).
Green, R. (2014) Collaborate or compete: How do landlords respond to the rise in coworking?. *Cornell Real Estate Review*, 12, 52–59. https://scholarship.sha.cornell.edu/crer (Accessed: 22 October 2019).
Greenwood, R., Smith, D.C., Daniels, D., & Bitner, D. (2018) *Coworking and Flexible Office Space: Additive or Disruptive to the Office Market?* www.cushmanwakefield.com/en/united-states/insights/2018-coworking-report (Accessed: 11 April 2021).
Harris, R. (2015) The changing nature of the workplace and the future of office space. *Journal of Property Investment & Finance*, 33(5), 424–435. https://doi.org/10.1108/JPIF-05-2015-0029
Heinzel, V., Georgiades, S., & Engstler, M. (2021) 'Corporate coworking – a catalyst for collaboration, creativity, and innovation', in Orel, M., Dvouletý, O., & Ratten, V. (eds.), *The Flexible Workplace. Human Resource Management*, Springer, Cham. https://doi.org/10.1007/978-3-030-62167-4_5

Hickey, S. (2015) The history of the office – why open-plan fell out of fashion. *The Guardian*. www.theguardian.com/small-business-network/2015/oct/15/history-office-open-small-business-workplaces (Accessed: 1 November 2023).

House of Commons (2022) *Building to net zero: Costing carbon in construction - First Report of Session 2022–23*. https://committees.parliament.uk/publications/22427/documents/165446/default/

IPF (2020) *Property Ownership in a Flexible World*. London. www.ipf.org.uk/static/uploaded/dce81df9-a1a8-4bc5-a1a8aff2b3a8a7f7.pdf (Accessed: 4 April 2021).

IPMS (2023) International property measurement standards: All buildings. *www.ipms.org*. International Property Measurement Standards Coalition (IPMSC). https://ipmsc.org/wp-content/uploads/2023/01/ipms-all-buildings-.pdf (Accessed: 10 March 2025).

IVSC (2024) *International Valuation Standards*. London: International Valuation Standards Council.

Jamal, A.C. (2018) Coworking spaces in mid-sized cities: A partner in downtown economic development. *Environment and Planning A: Economy and Space*, 50(4), 773–788.

JLL (2021) *The Future of Flex*. www.jll.co.uk/en/trends-and-insights/research/the-future-of-flex (Accessed: 8 February 2022).

JLL (2024) The global real estate transparency index 2024. *us.jll.com*. www.us.jll.com/en/trends-and-insights/research/global-real-estate-transparency-index (Accessed: 10 January 2025).

Johnston, W., & D'Souza, R. (2019) *How morningstar handles the unique risks we work tenants present to office properties*. www.morningstarcreditratings.com (Accessed: 5 April 2021).

Josef, B., & Back, A. (2019) Coworking aus unternehmenssicht-out of office, into the flow? *HMD*, 56, 780–794.

Kern, A. (2019) *How Landlords are Reaping the Benefits of Coworking, Commercial Property Executive*. www.cpexecutive.com/post/how-landlords-are-reaping-the-benefits-of-coworking/ (Accessed: 1 April 2021).

Knight Frank (2020) *Landlord-operated co-working Spaces: An Upcoming Trend?*

Knight Frank (2023) *ESG Property Investors Survey*.

Lorenz, D., & Lützkendorf, T. (2011) Sustainability and property valuation: Systematisation of existing approaches and recommendations for future action. *Journal of Property Investment & Finance*, 29(6), 644–676.

Lu, Z., Zhang, S., & Zhao, D. (2022) *Tax Me if You Can: Tax Evasion on Chinese Housing Market*. https://cre.mit.edu/wp-content/uploads/2023/03/Zhang-Sisi-Tax-Me-If-You-Can.pdf (Accessed: 10 January 2025).

Mei, L., & Ma, K. (2020, 18 December) PRC asset appraisal requirements and considerations. *moore.hk*. www.moore.hk/news/transaction-and-valuation/prc-asset-appraisal-requirements-and-consideration (Accessed: 10 November 2024).

Morgan Stanley (2018) *Coworking, Friend or Foe?* www.morganstanley.com/im/publication/insights/investment-insights/ii_coworkingfriendorfoe_us.pdf (Accessed: 22 October 2019).

Muldavin, S.R. (2010) *Value Beyond Cost Savings: How to Underwrite Sustainable Properties*. San Rafael: Green Building Finance Consortium.

Nareit (2024) *REIT Industry Sustainability Report 2024*. www.reit.com/investing/reits-sustainability/2024-reit-sustainability-report-resources

Nikiforou, P., Dimopoulos, T., & Sivitanides, P. (2022) Identifying how the time on the market affects the selling price: a case study of residential properties in Paphos (Cyprus) urban area. *Journal of European Real Estate Research*, 15(3), 368–386. https://doi.org/10.1108/JERER-11-2021-0051

Parker, T. (2020), For good measure. *Property Journal*, 34–35.

Porumb, V., Maier, G., & Anghel, I. (2020) The impact of building location on green certification price premiums: Evidence from three European countries. *Journal of Cleaner Production*, 272, 1–11. www.sciencedirect.com/science/article/pii/S0959652620321272

RICS (2024a). *RICS Valuation – Global Standards*. www.rics.org/content/dam/ricsglobal/documents/standards/Red-Book-Global-Standards-incorporating-IVS.pdf

RICS (2024b). *ESG Data List for Real Estate Valuations: A Practical Reference Document on Legislative, Market-Driven and Future ESG Requirements for Valuers and Financial Clients in the EU*. www.rics.org/content/dam/ricsglobal/documents/latest-news/WBEF-ESG-and-valuation-2024-data-list.pdf

Rumpfhuber, A. (2017) *Non-places of Immaterial Labour*, 1st edition. Verlag der Bauhaus-Universität.

Sankari, I. (2019) *Co-working Space as Workplace – Characteristics and User Experience*. Aalto University.

Sargent, K., Cooper, J., Mellwig, B., & McDonald, M. (2018) Coworking and the disruption of the current corporate real estate model. *Corporate Real Estate Journal*, 7(3), 267–276.

Savills (2024) *Spotlight: UK Flexible Offices – 2024*. www.savills.co.uk/research_articles/229130/370560-0

Seo, J., et al. (2017) Priorities of coworking space operation based on comparison of the hosts and users' perspectives. *Sustainability (Switzerland)*, 9(8). https://doi.org/10.3390/su9081494

Silva, F.A.B., Liu, N., & Hutchison, N. (2022) Flexible workspace providers as tenants: an analysis of the rental prices in the London market. *Journal of Property Investment & Finance*. https://doi.org/10.1108/JPIF-11-2021-0096

Spinuzzi, C. (2012) Working alone, together: Coworking as emergent collaborative activity. *Journal of Business and Technical Communication*, 26(4), 399–441.

Stoyanchev, A. (2024) *Occupancy Rates: What Numbers Indicate a Successful Coworking Space?* www.officernd.com/blog/occupancy-rates-guide/

Van Meel, J.J. (2000) *The European Office: Office Design and National Context*. Rotterdam: 010 Publishers.

Walsh, J. (2015) Designing work: A study of collaboration and concentration in open-plan offices open-plan offices. *Iterations*, (2), 44–49.

Weinbrenn, J. (2016) Coming of age: Is demand for flexible workspace changing the office market? *Corporate Real Estate Journal*, 5(3), 248–260.

The We Company (2019) *Form S-1: The We Company*. www.sec.gov/Archives/edgar/data/1533523/000119312519220499/d781982ds1.htm (Accessed: 22 October 2024).

Weijs-Perrée, M., van de Koevering, J., Appel-Meulenbroek, R., & Arentze, T. (2019) Analysing user preferences for co-working space characteristics. *Building Research and Information*, 47(5), 534–548. www.tandfonline.com/doi/full/10.1080/09613218.2018.1463750 (Accessed: 22 October 2019).

Wheaton, W.C., & Krasikov, A. (2019) *Will CoWorking Work?* https://papers.ssrn.com/sol3/papers.cfm?abstract_id=3784792 (Accessed: 19 January 2020).

World Green Building Council (2013) *The Business Case for Green Building: A Review of the Costs and Benefits for Developers, Investors and Occupants*. https://worldgbc.org/wp-content/uploads/2022/03/Business_Case_For_Green_Building_Report_WEB_2013-04-11-2.pdf

World Green Building Council (2019) *Bringing Embodied Carbon Upfront*. https://worldgbc.org/advancing-net-zero/embodied-carbon/

Chapter 14

Evaluating valuers

Introduction

Valuers play a pivotal role in safeguarding the integrity of real estate markets and bolstering economic stability. By providing unbiased assessments founded on established valuation methodologies, they support investors, lenders, government bodies, and other stakeholders in making informed decisions. Evaluating valuers therefore becomes essential for ensuring accurate, consistent, and transparent valuation outcomes. Core criteria include qualifications, adherence to professional standards, and continuous professional development – elements showcased through designations such as MRICS and REV, which underscore competence and commitment to ethical practice. Moreover, robust professional indemnity insurance (PII) requirements protect the public interest by guaranteeing that clients are safeguarded from losses arising from negligence or error. Peer reviews, regulatory oversight, and benchmarking further strengthen trust in the valuation process, mitigating risks like data inaccuracies or conflicts of interest. Ultimately, by upholding rigorous codes of conduct, staying current with market trends and technological advancements, and demonstrating methodological diligence, professional valuers significantly enhance transparency, confidence, and economic resilience in the property market.

Responsibilities of a valuer

A valuer is responsible for conducting thorough property inspections and market research to gather accurate data. The analysis of sales transactions, rental data, and market trends allows for a well-founded valuation. Additionally, valuers provide reports that comply with legal and professional standards, ensuring transparency in valuation methodology and justification of figures. Their expertise is also utilised in court cases, disputes, and arbitration proceedings, offering expert opinions that hold weight in legal matters. To maintain credibility, valuers must stay updated on changes in property laws, tax regulations, and economic conditions. So as mentioned in Chapter 4, the following are critical:

- Independence and integrity
- Compliance with standards
- Investigation and due diligence
- Client care and public interest protection
- Continuous professional development
- Legal and ethical obligations

DOI: 10.1201/9781003490357-17

By adhering to these responsibilities, valuers contribute to a transparent, efficient, and reliable property market, ensuring trust among investors, regulators, and clients.

Role of valuers in supporting a positive economic environment

Valuers play a crucial role in supporting a positive economic environment by ensuring transparency, stability, and confidence in the real estate market. Their assessments provide the foundation for financial decision-making, economic planning, and investment security, making them indispensable to economic growth and stability. Accurate and reliable property valuations contribute to efficient markets by determining fair prices for real estate transactions, preventing market distortions, and reducing the risks of speculative bubbles. Without their expertise, real estate markets could suffer from inefficiencies, leading to overpricing or undervaluation, which could, in turn, result in financial crises or economic downturns.

A key contribution of valuers to economic stability lies in their support for the banking and financial sectors. Banks rely on professional valuations to assess the collateral value of properties when granting loans, ensuring that lending decisions are based on accurate assessments of asset worth. This reduces the risk of loan defaults and financial instability, as witnessed in past real estate–driven economic crises. By providing objective and data-driven valuations, valuers help financial institutions manage credit risks effectively, protecting both lenders and borrowers from financial losses. Investor confidence is another critical area where valuers contribute to a healthy economic environment. Both domestic and international investors depend on accurate property valuations to make informed decisions about acquisitions, developments, and sales.

When investors trust the valuation process, they are more likely to commit capital to real estate projects, stimulating economic growth and job creation. Furthermore, institutional investors such as pension funds, real estate investment trusts (REITs), and asset managers require professional valuations to comply with financial regulations and to assess the performance of their property portfolios' compensation.

Real estate development also benefits significantly from professional valuations. Developers rely on valuation reports to determine project feasibility, profitability, and market demand. By assessing rental yields, potential growth, and market conditions, valuers reduce the risk of overbuilding or investing in projects that may not be financially viable.

Furthermore, by complying with anti-money laundering (AML) regulations, they help prevent real estate fraud and illicit financial activities. Another important contribution of valuers to the economic environment is their role in ensuring housing affordability and social equity. By assessing property values in the affordable housing sector, they help policymakers determine rent controls, subsidies, and housing initiatives that benefit lower-income groups.

Preventing speculative real estate practices that drive up prices beyond affordability is essential for social stability and economic inclusivity. Overall, valuers are integral to economic growth and stability. Their work supports the real estate sector, a major contributor to GDP in most economies, by ensuring financial security, investment confidence, and regulatory compliance. Accurate valuations facilitate construction, infrastructure development, and property market stability, all of which contribute to a robust economy. Moreover, valuers help attract foreign direct investment (FDI) in real estate, further stimulating economic expansion and development.

In essence, professional valuations serve as the backbone of a well-functioning property market, benefiting businesses, governments, and society as a whole. Through a process of accountability and evaluation, high professional and ethical standards can be maintained.

Key criteria for evaluating valuers

The importance of the property valuation profession is undisputable. Evaluating valuers requires a comprehensive approach that considers multiple factors to ensure their assessments are accurate, reliable, and professionally conducted. These criteria help determine whether a valuer has the necessary qualifications, knowledge, ethical integrity, and technical expertise to provide high-quality valuations. The following outline the primary aspects that should be considered when assessing the competency and performance of a valuer.

Qualifications and professional standards

Academic qualifications play a fundamental role in establishing the credibility of a valuer. A strong educational background in real estate, finance, economics, or a related discipline provides the theoretical foundation necessary for sound valuation practices. Formal education equips valuers with essential skills, such as financial analysis, property law comprehension, and market research proficiency. Furthermore, obtaining advanced degrees or specialised certifications demonstrates a commitment to professional development and enhances a valuer's ability to apply industry best practices. The integration of academic learning with practical experience ensures that valuers can navigate complex valuation scenarios with competence and confidence. A valuer's qualifications and adherence to professional standards serve as fundamental evaluation criteria. Membership in recognised professional bodies, such as the RICS, AI, American Standards Association (ASA), or TEGoVA, establishes credibility. Compliance with regulatory guidelines, including USPAP (2023) or IVS (IVSC, 2024), and possessing academic credentials in real estate, finance, or related fields further enhance a valuer's professional standing. CPD is essential for keeping up with industry trends, new regulations, and technological advancements.

CPD

CPD is vital for maintaining and enhancing professional competency, especially in a dynamic field like real estate and property valuation. As market conditions, regulations, and technological tools evolve, it is essential to stay current with the latest industry insights and best practices. CPD encompasses structured learning, formal training, and participation in conferences or seminars, all of which help professionals:

- *Stay updated.* Regular learning ensures familiarity with recent legislation, emerging trends, and new valuation methodologies.
- *Enhance skills and knowledge.* Engaging in targeted workshops or courses refines practical skills, sharpens analytical abilities, and deepens understanding of complex valuation scenarios.
- *Uphold professional standards.* Many professional bodies (such as RICS, TEGoVA, or national chambers) mandate CPD to ensure that members consistently meet high ethical and technical standards.
- *Adapt to innovation.* From advanced software tools to AI-driven mass appraisals, CPD equips professionals with the know-how to leverage innovative approaches that elevate the accuracy, efficiency, and value of their services.
- *Strengthen client trust.* Demonstrating a commitment to ongoing learning and expertise reinforces credibility and fosters confidence among clients, stakeholders, and the broader market.

By embracing CPD as part of a professional routine, valuation specialists and property consultants can ensure that their services remain relevant, reliable, and competitive in today's fast-evolving real estate environment.

Methodological competence

The ability to apply appropriate valuation methodologies is a key aspect of a valuer's competency. The *market approach*, also known as the comparative method, involves analysing recent property sales to determine value. The *income approach*, often referred to as the investment method, evaluates the present value of future income streams. The *cost approach*, which is based on the depreciated replacement cost, assesses property value by determining the cost of rebuilding minus depreciation. Justification of assumptions, adjustments, and data sources is necessary to ensure the reliability of valuation results. The use of reliable databases and analytical tools contributes to enhancing valuation accuracy.

Accuracy and consistency

A valuer's ability to deliver accurate and consistent valuations is crucial. Comparing valuations with actual transaction prices and third-party appraisals provides insights into reliability. Consistency in valuations across similar property types and market conditions further establishes credibility. Tracking historical performance helps in evaluating error rates and making necessary adjustments to improve valuation accuracy.

Ethical and legal compliance

Ethical conduct and legal compliance play a significant role in evaluating valuers. Adherence to ethical codes, such as those outlined by RICS, USPAP, AI, and TEGoVA, ensures professionalism and integrity in valuation practices. Avoiding conflicts of interest and maintaining impartiality in reporting strengthen the trustworthiness of valuations. Compliance with regulatory frameworks and industry best practices is essential, along with transparency in reporting limitations, potential biases, and data constraints.

Market knowledge and experience

A valuer must possess a deep understanding of local and international real estate markets. Experience in valuing various property types, including residential, commercial, industrial, agricultural, and specialised assets, is critical. Familiarity with macroeconomic factors, such as interest rates, inflation, zoning laws, and urban development trends, enables valuers to provide more accurate and relevant valuations.

Use of technology and innovation

The adoption of modern technology is essential for improving valuation accuracy. AVMs and big data analytics assist in precise valuation estimations. The integration of GIS, blockchain technology, and machine learning enhances valuation processes. The ability to utilise digital platforms for real-time data collection, market comparisons, and predictive analytics is beneficial.

Understanding the impact of sustainability and environmental considerations on property values is becoming increasingly important in valuation practices.

Client satisfaction and professional reputation

Client satisfaction and professional reputation are important measures of a valuer's competence. Positive feedback from clients, investors, and industry stakeholders indicates reliability and professionalism. Recognition and rankings in industry surveys and professional associations contribute to credibility. The resolution of disputes, complaints, and legal challenges related to valuation accuracy is a key aspect of maintaining a good reputation. The ability to provide clear, well-documented, and easily understandable reports enhances trust in valuation results.

Methods of evaluating valuers

Peer review and audit

Peer review and audit processes are essential for ensuring accuracy, compliance, and continuous improvement in the field of real estate and property valuation. By subjecting work to critical examination, whether by fellow professionals with external assessors or through structured internal audits, valuers and consultants can proactively identify potential errors, inconsistencies, and areas for refinement. Some key benefits include the following:

- *Enhanced quality control.* A structured peer review process ensures that valuation reports and analyses meet the highest standards. Feedback from experienced colleagues helps maintain consistency in methodologies, adherence to valuation standards (e.g. IVS or RICS guidelines), and clarity in reporting.
- *Compliance and risk management.* Audits, whether external (e.g. by professional bodies) or internal (within a firm), systematically verify that valuation procedures comply with legal regulations, ethical codes, and industry best practices. This mitigates the risk of legal challenges, regulatory fines, or reputational damage.
- *Transparency and accountability.* Both peer review and audit reinforce a culture of openness. Sharing work for review underpins accountability, as professionals must justify and document their approaches. This transparency ultimately fosters trust among stakeholders – clients, regulators, and peers alike.
- *Continuous learning and improvement.* Feedback from peers provides opportunities to identify gaps in knowledge and refine professional skills. Audits highlight systemic issues and guide process enhancements that improve overall efficiency and service quality.
- *Strengthening professional credibility.* Demonstrating a commitment to peer review and undergoing regular audits confirm that professionals adhere to rigorous standards. Such diligence bolsters credibility, reinforcing confidence in the quality and reliability of valuations and advisory services.

By embracing peer review and audit, real estate practitioners and valuation professionals not only safeguard the quality of their work but also uphold industry standards, foster an environment of collaboration, and enhance their professional reputation.

Benchmarking and performance metrics

Comparing valuations with market transactions and analysing valuation variances highlight accuracy and reliability. Industry benchmarks provide insight into how individual valuers perform relative to their peers, ensuring that valuation practices align with market trends.

Regulatory and institutional oversight

Professional bodies and regulatory authorities, such as RICS, USPAP, AI, TEGoVA, and national property appraisal boards, set standards, conduct inspections, and enforce compliance through disciplinary actions. Licensing requirements, periodic assessments, and adherence to codes of conduct contribute to maintaining valuation integrity.

Client and stakeholder feedback

Surveys and testimonials from clients, lenders, investors, and legal professionals offer insights into a valuer's reputation, responsiveness, and professionalism. Feedback mechanisms help identify strengths and areas for improvement in valuation services.

Challenges in evaluating valuers

Subjectivity and interpretation differences

Valuation is not an exact science, and different valuers may reach varying conclusions based on subjective interpretations of data. Disparities in approaches can lead to differing property valuations, affecting investor confidence and market transactions.

Data availability and reliability

Access to comprehensive and accurate data is crucial for valuations. However, restricted market data, unreliable transaction records, and outdated valuation models can affect valuation accuracy.

Conflict of interest and bias

Valuers may face pressure from clients, banks, or real estate firms to provide favourable valuations. Ethical breaches can undermine the credibility of valuations and distort market assessments.

RICS-regulated firms, MRICS and REV designations

RICS-regulated firms operate under the rigorous standards and ethical framework set out by the Institution, ensuring that clients receive professional, trustworthy, and high-quality services. These firms have voluntarily agreed to be monitored and guided by RICS, an internationally recognised body dedicated to maintaining consistent best practices in land, property, construction, and infrastructure. This commitment to regulation offers a range of benefits, particularly in terms of consumer protection, technical competency, and transparent conduct, in all surveying and property-related activities. One of the core advantages is the assurance that staff within RICS-regulated firms must fulfil ongoing obligations in terms of continuing professional development. In addition, many professionals within these firms hold the MRICS designation, which denotes individuals who have successfully demonstrated the requisite levels of education, practical experience, and adherence to professional and ethical standards. MRICS membership is recognised globally as a hallmark of professional excellence, underlining a surveyor's competence and dedication to the highest levels of service. Equally important is the stipulation to carry adequate professional indemnity insurance. In the event of any professional negligence or oversight, a client has the necessary safeguards to seek appropriate compensation or remediation. Such insurance is a built-in requirement for RICS membership and represents the institution's unwavering commitment to safeguarding the interests of clients and the broader public.

Beyond practical and financial safeguards, RICS-regulated firms must follow a code of conduct that emphasises integrity, independence, and accountability. This ethical foundation assures clients that any valuation, report, or piece of advice they receive is prepared with rigor and impartiality, whether it pertains to valuation assignments, project management, building pathology reports, or more specialised tasks. Furthermore, RICS-regulated firms are obliged to maintain clear and robust complaints-handling procedures, ensuring that clients know how to seek resolution if they are dissatisfied with any aspect of the service provided. These processes often involve an independent redress scheme, offering a fair and unbiased evaluation of the issue at hand. The global reputation of RICS also confers an added layer of trust across borders: From small-scale property valuations in local markets to complex portfolios spanning multiple countries, the standards enforced by RICS remain consistent worldwide. Periodic reviews and audits of RICS-regulated firms examine operational procedures, reporting methods, and overall professional competence, ensuring that these organisations are continually evolving to keep pace with emerging best practices.

When it comes to evaluating valuers specifically, clients may wish to look for professionals who hold the MRICS designation or the REV certification. Awarded by TEGoVA, the REV status is another layer of assurance that the valuer meets robust, pan-European standards, particularly in fields such as market analysis, valuation methodology, and ethical conduct. Together, MRICS and REV represent a powerful combination of qualifications, reflecting comprehensive training and thorough competence in the complexities of property valuation. By choosing a valuer or firm that holds these accreditations and operates under the umbrella of RICS regulation, clients can be confident that the advice received is grounded in best practice guidelines and fortified by strict ethical requirements. Ultimately, engaging a RICS-regulated firm – especially one staffed by MRICS- and possibly REV-qualified professionals – significantly reduces the risks that can accompany real estate or property-related transactions, fosters transparent and ethical decision-making, and ensures that each client's best interests remain paramount at every stage of the process.

PII

PII is a cornerstone of RICS regulation, specifically aimed at protecting public interest and safeguarding clients from potential financial harm. By requiring RICS-regulated firms to maintain adequate PII coverage, RICS ensures that if a client suffers losses due to professional negligence, errors, or omissions, there is a dependable safety net to address any compensation needs. This arrangement fosters greater accountability among property professionals, as it compels them to uphold the highest standards of diligence and care in their advice and service delivery. Moreover, the existence of robust PII coverage strengthens public confidence in the surveying and valuation professions, reassuring stakeholders that any serious mistakes will be promptly and fairly remedied. Whether engaging a RICS-regulated firm for a simple valuation or a complex development appraisal, clients can trust that their interests are safeguarded, thereby reinforcing the integrity and reliability of the real estate industry as a whole.

Conclusion

Evaluating valuers is essential to maintaining confidence in the property market and ensuring that valuation reports are credible and reliable. A combination of professional qualifications, methodological rigor, ethical conduct, market expertise, and technological integration forms the foundation of an effective valuer evaluation system. Continuous monitoring, stakeholder feedback, and adherence to international best practices help improve valuation standards and maintain trust in the profession. By addressing challenges and promoting innovation, the valuation profession can evolve to meet the growing demands of an increasingly complex real estate market, ensuring transparency, accuracy, and reliability in property valuation assessments.

Questions

1. Why is evaluating valuers important for the integrity of real estate markets?
2. What are the key responsibilities of a valuer?
3. How do valuers contribute to a positive economic environment?
4. What criteria should be considered when evaluating a valuer's performance?
5. Why is continuous professional development (CPD) essential for valuers?
6. What are the common methods used to evaluate valuers?
7. How do RICS-regulated firms and professional designations like MRICS and REV impact valuation credibility?
8. What role does professional indemnity insurance (PII) play in valuation practice?

References/Further Reading

American Society of Appraisers (ASA) (2021). ASA Business Valuation Standards.
Appraisal Institute (AI) (2022). The Appraisal of Real Estate, 15th Edition.
European Group of Valuers' Associations (TEGoVA) (2022). European Valuation Standards (EVS 2022).
International Valuation Standards Council (IVSC) (2024). International Valuation Standards (IVS 2024).
Uniform Standards of Professional Appraisal Practice (USPAP) (2023). USPAP Manual.
World Bank (2023). *Real Estate Valuation and Market Stability: Global Insights*.

Part IV

Case studies, illustrative reports, and solutions

In Part IV, we provide two additional case studies from practice, with the aim of bringing to life the income approach in two very different valuation methods and contexts.

- *Case study 1.* A comprehensive development scenario of a strategic master-planned community in the UK, with an illustrative residual land valuation and an explanation of the value-added elements of site assembly, preparation, and infrastructure provision.
- *Case study 2.* A profits-based valuation of a historic windmill in the Netherlands.

In addition, in Appendices A and B, we provide two illustrative valuation reports

- **Appendix A:** Illustrative investment property purchase report
- **Appendix B:** Illustrative development site appraisal report

In **Appendix C,** we provide the solutions to the questions posed within numerous chapters throughout the book and to Case Study 2.

DOI: 10.1201/9781003490357-18

Case Study 1

Introduction

This case study is produced from materials provided by and discussions with Gareth Thomas, Development Director (Midlands); James Gibson-Flowith, Development Manager; and Soraya Hemming, Surveyor (Degree Apprenticeship), from Harworth. We are very grateful for their contribution.

Harworth is one of the UK's leading land and property regeneration businesses, owning and managing over 14,000 acres, spread across 100 sites in the North of England and the Midlands. Based in Rotherham, South Yorkshire, they also have regional offices in Birmingham, Leeds, and Manchester. Their mission is to invest to transform land and property, both former industrial and urban areas, into vibrant sustainable places where people want to live and work, thereby supporting new homes (including affordable options), jobs, and communities across the regions and delivering long-term value for all their stakeholders. As a master developer, Harworth creates long-term value by acquiring and assembling sites that are large, are complex, and often had former industrial uses, transforming them into sustainable residential and employment spaces with a focus on placemaking.

Harworth's motivation for getting involved in this case study is that, as a master developer, they feel that hands-on experience provides invaluable real-world insight into development site valuation. By contributing a case study residual appraisal, Harworth aims to bridge the gap between academic theory and practical application. They also hope to help equip the next generation of surveyors with some of the knowledge needed to appraise and develop land effectively in real-world scenarios.

Important note: *Whilst based on an actual major redevelopment scheme, the development appraisal provided is based on data produced for illustration and academic purposes only.*

Context of the appraisal

Harworth's role is to acquire challenging sites, often requiring substantial remediation, and add value. This will frequently involve site clearance and demolition, contamination

DOI: 10.1201/9781003490357-19

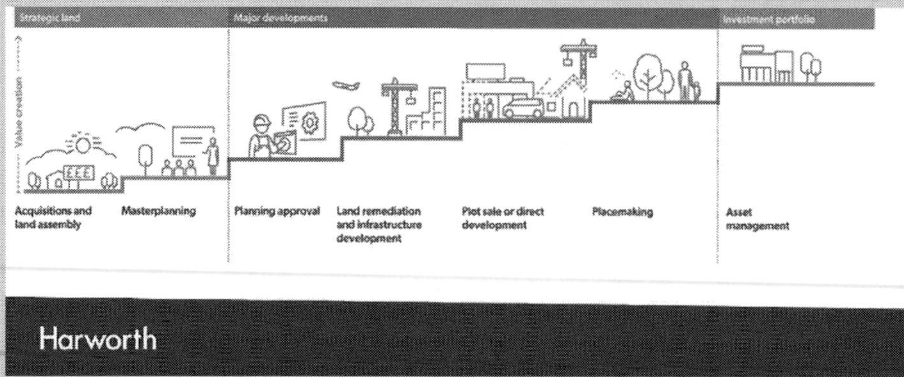

Figure CS1 The big picture of value creation in a large-scale development project.

remediation, infrastructure investment, master planning and placemaking, and planning negotiations, including achieving permission(s) for the highest and best use of the many parcels within their masterplan. Their focus is on future potential rather than past performance – for example, ensuring the delivery of the masterplan reflects the anticipated product, typically ten years ahead. This requires addressing timelines, risks, and strategic planning at each stage to unlock the site's full potential. This is well illustrated in Figure CS1.

The strategic development process

Site acquisition stage

The site which forms the subject of this case study was originally known as Ironbridge. It contained two power stations, constructed in the early 1930s (also known as the Buildwas power stations), that occupied a site on the banks of the river Severn at Buildwas in Shropshire, England. The Ironbridge B Power Station was operated by E.ON UK before being acquired by the Harworth Group in 2018.

The site comprises a gross development area of 359 ac (145 ha) and a net development area of 100 ac (40.5 ha). The site was bought without planning permission, so considerable work was required to formulate a strategic masterplan and negotiate approval of an application in September 2021.

The initial site valuation on purchase reflected the significant demolition and remediation costs, plus the challenges of securing a planning permission acceptable to all stakeholders.

This valuation reflected the following constraints (typical in a major land assembly context):

- *Access.* Limited or challenging access points.
- *Planning status.* Land without pre-approved planning permission.

Figure CS2 The site on acquisition.

- *Topography.* Challenging terrain, irregular landforms, and complexities of previous use.
- *Legal issues.* Encumbrances, including rights of way and other easements.

Planning stage

The key activities at this stage involve the securing of outline planning permission, addressing S106 contributions (which is the name given in the UK for planning gain, where developers are expected to contribute to infrastructure provision, affordable housing, and/or community benefits), and agreeing on any reserved matters or conditions of the permission.

There are considerable cost implications at this stage. Planning activities may incur significant costs in the region of £300,000, dependant on the size and scale of the site. These costs include appointment of consultants; development of documentation, plans, and surveys; and legal advisors and their associated fees.

An initial outline planning permission application was submitted in December 2019, and permission was formally granted in September 2022, as well as an agreed and endorsed S106 agreement. This permission allowed consent for 1,000 new dwellings, retirement village, employment land, retail, leisure use, and a new primary/nursery school.

Impact on adding value
Obtaining planning permission significantly increases the site's value by:

- Validating its potential for development
- Expanding market interest by unlocking the land's use
- Reducing risk perception among stakeholders

Figure CS3 The site following demolition and initial preparation.

At this point, the site value had increased significantly since the 2018 acquisition value, reflecting the demolition of structures on-site, remediation, initial infrastructure provision, and the imminent completion of the first plot sale to a housebuilder. This also reflects the significant reduction of risk from acquisition to this point.

Market analysis

As part of the ongoing development and valuation processes, detailed market analysis is undertaken, not only to assess land values, but also to help inform the type of residential products desired by customers and which will provide the highest and best use for each plot to be sold. This understanding of both product specifications and value is an essential component of the appraisal process.

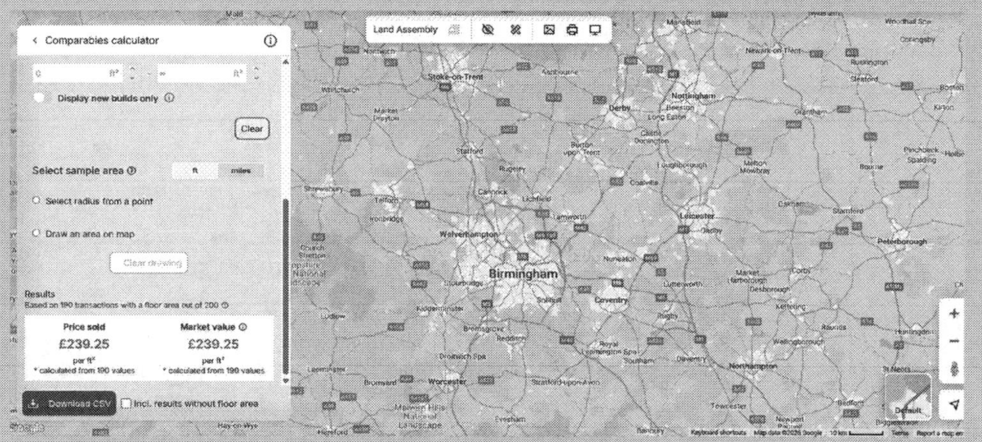

Figure CS4 Land values in the Midlands area taken from the LandInsights software.

Known as land trader analysis, a detailed assessment is conducted to outline potential costs and anticipated value enhancements through development. This serves as a baseline for decision-making during subsequent stages. Amongst other tools, LandInsights software can be used to examine prices per square foot for different house types. A screenshot of their analysis is provided in Figure CS4.

Infrastructure provision

Post-planning approval, the focus shifts to delivering essential infrastructure, which typically includes:

- Site remediation
- Levelling the land to create saleable plots
- Securing National House Building Council (NHBC) Land Quality Endorsement (LQE) certificates

Contractors are chosen through a tender process to ensure cost-efficiency and compliance with timelines.

Impact on adding value

Infrastructure investments further enhance value by:

- De-risking plots to end user developers, that is, housebuilders or logistics providers, Harworth ensures key roadways are delivered and that power, water, and drainage are provided for.

- Providing a cohesive environment across the whole development, including specific placemaking initiatives, such as schools and local quarters.
- This attracts potential buyers (e.g. housebuilders and/or industrial occupiers, if Harworth wishes to retain the asset).

The duration of this stage can vary significantly based on project scale and complexity.

Disposal of parcels to housebuilders and units sold

The following image shows another Harworth site, South East Coalville, to demonstrate this phase of the strategic development process, where the remaining plots, shown in white, are being prepared for disposal to regional and national housebuilders. The parcels shown in light grey indicate the completed phases.

Impact on adding value

When a housebuilder acquires a parcel, it demonstrates confidence in the location and market demand, which can enhance the value of the remaining land.

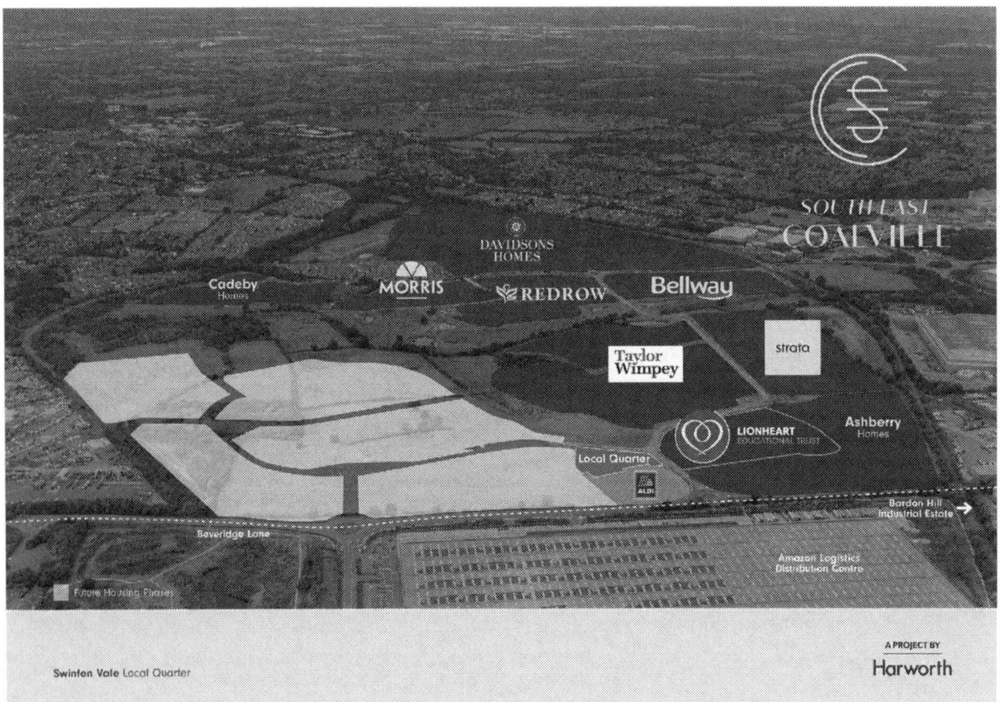

Figure CS5 An example of phased preparation and disposal of sites.

Placemaking and community involvement

Placemaking is a critical component of any major development, adding long-term value through community-focused enhancements. These can include:

- *Primary schools.* Establishing educational infrastructure.
- *Pocket parks and play areas.* Creating accessible green spaces and play equipment for all ages.
- *Amenities to support new residents and the community.* Such as supermarkets, medical centres, nurseries, etc., to support local needs.

Impact on adding value
Enhances the appeal of the development to potential buyers. Creates a sense of community and long-term viability and increases overall property values in and around the development.

Concluding remarks

The life cycle of property development demonstrates a strategic increase in valuation through targeted activities at each stage. From acquisition through placemaking to phased disposal, every phase contributes to unlocking the land's full potential, ensuring the highest and best use while navigating constraints and risks effectively. Throughout this process, robust development appraisals are required to monitor value enhancement, but also specifically to assess the value on disposal of an enhanced site to residential developers. Harworth's strategic approach to the development process is effectively summarised in Figure CS6.

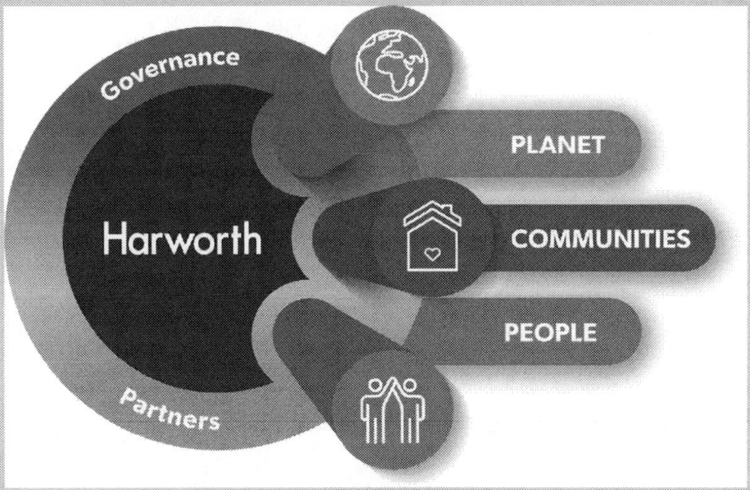

Figure CS6 The strategic development vision of Harworth.

We present an appraisal of one of the improved parcels of land for disposal in the following section.

An illustrative appraisal produced for disposal of an enhanced site

The appraisal summary
(produced using Argus Developer Version 8.20.003)

APPRAISAL SUMMARY **HARWORTHESTATES**

Appraisal Summary for Merged Phases 1 2

Currency in £

REVENUE					
Sales Valuation	**Units**	**ft²**	**Sales Rate ft²**	**Unit Price**	**Gross Sales**
Private Units	525	735,000	300.00	420,000	220,500,000
Affordable Units	225	189,000	180.00	151,200	34,020,000
Totals	**750**	**924,000**			**254,520,000**
NET REALISATION				**254,520,000**	
OUTLAY					
ACQUISITION COSTS					
Residualised Price (53.00 Acres @ 858,096.32 /Acre)			45,479,105		
				45,479,105	
Stamp Duty		5.00%	2,273,955		
				2,273,955	
CONSTRUCTION COSTS					
Construction	**ft²**	**Build Rate ft²**	**Cost**		
Private Units	735,000	120.00	88,200,000		
Affordable Units	189,000	120.00	22,680,000		
Totals	**924,000 ft²**		**110,880,000**		
Developer's Contingency		10.00%	1,387,500		
Plot Abnormals			6,000,000		
Road/Site Works			7,875,000		
				126,142,500	
Section 106 Costs					
Section 106 Costs			7,500,000		
				7,500,000	
PROFESSIONAL FEES					
		7.00%	7,761,600		
				7,761,600	
DISPOSAL FEES					
Sales Agent Fee		2.00%	5,090,400		
Sales Legal Fee			450,000		
				5,540,400	
FINANCE					
Debit Rate 7.000%, Credit Rate 0.000% (Nominal)					
Total Finance Cost				8,918,440	
TOTAL COSTS				**203,616,000**	
PROFIT					
				50,904,000	
Performance Measures					
Profit on Cost%		25.00%			
Profit on GDV%		20.00%			
Profit on NDV%		20.00%			
IRR% (without Interest)		76.84%			
Profit Erosion (finance rate 7.000)		3 yrs 3 mths			

An explanation of the inputs and assumptions to the summary appraisal

Revenue

The number of bedrooms per unit is determined by the housebuilder, taking into consideration the demographic of the area. Typically, developments feature a mix of two-, three-, and four-bedroom houses, while one-, five-, and six-bedroom homes are less common due to prevailing demographic trends and typical family structures.

Market and customer analysis demonstrates that it is common for new four-, five-, and six-bedroom homes to include a downstairs study, accommodating the increasing demand for working from home. Housebuilders offering larger properties (four, five, and six bedrooms) generally price these homes at a premium to reflect their size and additional features.

In this scenario, the appraisal is for a total of 750 houses, split to provide 30% affordable homes (225 houses) and 70% private (525 houses) in accordance with local authority requirements. The market demand for private sale in terms of size depends on the market average for that area and comparable evidence. Likewise with the bedrooms mix, it depends on the demographics of the area as well as the local planning authorities' requirements. Affordable homes generally favour smaller housing options. As a result, the design and development of house types, as well as their square footage, are tailored to align with this preference, ensuring that they remain both cost-effective and accessible to a broader range of buyers.

Type	*Housing for private sale*	*Affordable housing allocation*
Number of houses	525	225
Average area per unit	1,400 ft^2 (130 m^2)	840 ft^2 (78 m^2)
Price per ft^2/m^2 *(Market rate per ft^2 based on comparable evidence and led by market conditions)*	£300/ft^2 per unit £28/m^2 per unit	£180/ft^2 per unit (60% of private housing price) £17/m^2 per unit

Note: These numbers are for illustrative purposes only and do not represent Harworth's view of current market conditions.

Affordable housing requirements vary across local authorities, as they are determined based on local demand, housing needs assessments, and policy objectives. To promote social integration and avoid the concentration of affordable units in specific areas, these homes will be 'pepper-potted' throughout the scheme. This approach ensures a balanced, mixed-tenure community, aligning with best practices in urban planning and housing policy. Affordable homes are typically sold or rented through a registered provider.

As affordable housing units are sold at a discounted rate compared to private market units, we have applied, in this scenario, a 60% discount to the sales rate (£/ft^2/m^2) to reflect this pricing adjustment. This discount accounts for affordability requirements while maintaining financial viability for the development.

Land value: residualised price for 53 ac at £858,096.32/ac

Land value is determined by market dynamics and varies from site to site based on a range of factors. These include the specific characteristics of the surrounding area, site-specific constraints, development costs, and the overall financial viability of the project, including projected expenditures and profit margins.

Stamp duty: 5%

Stamp duty (a land tax paid on disposal) is set by the UK government.

Construction costs: private and affordable units £120/ft^2, £11.2/m^2

Construction costs are typically determined based on a cost plan prepared by a quantity surveyor. Their assessments are informed by market and industry standards, which are subject to change in response to fluctuations in material and supplier demand, as well as variations in interest and borrowing rates. This includes foundations, external and internal structural framework, fit-outs for bathrooms, kitchen, etc.

Developers' contingency: 10%

This is a standard addition to reflect the uncertainties and risks inherent in construction projects.

Plot abnormals: £8,000/plot

Plot abnormals are additional allowances made in the earlier stages of the appraisal for unknowns, such as specific house type finishes, foundation types, etc.

Road/site works: £7,875,000 (£10.5k per plot)

These costs would be determined through a comprehensive technical review conducted by ground and drainage engineers. The allocated budget would account for various infrastructure elements, including highway roads, paving, and drainage systems, such as sustainable urban drainage systems (SuDS). Additionally, other essential external works, such as road lighting and associated site infrastructure, would be incorporated to ensure compliance with regulatory standards.

S106 agreement: £10,000/plot

S106 agreements are legally binding agreements between local planning authority and the developer/landowner. These detail contributions to the provision of public services and infrastructure like highways, recreational facilities like sports pitches and parks, public open space, education, health, and affordable housing. The negotiation and finalisation of these agreements can take a long time to agree but are fairly and reasonably related to the development.

These contributions are usually indexed, linked, and will have a trigger attached to the obligations, such as:

> 'Not to commence development on site until owners have paid *X* for the travel plan monitoring contribution to the county council'.
>
> 'Prior to the occupation of the 100th dwelling, *X* shall be contributed in instalments to the county council for the construction of a new nursery'.

Professional fees: 6%

Fees for consultants, including planning specialists, architects, engineers, and environmental experts, engaged to assess the site and inform decision-making.

Marketing and leasing: 2.5%

Allowance made for photography, drone footage, marketing packs, and advertisements to attract potential buyers.

Disposal: sales agent fee, 2%, and legal fee, £600/plot

Selling agent fees typically vary between 1% and 3% of the sale price. The lower rate generally applies to straightforward transactions, often handled by local agents, while the higher rate is usually charged by multiple national agents for more complex cases. Additionally, fees depend on the anticipated workload, for example, if the property has intricate lot divisions, has multiple buildings, or requires planning consents, or if the agent must conduct all viewings, the fee will be higher, as their involvement in the site is greater and more time-consuming.

On this site it has been assumed that two selling agencies have been instructed.

Legal consultancy is an essential component of property development, ensuring that all aspects of the project comply with applicable laws and regulations while mitigating potential risks and liabilities for all stakeholders involved. Solicitors play a critical role in this process by drafting, reviewing, and finalising sales agreements and contractual documents when transferring ownership of land parcels or, in this instance, the residual land. Their expertise ensures that transactions are legally sound, transparent, and structured to protect the interests of all parties involved.

Finance: debit rate, 7%

A 7% borrowing rate has been applied in this appraisal to reflect prevailing finance rates.

Performance measures

Profit on cost: 25%
Profit on gross development value: 20%
Profit on net development value: 20%

These performance measures are set as an industry standard and are higher in earlier stages of the development and then reduced slightly as risks become known.

We hope that this case study helps bring the theory explored in Chapter 10 to life and explain the added complexities of projects that are not simply a single site bought, developed, and sold. Many developments such as this take many years and require strategic planning, significant capital, and execution of both the masterplan and the placemaking activities, all of which must be informed, not only by comprehensive value data, but also by market data, ensuring the valuations reflect the expectations of commercial end users and residents.

Case Study 2

The restaurant De Jonge Dikkert

Introduction

This case study is produced from materials provided by and from discussions with Philip Bacon, Senior Director; Dorus Helsen, Managing Partner; and especially Joris Blommaert, Consultant, from Horwath HTL. We are very grateful for their considerable enthusiasm for the project and their significant contribution.

Horwath HTL is part of the Crowe Global network, a top-ten accounting and financial services network. Their focus is 100% on hospitality, tourism, and leisure consulting, covering every aspect of hotel real estate, tourism, and leisure development. With over 100 years of history, 52 offices in 38 countries, and a track record of successfully completing over 50,000 assignments for private and public clients, they are a global leader in hospitality, tourism, and leisure consulting. We are delighted that they have made this very interesting and different case study available as a learning opportunity.

Disclaimer

Here we present the standard disclaimer and copyright statement used by Horwath, both as a legal request and also as an illustration of how their reports are prefaced.

Horwath HTL

The conclusions in the reports we prepare, including the financial projections, will never be guaranteed in any way. Our conclusions and recommendations are based on our analyses. We will never guarantee in advance that these analyses will lead to a positive outcome. We are not responsible for adjusting the reports based on events or circumstances that become known after publication. However, we are available to discuss the possibilities for adjustments. The General Terms and Conditions in the appendices apply to this report.

Horwath HTL is a member of Crowe Global, a Swiss organisation. Each member of Crowe Global is a separate and independent legal entity. Horwath HTL and its affiliated

DOI: 10.1201/9781003490357-20

companies are not responsible or liable for any acts or omissions of Crowe Global or any other Crowe Global member. Crowe Global does not provide professional services and does not have any ownership or partnership interest in Horwath HTL.

Our reports are based on assumptions, estimates, and other information derived from market research and our industry expertise. Some assumptions may not materialise, and unexpected events or circumstances may occur. As a result, actual outcomes may differ from the projections in our reports, and these deviations could be substantial.

Copyright

Context of the Case Study

We have worked with Horwath to present this case study as a discussion piece, encouraging readers to think about the assumptions and variables and make their own judgements and assessments. You are encouraged to try to formulate your own solutions using the three valuation methods presented before examining the solutions presented at the end of Part IV.

The historic and cultural characteristics of the property

Restaurant De Dikkert is housed in an impressive, historic *stellingmolen*, or scaffold mill, dating back to the 17th century. This type of windmill, built on an elevated platform, was traditionally used for grinding grain. Originally constructed in 1672 in Zaandam, it was known as *De Walvisch*. In the early 20th century, the mill was dismantled and rebuilt in Amstelveen in 1929, where it was renamed De Dikkert.

The windmill is a national monument and an important piece of cultural heritage, reflecting the rich history of Dutch windmills. Its characteristic wooden structure, large sails, and sturdy appearance make it an iconic landmark in the Amstelveen region, as can be seen in Figure CS2.1.

Architecture and features

- Octagonal wooden windmill, built in the traditional Zaan-style architecture.
- Large *stelling* (platform), previously used to manually adjust the sails to the wind.
- Authentic thatched roof, contributing to its historical character.
- Large wooden beams and original mill components, still visible in the interior.

Figure CS2.1 Photograph of the Restaurant De Jonge Dikkert

- Thoughtfully restored, combining historic elements with a warm, modern interior that suits the restaurant concept.

Cultural heritage and significance

De Dikkert is a living monument that represents the centuries-old Dutch windmill culture. Its reconstruction and repurposing as a restaurant have ensured the preservation of this heritage while giving it a new role within the community. The restaurant contributes to the conservation of historic landmarks while offering visitors a unique blend of history and gastronomy. Furthermore, the windmill is a defining feature of Amstelveen, cherished by both locals and tourists. It symbolises traditional Dutch craftsmanship and architectural heritage. With its unique setting in the windmill, De Dikkert offers not only a fine dining experience but also a historical journey in one of Amstelveen's most remarkable national monuments.

From functional to symbolic

When a windmill loses its function as a working mill, market demand changes. The value of a mill that still operates in its traditional role, for example, as a grain mill, is often much higher than that of a mill without working mechanics. In this case, the value of the object depends largely on its iconic appearance and historical significance, while its economic value is primarily based on the land and location rather than the functional usability of the mill itself.

If alternative use is permitted, the dynamics of market transactions change entirely. When a windmill is repurposed for hospitality, such as a restaurant, the *highest and best use* of the object is considered. This concept means that a real estate asset is utilised in the most profitable and efficient way. In most cases, this means that the mill is repurposed as a hospitality venue, where the historical ambiance of the building serves as the backdrop for a restaurant, hotel, or even a brasserie. This transformation brings several advantages: the preservation of heritage, attraction for tourists, and the creation of new employment opportunities. However, challenges also arise, such as compliance with modern building and safety regulations, which may sometimes conflict with the protected status of the monument.

Useful background information and data to be collected to prepare for the valuation

Physical location

Restaurant De Jonge Dikkert is located at Amsterdamseweg 104(a) in Amstelveen, a busy street connecting the city with other parts of the region. The restaurant is situated in an area that blends urban and green elements, providing a varied environment.

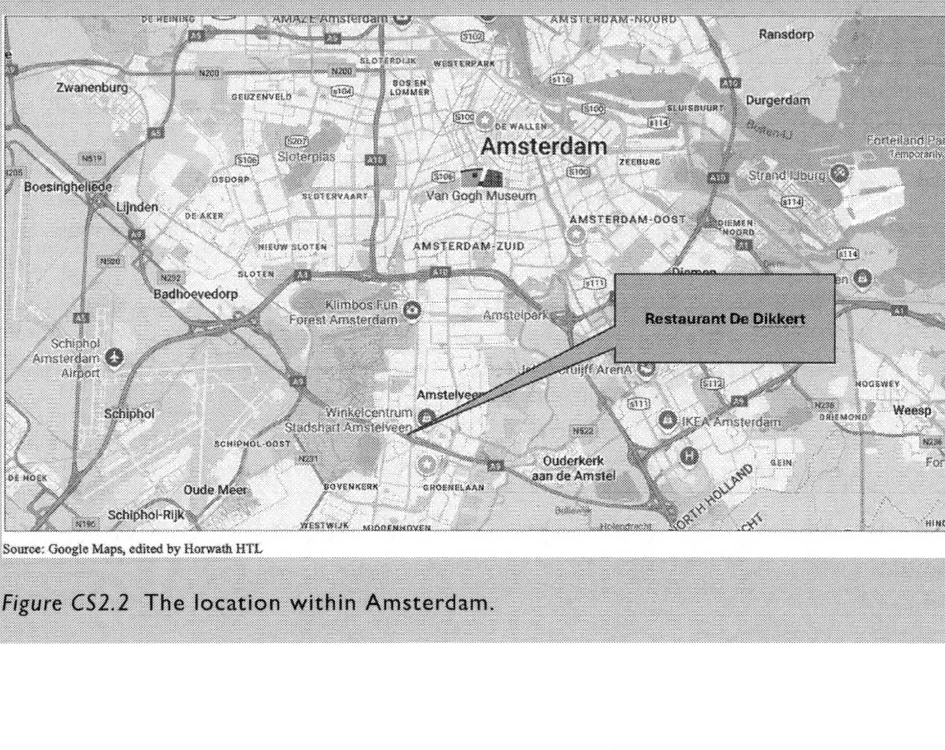

Source: Google Maps, edited by Horwath HTL

Figure CS2.2 The location within Amsterdam.

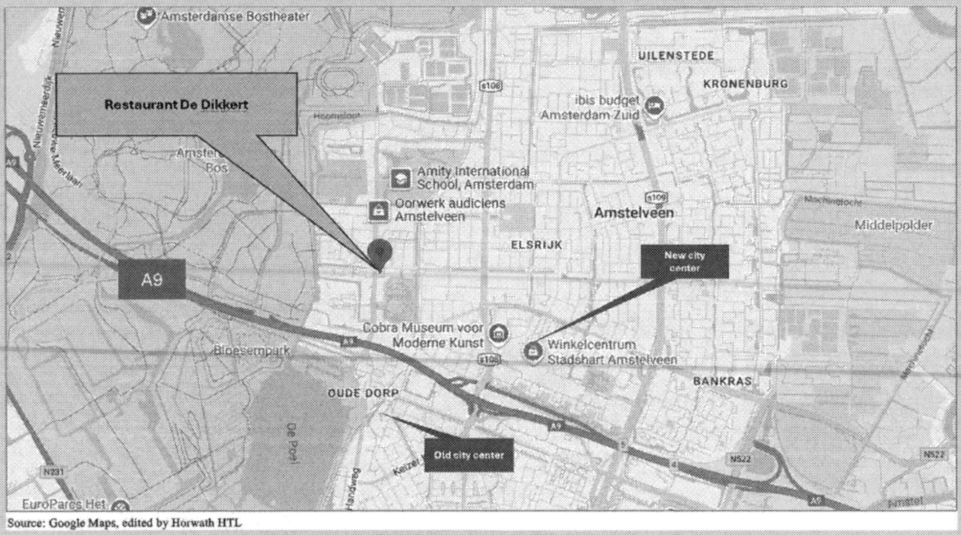

Figure CS2.3 The location within Amstelveen.

Visibility and recognisability

The location is clearly visible from the street, primarily due to the presence of the historic windmill, which serves as a landmark. The windmill is a striking feature in the landscape, making the restaurant relatively easy to find for passersby. Its visibility is further enhanced by the way the windmill stands out above the surrounding area, making it identifiable from a distance.

Accessibility

The restaurant is easily accessible by both car and public transport.

- *By car.* Amsterdamseweg is an important traffic route in the region, providing connections to the A9 and other major roads. As a result, the restaurant is easily reachable from both Amstelveen and Amsterdam, as well as other nearby cities. There is ample parking around the restaurant and in the vicinity, making it convenient for guests arriving by car.
- *By public transport.* The restaurant is accessible via various bus and tram stops in the area. The proximity to the centre of Amstelveen also makes it easy to reach by public transport.

Comparable offerings in the area

There are several other dining establishments in the immediate area, but none with the same historical setting as De Dikkert. Other restaurants and cafés can be found nearby, but the

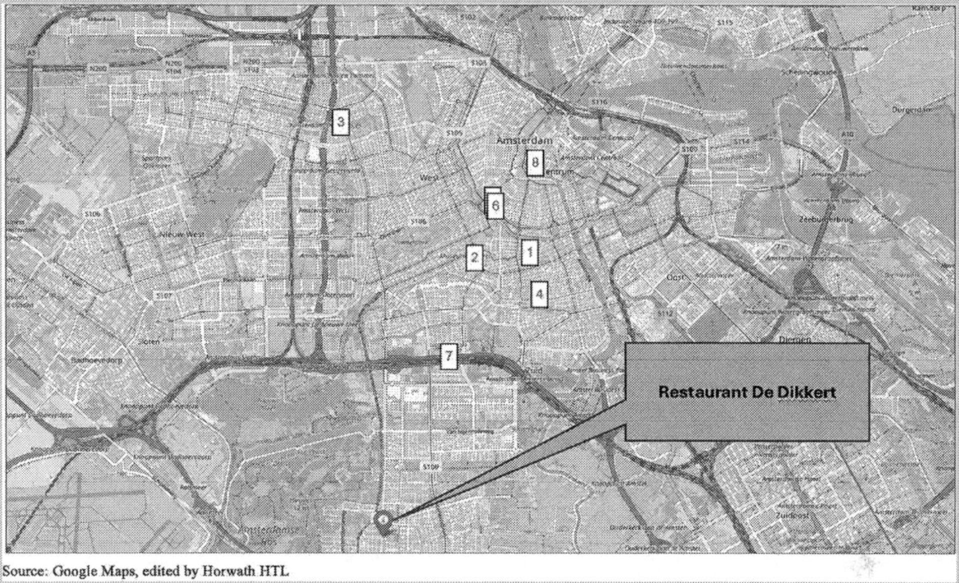

Figure CS2.4 Locations of other restaurants within the vicinity of the subject property.

combination of fine dining in a historic windmill makes this restaurant unique. Additionally, commercial centres and shopping areas, such as Amstelveen's Stadshart, are in close proximity, which adds to the location's appeal. The greater region of Amsterdam has more comparable establishments in the area. A list with relevant references has been listed in what follows, together with their location on the map provided as Figure CS2.4.

1. Albert Cuypstraat 31H, Amsterdam
2. Van Baerlestraat 128h, Amsterdam
3. Bos en Lommerweg 191, Amsterdam
4. Cornelis Troostplein 2–4, Amsterdam
5. Lange Leidse-dwarsstraat 93H, Amsterdam
6. Korte Leidse-dwarsstraat 62, Amsterdam
7. Beethovenstraat 27-a, Amsterdam
8. Rokin 93, Amsterdam

Surrounding area and attractions

Amsterdamseweg is situated in an area that mixes urban and natural elements. The restaurant is located near the Amsterdamse Bos, a large park that is frequently visited by both locals and tourists. This green space allows visitors to combine a meal at the restaurant with a walk or other outdoor activities.

Furthermore, the restaurant is close to business and tourist centres, including the Zuidas and Schiphol, making it attractive to business travellers and visitors to the region.

Competitive factors

While there are several other dining options in the area, De Dikkert stands out primarily due to the historic windmill that houses the restaurant. The location on Amsterdamseweg and its proximity to other commercial and recreational facilities make it easily accessible to a broad audience, though the presence of other dining establishments in the area may create some competition.

The subject property: restaurant De Jonge Dikkert

The restaurant is a stylish and gastronomic restaurant in Amstelveen, housed in a historic 17th-century windmill. The restaurant combines a warm, atmospheric ambiance with refined gastronomy, blending classic and modern culinary techniques.

Concept and cuisine

De Dikkert is known for its French–Dutch cuisine with influences from international culinary trends. The dishes are elegantly presented and prepared using seasonal, fresh, and sustainable ingredients. A strong focus on local products and artisanal preparation methods results in refined and surprising flavours.

Atmosphere and experience

- *Authentic setting.* The windmill provides a unique, warm, and intimate atmosphere, combining rustic charm with contemporary elegance.
- *Fine dining with excellent service.* The staff is professional and welcoming, paying close attention to details and wine pairings.
- *Extensive wine selection.* A carefully curated selection of quality wines, perfectly paired with the dishes.

Menu and specialties

The menu changes seasonally and offers both à la carte options and elaborate, multi-course menus, emphasising refined flavour combinations. Signature aspects of De Dikkert include:

- Classic French preparations with a modern twist
- Use of fresh fish, meat, and vegetarian ingredients
- Culinary creativity in flavour, texture, and presentation

Target audience

The restaurant attracts a diverse clientele, including business guests, food enthusiasts, and couples looking for a special dining experience. Its ambiance and quality also make it ideal for special occasions, such as anniversaries and business dinners. In short, De Dikkert offers a luxurious culinary experience in a unique setting, where gastronomy and history come together.

The valuation scenario and methods of valuation

You are a potential restaurant operator who is considering taking over a restaurant in Amstelveen. The restaurant, *De Dikkert*, is located within a historic windmill. The windmill is fully owned by a third party and is currently vacant in terms of rental.

To make a decision as to whether or not you want to go ahead with operating this restaurant, you need to conduct a feasibility analysis and comparative analysis.

In this case study, you will apply three key valuation methods to estimate the value of the restaurant:

Method 1: *rental value capitalisation*
Method 2: *economic rental value*
Method 3: *profits method*

Summary of the three valuation methods to be applied

Method 1: rental value capitalisation

This method determines the value of the restaurant by capitalising its potential rental income. The annual rent that could be charged for the property is multiplied by a capitalisation factor, which is derived from market data and reflects investment risk. This comparative approach is particularly relevant for properties where rental income is a key determinant of value.

Method 2: economic rental value

The economic rental value represents the rent that a tenant in the hospitality sector could reasonably afford to pay based on industry benchmarks and expected revenue. This method considers the financial sustainability of the rent in relation to the restaurant's business model, ensuring that the lease terms are viable for a long-term operation.

Method 3: profits method

This approach values the restaurant based on its earning potential. It involves calculating the expected annual profit and applying a capitalisation rate to determine the business's market value. This method is widely used in hospitality valuations as it reflects the restaurant's financial performance and operational efficiency.

Combining the methods

By combining these three valuation techniques, you will gain a well-rounded perspective on the financial feasibility of taking over De Jonge Dikkert, enabling you to make an informed investment decision.

To ensure consistency in this case study, a gross rent multiplier **(GRM) of 9.1%** should be used for both the *rental value capitalisation* and the *economic rental value* methods. In real-world valuations, this multiplier would typically be determined based on market conditions, comparable transactions, and investor expectations. However, for the purposes of this case study, the GRM is provided as a fixed value to allow for a more structured analysis and focus on the valuation process itself. Additionally, purchaser costs are set at a **fixed rate of 11%** and will be applied across all valuation methods. The final market value will be presented both **including and excluding** purchaser costs to ensure a comprehensive understanding of the total investment required.

Undertaking the valuations using all three methods

Valuation method 1: rental value capitalisation

In this method, you will determine the rental capitalisation value of the restaurant by assessing its rental potential and applying a capitalisation factor or gross rent multiplier (GRM). This process involves two key components:

- *A comparative analysis (refer to Figure CS2.).* Analysing comparable rental properties to determine the market positioning of the restaurant.
- *Production of a rental calculation table.* Assigning weights, calculating the adjusted rental area, and applying the GRM to derive the rental capitalisation value.

Step 1: *Comparative analysis*

To estimate a fair rental value, it is crucial to position De Dikkert relative to comparable restaurants. Figure CS2.1 contains a pre-filled reference table with market data on various restaurants. Using this table, you will:

- *Identify the restaurant's competitive positioning.* Compare De Dikkert with similar establishments based on factors such as location, accessibility, visibility, and size.
- *Assess the importance of each factor.* Not all elements have the same impact on a restaurant's rental value. Consider which aspects are more valuable for a restaurant than for other types of commercial properties.

- Using Figure 1, determine where De Jonge Dikkert stands in comparison to the market.
- This positioning will help you determine the preliminary rent per square metre.
- You should be familiar with this process from reading Chapter 7.

Table CS2.1 Comparative analysis

Address	*Surface (m^2)*	*Monument (yes/no)*	*Terrace*	*Maintenance*	*Location and accessibility*	*Socio-economic level*	*Rental price (per m^2)*
Albert Cuypstraat 31H Amsterdam	165 m^2	N	−−	+/−	+	+/−	€378,−
Van Baerlestraat 128h Amsterdam	217 m^2	N	−−	+	+	+	€392,−
Bos en Lommerweg 191 Amsterdam	200 m^2	Y	+/−	+	+/−	−	€420,−
Cornelis Troostplein 2–4 Amsterdam	180 m^2	Y	−	+/−	+/−	+/−	€428,−
Lange Leidse-dwarsstraat 93H Amsterdam	73 m^2	Y	−−	−	++	−	€461,−
Korte Leidse-dwarsstraat 62 Amsterdam	80 m^2	N	−−	+/−	++	−	€613,−
Beethovenstraat 27-a Amsterdam	155 m^2	N	−−	++	+	+	€619,−
Rokin 93 Amsterdam	220 m^2	Y	++	+	++	+	€659,−

Step 2: *Determining the rental capitalisation value*

After determining the market position of De Dikkert, the next step is to calculate the rental capitalisation value using the calculation table on the next page. This involves assigning values to different areas of the restaurant, determining the annual rental income, and applying a capitalisation factor.

Step 2.1: *Assigning weights to spaces*

Not all areas within the restaurant contribute equally to its operational value. Some spaces, such as the dining area and kitchen, are more essential to the business than others, like storage rooms or parking.

Using Table CS2.2 students should attempt to assign a weight factor to each space based on its relevance to the restaurant's operations. Higher weights should be given to spaces that are crucial for generating revenue, while lower weights should be assigned to areas with less impact on the business.

Step 2.2: *Calculate the corrected surface area*

Each space's original surface area is multiplied by its assigned weight factor to determine the corrected surface area.

$$Corrected\ Surface\ Area = Surface\ Area \times Weight\ Factor$$

The total corrected surface is then used for further calculations.

Step 2.3: *Determine the yearly rent*
Once all corrected surface areas are determined, you can calculate the annual rent using the following formula. The rent per square metre is based on the previous comparative analysis.

$$Annual\ Rent = Total\ Corrected\ Surface \times Rent\ per\ m^2$$

Step 2.4: *Applying the GRM*
After calculating the yearly rent, the rental capitalisation value is determined by applying the GRM. This multiplier represents the relationship between rental income and property value, providing an estimate of what the restaurant is worth based on its rental potential.

$$Rental\ Capitalisation\ Value = \frac{Annual\ Rent}{GRM}$$

Step 2.5: *Accounting for purchaser costs*
Once the rental capitalisation value has been established, the final step is to account for purchaser costs, which represent additional acquisition expenses, such as taxes, legal fees, and valuation costs. These costs must be included in the final valuation to ensure a full overview of the investment required. The final outcome of this step is a rental capitalisation value, presented both including and excluding purchaser costs.

Table CS2.2 Calculation table for method 1

A. Rental value capitalisation		*Price per m²*	
Ground floor	*Surface in m²*	*Weight*	*Surface after correction*
Entrance area	15		0.00
Wardrobe	10		0.00
Bar	40		0.00
Restaurant	120		0.00
Kitchen	60		0.00
Storage	25		0.00
Toilet	35		0.00
Garden	125		0.00
Terrace	100		0.00
Parking	20		0.00
Subtotal	550		–

(*Continued*)

(Continued)

A. *Rental value capitalisation*		*Price per m²*	
First floor	*Surface in m²*	*Weight*	*Surface after correction*
Private room	25		0.00
Toilet	15		0.00
Office	30		0.00
			0.00
Subtotal	70		–
Basement	*Surface in m²*	*Weight*	*Surface after correction*
Storage	125		0.00
			0.00
			0.00
			0.00
Subtotal	125		–
TOTAL	–	€ –	€
Maximum investment	Rental value GRM	€ –	Incl. buyer costs Excl. buyer costs

Valuation method 2: economic rental value

The *economic rental value* (ERV) method estimates the rental value based on the restaurant's fair maintainable turnover (FMT). This approach ensures that the rent is financially sustainable in relation to the business's expected turnover. Unlike the rental value capitalisation method, which is based on market rental benchmarks, the **ERV method** considers what a restaurant operator can reasonably afford to pay in rent while maintaining profitability.

The following steps are required to determine the **ERV.**

Step 1: *Determining the fair maintainable turnover*

The first step is to assess the restaurant's expected revenue by analysing the **profit and loss statement** provided in what follows. This statement contains financial data from the **previous three years**, allowing for a trend analysis.

You must now evaluate:

- Whether revenue has been stable, increasing, or declining
- The most realistic projection based on past performance
- Any potential external factors that could impact future revenue

The FMT will serve as the basis for calculating the **maximum sustainable rent** the restaurant could afford.

Step 2: *Determining the rent-to-revenue ratio*

The rent-to-revenue ratio of the ERV represents the percentage of a restaurant's revenue that is allocated to rent. This ratio helps determine whether the rent is financially sustainable in relation to the restaurant's expected turnover.

- **Higher** rent-to-revenue ratio is common in **prime locations**, where property demand is high, rental supply is limited, and the restaurant benefits from a strong market presence or a luxury positioning. Restaurants in **high-traffic areas, city centres, or exclusive dining districts** often operate with a higher ratio, as the location itself adds significant value to the business.
- **Lower** rent-to-revenue ratio are typically found in **less-desirable areas** with lower market demand, a larger supply of available rental spaces, or when

De Jonge Dikkert

Profit and Loss Statement

REVENUE	**2023**	**%**	**2022**	**%**	**2021**	**%**
Food revenue	€ 945.000	69,3%	€ 876.000	67,8%	€ 769.000	77,5%
Beverage revenue	€ 416.000	30,5%	€ 414.000	32,0%	€ 222.000	22,4%
Miscellaneous revenue	€ 2.000	0,1%	€ 2.000	0,2%	€ 1.000	0,1%
TOTAL REVENUE	**€ 1.363.000**	**100%**	**€ 1.292.000,00**	**100%**	**€ 992.000,00**	**100%**
Purchase Costs	**2023**	**%**	**2022**	**%**	**2021**	**%**
C.O.S. food	€ 244.000	17,9%	€ 258.000	20,0%	€ 211.000	21,3%
C.O.S. beverage	€ 66.000	4,8%	€ 75.000	5,8%	€ 61.000	6,1%
Direct revenue costs	€ -	0,0%	€ -	0,0%	€ 12.000	1,2%
TOTAL Purchase Costs	**€ 310.000**	**22,7%**	**€ 333.000,00**	**25,8%**	**€ 284.000,00**	**28,6%**
GROSS OPERATING PROFIT	**€ 1.053.000,00**	**77,3%**	**€ 959.000,00**	**74,2%**	**€ 708.000,00**	**71,4%**
Company expenses	**2023**	**%**	**2022**	**%**	**2021**	**%**
Staff wages	€ 547.000,00	40,1%	€ 484.000,00	37,5%	€ 364.000,00	36,7%
Depreciation	€ 17.000,00	1,2%	€ 18.000,00	1,4%	€ 17.000,00	1,7%
Housing costs	€ 201.000,00	14,7%	€ 186.000,00	14,4%	€ 136.000,00	13,7%
General & Administrative	€ 85.000,00	6,2%	€ 73.000,00	5,7%	€ 68.000,00	6,9%
Selling costs	€ 48.000,00	3,5%	€ 47.000,00	3,6%	€ 41.000,00	4,1%
Financial costs	€ 10.000,00	0,7%	€ 10.000,00	0,8%	€ 5.000,00	0,5%
TOTAL EXPENSES	**€ 908.000,00**	**66,6%**	**€ 818.000,00**	**63,3%**	**€ 631.000,00**	**63,6%**
EBITDA	**€ 145.000,00**	**10,6%**	**€ 141.000,00**	**10,9%**	**€ 77.000,00**	**7,8%**
Financial income and expenses	**2023**	**%**	**2022**	**%**	**2021**	**%**
Interest income and similar revenues	€ 3.000,00	0,2%	€ -	0,0%	€ -	0,0%
Interest expenses and similar costs	€ -	0,0%	€ 3.000,00	0,2%	€ 2.000,00	0,2%
TOTAL FINANCIAL INCOME & EXPENSES	**€ 3.000,00**	**0,2%**	**€ -3.000,00**	**-0,2%**	**€ -2.000,00**	**-0,2%**
Extraordinary income and expenses	**2023**	**%**	**2022**	**%**	**2021**	**%**
Extraordinary income	€ -	0,0%	€ 49.000,00	3,8%	€ 143.000,00	14,4%
Extraordinary costs	€ -	0,0%	€ -	0,0%	€ -	0,0%
TOTAL FINANCIAL INCOME & EXPENSES	**€ -**	**0,0%**	**€ 49.000,00**	**3,8%**	**€ 143.000,00**	**14,4%**
NET OPERATING PROFIT	**€ 148.000,00**	**10,9%**	**€ 187.000,00**	**14,5%**	**€ 218.000,00**	**22,0%**

Figure CS2.5 The profit and loss statement.

Economic Rental Value				
	x 1000		Rent-to-revenue ratio	
Fair maintainable turnover (FMT)				
			Economic rental value	
Maximum investment	Rental value	€ -	€ -	Incl. Buyer Costs
	GRM		€ -	Incl. Buyer Costs

Figure CS2.6 Calculation table for method 2.

a restaurant is able to operate with greater cost-efficiency. These restaurants may have **lower overhead costs**, **more flexible business models**, or **less dependence on location-driven foot traffic**.

For this case study, you must select an appropriate rent-to-revenue ratio typically within the **7–12% range**, based on the restaurant's expected financial performance, market conditions, and positioning.

Step 3: *Calculating the ERV*

Once the FMT and an appropriate rent-to-revenue ratio have been determined, the ERV can be calculated using the following formula:

$$ERV = \frac{Projected\ \ Revenue}{Rent - to - Revenue\ \ Ratio}$$

Since the ratio is expressed as a percentage, it must be converted into decimal form (e.g. 10% = 0.10). This calculation determines the maximum sustainable rent the restaurant could afford based on its expected revenue.

Valuation method 3: profits method

The *profits method* estimates the restaurant's value based on its earning potential. Unlike the previous methods, which focus on rental income, this approach capitalises the **operational profitability** of the business to determine its market value. This method is particularly relevant for hospitality businesses, where profitability directly impacts investment decisions.

Whilst the *rental value capitalisation method* and *economic rental value method* primarily assess the financial feasibility of renting the property, the *profits method* goes further by evaluating the **intrinsic value of the operational business itself**. This method helps determine whether the restaurant generates sufficient profit to justify its market valuation beyond just its rental conditions. By capitalising its net operating profit, we gain insight into the **long-term earning potential of the business** and whether it holds additional value beyond the real estate or rental agreement.

The following steps outline how to apply the **profits method**:

Step 1: *Completing the table and calculating the fair market operating profit (FMOP)*
The FMT, determined in **method 2,** *economic rental value*, serves as the foundation for the following method. Since food and beverage (F&B) sales are the restaurant's primary income source, this revenue projection ensures consistency across all valuation approaches.

To gain a complete financial picture, you must complete the missing fields in the following simplified **profit and loss table** (Table CS27), using the historical profit and loss statements as a frame of reference.

By filling in the missing fields based on historical trends and useful information, you can create a reliable financial projection for the valuation process. The completed **table** will then be used to calculate the **FMOP** using the following formula:

FMOP = FMT – Purchase Costs – Operating Expenses

Step 2: *Determining the annual capital reserve*
A portion of the FMOP must be set aside for reinvestment in the business. This *annual capital reserve* accounts for ongoing maintenance, equipment replacement, and other long-term operational costs.

- *A lower reserve percentage (3%).* Used for **low-risk businesses** with stable revenue, minimal fluctuations, and predictable maintenance costs.
- *A higher reserve percentage (5%).* Applied to **higher-risk businesses** that operate in volatile markets, require significant ongoing investments, or face potential operational disruptions.

The net operating profit is adjusted using the following table:

Net Operating Profit = FMOP – Annual Capital Reserve

Complete the profit and loss calculation table (Figure CS2.7) using the information provided and your own assumptions and analysis.

Step 3: *Determining the capitalised profit value*
To determine the capital value (also known as capitalised profit value), the **net operating profit** is capitalised using a fixed rate of return of **12.5%**. This capitalisation rate reflects the expected return on investment for a business of this type and accounts for both operational risk and market conditions.

Profits method		
NORM	*x 1000*	
Fair maintainable turnover (FMT)		**100,0%**
Purchase costs	€	%
Gross profit	€	%
Wages	€	%
Other costs	€	%
Fair market operating profit	€	
Annual capital reserve	€	%
Net operating profit	€	%
Rate of return	12,5%	
Capital value	€ -	
Value incl. buyer costs	€	
Value excl. buyer costs	€	
Value of operational business	€	

Figure CS2.7 Calculation table for the profits method.

The **capital value** is then calculated as follows:

$$\textit{Capital Value} = \frac{\textit{Net Operating Profit}}{\textit{Rate of Return}\,(\%)}$$

Step 4: *Determining the operational business value*

Once the capitalised profit value has been calculated, the next step is to determine the **value of the operational business**. This is done by **deducting the average market value** derived from the first two valuation methods (*rental value capitalisation* and *economic rental value*) from the capital value. Make sure to use the market value that includes the purchaser costs to provide a realistic representation of the investment costs.

$$\textit{Value Operational Business} = \textit{Capital Profit Value} - \textit{Market Value}$$

This final step isolates the **business operation's intrinsic value**, separate from the underlying property or rental value. A **positive result** suggests that the restaurant is profitable beyond its location and rental conditions, while a **low or negative value** may indicate that the business has limited profitability despite its rental position.

Questions

- What additional data and information might you need to complete the valuation?
- What are the variables that have the most impact in terms of sensitivity of your valuation?
- How have you reconciled the different methods to arrive at your valuation?
- How does the location of the building influence the valuation?
- How does the history and culture of the building influence the valuation?
- How confident are you of your projection of the business elements? How do you account for uncertainty in the business environment?

Using the data provided in the text and tables, follow the step-by-step instructions to value the restaurant and determine if you consider it as a viable business and how much a new operator should pay for it.

Try to undertake your own valuations before looking at the solutions at the end of Part IV.

Appendix A

Illustrative investment property purchase report

The report set out here is used solely as an illustration for students who are sometimes asked on their courses to write valuation reports and have very little material to access. We hope it helps, but do not grade it in academic or practice terms; the RICS (through the *Red Book*), clients, and the investment industry in general are constantly changing, and as a result it is the responsibility of the valuer to constantly update his or her approach to suitably reflect contemporary needs.

Students are referred to VPS 6 Valuation Reports in the *Red Book* (see also Chapter 5) for a list of 'report contents'; all *Red Book* reports must also contain a copy of the terms of engagement which must be addressed within the report. Needless to say, the valuer's records should meet the requirements of VPS 4 on valuation records.

A full investment report, complete with all background information and descriptive material, may well run to 50 pages and longer for the more complex multi-million-pound, multi-tenanted properties.

The report

Whilst each valuer and/or company may have a different style, the report will generally begin with an executive summary to highlight the salient points, but with a proviso (see following) and may take the following form. Such reports will usually be fully supported with photographs, location plans, floor plans with measurements as per the RICS Code of Measuring Practice (or the more recent International Property Measurement Standards), and market evidence. A key reason for this is to take the reader to the logical conclusion arrived at by the valuer and will not be ambiguous, be misleading, or create a false impression.

1.0 Executive summary

Terms have been agreed on behalf of XYZ pension fund to purchase the freehold shop investment known as 20 Market Street, Market Town, AB1 2CD from (name and address of the current owner) for £525,000 (five hundred and twenty-five thousand pounds). In brief:

Tenure: Freehold
Net internal area: 100 m^2
User: High-quality retail outlet
Tenancy: Let to AAA and BBB trading as ZZZ bakers on an assignment from XXXX (UK) Ltd for a term of 25 years, from 00/00/00 to 00/00/00, on FRI terms at a starting rent of £24,000 pa payable quarterly in advance on the normal quarter days.

Rent reviews: Every five years upward only, the next review being on 00/00/00.
Estimated market rent: £30,000 pa on similar terms.
Estimated yields: 4.32% net initial, rising to 5.40%, on reversion in 00/00/00; net equivalent yield 5.27%.

In conclusion, it is our opinion that the price agreed reflects the current market, and we recommend the acquisition at this price to XYZ pension fund. However, this summary must only be read in conjunction with the detailed comments which follow.

2.0 Client

This report has been prepared in accordance with the agreed terms of engagement (scope of work) as set out in annex 1 for the XYZ pension fund, Fund House, Edinburgh, Scotland, E1 1AZ.

3.0 Valuation purpose

The purpose of this valuation is to advise XYZ on the acquisition of the freehold interest in 20 Market Street, Market Town, AB1 2CD, at an agreed price of £525,000 (five hundred and twenty-five thousand pounds) subject to contract and survey. It should be noted that whilst we have inspected the property for the purpose of the valuation, we are not instructed to report on its structural condition. We understand that a full survey has been carried out on your behalf by ____________________ and that you are satisfied with the result. Further, that you have commissioned an environmental report which we have seen, and we note that it has paid full regard to current RICS and industry guidelines, any aspects of this report relevant to our valuation noted in Section 17.0.

4.0 Subject matter

This report relates solely to the property known as 20 Market Street, Market Town AB1 2CD, as identified on location plan in annex 2.

5.0 Interest

The interest to be valued is the freehold, land registration title no.12345, being offered for sale subject to the lease dated 00/00/00 as detailed in section 14. We understand from your solicitors, S, S, and S, that there is a right of way in fee simple to the owners and occupiers of the adjoining property, 22 Market Street, over the passageway along the side of the subject property. This gives access to the upper floors of the adjoining property. This is marked in green on the plan in annex 2.

6.0 Description and construction

The property is a Grade II listed building constructed circa 1850 comprising a ground-floor shop unit with single-storey rear storage and three floors of disused residential space above *Appendix A* 357 having access from a rear yard. Access to the yard is provided via a passageway to the side of the shop (see 5.0 earlier). The shop has limited internal width averaging 3.2 m (see annex 2).

The building is constructed of brick and sandstone quoin blocks and eaves cornice under a pitched slate roof. The windows are single-glazed, wooden sash. The shop has no central heating but is warmed by wall-mounted electric heaters. The upper floors are heated by a coke burning boiler serving radiators; this boiler and the radiators are in need of replacement.

The separate access to the upper floors means that the three floors of disused space could be brought back in to use subject to any necessary consents. Similar space in 22 Market Street was recently converted to office use.

(Key features are shown in annex 6 photographs.)

7.0 Basis of value

The basis of value is market value, defined in annex 3.

8.0 Date of valuation

The property has been valued as at 00/00/00. Values may change over time, particularly at the present time, with difficult borrowing conditions.

9.0 Material involvement

Neither the valuer, whose signature appears below, nor BBBB Chartered Surveyors have had any material involvement with this property or its owners.

10.0 Status of the valuer

This report and valuation has been prepared by Mr A. Proff FRICS of Proff and Partners. A. Proff is a fellow of the Royal Institution of Chartered Surveyors with over 20 years' experience of retail property in the local and national market.

11.0 Currency

All values are in £GBP.

12.0 Assumptions and special assumptions

[*Here, the valuer will list any assumptions or 'special assumptions' that they have relied upon (or been asked to report on). These may include a number of scenarios, such as vacant possession or in relation to planning situations. Current RICS guidance is that any special assumption can be reported on, so long as it is 'reasonable' (and, in each case, then the valuer has the skills, understanding, and knowledge to report under the assumption).*]

13.0 Investigations

In reporting our opinion of value, we have relied upon a number of sources including (but not limited to) the Land Registry and local planning databases.

13.1 Location

The property is located in a historic city with a population of 100,000 persons, which is substantially increased by tourists and students. The estimated shopping catchment is some 200,000 persons (see annex 4). The city is situated close to the junction of the MO and AOO, this providing good access to London (50 mi), XXX (00 mi), and YYY (00 mi). There are good road connections to all parts of the region.

A main line rail service operated by TTTTTTT connects to London King's Cross (the fastest journey time being 1 hour, 50 minutes). The Market Town international airport is some 20 minutes by road from the city centre and provides regular flights to Paris, Brussels, and Amsterdam but is primarily a holiday destination airport with flights by UUU, PPP, and EEE.

The property is located within the established shopping area in the city centre on the south side of Market Street close to all the major retailers (see goad plan, annex 2).

13.2 Site

The site slopes up to the rear and is roughly rectangular in shape. It has a frontage to Market Street of 4.35 m and a depth of 31.6 m. The site extends to an area of approximately 126 m^2 and is outlined in red on the attached Ordnance Survey extract (annex 2). [*In many valuation reports, imperial measures will be given first, followed by metric.*]

13.3 Accommodation

The property comprises a ground floor shop unit with single-storey rear storage and three floors of disused residential space above having access from a rear yard. Please see appendix X for a full schedule of accommodation.

13.4 Services

Whilst we have not inspected, for the purposes of this valuation, we have assumed that all services to the property are fully functioning. All information necessary for a valuation of this nature has been obtained, and its sources are detailed in annex 4. We suggest you seek expert professional advice in relation to this.

13.5 Town planning

The property is located within the borough of XXXX and therefore conforms to the XXXX development plan (a copy of which is available at www.XXXXXboroughcouncil.co.uk/planning).

13.6 Rateable value

We have checked the RV on the Valuation Office website and have noted that this property is currently listed (as at 00/00/00) with an RV of £28,000. The current Uniform Business Rate is 00.0p in the £. The tenant is liable for paying the UBR.

14.0 Tenancy

The entire property is held on a lease granted in 00/00/00 to XXXX for a term of 25 years, ending on 00/00/00. It is subject to upward-only rent reviews every five years; the next

359 review is due on 00/00/00. The lease is on full repairing and insuring terms. We understand that you are satisfied that your survey has confirmed the premises are being properly maintained and decorated. This lease was formally assigned with the existing owners' consent in 00/00/00 to the current occupiers.

The upper floors are included in the lease but are no longer used for residential purposes. In terms of our valuation, we have included this space as ancillary to the retail use and not as unimproved residential.

15.0 Sources of information

All the information necessary for a valuation of this nature has been obtained, and its sources are detailed in annex 4.

16.0 Market rent

Market Town is a cathedral city with a substantial catchment area and is much sought after by national multiple retailers. We are currently aware that AA, BB, CC, and DD are, even in the current weak market, seeking shop units in Market Street. Market rents have remained stable over the last three years, having dipped from their peak in 2007. Relevant extracts from our retail report for this year are included in annex 5, together with evidence of recent lettings and market rents agreed at rent review and lease renewal. Whilst not conclusive in today's uncertain market, we feel that there is sufficient evidence to support a zone A rent of £400/m^2 pa, which suggests a current market rent in the region of £30,000 pa for the premises. This might be increased if planning consent could be obtained for a change of use of the upper floors to offices.

17.0 Valuation

Our valuation has been prepared in accordance with the latest edition of the *RICS Valuation – Global Standards 2017*, which came into effect on 1 July 2017. The evidence of current market rents suggests a market rent of £30,000 pa.

Terms have been agreed for the purchase of the freehold interest in 20 Market Street for £525,000 (five hundred and twenty-five thousand pounds) subject to contract. Our own valuation set out in annex 6 using the ABC investment software supports this sum, and in our opinion, the market value of the freehold interest, subject to the right of way and tenancy, is £525,000 (five hundred and twenty-five thousand pounds).

A purchase at this price will give an initial yield of 4.32%, rising to 5.40%, on reversion in 00/00/00. This represents a net equivalent yield of 5.27%. These yields are after allowance for acquisition costs of 5.75% to cover fees and SDLT.

18.0 Exclusions of liability and publication

This valuation has been prepared in accordance with the terms of engagement agreed on 00/00/00 and set out in annex 1 together with an agreed statement on publication and liability.

The property was inspected and valued by Mr A. Proff of Proff and Partners, Chartered Surveyors. A. Proff is a fellow of the Royal Institution of Chartered Surveyors with over 20 years' experience of retail property in the local and national market and has the knowledge, skill, and understanding to undertake this valuation.

19.0 Signature

A. Proff
00/00/00

Annexes

1. Terms of engagement
2. Ordnance survey extracts, floor plans, goad plans
3. Market value definition
4. Data used and sources
5. Company retail report
6. Photographs

Appendices

1. Schedule of accommodation

Other reports

Students should note that clients may also ask for reports on investment value (worth). These are normally prepared after agreeing with the client a number of key variables, such as holding period, discount rate, rates of rental growth, terminal capitalisation rate, etc. Further information on investment value reporting can be found in Chapter 10.

Appendix B

Illustrative development site appraisal report

The following report is based on the standard report produced by ARGUS Developer Version 4.05.001 using the data from Example 10.1 on p. 295. Note that other software, including EstateMaster, which we demonstrate in Chapter 10, is available to undertake similar analysis.

Also, to be complete, we have included the exhaustive list of assumptions that the user can change in the Argus programme. Readers should note the detail and complexity that the system can manage, including the management of taxation, such as VAT in the UK, and in arrears and advance payments and receipts. The data used in the input screens and assumptions will form the framework of the cash flow basis for the appraisal, and therefore, it is essential that the valuer *understands*:

- The implications of the selections made in the assumptions settings
- How those implications will direct the construction of the cash flow
- What assumptions are the default settings and whether or not these are applicable to the subject valuation
- Any change in tax or other legislation and ensures the system has been updated if necessary or the assumptions adjusted in accordance with the new requirements.

Note: The land value calculated by the Argus system is not exactly the same as in the manual calculation due to rounding and minor variations within the assumptions made in the programme.

Timescale and assumptions

Timescale (duration in months)

Project commences September 2017

Phase 1

Stage name	*Duration (months)*	*Start date*	*End date*
Phase start		Sept 2017	
Construction	24	Sept 2017	Aug 2019
Phase end		Sept 2019	
Phase length	**24**		
Project length	**25**	**(Includes exit period)**	

This first part shows the timescales and durations of the project. The development may be split into a number of phases, which would be shown in this section. Phasing allows better manipulation and control of the cash flow. In this simple calculation, there is no void period; this is unrealistic, unless the development is fully pre-let.

Assumptions

Expenditure

Professional fees are based on construction.
Purchaser's costs are based on gross capitalisation.
Purchaser's costs are deducted from sale (not added to cost).
Sales fees are based on net capitalisation.
Sales fees are added to cost (not deducted from sale).

Receipts

Show tenant's true income stream	On
Offset income against development costs	Off
Rent payment cycle	Annual (adv.)
Apply rent payment cycle to all tenants	On
Renewal void and rent-free apply to first renewal only	Off
Initial yield valuation method	Off
Default capitalisation yield	0.0000%
Apply default capitalisation to all tenants	Off
Default stage for sale date	Off
Align end of income stream to sale date	Off
Apply align end of income stream to all tenants	On
When the capital value is modified in the cash flow	Recalculate the yield
Valuation tables are	Annually in arrears
Rent-free method	Defer start of tenant's rent

Finance

Financing method	*Basic (interest sets)*
Interest compounding period	Annual
Interest charging period	Annual
Nominal rates of interest used	
Calculate interest on payments/receipts in final period	Off
Include interest and finance fees in IRR calculations	Off
Automatic inter-account transfers	Off
Manual finance rate for profit erosion	Off

Calculation

Site payments	In advance	
Other payments	In advance	
Negative land	In arrears	
Receipts	In advance	
Initial IRR guess rate		8.00%
Minimum IRR		−100%
Maximum IRR		99999%
Manual discount rate		Off
IRR tolerance		0.001000

General assumptions

Letting and rent review fees are calculated on	Net of deductions
Development yield and rent cover are calculated on	Rent at sale date(s)
Include tenants with no capital value	On
Include turnover rent	Off
Net of non-recoverable costs	On
Net of ground rent deductions	On
Net of rent additions/costs	On

Value-added tax

Global VAT rate	0.00%
Global recovery rate	0.00%
Recovery cycle every	2 months
1st recovery month	2 (Oct 2010)
VAT calculations in cash flow	On
GST margin calculations in cash flow	Off

Residual

Land cost mode		Residualised land value	
Multi-phasing		Separate land residual for each phase	
Target type		Profit on GDV	
Phase number	Target value	Locked Value	Treat neg land as revenue
Phase 1	15.00%	No	No

Distribution of payments and receipts

Construction payments are paid on	S-curve
Sales receipts are paid on	Single curve
Sales deposits are paid on	Monthly curve

Interest sets

Site

Debit rate	*Credit rate*	*Months*	*Start date*
15.000%	0.000%	24	Sept 2010
0.000%	0.000%	Perpetuity	Sept 2012

Construction

Debit rate	*Credit rate*	*Months*	*Start date*
15.000%	0.000%	24	Sept 2017
0.000%	0.000%	Perpetuity	Sept 2017

This exhaustive set of assumptions gives the valuer many options and the flexibility to ensure the system mirrors the individual circumstances of the development project. As we said earlier, however, great care must be exercised in ensuring that the assumptions, especially the default settings, are what are expected by the client.

Appraisal summary

Summary appraisal for phase 1

REVENUE				
Rental area summary	*m²*	*Rate (m²)*	*Gross MRV*	
Office	6,000.00	£120.00	720,000	
Investment valuation office				
Current rent	720,000	YP @	7.0000%	14.2857
NET REALISATION				**10,285,714**
OUTLAY				
ACQUISITION COSTS				
Residualised price			3,533,726	
Stamp duty		1.00%	35,337	
Agent fee		1.00%	35,337	
Legal fee		1.00%	35,337	
				3,639,738
CONSTRUCTION COSTS				
Construction	*m²*	*Rate (m²)*	*Cost*	
Office	7,000.00	£400.00	2,800,000	**2,800,000**
PROFESSIONAL FEES				
Architect		9.00%	252,000	
Quantity surveyor		3.50%	98,000	
				350,000
MARKETING and LETTING				
Marketing			50,000	
Letting agent fee		20.00%	144,000	
				194,000
DISPOSAL FEES				
Sales agent fee		1.00%	102,857	
				102,857
FINANCE				
Multiple finance rates used (see 'Assumptions')				
Debit rates varied throughout the cash flow				
Total finance cost				1,656,262
TOTAL COSTS				**8,742,857**
PROFIT				
				1,542,857

Performance measures

Profit on cost %	17.65%
Profit on GDV %	15.00%
Profit on NDV %	15.00%
Development yield % (on rent)	8.24%
Equivalent yield % (nominal)	7.00%
Equivalent yield % (true)	7.32%
Gross initial yield %	7.00%
Net initial yield %	7.00%
IRR	27.79%
Rent cover	2 years, 2 months
Profit erosion (finance rate 15.000%)	1 year, 2 months

The system can be used for capital sales (e.g. residential development), as well as capitalised rental income, as in this example, or a mixture of the two.

The expenditure summary shows additional items not considered in the simple manual calculation on page 295.

This page also shows some useful performance measures for the development, including the following:

The profit on GDV

In this case, the profit on GDV has been stipulated by the user at 15% in order to calculate the residual site value. Where the system is used for calculating the developer's profit, the land value would be provided and the profit calculated by the system.

Profit erosion

The profit erosion cover shows the period of time over which the profit would be completely eroded by continuing interest payments if the development remained un-let.

Rent cover

The rent cover is calculated by dividing the residual profit by the rental value. If the developer is guaranteeing rent as part of a funding arrangement, a profit will still be made (decreasing over this period), provided the development is let within the period of years shown, in this case two years and two months.

Development yield

The development yield is calculated by dividing the rental income by the total costs.

The other yields are explained in Chapter 11.

Sensitivity analysis

Table of residual land price

Construction Rate (step and amount) £/m²					
Rent rate (step and amount) £ pm²	*step* **−£20.00 m²** *amount* **£380.00 m²**	*step* **−£10.00 m²** *amount* **£390.00 m²**	*step* **£0.00 m²** *amount* **£400.00 m²**	*step* **+£10.00 m²** *amount* **£410.00 m²**	*step* **+£20.00 m²** *amount* **£420.00 m²**
step **−£5.00 m²** amount **£115.00 m²**	£3,407,180	£3,340,514	£3,273,848	£3,207,181	£3,140,515
step **−£2.50 m²** amount **£117.50 m²**	£3,537,120	£3,470,453	£3,403,787	£3,337,121	£3,270,454
step **£0.00 m²** amount **£120.00 m²**	£3,667,059	£3,600,393		£3,467,060	£3,400,394
step **+£2.50 m²** amount **£122.50 m²**	£3,796,998	£3,730,332	£3,663,665	£3,596,999	£3,530,333
step **+£5.00 m²** amount **£125.00 m²**	£3,926,937	£3,860,271	£3,793,605	£3,726,938	£3,660,272

Sensitivity analysis: assumptions for calculation

Construction rate

Heading	*Phase*	*Original Value*
Office	1	£400.00 pm^2

These fields varied in fixed steps of £10.00 pm2.

Rent rate

Heading	*Phase*	*Original Value*
Office	1	£120.00 pm^2

These fields varied in fixed steps of £2.50 pm^2.

The system can be used to generate a sensitivity analysis matrix.

In this case, the matrix shows the effect on the residual land value with changes in two variables: the construction cost rate and the rental rate. The rental rate changes in steps of £2.50, and the building costs in steps of £10.

The 'worst'-case scenario is therefore in the top right-hand corner of the matrix, and the 'best'-case scenario in the bottom left-hand corner. Sensitivity here is not that significant, although when analysing yields, very small changes in the yield may result in significant changes in the residual value or profit of a scheme.

The original residual value (£3,553,726), calculated with the unadjusted values for rent and construction, is shown in the centre of the table.

When using the system for appraisals, the matrix would show the amount of profit in £ (or other currency) and as a percentage of either cost, GDV, or NDV, as selected by the user. The sensitivity can also produce a three-dimensional matrix by using five layers or windows which are printed separately.

Report produced by
ARGUS Developer Version: 4.05.001

Appendix C

Solutions to questions set in the text

Chapter 1

1. *Property* is a broad legal and economic concept that includes both tangible and intangible assets, while *real estate* specifically refers to land and immovable structures. Real estate is a subset of property focused on land and the permanent changes made to it.
2. Aristotle viewed property as natural and essential for personal well-being and societal harmony, supporting private ownership with moral responsibilities. Plato, in contrast, advocated for communal ownership among the ruling class to prevent corruption and division, although he later accepted limited private ownership with regulations.
3. Marx argued that private property in the means of production led to exploitation, as it enabled the bourgeoisie to profit from workers' labour. He viewed private property as a root cause of class struggle and advocated for its abolition in favour of collective ownership to benefit society as a whole.
4. Property valuation is essential for various financial decisions, including secured lending, financial reporting, buying and selling properties, development appraisals, taxation assessments, and compensation claims. It ensures fair market transactions and informed investment decisions.
5. The main methods of property valuation include:

   a. *Sales comparison method (market approach).* Compares similar properties to determine value.
   b. *Income capitalisation method.* Uses income potential to assess value.
   c. *Cost approach.* Estimates value based on reconstruction costs minus depreciation.
   d. *Profits method.* Assesses properties generating income from business operations.
   e. *Residual method.* Evaluates development potential by estimating future value minus costs.

6. Different bases of value influence how properties are assessed. *Market value* reflects open market conditions, *investment value* considers specific investor objectives, *synergistic value* accounts for added worth in combination, and *liquidation value* estimates sale price in forced situations. The choice of basis depends on the valuation's purpose.
7. Ethical considerations in property valuation include integrity, objectivity, confidentiality, and due diligence. Valuers must avoid conflicts of interest, ensure transparency in their reports, and adhere to professional standards set by organisations such as RICS and IVSC to maintain public trust in property markets.

8. Property valuation standards, such as those set by RICS, IVSC, and IFRS, ensure transparency, consistency, and reliability in valuations. This helps prevent overvaluation and undervaluation, reduces financial risk, supports informed decision-making, and enhances investor confidence in real estate markets.

Chapter 3

1. *Real estate* is a distinctive investment due to its tax efficiency, stability, potential for appreciation, and capacity for income generation. It also offers tangible characteristics, versatility, and in ideal cases, resilience to market volatility.
2. According to RICS (2022), *investment property* is defined as land or buildings held for earning rentals or capital appreciation, rather than for use in production, supply of goods or services, administrative purposes, or sale in the ordinary course of business.
3. Location is crucial because it determines the accessibility and desirability of a property. Traditionally, physical location was the most important factor, but in recent years, connectivity, particularly internet infrastructure, has gained significance, impacting property values.
4. The four main phases of a property cycle are:

    a. *Boom/expansion.* Rising demand, increasing rents, and high property values.
    b. *Peak.* Prices and rents reach their highest point, often leading to speculative investment.
    c. *Recession/contraction.* Decreasing demand, falling rents, and declining property values.
    d. *Recovery.* Stabilisation of demand, rents, and values, leading to renewed investment activity.

5. Property cycles affect investment decisions by influencing risk, return expectations, and market timing. Investors aim to buy during the recovery phase and sell during the boom phase to maximise returns, while avoiding overpaying during peak markets.
6. The *capitalisation rate* (cap rate) is the ratio of net operating income to property value or purchase price. It is used to compare investments and assess market perceptions of risk, playing a key role in property valuation. It is used to capitalise the NOI in order to convert it to capital value, normally at the exit year of a DCF.
7. Property market cycles are influenced by factors, such as:

    a. *Economic conditions.* GDP growth, employment rates, and consumer confidence.
    b. *Interest rates.* Higher rates reduce affordability and investment activity.
    c. *Supply and demand.* Overbuilding can lead to market downturns.
    d. *Government policies.* Taxation, planning regulations, and incentives impact investment trends.

8. 'Lumpiness' refers to the substantial capital requirements of real estate investments. Unlike stocks or bonds that can be purchased in small increments, real estate transactions typically involve large, indivisible assets, affecting liquidity and investment flexibility.
9. Leverage allows investors to finance property acquisitions using borrowed funds, potentially increasing returns. However, it introduces financial risks, including interest rate fluctuations, refinancing risks, and potential losses if property values decline.
10. Real estate is considered illiquid because transactions involve long due diligence processes, legal complexities, and significant costs, such as taxes and agent fees. During market downturns, selling properties quickly at fair market value can be challenging.

Chapter 4

1. The real estate system consists of the space market, asset market, and development industry. The *space market* (or occupier market) is where properties are leased, determining rental income and cash flows. These cash flows affect property values in the *asset market* (or investment market), where properties are bought and sold while competing with other assets, such as bonds and equities. Investment decisions in the asset market drive activity in the *development industry*, which converts financial capital into physical assets, which shape future real estate supply.
2. Space market: Rental levels, vacancy rate, occupancy rate, take-up levels.

   Asset market: Price, cap rate, investment volumes.
   Development industry: Construction activity, development pipeline.
3. Endogenous factors, that is, actual prices and rents, influence real estate demand through direct market interactions, leading to movements *along* the demand curve (see Figure 4.1). For instance, higher rents may reduce demand, as fewer tenants can afford the space. In contrast, exogenous factors, including market size, income levels, the price of substitutes, consumer expectations, and interest rates, cause *shifts* in the demand curve (see Figure 4.2). For example, population growth can increase demand for housing and commercial properties.
4. Economic crises and pandemics necessitate a wider temporal perspective that goes beyond just recent market conditions. While valuation reports typically focus on data from the last few quarters or the previous year, significant disruptions can result in structural changes in occupier demand, lease agreements, and investment strategies. These events can speed up existing trends, shift risk perceptions, and redefine space requirements, ultimately transforming long-term market dynamics. Therefore, market analysis should consider both short-term fluctuations and long-term changes to provide a clearer understanding of real estate performance, pricing trends, and investment risks.
5. *Short-term supply inelasticity* means that real estate supply cannot quickly adjust to changes in demand because of lengthy development processes, regulatory hurdles, and high construction costs. As a result, sudden increases in demand can lead to sharp rises in rents and property prices, while abrupt demand declines may lead to rising vacancy rates and downward pressure on values. For valuation, this inelasticity contributes to market volatility, requiring a thorough assessment of rental trends, occupancy levels, and investor sentiment to gauge potential risks and opportunities.

Chapter 5

1. Valuers play a critical role in ensuring accurate, transparent, and consistent property assessments, which help maintain financial stability by preventing overvaluation or undervaluation of assets. Their work supports risk management, investor confidence, regulatory compliance, and systemic financial stability.
2. Valuation results may differ due to the purpose of the valuation, choice of methodology, market conditions at the time of valuation, data availability and quality, professional judgement and assumptions, and regulatory or legal factors. These variations highlight the importance of clear explanations and adherence to professional standards.
3. The three primary approaches to property valuation are:

   a. *Market approach (sales comparison).* Based on recent sales of comparable properties.
   b. *Income approach.* Determines value based on income generation potential.
   c. *Cost approach.* Estimates value by considering replacement or reconstruction costs.

4. Valuation standards, such as those established by RICS, IVS, TEGoVA, and IFRS, ensure accuracy, consistency, and transparency in property valuations. They provide a structured methodology, enhance credibility, and facilitate comparability across international markets.
5. Professional bodies, such as RICS and IVSC, establish valuation standards, provide regulatory oversight, offer education and certification programmes, resolve disputes, and promote public trust. They ensure valuers adhere to ethical and professional guidelines, maintaining integrity in the valuation process.
6. Transparency in property valuation ensures that assessments are conducted fairly, fostering trust among investors, lenders, and regulatory authorities. It involves standardised methodologies, disclosure of assumptions, compliance with ethical standards, reliance on market data, and regulatory adherence.
7. Measurement inconsistencies can arise from differences in measurement standards (e.g. IPMS, RICS Code), methodology variations, unclear definitions, human error, and differences in technology. Standardising measurement practices ensures uniformity, comparability, and accuracy in valuations.
8. Valuers must maintain integrity, objectivity, and confidentiality while avoiding conflicts of interest. They should exercise professional scepticism, adhere to ethical codes, ensure legal compliance, and provide unbiased assessments based on reliable data to uphold the credibility of the valuation industry.

Chapter 6

1. $61,391.33
2. $1,200,000 in two years from now (PV of it is $1,088,436.37, which is higher than $1,000,000).
3. £7,561.33
4. $851,140.28
5. $40,556.25
6. $341,590.53
7. €35,676.30
8. €921,685.07
9. R$4,331,940.32
10. £12,841.52
11. €81,939.67
12. 25.378 years
13. R$1,013,853.57
14. £16,605.89
15. Bank B (EAR 7.33%)

Chapter 7

1a. €275,000
1b. €266,000
1c. Comp A: €705.13/m²
Comp B: €633.33/m²
1d. €272,478.63 (using linear interpolation using the price per square metres calculated in the previous question)

2. Indicative answer:

Comp	*Price per m² (£)*	*Comment*
A	3,393	The lowest price per square metre, reflecting its older age, fewer en-suites, no pool, and only one parking space. The subject property offers superior amenities, justifying a higher value.
B	3,750	The highest price per square metre, driven by its newer age, additional bedroom, and similar amenities. The subject property is slightly older and has fewer bedrooms, suggesting a lower value.
C	3,621	A newer build but lacks a pool and has fewer parking spaces. The subject property's additional amenities suggest a higher price.
D	3,710	Very similar in features but slightly newer. The subject property is expected to be valued slightly lower.
E	3,448	Same bedroom configuration and same number of parking spaces as the subject property but is older and lacks a pool, which would justify a higher price for the subject property.

From the comparable evidence, comparables B and D are the most similar to the subject property in terms of amenities and parking, though both are newer. The subject property is expected to achieve a price per square metre below B (£3,750/m^2) and D (£3,710/m^2) but above A (£3,393/m^2), C (£3,621/m^2), and E (£3,448/m^2), which offer fewer amenities. Given these factors, it is our opinion that the market value of the subject property is £1,107,000 (£3,690/m^2).

3. $144,000 pa
4. €79,495.61
5. WAULT area = 38.40 months ~ 3.2 years
   WAULT rent = 37.41 months ~ 3.12 years
   Since WAULT weighted by rent (37.4 months) is lower than WAULT by area (38.4 months), the income profile appears slightly less secure than the physical occupancy. While the WAULT figures do not vary enormously, a WAULT of around 3 years does not provide significant income security. A property valuer might therefore increase the ARY if comparable evidence shows properties with longer WAULTs.

Chapter 9

1. €6,000,000
2. €29,438,839.29
3. £5,625,000
4. €7,433,333.33
5. £275,689.07

6a. €652,627
6b. €2,558,700 and €1,747,271
6c. €2,399,898
7a. Personal expenses are part of the NOP but should not be included in the calculation of value; therefore, adjusted NOP is €250,000.
7b. €2,500,000

8. Objective modelling is difficult for unique trading establishments because:

   - They have distinct business models and operational structures.
   - Revenue streams can fluctuate significantly due to market conditions.

- Lack of direct comparables makes traditional valuation methods ineffective.
- Financial accounts may include non-standard expenses, such as personal expenditures.

That is why benchmarking helps:

- Benchmarking provides industry averages for expenses, profit margins, and revenue per unit.
- It enables valuers to compare the financial performance of a business against similar businesses.
- It ensures fair adjustments to financial statements, making valuations more standardised.

9a. €70,000
9b. €1,000,000
10. Comparables are critical in the profits method because:

- They provide a real-world benchmark for revenue, expenses, and profitability.
- They ensure valuations reflect market conditions rather than relying purely on assumptions.
- They help in assessing rental values, where market-based rents are difficult to determine.

Adjustments for unique trading styles:

- Consider differences in business efficiency and operational costs.
- Adjust for location, target market, and customer base.
- Factor in lease conditions, especially turnover rents or franchise agreements.

11. a. €1.08 mil.
    b. €1.4 mil.
    c. Approx. €17.95 mil.
12.

- *Tax optimisation.* Businesses may minimise taxable profits by inflating expenses, which can distort real profitability.
- *Personal expenses.* Audited accounts often include non-business-related expenses, affecting net operating profit.
- *Inconsistencies across businesses.* Different businesses use different accounting policies, making direct comparisons difficult.
- *Undisclosed revenue streams.* Cash transactions or unreported income can lead to undervaluation.
- *Variability in operating costs.* Expenses like staffing, rent, and energy costs vary widely, complicating benchmarking.
- *Adjustments required.* Valuers must normalise financial data, remove extraordinary expenses, and compare against market benchmarks.

Chapter 11

1a. 1.4286%
1b. $357,142.86
1c. $2,142,857.14
2. $4,429,211.90
3a. $1,875,000
3b. $1,125,000
3c. $1,575,000

4a. $2,400,000
4b. $4,800,000
4c. $144,000
5. New properties under construction

- Developers may face higher costs, leading to increased prices for buyers.
- Some projects might be delayed or cancelled if funding cannot cover the cost increase.
- Developers may seek higher pre-sales or pre-leases to secure financing.

Newly completed properties

- Properties completed before the cost increase benefit from lower construction costs, making them more attractive.
- Sellers might increase their prices due to higher replacement costs.
- Investors might prefer existing properties over new construction due to cost stability.

Resales (second-hand market)

- The rise in new construction costs pushes demand toward existing properties.
- Property owners may increase prices in response to rising new-build costs.
- Buyers may seek older properties as they become a more affordable alternative.

6. Why is DRC not acceptable?

- *Market-based lending.* Banks prefer market evidence rather than cost-based estimates.
- *Lack of comparables.* DRC does not consider market trends or demand-driven pricing.
- *Liquidity risk.* If a lender repossesses a property, they need a realistic resale value, which DRC may not reflect.
- *Depreciation and obsolescence.* DRC considers cost replacement, but buyers may not pay for outdated designs or inefficient layouts.

When can DRC be accepted?

- *Specialised properties.* Public buildings, hospitals, or infrastructure projects that have no comparable sales data.
- *Government or institutional lending.* Some public sector loans or insurance-backed financing may use DRC.
- *Legal or tax requirements.* Some courts or tax authorities mandate a cost-based valuation.

7. There are three issues with the approach taken:

a. The allowable density was not considered.

- The site has 140% density – meaning 770 m^2 can be developed (550 × 140%).
- Only 250 m^2 has been built, leaving 520 m^2 of unused potential.
- The valuation does not account for the full development potential.

b. The land value may be overstated.

- If full density is not used, the land's effective value should be discounted.
- A site with underutilised density is worth less than one that is fully developed.

c. Better alternative uses not considered.

- If the site allows multiple units, its value should reflect higher potential income.
- The valuation ignores what a developer or investor would pay for a fully utilised site.

Revised valuation approach is recommended:

- If full density is not usable, the land value should be adjusted based on actual use.
- If full density is buildable, then a residual valuation or highest and best use approach should be used.
- The valuer must clarify if and how much of the unused density can be realistically developed.

Conclusion:

- The approach ignores development potential, making it flawed.
- The valuer should adjust the land value to reflect realistic use.
- If full density is feasible, a market-based valuation method should be considered instead.

Chapter 12

1. Mass appraisal is the systematic valuation of groups of properties at a specific date using standardised procedures and statistical testing. Unlike traditional single-property valuation, which evaluates one property at a time based on individual characteristics, mass appraisal applies uniform methodologies to large datasets to ensure consistency and efficiency.
2. Mass appraisals and AVMs have gained prominence due to advancements in AI and machine learning, the increasing need for rapid and cost-effective valuations, the digitisation of property data, and regulatory interest in ensuring transparency and consistency in property valuations.
3. Key beneficiaries include tax authorities, financial institutions, lenders, investors, insurance companies, real estate professionals, and government agencies. These stakeholders rely on AVMs for accurate, quick, and scalable valuations in lending, investment decisions, taxation, and portfolio management.
4. Both CAMA and AVMs are automated valuation tools. CAMA is mainly used for property taxation and emphasises consistency over accuracy, whereas AVMs are designed for real estate markets, mortgage lending, and investment analysis. AVMs integrate AI, big data, and predictive analytics, making them more adaptive to market conditions compared to CAMAs.
5. AVMs rely on high-quality, accurate, and up-to-date data to generate reliable valuations. Poor data quality, including outdated or incorrect property attributes, missing transaction data, or biased datasets, can result in incorrect valuations, increased financial risks, and decreased confidence in AVMs.
6. AVMs use transaction data, property characteristics, geographic information, and economic indicators. Challenges that can undermine reliability include missing or outdated data, incorrect property attributes, market volatility, and biases within the dataset that may distort valuation outcomes.
7. Professional valuation standards emphasise transparency, documentation, and human oversight when using AVMs. IVS 105, for example, specifies that AVMs alone cannot provide IVS-compliant valuations without professional judgement to verify assumptions, model inputs, and market conditions.
8. Valuers are required to verify AVM assumptions, validate the data, interpret the outputs in market context, and ensure compliance with legal and professional standards. AVMs lack the ability to assess qualitative factors, such as property condition, desirability, and legal considerations.

9. Advantages: Speed, cost-efficiency, consistency, scalability, and ability to handle large datasets.
   Limitations: Limited ability to assess unique or complex properties, reliance on data quality, potential biases, and lack of in-depth property-specific analysis.
10. Ethical concerns include algorithmic bias, lack of transparency in valuation processes, potential manipulation of valuation outcomes, and displacement of traditional valuers. There is also the risk of systematic errors affecting financial decisions on a large scale.
11. Regulators and professional bodies can enforce transparency standards, require regular audits, ensure human oversight in AVM valuations, establish compliance protocols, and introduce ethical guidelines for AVM usage to prevent market manipulation and unfair assessments.
12. Emerging technologies include blockchain for secure property records, IoT sensors for real-time property monitoring, AR/VR for remote property inspections, machine learning for improved predictive analytics, and deep learning models for enhanced valuation accuracy.
13. Steps include integrating high-quality data sources, ensuring transparency in valuation methodologies, providing public education on AVMs, implementing human oversight, conducting regular audits, and maintaining compliance with legal and ethical valuation standards.
14. Valuers can argue that AVMs lack contextual understanding, cannot assess qualitative property attributes, struggle with unique properties, may misinterpret market trends, and require professional oversight to ensure accuracy and compliance with valuation standards.
15. While AVMs provide efficiency and cost savings, reliance on flawed data or inadequate modelling can introduce systemic errors. A balanced approach that integrates AVM insights with human verification ensures optimal results, maintaining accuracy while benefiting from automation.

Chapter 14

1. Evaluating valuers ensures accurate and reliable property valuations, which are crucial for maintaining transparency, financial stability, and investor confidence in real estate markets. It helps prevent misleading valuations that could lead to financial risks and market distortions.
2. A valuer must conduct thorough property inspections, adhere to professional and legal standards, perform detailed market analysis, provide transparent valuation reports, and stay updated with market trends and regulations. They also offer expert opinions in disputes and legal cases.
3. Valuers support economic stability by providing accurate valuations that help prevent real estate bubbles, inform lending decisions, support investment confidence, and ensure fair market transactions. Their work also assists in infrastructure development and regulatory compliance.
4. Key criteria include academic qualifications, professional certifications (such as MRICS and REV), compliance with valuation standards, methodological competence, accuracy and consistency of valuations, ethical conduct, and ongoing professional development (CPD).
5. CPD ensures that valuers stay updated with evolving market conditions, valuation methodologies, and regulatory changes. It helps enhance skills, improve valuation accuracy, adapt to new technologies, and maintain high professional standards in the valuation industry.

6. Common evaluation methods include peer reviews, audits, benchmarking against market transactions, regulatory oversight by professional bodies, and feedback from clients and stakeholders. These methods help ensure accuracy, compliance, and professional integrity.
7. RICS-regulated firms operate under strict ethical and professional standards, ensuring transparency and reliability in valuations. Designations like MRICS and REV indicate a valuer's expertise, adherence to global best practices, and commitment to ethical valuation processes.
8. Professional indemnity insurance (PII) protects clients from financial losses due to valuation errors or negligence. It ensures accountability among valuers, fosters public trust, and reinforces the credibility of the valuation profession.

Case Study 2

Rental Value Capitalisation		Price per m²	€ 400,00	
Ground floor	Surface in m²	Weight	Surface after correction	
Entrance area	**15,00**	**75%**	11,25	
Wardrobe	**10,00**	**75%**	7,50	
Bar	**40,00**	**100%**	40,00	
Restaurant	**120,00**	**100%**	120,00	
Kitchen	**60,00**	**75%**	45,00	
Storage	**25,00**	**75%**	18,75	
Toilet	**35,00**	**75%**	26,25	
Garden	**125,00**	**5%**	6,25	
Terrace	**100,00**	**15%**	15,00	
Parking	**20,00**	**15%**	3,00	
Subtotaal	550		293	
First Floor	Surface in m²	Weight	Surface after correction	
Private room	**25,00**	**75%**	18,75	
Toilet	**15,00**	**50%**	7,50	
Office	**30,00**	**50%**	15,00	
			0,00	
Subtotaal	70		41	
Basement	Surface in m²	Weight	Surface after correction	
Storage	**125,00**	**25%**	31,25	
			0,00	
			0,00	
			0,00	
Subtotaal	125		31	
Total	366	€ 400,00	€ 146 200,00	
Maximum investment	**Rental value**	€ 146 200	€ **1 606 593,41**	**Incl. Buyer Costs**
	Investment Yield (GRM)	**9,1%**	€ **1 447 381,45**	**Excl. Buyer Costs**

$\textit{Value including Buyer Costs} = \textit{approximately} \text{ €} 1.605.000$

$\textit{Value excluding Buyer Costs} = \textit{approximately} \text{ €} 1.445.000$

Economic Rental Value

	x1000			Rent-to-revenue ratio	
Fair Maintainable Turnover (FMT)	€ 1 400,00			10,0%	
Maximum investment	Rental value	€	140 000	€ 1 538 461,54	**Incl. Buyer Costs**
	Investment Yield		9,1%	€ 1 386 001,39	**Excl. Buyer Costs**

$Value\ including\ Buyer\ Costs = approximately\ € \mathbf{1.540.000}$

$Value\ excluding\ Buyer\ Costs = approximately\ € \mathbf{1.385.000}$

Profits Method

NORM	*x 1000*		
Fair Maintainable Turnover (FMT)	€	1 400,00	100,0%
Purchase costs	€	350,00	25,0%
Gross Profit	€	1 050,00	75,0%
Wages	€	525,00	37,5%
Other costs	€	224,00	16,0%
Fair Market Operating Profit (FMOP)	€	301,00	21,5%
Annual Capital Reserve	€	60,00	4,3%
Net Operating Profit	€	241,00	17,2%
Rate of Return		**12,5%**	
Capital Value	€	1 928 000,00	
Value incl. buyer costs	€	1 573 000,00	
Value excl. buyer costs	€	1 417 000,00	
Value Operational Business	€	355 000,00	

$Capitalised\ Profit\ Value = approximately\ € \mathbf{1.928.000}$

$Value\ Operational\ Business = approximately\ € \mathbf{355.000}$

Index

Note: Page numbers in *italics* indicate figures, and page numbers in **bold** indicate tables in the text

FSC
www.fsc.org
MIX
Papier | Fördert gute Waldnutzung
FSC® C083411

Kolibri 360 GmbH
Ferdinand-Jühlke-Straße 7
99095 Erfurt, Deutschland
produktsicherheit@kolibri360.de